A Survey of Numerical Mathematics

by David M. Young
Center for Numerical Analysis
The University of Texas at Austin

and Robert Todd Gregory
Formerly at the Center for Numerical Analysis
The University of Texas at Austin

In Two Volumes
Volume I

Dover Publications, Inc., *New York*

Copyright © 1972, 1973 by David M. Young, Jr., and Robert T. Gregory.
All rights reserved under Pan American and International Copyright
Conventions.

Published in Canada by General Publishing Company, Ltd., 30 Lesmill
Road, Don Mills, Toronto, Ontario.
Published in the United Kingdom by Constable and Company, Ltd., 10
Orange Street, London WC2H 7EG.

This Dover edition, first published in 1988, is an unabridged, corrected
republication of the work originally published by the Addison-Wesley Publishing Company (Addison-Wesley Series in Mathematics), Reading, Mass., 1972
(Vol. I) and 1973 (Vol. II). In the Dover edition the Bibliography and Index of
Volume I, given in incomplete form in the original edition, have been replaced
by the complete versions that first appeared in Volume II. Appendix B and
Appendix C, newly added to the Dover edition, include revisions of two
passages in Volume I.

Manufactured in the United States of America
Dover Publications, Inc., 31 East 2nd Street, Mineola, N.Y. 11501

Library of Congress Cataloging-in-Publication Data

Young, David M., 1923–
A survey of numerical mathematics.

Reprint. Originally published: Reading, Mass. : Addison-Wesley,
c1972–c1973. (Addison-Wesley series in mathematics)
Includes bibliographies and index.
1. Numerical analysis—Data processing. I. Gregory, Robert Todd,
1920–1984. II. Title. III. Series: Addison-Wesley series in mathematics.
QA297.Y63   1988   519.4   88-3630
ISBN 0-486-65691-8 (pbk. : v. 1)
ISBN 0-486-65692-6 (pbk. : v. 2)

519.4
Y84
v.1

91-608

MW-91-5/4

To the memory of

GEORGE E. FORSYTHE

PREFACE

Since the advent of automatic digital computers there has been a rapid development of the branch of applied mathematics known as *numerical analysis*. Numerical analysis is concerned with the development, analysis, and evaluation of numerical algorithms which can be carried out on a computer (usually an automatic digital computer) for obtaining numerical solutions to mathematical problems. Although the methods and results of classical analysis can often be helpful, they usually provide only background and/or a starting point for the numerical analyst. For example, the pure mathematician may be fully satisfied if he can prove that a unique solution to a given problem exists, but it is up to the numerical analyst to devise a procedure for actually computing a solution to within a specified accuracy and within a reasonable time. In developing an algorithm to use in solving a given problem, the numerical analyst must be concerned not only with the number of arithmetic operations and the theoretical accuracy but also with the cumulative effects of rounding errors which are made when the algorithm is implemented on a computer.

The purpose of this book (Volume I and Volume II) is to study computer-oriented numerical algorithms for solving various types of mathematical problems. An attempt has been made to provide a judicious mixture of mathematics, numerical analysis, and computation. It is intended that the reader should not only acquire a working knowledge of practical techniques for solving real problems, but that he should also be prepared for deeper studies of particular topics at the graduate level.

In Chapter 1, we try to define numerical analysis as a subject area and to distinguish it from classical mathematical analysis. We attempt to give an idea of some of the kinds of problems encountered by the numerical analyst and the kinds of techniques used to solve these problems. In Chapter 2, we describe some of the characteristics of automatic digital computers which are of concern to the numerical analyst, and we show, by way of illustration, how accurate and reliable procedures can be developed for evaluating many of the elementary mathematical functions. Chapter 3 presents methods for monitoring and controlling the size of numbers involved in certain types of problems. Chapter 4 is concerned with the solution of the single nonlinear equation $f(x) = 0$ and systems of such equations. The special case of polynomial equations is treated in Chapter 5. Interpolation and approximation are examined in Chapter 6 and numerical differentiation and numerical quadrature in Chapter 7.

Chapters 8, 9, and 10 deal with methods for solving ordinary differential equations. Chapter 8 describes some of the basic methods for solving initial-value problems for a single equation or for systems of equations, and this portion of the material is included in Volume I. Chapter 9, which begins Volume II, is concerned with stability, convergence, and accuracy of the various methods. Two-point boundary-value problems involving second-order differential equations are treated in Chapter 10. Much of the analysis of Chapter 10 is applicable to the study of boundary-value problems involving elliptic partial differential equations in Chapter 15.

Chapter 11 contains a review of the elements of linear algebra and matrix theory, along with an introduction to vector and matrix norms (with emphasis on those results which are used in this book). Arbitrary systems of linear algebraic equations are discussed in Chapter 12. The discussion is restricted to Gaussian elimination with pivoting (and the equivalent triangular decomposition) followed by iterative improvement. Difficulties associated with numerical instability and with ill-conditioned systems are discussed.

We survey the elements of residue arithmetic in Chapter 13 and describe an algorithm (a modified version of Gauss-Jordan elimination) for solving systems of linear algebraic equations *exactly*, using residue arithmetic. This chapter is a modified version of a University of Texas Computation Center report, TNN 82 (revised) by Howell and Gregory [1969a]. Moreover, the report itself is a revised version of a Master's thesis written by Mrs. Howell under the direction of the second author.

The algebraic eigenvalue–eigenvector problem is presented in Chapter 14, and various algorithms are described for handling Hermitian matrices as well as non-Hermitian matrices. Again, the difficulties associated with numerical instability and with ill-conditioned matrices are discussed.

In Chapters 15, 16, and 17 we deal with the numerical solution of partial differential equations. Finite difference methods for solving elliptic equations are studied in Chapter 15. It is shown that the problem can be reduced to that of solving a large system of linear algebraic equations. Iterative methods, such as the successive overrelaxation method for solving these systems, are described in Chapter 16. Initial-value problems, such as the heat equation in two and three (space) dimensions, are considered in the final chapter.

The mathematical and computational background required of the reader varies over different parts of the book. He should have had a course on the elementary aspects of real analysis as covered in a good course in "advanced calculus" or "elementary analysis," and he should also be familiar with elementary complex analysis to some extent. In Appendix A, we have included the basic theorems which we use. In addition to the mathematical background required, the reader should have some knowledge of automatic digital computers and be able to program in an algebraic compiler language such as Fortran or Algol.

For the study of Chapters 8, 9, and 10 the reader should have some knowledge of the elementary theory of ordinary differential equations. Similarly, although the treatment in Chapters 11, 12, 13, and 14 is largely self-contained, a beginning course in linear algebra is almost essential. For Chapters 15, 16, and 17 some background in partial differential equations would be helpful.

The book is designed as a text for a sequence of three one-semester courses beginning at the senior year and continuing into the first year of graduate work. The first course would cover Volume I. The second course dealing with computational methods in linear algebra would cover Chapters 11–14 included in Volume II. The third course would cover ordinary differential equations (Chapters 9 and 10 along with a review of Chapter 8) and partial differential equations (Chapters 15, 16, and 17).

Exercises are given for most of the sections. Some of them are designed to extend the theory, but the primary objective of the exercises is to enhance the student's understanding of the material. In some cases he is asked to work numerical problems, whereas in other cases he is required to fill in gaps in the discussion in the text. A few problems can be worked with the aid of a slide rule or a desk calculator and tables, but many have to be solved on an automatic digital computer.

Decimal notation is used for the numbering of the sections and chapters. For example, the third section of Chapter 6 is numbered 6.3 and referenced as Section 6.3. Equations, theorems, lemmas, figures, and tables are numbered consecutively as items within sections, e.g., the tenth item in Section 6.3 is numbered 3.10. This item might be referenced in several ways. For instance, if it is an *equation*, it is referred to as (3.10), but if it is any other kind of item (such as a theorem), it is called Theorem 3.10. If it is referenced outside Chapter 6, then the notation is 6-(3.10) or Theorem 6-3.10. At this point we should add that A-(3.10) refers to the tenth item in the third section of Appendix A.

We decided to divide the material into two volumes because of its length. The most logical point of division appears to be between Chapters 8 and 9, primarily because the material in the first eight chapters seems suitable for a one-semester introductory course. Appendix A and the complete bibliography are included in both volumes.

Austin, Texas   D.M.Y.
April 1972   R.T.G.

ACKNOWLEDGMENTS

We wish to express our appreciation to the many people who helped make this book a reality. We are especially grateful to Richard Varga who reviewed the entire manuscript. We are also grateful to A. E. McDonald, Y. Ikebe, J. Dauwalder, T. Lyche, Jo Ann Howell, and V. Benokraitis, each of whom reviewed substantial portions of the material. All of the reviewers made excellent suggestions for improving the book. We benefited greatly from our discussion with G. W. Stewart, who generously allowed us to read the manuscript of his forthcoming book on numerical linear algebra. Jo Ann Howell's greatest contribution —over and above her contribution in reviewing some of the material— stems from the fact that Chapter 13 is essentially a revised version of her Master's thesis.

An early draft of portions of this book was prepared in the form of class notes taken by Sylvia Goodrich from lectures given by the first author.

Finally, we must thank Dorothy Baker for her complete dedication to this effort. Her superb job of preparing the manuscript enabled us to bring this project to completion. She was aided at various stages by Barbara Allen, Marge Dragoo, and Linda Brothers.

We wish to dedicate this book to the memory of Professor George E. Forsythe, who was a pioneer and an inspiration in the development of the field of numerical mathematics.

CONTENTS

Chapter 8   Ordinary Differential Equations

VOLUME II

Chapter 9   Ordinary Differential Equations: Stability, Convergence, and Accuracy

Chapter 10   Ordinary Differential Equations: Boundary Value and Eigenvalue Problems

Chapter 11   Vectors, Matrices, and Norms

Chapter 12   The Solution of Systems of Linear Algebraic Equations by Direct Methods

Chapter 13   Solving Systems of Linear Algebraic Equations Using Residue Arithmetic

Chapter 14   The Algebraic Eigenvalue–Eigenvector Problem

Chapter 15   Partial Differential Equations: Elliptic Boundary Value Problems

Chapter 16   Iterative Methods for Solving Large Linear Systems

Chapter 17   Partial Differential Equations: Initial Value Problems

CHAPTER 1

NUMERICAL ANALYSIS AS A SUBJECT AREA

1.1 INTRODUCTION

Numerical analysis is concerned with the application of mathematics to the development of constructive, or algorithmic, methods which can be used to obtain numerical solutions to mathematical problems. Quite often (but not always) we find that certain results of classical analysis are not useful to the numerical analyst. For example, existence and uniqueness theorems for certain classes of problems are often proved by non-constructive methods such as, for instance, showing that the nonexistence of a solution leads to a contradiction. A proof of this kind might not provide any useful information as to how a solution (which has been shown to exist) can actually be found.

Even if an *analytical* solution to a given problem is available, it may not help us in obtaining a *numerical* solution. For example, it can be shown that the series

1.1
$$1 + x + \frac{x^2}{2!} + \frac{x^3}{3!} + \cdots$$

converges absolutely to the function e^x for any value of x. However, the use of the series to evaluate e^{-100} would be completely impractical. Indeed, we would need to compute about 100 terms of the series before the size of the terms would begin to decrease. The amount of work involved for hand computation would be prohibitive. Even for computation using an automatic digital computer, the amount of time required would be excessive. Much worse, there would be an unacceptably high loss of accuracy which would result from the addition of terms of large magnitudes and alternating signs. We remark that the function e^x can be evaluated easily and accurately, for $x = -100$, by other methods such as, for example, by the use of logarithms (see Chapter 2).

Another example, illustrating the fact that an analytical solution to a given problem may not be of practical use, is the following. Given the real numbers a_{ij} $(i, j = 1, 2, \ldots, N)$ and $b_1, b_2, \ldots, b_N$, find real numbers $u_1, u_2, \ldots, u_N$ such that the system of linear algebraic equations

1.2
$$\sum_{j=1}^{N} a_{ij}u_j = b_i, \qquad i = 1, 2, \ldots, N$$

is satisfied. When a unique solution exists it can be described analytically by

Cramer's rule.* However, as we show in Section 12.6, Cramer's rule is completely impractical except for small values of N (for example, $N \leq 3$).

1.2 SOME PITFALLS IN COMPUTATION

During the pre-computer era, numerical calculations were performed by hand, sometimes with the aid of tables of logarithms and other functions. Desk calculating machines or, when limited accuracy would suffice, slide rules were sometimes used. A great deal of numerical computation was done in connection with problems in astronomy and in ballistics. Extensive tables of mathematical functions were computed by hand during the 1930's under the sponsorship of the United States Government through the Works Progress Administration.

The advent of automatic digital computers with their almost infinitely greater speed than the human computer has resulted in the development of many computational algorithms which are largely different from those used in hand computation. Usually, computer-oriented algorithms are designed to be simple and as foolproof as possible. Frequently, economy of the number of arithmetic operations is sacrificed for simplicity. Emphasis is usually placed on the control of the accumulation of rounding errors. The problem of rounding-error accumulation is present also in hand computation but is usually less serious, since there are fewer operations involved and since the human computer can often apply corrective action where appropriate. For example, consider the evaluation of the function

2.1
$$f(x) = \sqrt{x^2 + 1} - 1$$

for $x = 0.25$. Let us assume that the calculations are being carried out to three significant digits, that is, each calculated result is rounded, where appropriate, so that there are no more than three digits after and including the first nonzero digit. (Of course, in the case of addition and subtraction we may produce fewer than three significant digits and so there would be no rounding.) Proceeding in a straightforward way, we obtain†

2.2
$$\begin{aligned}
f(0.25) &= \sqrt{(0.25)^2 + 1} - 1 \\
&\doteq \sqrt{1.06} - 1 \\
&\doteq 1.03 - 1 \\
&= 0.03
\end{aligned}$$

On the other hand, the answer, correct to three significant digits, is 0.0308.

---

* See Theorem A-7.2 and Theorem 12-4.7.

† The symbol $\doteq$ means "is approximately equal to" throughout this book.

Hopefully, a human computer would observe the loss of significant digits resulting from the subtraction of two nearly equal quantities and would transform $f(x)$ into the form

2.3
$$f(x) = \frac{x^2}{\sqrt{1 + x^2} + 1}$$

obtained by multiplying and dividing $f(x)$ by $\sqrt{1 + x^2} + 1$. Computing to three significant digits with the transformed form, we now obtain

2.4
$$f(0.25) = \frac{(0.25)^2}{\sqrt{(0.25)^2 + 1} + 1}$$
$$\doteq \frac{0.0625}{1.03 + 1}$$
$$\doteq 0.0308,$$

a much more accurate result.

The difficulty in evaluating $f(x)$ and the procedure for avoiding the difficulty are special cases of phenomena which occur in the solution of the quadratic equation

2.5
$$Az^2 + Bz + C = 0,$$

whose solution, in general, is written

2.6
$$z = \frac{-B \pm \sqrt{B^2 - 4AC}}{2A}.$$

As an illustration, we let $A = 1$, $B = 2$, $C = -x^2$, and obtain, as roots of (2.5),

2.7
$$\begin{cases} r_1 = -1 - \sqrt{1 + x^2} \\ r_2 = -1 + \sqrt{1 + x^2}. \end{cases}$$

In the case $x = 0.25$, we obtain $r_1 \doteq -2.03$, which is correct to three significant digits. However, if we use (2.7), we obtain $r_2 \doteq 0.03$, which is correct to only one significant digit. This is our result (2.2), and we observed above that the loss of significant digits is due to the subtraction of two nearly equal quantities.

It is desirable to rewrite (2.6) in a form which enables us to avoid the algebraic addition of numbers of opposite sign. Thus, if $B \geqq 0$ and $B^2 - 4AC \geqq 0$, we should use the formulas

2.8
$$
\begin{cases}
r_1 = \dfrac{-B - \sqrt{B^2 - 4AC}}{2A} \\[3mm]
r_2 = \dfrac{2C}{-B - \sqrt{B^2 - 4AC}}.
\end{cases}
$$

Similarly, if $B < 0$ and $B^2 - 4AC \geqq 0$, we should use the formulas

2.9
$$
\begin{cases}
r_1 = \dfrac{-B + \sqrt{B^2 - 4AC}}{2A} \\[3mm]
r_2 = \dfrac{2C}{-B + \sqrt{B^2 - 4AC}}.
\end{cases}
$$

In the first case both B and $\sqrt{B^2 - 4AC}$ have the same sign. In the second case both $-B$ and $\sqrt{B^2 - 4AC}$ have the same sign. Hence, in each case, there is no possibility that two nearly equal quantities could be subtracted, with a subsequent loss of significant digits.

The derivation of (2.8) and (2.9) can be motivated by noting that the roots r_1 and r_2 of the quadratic equation (2.5) satisfy

2.10
$$
r_1 r_2 = \frac{C}{A},
$$

or, equivalently,

2.11
$$
r_2 = \frac{C}{A r_1}.
$$

If we wish to develop a computer program which can solve (2.5) for *any* given values of A, B, and C, then we must first analyze certain special cases. This is necessary if the computer program is to behave with the same intelligence as a human computer.

The first obvious possible difficulty would be the case where $A = 0$ but $B \neq 0$. In hand computation, the human computer would recognize immediately that he should look at the simpler problem $Bz + C = 0$, and he would compute $z = -(C/B)$. If A and B were both zero, then he would certainly know enough to refrain from computing $-(C/B)$ and, instead, he would look at C. If $C = 0$, every value of z satisfies (2.5), whereas if $C \neq 0$, there is no solution.

It is desirable for a computer program to do as well as the human computer, and so it must be designed to take any set of values of A, B, and C and to solve (2.5) for z. Thus, the program must first test to see if A vanishes, and if $A = 0$, it must attempt to solve $Bz + C = 0$. Here again, it must test to see whether or not $B = 0$. If so, then it should tell us that there is no solution, if $C \neq 0$, and it should tell us that every value of z is a solution, if $C = 0$.

Difficulties arise in the above procedure if some or all of the quantities, A, B, C, are not exactly zero but are very "small." If a quantity is "sufficiently small," the program should replace it by zero. However, the decision as to what is sufficiently small is by no means an easy decision to make. We shall consider this question in more detail in Chapter 3.

In view of these comments, then, it appears that an elementary computer program for solving (2.5) should have as input parameters three positive constants ε_A, ε_B, and ε_C. If $|A| < \varepsilon_A$, then A should be replaced by zero. Similarly, for B and C. If A vanishes, then the program should attempt to solve $Bz + C = 0$. If $A \neq 0$ and $B^2 - 4AC \geq 0$, then the program should compute both roots to as high a degree of accuracy as possible using the procedure previously described. If $B^2 - 4AC < 0$, then it should compute both the real and the imaginary parts of the complex conjugate pair

2.12
$$\begin{cases} r_1 = \dfrac{-B}{2A} + i\dfrac{\sqrt{4AC - B^2}}{2A}, \\[2ex] r_2 = \dfrac{-B}{2A} - i\dfrac{\sqrt{4AC - B^2}}{2A}. \end{cases}$$

We have mentioned some of the difficulties which arise in an attempt to develop an effective and foolproof algorithm for solving a quadratic equation. One difficulty we have not yet mentioned is the loss of accuracy which occurs when $B^2 - 4AC$ is very small. In Chapter 3 we discuss, in some detail, this and other difficulties involved in the solution of a quadratic equation.

1.3 MATHEMATICAL AND COMPUTER ASPECTS OF AN ALGORITHM

We have already mentioned that for many types of problems there is a substantial difference between algorithms which are adapted for hand computation and those which are well adapted for machine computation. Unfortunately, many of the books on numerical analysis (and also many university courses) emphasize the former and do not give much consideration to the latter. Although an emphasis on mathematics which causes a neglect of the automatic digital computer is undesirable, the opposite emphasis should also be avoided. A number of books have been written which emphasize the use of automatic digital computers without much regard for the mathematical aspects of numerical analysis. This is unfortunate, since the subject of numerical analysis has not become solidified to the point where there is a universally accepted best algorithm for solving each type of problem. Hence, the problem solver should be aware of the *mathematical* aspects, as well as the *computer* aspects, of selecting an algorithm to solve his problem.

A simple example which illustrates the need for understanding the mathematical as well as the computer aspects of a computational algorithm is the problem of evaluating the function

3.1 $$f(x) = \tan x - \sin x$$

for small values of x. Suppose we let $x = 0.1250$ and perform all calculations correct to four significant digits. In four-place tables of $\sin x$ and $\tan x$, tabulated at increments of 0.0001 in x, we have

3.2 $$\begin{cases} \tan(0.1250) \doteq 0.1257 \\ \sin(0.1250) \doteq 0.1247 \end{cases}$$

and so $f(0.1250) \doteq 0.0010$. However, the value of the function correct to four significant digits is 0.0009804. The result, 0.0010, is correct to two significant digits at most.*

We can obtain a better answer by finding $\tan x$ and $\sin x$ to greater accuracy,† but the number of significant digits needed for the intermediate quantities, namely $\tan x$ and $\sin x$, remains large relative to the number of significant digits in the answer. In such a situation it may be best to search for an equivalent mathematical formulation of the function such that the level of accuracy required for the intermediate calculations will be reasonably in line with the accuracy of the result. One such formulation is obtained by expanding $\tan x$ and $\sin x$ in Taylor's series, obtaining

3.3 $$\begin{cases} \tan x = x + \tfrac{1}{3}x^3 + \tfrac{2}{15}x^5 + \tfrac{17}{315}x^7 + \cdots \\ \sin x = x - \tfrac{1}{6}x^3 + \tfrac{1}{120}x^5 - \tfrac{1}{5040}x^7 + \cdots \end{cases}$$

so that the function becomes

3.4 $$f(x) = \tfrac{1}{2}x^3 + \tfrac{1}{8}x^5 + \tfrac{13}{240}x^7 + \cdots$$

If we use this formulation, we get $f(0.1250) \doteq 0.0009804$.

This is just one very elementary example of the need to examine a problem and, if necessary, make a mathematical reformulation in order to obtain accurate answers with a reasonable expenditure of machine time. Other (more sophisticated) examples will be presented throughout the book.

---

* The relative error is about $0.00002/0.00098 \doteq 2/100$. Hence, we say that we have about *two significant digits* of accuracy.
† If we use eight-place tables then we get $\tan(0.1250) \doteq 0.12565514$, $\sin(0.1250) \doteq 0.12467473$, and hence $f(0.1250) \doteq 0.00098041$.

3.5 *Remark.* When a computational algorithm is applied in practice there are considerations over and above the usual mathematical considerations which must be taken into account. These additional considerations are necessary due to the fact that the computations required are carried out on an automatic digital computer. As we show in Section 2.6, this means that our mathematical universe consists of a *finite* subset of the rational numbers (called machine representable numbers) rather than the *infinite* set of real numbers.* Moreover, the arithmetic cannot be carried out exactly, in general (see Chapter 13 for a notable exception), and so rounding errors contaminate the computed results. Consequently, the usual properties of an algebraic field (closure, associativity etc.) cannot be assumed. For example, we cannot guarantee that

$$a\left[\frac{b}{c}\right] = b\left[\frac{a}{c}\right] = \frac{ab}{c}$$

due to rounding errors which could be introduced.

1.4 NUMERICAL INSTABILITY OF ALGORITHMS AND ILL-CONDITIONED PROBLEMS

If we compute numerical values for the two roots of the quadratic equation (2.5) by using (2.6) we find that, for certain values of the coefficients A, B, and C, one of the roots is not computed with the same relative accuracy as the other root. However, by using (2.8) and (2.9) instead of (2.6) we avoid the possible loss of significant digits by cancellation and improve our chances of obtaining accurate answers.

This brings us to the point where we wish to separate the difficulties associated with a *problem* which requires a numerical solution from those associated with a *computational algorithm* which is used to obtain the numerical solution. In the previous paragraph the problem is to find the roots of a quadratic equation. The two† computational algorithms mentioned above involve the use of either (2.6) or (2.8) and (2.9) in obtaining the roots.

We say that the algorithm using (2.6) is numerically *unstable* because there are choices of the coefficients A, B, and C for which there is a loss of significant digits due to cancellation. On the other hand, we call the algorithm using (2.8) and (2.9) numerically *stable* because no loss of significant digits can occur due to cancellation. By the same token, the algorithm based on (3.4) is numerically stable.

---

* In the case of complex arithmetic we are restricted to working with complex numbers whose real and imaginary parts are machine representable.

† There are other algorithms, of course.

Now we switch our attention from algorithms to problems. Consider the problem of solving the systems of two linear algebraic equations.*

4.1
$$\begin{cases} (2.000)x + (0.6667)y = 2.000 \\ (1.000)x + (0.3333)y = 1.000. \end{cases}$$

If we subtract twice the second equation from the first equation, we obtain $(0.0001)y = 0.0000$ which yields the *unique* solution

4.2
$$\begin{cases} x = 1.000 \\ y = 0.000. \end{cases}$$

On the other hand, consider the system**

4.3
$$\begin{cases} (2.000)x + (0.6666)y = 2.000 \\ (1.000)x + (0.3333)y = 1.000 \end{cases}$$

which differs from (4.1) only in the fourth significant digit of the coefficient 0.6667. As any high school algebra student knows, these equations have *a single infinity of solutions*.† (The coefficients of the first equation are twice the coefficients of the second equation.) Obviously, the equations

4.4
$$\begin{cases} x = 1.000 - (0.3333)k \\ y = k \end{cases}$$

provide us with a solution for *every possible value of k*, including‡ $k = 0$.

The thing to observe here is that a "small perturbation" in a single coefficient of (4.1) changes the problem from one with a unique solution to one with a single infinity of solutions. This is a *mathematical property of* the systems (4.1) which is completely independent of any algorithm used.

If a problem has the property that "small" perturbations in any (or all) of the data produce "small" perturbations in the mathematical solution, we call the problem *well-conditioned*. On the other hand, if "small" perturbations in even part of the data produce "large" perturbation in the mathematical solution, we call the problem *ill-conditioned*. Consequently, if we view the systems (4.1) and (4.3) as differing only by a small perturbation in a single coefficient, then obviously both systems are ill-conditioned.

---

* The coefficients in (4.1) are assumed to be *exact*.

** Again, we assume that the coefficients are *exact*.

† This concept is discussed, in detail, in Chapter 12.

‡ Thus, the unique solution to (4.1) is also one of the solutions to (4.3).

Suppose we introduce the symbol D to represent the exact data which characterize a problem and the symbol F to represent a mathematical function for obtaining the exact solution $F(D)$. We can write this as the mapping

4.5 $$D \xrightarrow{\quad F \quad} F(D).$$

If the data are perturbed so that we must work with D^*, rather than D, then we write the exact solution of the perturbed problem

4.6 $$D^* \xrightarrow{\quad F \quad} F(D^*).$$

4.7 Definition. The problem characterized by the data D is well-conditioned if the fact that D^* is close to D guarantees that $F(D^*)$ is close to $F(D)$, in some sense, and it is ill-conditioned, otherwise.

The sense in which D^* is close to D and $F(D^*)$ is close to $F(D)$ cannot be made precise without knowing the form of the data and the form of the solution. For example, if both D and $F(D)$ are real or complex numbers, then we merely examine $|D - D^*|$ and $|F(D) - F(D^*)|$. On the other hand, if they are vectors or matrices we examine the norms† $\|D - D^*\|$ and $\|F(D) - F(D^*)\|$.

4.8 Definition. Let F^* be a computational algorithm for solving the problem characterized by the data D. Then F^* is a stable algorithm if there exists D^* close to D such that $F(D^*)$ is close to $F^*(D)$, in some sense, and it is unstable, otherwise.

The thing that characterizes a stable computational algorithm is the fact that a solution obtained, using the algorithm, is close, in some sense, to the exact solution of a slightly perturbed problem.

4.9 *Remark.* Certainly, we cannot expect a stable algorithm to solve an ill-conditioned problem more accurately than is warranted by the data. On the other hand an unstable algorithm, when applied to a well-conditioned problem, can produce poor results (see Section 12.13 for an excellent example of this) and obviously, we would not wish to apply an unstable algorithm to the solution of an ill-conditioned problem.

4.10 *Remark.* If we take the point of view that any *computed solution* to a problem is the *exact solution* to a slightly perturbed problem, then the *details* of what caused the computed solution to differ from the exact solution are not too important. This is the motivation for what is called *backward error analysis* as opposed to the usual *forward error analysis*, where we try to keep track of errors as they are introduced and develop bounds on the accumulated errors.

---

† See Chapter 11.

Consider the following diagram

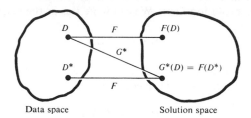

Data space   Solution space

If D represents the exact data, if $F(D)$ represents the exact solution, and if $G^*(D)$ represents the solution obtained by a computer (with all sources of error included), then we assume the existence of perturbed data D^* for which the exact solution $F(D^*) = G^*(D)$. At this point, then, we try to express $\|F(D^*) - F(D)\|$ in terms of $\|D^* - D\|$. In other words, we try to relate a perturbation in the solution to a perturbation in the problem. Once this relationship is known, the error analysis consists of finding out how all possible sources of error lumped together can be translated into a perturbation in the original problem.

Note that G^* is slightly more general than F^* (used above). For example, if D must be perturbed in order to get the data into the computer, then F^* operates on perturbed data (rather than on D) to produce $G^*(D)$.

Backward error analysis was used by Givens [1954] and Lanczos [1956], but its greatest advocate is Wilkinson, and his writings (for approximately fourteen years) reflect this point of view.

1.5 TYPICAL PROBLEMS OF INTEREST TO THE NUMERICAL ANALYST

In this book we attempt to provide a judicious mixture of theoretical numerical analysis on the one hand and practical computer techniques on the other. Thus, we endeavor to combine the mathematical aspects with the computer aspects of numerical analysis. However, we try to avoid undue concern with computer properties which are not likely to have lasting significance. For example, we expect that memory size will increase greatly in the next few years and so we do not emphasize techniques designed to use a small amount of computer memory, especially when this requires a sacrifice of some sort, such as computer time. Of course, large reductions in memory requirements are to be desired. In general, if we refer to a computer language in our discussions, we usually use an algebraic compiler language rather than machine or assembly language. However, in Chapters 2 and 3 we illustrate how certain frequently used computer programs can more effectively be written in assembly language.

As a working definition, let us define a numerical analyst as a mathematician who develops, analyzes, and evaluates computational algorithms for obtaining approximate numerical solutions to mathematical problems. Many of these problems arise in fields such as engineering, the physical and biological sciences, the social sciences, and the behavioral sciences.

In applying mathematical analysis to analyze and evaluate computational algorithms, the numerical analyst is often required to develop new mathematical results and to adapt these new results for effective use with automatic digital computers. The evaluation of algorithms so developed is done by analysis based on mathematical considerations as well as computer considerations (for example, considerations of rounding error accumulation, computation time, etc.). Evaluation of algorithms is also done by solving test problems on the computer. We remark that, following Hartree [1952], we do not consider the actual carrying out of a numerical procedure as numerical analysis except insofar as it might constitute a test of the algorithm.

Some typical problems, which a numerical analyst might be asked to develop procedures for solving, include the following:

i) Evaluate the function

$$f(x) = \frac{\tan x - \sin x}{\sqrt{1 + x^3}}$$

for $x = 0.0, 0.1, 0.2, \ldots, 1.0$. A more important and related problem is to select from among a given class of functions (for example, the class of all polynomials of degree 3 or less) the one which (in some sense) gives a best approximation to $f(x)$ in a prescribed range of the independent variable. Alternatively, we might select a satisfactory approximation rather than the best approximation. Frequently the Taylor series expansion of a function yields a satisfactory approximation, but other functions, for example, rational functions, often yield better approximations. Various kinds of approximation are considered in Chapter 6.

ii) Solve the equation

$$\log z + 2 = z.$$

iii) Given values of $f(x)$, for $x = 0.0, 0.1, \ldots, 1.0$, find an approximate value of $f(x)$ for $x = 0.175$. This is an *interpolation problem* and such problems are very rarely encountered in practice. Related, and more frequently encountered problems for this function are:

     a) Find an approximate value for the derivative of $f(x)$ for $x = 0.2$. This is a *numerical differentiation problem*.

     b) Find an approximate value for $\int_{0.0}^{1.0} f(x)\,dx$. This is a *numerical integration (quadrature) problem*.

iv) Given a set of values $y_1, y_2, \ldots, y_N$ of the dependent variable y, (perhaps based on experimental observations) corresponding to the values $x_1, x_2, \ldots, x_N$ of the independent variable x, find a function $f(x)$ from among a given class of functions which is (in some sense) optimal. For example, a given class of functions might include all polynomials of degree k, or less, and we might choose $f(x)$ from this class so that

$$\left[\sum_{i=1}^{N} [f(x_i) - y_i]^2 \right]^{1/2}$$

is minimized. Such a problem frequently arises in the analysis of experimental data. This is a *least squares problem*.

v) Find a function $y(x)$ which satisfies the ordinary differential equation

$$\frac{dy}{dx} = x + y^2$$

in the interval $0 \leq x \leq 2$ and, at the same time, satisfies the *initial condition* $y(0) = 1$. This is an *initial value problem* in ordinary differential equations.

vi) Given a set of real or complex numbers a_{ij} $(i, j = 1, 2, \ldots, N)$, find a real or complex number λ and a set of real or complex numbers $u_1, u_2, \ldots, u_N$ (not all zero) such that

$$\sum_{j=1}^{N} a_{ij} u_j = \lambda u_i, \qquad i = 1, 2, \ldots, N.$$

This is *the algebraic eigenvalue–eigenvector problem*.

vii) Find a function $u(x, y)$, defined and continuous in the region shown in Fig. 5.1, which satisfies Laplace's equation

$$\frac{\partial^2 u}{\partial x^2} + \frac{\partial^2 u}{\partial y^2} = 0$$

in the interior and satisfies the condition

$$u(x, y) = \sin \pi x + y^3$$

on the boundary. This is an *elliptic boundary value problem* in partial differential equations.

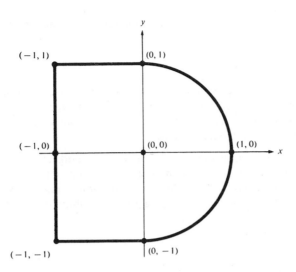

5.1 Fig.

1.6 ITERATIVE METHODS

The availability of an automatic digital computer makes it feasible to consider the use of *iterative methods* (as opposed to *direct methods*) for solving certain types of problems. With an iterative method, we choose an (almost) arbitrary initial approximation to the solution of the problem (a number, a function, etc.) and successively improve this approximate solution (that is, we iterate) in such a way that the sequence of improved solutions converges to the solution of the given problem. Such methods are frequently very simple, and consequently very attractive for use with digital computers, in spite of the many arithmetic operations normally involved, because of the high computation speeds of most computers. Iterative methods have the very desirable feature that they are self-correcting.

Whereas direct methods (theoretically, at least) converge in a *finite number* of steps, iterative methods require an *infinite number* of steps for convergence. However, an iterative method is seldom used unless its convergence is rapid enough so that a (relatively small) finite number of iterations produces an acceptable numerical approximation to a solution, whenever the initial approximation is reasonably close to a solution.

Iterative methods are frequently used to solve equations of the form

6.1 $$f(x) = 0,$$

where $f(x)$ is a nonlinear function of x. We select an initial approximation x_0 to a solution x and successively compute $x_1, x_2, \ldots$, according to some iteration algorithm. For example, we might choose a function $\phi(x)$ and determine the x_i by using the iteration algorithm

6.2 $$x_{n+1} = \phi(x_n), \qquad n = 0, 1, 2, \ldots .$$

In developing iterative methods, the numerical analyst is faced with such tasks as proving convergence, determining the rate of convergence, and establishing criteria for determining when to stop the iteration with an acceptable approximation to the solution of a given problem.

6.3 EXAMPLE. To illustrate the use of an iterative method, let us consider, in some detail, the problem of determining the square root of a nonnegative number N. Actually, the algorithm is useful for hand computation (using a desk computer) as well as for automatic computation (using a digital computer). The algorithm is described by the formulas

$$\begin{cases} x_0 = \dfrac{1 + N}{2} \\[2ex] x_{n+1} = \dfrac{1}{2}\left[x_n + \dfrac{N}{x_n} \right], \qquad n = 0, 1, 2, \ldots . \end{cases}$$

where x_0 is the initial approximation. As shown in Chapter 4, the above formulas can be derived by applying the Newton method to the solution of the equation $x^2 - N = 0$.

For instance, consider the case $N = 0.5$. If we substitute in the formulas above, we generate the approximations

$$x_0 = 0.75,$$

$$x_1 = (0.5)\left[0.75 + \frac{0.5}{0.75} \right] \doteq 0.70833,$$

$$x_2 = (0.5)\left[0.70833 + \frac{0.5}{0.70833} \right] \doteq 0.70711.$$

Continuing in this manner, we obtain

$$x_3 \doteq 0.707107,$$

$$x_4 \doteq 0.7071068,$$

$$x_5 \doteq 0.7071068. \ \text{-}$$

Since $(0.7071068)^2 \doteq 0.50000003$ and $(0.7071067)^2 \doteq 0.49999988$ it follows that the final value 0.7071068 is correct to within one unit in the seventh decimal place.

To prove that this iterative method converges, we first observe that

6.4
$$x_0 - \sqrt{N} = \frac{1+N}{2} - \sqrt{N}$$

$$= \frac{1 - 2\sqrt{N} + N}{2}$$

$$= \frac{(1 - \sqrt{N})^2}{2}$$

is nonnegative and so $x_0 \geq \sqrt{N}$. In the same way we show that

6.5
$$x_{n+1} - \sqrt{N} = \frac{1}{2x_n}[x_n - \sqrt{N}]^2$$

and so $x_{n+1} \geq \sqrt{N}$. Finally,

6.6
$$x_{n+1} - x_n = \frac{1}{2x_n}[N - x_n^2].$$

Consequently, $x_n \geq \sqrt{N}$ for all n and $x_0 \geq x_1 \geq x_2 \geq \ldots$. Thus, the sequence $x_0, x_1, x_2, \ldots$ is a monotone nonincreasing sequence and is bounded below by $\sqrt{N}$. Consequently, by Theorem A-2.8, the sequence converges to a limit z.

Suppose

$$z = \lim_{n \to \infty} x_n = 0.$$

Then, since the sequence converges to 0, and since $x_n \geq \sqrt{N}$ for all n, it follows that $N = 0$. On the other hand, suppose

$$z = \lim_{n \to \infty} x_n > 0.$$

Then we can take limits of both sides of the iteration formula and obtain

6.7
$$z = \frac{1}{2}\left[z + \frac{N}{z}\right],$$

which may be written

6.8
$$z^2 = N.$$

Both $z = +\sqrt{N}$ and $z = -\sqrt{N}$ satisfy this equation but we have assumed that $z > 0$. Hence $z = \sqrt{N}$. This proves the convergence.

One of the attractive features of the algorithm for use on an automatic digital computer is the fact that if $N < 1$, then $\sqrt{N}$ can be computed without ever encountering numbers which are greater than one. This makes it convenient, as we shall see in Chapter 2, to use fixed-point arithmetic. To show that no intermediate value exceeds one let $N < 1$ and write the formulas of (6.3) in the form

6.9
$$\begin{cases} x_0 = \dfrac{1}{2} + \dfrac{N}{2} \\[2ex] x_{n+1} = \dfrac{x_n}{2} + \dfrac{N/2}{x_n}. \end{cases}$$

Note that $N < 1$ guarantees that $x_0 < 1$ in (6.9) whereas, if we use $x_0 = (1 + N)/2$ the numerator $1 + N$ can be greater than one and this defeats our purpose. Similarly, in (6.9) we know that $x_n/2 < 1$, since $x_0 < 1$ and the x_n form a nonincreasing sequence. Moreover, $x_n > \sqrt{N}$ so that $N/(2x_n) < \sqrt{N}/2 < 1$. However, we want to avoid dividing N by $2x_n$, since $2x_n$ may exceed one, so we choose the form

6.10
$$\frac{N/2}{x_n},$$

where N is first divided by 2 and the result is then divided by x_n.

The algorithm converges slowly for some values of N. To take an extreme case, suppose that $N = 0$. In this case we have $x_0 = \frac{1}{2}$, $x_1 = [\frac{1}{2}]^2$, $x^2 = [\frac{1}{2}]^3, \ldots$, $x_n = [\frac{1}{2}]^{n+1}$. To converge to within ε of the true answer, $\sqrt{N} = 0$, we would require p iterations where

6.11
$$2^{-(p+1)} \leqq \varepsilon.$$

If $\varepsilon = 10^{-8}$, we would require about 24 iterations. Even more iterations would be required for values of $N \gg 1$; for instance, $N = 10^{200}$.

By limiting N to a prescribed range, however, we can limit the number of iterations needed for convergence. For example, if we require that N lie in the range

6.12
$$\tfrac{1}{4} \leqq N < 1,$$

then the number of iterations required to reduce the error $|x_n - \sqrt{N}|$ to no more than 2^{-48}, does not exceed 4. To show this, we note that since $\sqrt{N} \geqq \frac{1}{2}$ we have $x_n \geqq \sqrt{N} \geqq \frac{1}{2}$ for all n. By (6.5) we have

6.13
$$|x_{n+1} - \sqrt{N}| \leqq |x_n - \sqrt{N}|^2$$

and

6.14
$$|x_n - \sqrt{N}| \leqq |x_0 - \sqrt{N}|^{2^n} \leqq [\tfrac{1}{8}]^{2^n}.$$

Thus if $n = 4$, we have

6.15 $$|x_n - \sqrt{N}| \leqq 2^{-48}.$$

and the result is proved.

As we shall see in Chapter 2 (where the procedure for handling the case of an arbitrary $N > 0$ is described in detail) numbers m and p can be determined so that

6.16 $$N = m2^p,$$

where p is an even integer and $\frac{1}{4} \leqq m < 1$. We then determine $\sqrt{m}$ as described above, and obtain the result

6.17 $$\sqrt{N} = \sqrt{m}2^{p/2}.$$

EXERCISES

1. How many terms of the series

$$e^x = 1 + x + \frac{x^2}{2!} + \frac{x^3}{3!} + \cdots$$

must be used to get e^x correct to 4 significant figures if $x = -2$? If $x = 0.6$? In each case evaluate the sum of the first 6 terms and compare the observed error with an appropriate bound for the error.

2. Solve the following system by Cramer's rule

$$\begin{cases} 2u_1 - u_2 = 1, \\ -u_1 + 2u_2 - u_3 = 2, \\ -u_2 + 2u_3 = 1. \end{cases}$$

3. Describe a method for evaluating a determinant and estimate the number of multiplications required to evaluate a 10 by 10 determinant.

4. Evaluate $(\tan x - \sin x)x^{-3}$, for $x = 0.01$, correct to four significant digits.

5. Evaluate $f(x) = \sqrt{1 + x^2} - 1$, for $x = 0.27$, correct to three significant digits, carrying three significant digits at each intermediate stage of the computation.

6. Find, correct to three significant digits, the smaller root of the equations

$$x^2 - 18x + 1 = 0$$

and

$$x^2 - 20.5x + 1.25 = 0.$$

Each intermediate operation should be carried out to three significant digits.

7. Write a computer program for solving the quadratic equation

$$Az^2 + Bz + C = 0,$$

where A, B, and C are given real numbers. The program should handle special cases (for example, where $A = 0$) and should minimize cancellation and the effects of rounding errors. Apply the program for the following cases

$$
\begin{array}{lcccccccc}
A: & 2 & 0 & 0 & 1 & 0 & 1 & 1 & 1 \\
B: & 10 & 0 & 1 & 0 & 0 & 1 & -2 & -8.001 \\
C: & 1 & 1 & 0 & 0 & 0 & 10 & 8 & 16.004
\end{array}
$$

8. Carry out two iterations of the algorithm described in (6.3) and compute an approximation to $\sqrt{1/3}$. Compare the actual error with the bound for the error given in the text. (First compute x_2 as a rational number.)

9. Find a number λ such that for some u_1 and u_2, not both zero, we have

$$
\begin{cases}
2u_1 - u_2 = \lambda u_1 \\
-u_1 + 2u_2 = \lambda u_2.
\end{cases}
$$

CHAPTER 2

ELEMENTARY OPERATIONS
WITH AUTOMATIC DIGITAL COMPUTERS

2.1 INTRODUCTION

In this chapter we discuss some of the characteristics of automatic digital computers. We include a brief description of number representation using bases other than ten (with an emphasis on base two representation). Fixed- and floating-point arithmetic (in binary) are discussed and we include an analysis of rounding errors. We close the chapter with a discussion of the calculation of some of the elementary functions using an algorithmic language.

The reader is assumed to be familiar with some version of an algorithmic language such as Fortran or Algol and to know how to read a simple logical flow diagram.

2.2 BINARY ARITHMETIC

In order to understand how most scientific computers work, it is necessary to understand the elements of binary arithmetic. Although the reader may have had some introduction to binary arithmetic in previous courses, we shall review the subject, briefly, and introduce some basic notation.

We begin by describing our standard positional notation, whereby the meaning of a digit depends on its position relative to other digits in a number representation. Let B be any positive integer greater than unity and let d_i, for any integer i, be one of the numbers $0, 1, 2, \ldots, B - 1$. Then any positive *integer* can be written (uniquely)

2.1 $$d_n d_{n-1} \ldots d_2 d_1 d_0 = d_0 B^0 + d_1 B^1 + d_2 B^2 + \cdots + d_{n-1} B^{n-1} + d_n B^n$$

$$= \sum_{i=0}^{n} d_i B^i.$$

Likewise, any positive *fraction** can be written

2.2 $$0 . d_{-1} d_{-2} \ldots d_{-j} \ldots = d_{-1} B^{-1} + d_{-2} B^{-2} + \cdots + d_{-j} B^{-j} + \cdots$$

$$= \sum_{j=1}^{\infty} d_{-j} B^{-j}.$$

---

* In this book we define a fraction to be a real number x such that $|x| < 1$.

(The proof of these two statements is left as an exercise for the reader. See Exercise 3.)

It should be pointed out that for many rational fractions (but, for a given B, not all rational fractions) the series (2.2) reduces to a finite sum. When this is the case there is also an infinite series which represents the same fraction. For example, 0.1 and 0.0111 ... both represent one-half, if B is two, and 0.5 and 0.4999 ... both represent one-half if B is ten. A rational fraction may have a finite representation for some values of B and no finite representation for most other values of B. For example, one-tenth is 0.1 if B is ten and 0.000110011001100 ... if B is two.

It follows from (2.1) and (2.2) that any positive real *number* can be written

2.3 $$d_n d_{n-1} \ldots d_2 d_1 d_0 . d_{-1} d_{-2} \ldots d_{-j} \ldots = \sum_{i=0}^{n} d_i B^i + \sum_{j=1}^{\infty} d_{-j} B^{-j}.$$

The integer $B > 1$ is called the *base* of the representation. The familiar decimal system uses the base ten. If the base is two or eight, we use the terms *binary* or *octal*, respectively.

For example, if B is ten, we know that

2.4 $347.29 = [7 \cdot 10^0 + 4 \cdot 10^1 + 3 \cdot 10^2] + [2 \cdot 10^{-1} + 9 \cdot 10^{-2}],$
            $= [7 + 40 + 300] + [2/10 + 9/100],$
            $=$ three hundred forty-seven and twenty-nine hundredths.

In order to avoid confusion when more than one number base is used, let us introduce the notation

2.5 $$d_n d_{n-1} \ldots d_1 d_0 . d_{-1} d_{-2} \ldots d_{-j} \ldots = \phi_B(x)$$

to represent the number x to the base B, where the nonnegative integers d_i and d_{-j} are less than B. *Whenever numerical values are assigned to the letters B and x in $\phi_B(x)$ we shall always use decimal notation for those two letters.* For example,

2.6
$$\begin{cases} \phi_{10}(1/3) = 0.3333 \ldots \\ \phi_2(1/3) = 0.010101 \ldots \\ \phi_8(1/3) = 0.252525 \ldots \\ \phi_{10}(3/2) = 1.5 \\ \phi_2(3/2) = 1.1 \\ \phi_8(3/2) = 1.4 \\ \phi_2(1/10) = 0.000110011001100 \ldots \\ \phi_{10}(21/2) = 10.5 \\ \phi_{12}(21/2) = T.6 \end{cases}$$

where ten and eleven are represented by the symbols T and E, respectively, if $B = 12$.

Since most scientific computers use $B = 2$, then most scientific computers perform arithmetic operations on numbers which have finite *binary* representations. These computers, in effect, store the *binary* addition table

+	0	1
0	0	1
1	1	10

and the binary multiplication table

×	0	1
0	0	0
1	0	1

Operations can be performed using the binary representation of numbers in a manner analogous to the procedures using the decimal representation. The same rules for "carrying" in addition and multiplication, and for "borrowing" in subtraction are used. Division is easy to carry out using a procedure similar to that for long division of decimal numbers. However, division in binary is much easier in the sense that at each stage we can tell by inspection whether a "trial divisor" of one will go into the dividend. Thus, for example, we can obtain $\phi_2(1/10)$ above by dividing one by ten using binary notation as follows:

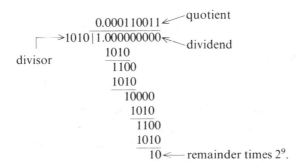

In other words, if we stop at this point, we can write

2.7
$$\phi_2(1/10) = \frac{1}{1010}$$

$$= 0.000110011 + \frac{0.000000010}{1010}$$

and so the quotient and the remainder are 0.000110011 and 0.000000010, respectively.

However, one-tenth is a rational number, and if we continue the division we must get a repeating pattern of binary digits. It is obvious, from the analysis above, that

2.8
$$\frac{1}{1010} = 0.000110011001100\ldots .$$

Obviously, in a fixed-word-length binary computer we can store only a finite number of binary digits and so we cannot store one-tenth exactly.

Notice that we must be very careful with our notation when we are not using the decimal numeral system exclusively. To begin with, we are discussing the *representation* of a number and not the *concept* of a number. The concept of seven, for example, has many representations, including

seven, VII, $[111]_{two}$, 7.

We choose different representations to suit different uses, but the concept of seven is unaffected by its representation. The English word for the concept is unambiguous (to those who understand English), whereas 111 usually must be written $[111]_{two}$ unless it is clearly understood that this is the binary representation of seven.

Table 3.0 gives different representations of the first twenty positive integers.

EXERCISES 2.2

1. Express 3 and 10 in binary and compute 3/10. Show that the result has a repeating pattern. By summing a suitable geometric series verify that the quotient is indeed 3/10.

2. Show that if y is an integer, then the representation of $1/y$ to the base $B > 1$ is repeating. Also show that this is true of x/y where x is also an integer.

3. Verify the representations (2.1) and (2.2).

2.3 CONVERSION FROM BASE D TO BASE B REPRESENTATION

It is easy to describe an algorithm for converting from $\phi_D(x)$ to $\phi_B(x)$ and vice versa. We usually break the procedure into two parts and convert the *integer part* of x and the *fraction part* of x separately. The two parts can then be combined.

3.0 Table

Different Representations of the First Twenty Positive Integers

Number	Base Two	Base Three	Base Eight	Base Ten	Base Twelve
one	1	1	1	1	1
two	10	2	2	2	2
three	11	10	3	3	3
four	100	11	4	4	4
five	101	12	5	5	5
six	110	20	6	6	6
seven	111	21	7	7	7
eight	1000	22	10	8	8
nine	1001	100	11	9	9
ten	1010	101	12	10	T
eleven	1011	102	13	11	E
twelve	1100	110	14	12	10
thirteen	1101	111	15	13	11
fourteen	1110	112	16	14	12
fifteen	1111	120	17	15	13
sixteen	10000	121	20	16	14
seventeen	10001	122	21	17	15
eighteen	10010	200	22	18	16
nineteen	10011	201	23	19	17
twenty	10100	202	24	20	18

Consider the algorithm for converting *integers* from base *D* to base *B*, where *D* and *B* may be arbitrary bases. Let the integer *N* have the two equivalent representations

3.1
$$N = \sum_{i=0}^{r} a_i D^i,$$

and

3.2
$$N = \sum_{i=0}^{s} c_i B^i,$$

where the digits a_i are known and the digits c_i are to be found. If we divide by *B*, we obtain

3.3
$$\frac{N}{B} = \sum_{i=1}^{s} c_i B^{i-1} + \frac{c_0}{B}$$

$$= N_1 + \frac{c_0}{B},$$

where N_1 is the integral quotient and c_0 is the remainder. Notice that the remainder is the least significant digit of the base B representation.

Now divide N_1 by B.

3.4
$$\frac{N_1}{B} = \sum_{i=2}^{s} c_i B^{i-2} + \frac{c_1}{B}$$

$$= N_2 + \frac{c_1}{B},$$

where N_2 is the integral quotient and c_1 is the remainder. This time the remainder is the next significant digit of the base B representation. It is not difficult to show that if we perform $s + 1$ such steps, we obtain the digits $c_0, c_1, \ldots, c_s$, in that order.

3.5 PROBLEM. If $\phi_{10}(x) = 75$, find $\phi_2(x)$.

Solution. The arithmetic will be carried out using decimal notation.

$$\begin{aligned}
\tfrac{75}{2} &= 37 + \tfrac{1}{2}, & c_0 &= 1, \\
\tfrac{37}{2} &= 18 + \tfrac{1}{2}, & c_1 &= 1, \\
\tfrac{18}{2} &= 9 + \tfrac{0}{2}, & c_2 &= 0, \\
\tfrac{9}{2} &= 4 + \tfrac{1}{2}, & c_3 &= 1, \\
\tfrac{4}{2} &= 2 + \tfrac{0}{2}, & c_4 &= 0, \\
\tfrac{2}{2} &= 1 + \tfrac{0}{2}, & c_5 &= 0, \\
\tfrac{1}{2} &= 0 + \tfrac{1}{2}, & c_6 &= 1.
\end{aligned}$$

Hence,

$$\phi_2(75) = 1001011.$$

To convert *fractions* from base D to base B we proceed in similar fashion. Let $0 \leqq F < 1$ and let F have the equivalent representations

3.6
$$F = \sum_{i=1}^{\infty} f_{-i} D^{-i}$$

and

3.7
$$F = \sum_{i=1}^{\infty} g_{-i} B^{-i},$$

where the digits f_{-i} are known, and the digits g_{-i} are to be found.

This part of the algorithm involves multiplying by B instead of dividing by B as we did above. Thus,

3.8
$$BF = g_{-1} + \sum_{i=2}^{\infty} g_{-i}B^{1-i}$$

$$= g_{-1} + F_1,$$

where g_{-1} is an integer and $F_1 < 1$. Notice that the integral part of the product is the most significant digit of the base B representation.

Next, multiply F_1 by B. This gives us

3.9
$$BF_1 = g_{-2} + \sum_{i=3}^{\infty} g_{-i}B^{2-i}$$

$$= g_{-2} + F_2,$$

where g_{-2} is an integer and $F_2 < 1$. This time the integral part of the product is the next most significant digit of the base B representation of F. It is easy to show that if we perform t such steps we obtain the digits $g_{-1}, g_{-2}, \ldots, g_{-t}$, in that order.

3.10 PROBLEM. If $\phi_{10}(x) = 0.8$, find $\phi_2(x)$.

Solution. The arithmetic will be carried out using decimal notation. Thus,

$$\begin{array}{ll}
(0.8)(2) = 1.6, & g_{-1} = 1, \\
(0.6)(2) = 1.2, & g_{-2} = 1, \\
(0.2)(2) = 0.4, & g_{-3} = 0, \\
(0.4)(2) = 0.8, & g_{-4} = 0.
\end{array}$$

At this point we observe that we are back where we started and so the digits are periodic, with a period of four. Hence,

$$\phi_2(0.8) = 0.110011001100\ldots .$$

This problem illustrates the fact that a rational fraction with a finite representation with respect to one base may have an infinite (but periodic) representation with respect to another base.

If we combine (3.5) with (3.10) we obtain

3.11
$$\phi_2(75.8) = 1001011.110011001100\ldots .$$

It is sometimes simpler to use the definition of our positional notation, (2.3), to carry out these conversions than to use the algorithm described above, especially when the conversion is *to* the decimal numeral system from some other system.

3.12 PROBLEM. If $\phi_2(x) = 1110.101$, find $\phi_{10}(x)$.

Solution. From (2.3), using decimal notation, we obtain

$$\begin{aligned}\phi_{10}(x) &= (2^3 + 2^2 + 2) + (2^{-1} + 2^{-3})\\ &= (8 + 4 + 2) + (1/2 + 1/8)\\ &= 14 + 5/8\\ &= 14.625.\end{aligned}$$

3.13 PROBLEM. If $\phi_8(x) = 27.43$, find $\phi_{10}(x)$.

Solution. From (2.3), using decimal notation, we obtain

$$\begin{aligned}\phi_{10}(x) &= (2 \cdot 8^1 + 7 \cdot 8^0) + (4 \cdot 8^{-1} + 3 \cdot 8^{-2})\\ &= 23 + 35/64\\ &= 23.546875.\end{aligned}$$

The reader should carry out these conversions using the algorithm described previously, in order to see the difference between the two procedures.

Notice that when we use the conversion algorithm described above we do the arithmetic in the notation of the base *from which* we are converting. This is necessary because the number being converted has a given representation in that numeral system and presumably we don't know its representation in any other numeral system, at this point.

On the other hand, when we use the definition of our positional notation, (2.3), we do the arithmetic in the notation of the base *to which* we are converting. This is particularly useful when we convert to the decimal numeral system. For example, consider (3.11) again, where (2.3) will be used and the arithmetic will be done in binary.

3.14 PROBLEM. If $\phi_{10}(x) = 75.8$, find $\phi_2(x)$.

Solution.

$$\begin{aligned}\phi_2(x) &= [(111)(1010) + (101)] + [(1000)(1010)^{-1}]\\ &= (1000110 + 101) + [(1000)(0.000110011001100\ldots)]\\ &= 1001011.110011001100\ldots.\end{aligned}$$

One final example illustrates how the algorithm described earlier can be carried out when the arithmetic is binary.

3.15 PROBLEM. If $\phi_2(x) = 1001011.11001100$, find $\phi_{10}(x)$.

Solution. Notice that this is approximately the inverse of Problem 3.14. First we convert the integer part of x.

$$\frac{1001011}{1010} = 111 + \frac{101}{1010} \qquad c_0 = 5,$$

$$\frac{111}{1010} = 0 + \frac{111}{1010} \qquad c_1 = 7.$$

Next, we convert the fraction part of x.

$$
\begin{aligned}
(0.110011)(1010) &= 111.111110, & g_{-1} &= 7, \\
(0.111110)(1010) &= 1001.101100, & g_{-2} &= 9, \\
(0.101100)(1010) &= 110.111000, & g_{-3} &= 6, \\
(0.111000)(1010) &= 1000.110000, & g_{-4} &= 8, \\
(0.110000)(1010) &= 111.100000, & g_{-5} &= 7, \\
(0.100000)(1010) &= 101.000000, & g_{-6} &= 5.
\end{aligned}
$$

Hence,

$$\phi_{10}(x) = 75.796875.$$

Conversion from binary to a base which is a power of two (for example, eight) is relatively simple and can be done by inspection. To illustrate this, consider the binary representation of the positive integer N,

3.16
$$b_{3k+2}b_{3k+1}b_{3k}\ldots b_5b_4b_3b_2b_1b_0 = \sum_{i=0}^{3k+2} b_i2^i,$$

where at least one of the digits b_{3k+2}, b_{3k+1}, or b_{3k} is different from zero. This may be rearranged and written

3.17
$$
\begin{aligned}
N = \;& 2^{3k}(b_{3k+2}\cdot 2^2 + b_{3k+1}\cdot 2^1 + b_{3k}\cdot 2^0) + \cdots \\
& + 2^3(b_5\cdot 2^2 + b_4\cdot 2^1 + b_3\cdot 2^0) \\
& + 2^0(b_2\cdot 2^2 + b_1\cdot 2^1 + b_0\cdot 2^0).
\end{aligned}
$$

If we set

3.18
$$
\left\{
\begin{aligned}
d_0 &= b_2\cdot 2^2 + b_1\cdot 2^1 + b_0\cdot 2^0, \\
d_1 &= b_5\cdot 2^2 + b_4\cdot 2^1 + b_3\cdot 2^0, \\
&\;\;\vdots \\
d_k &= b_{3k+2}\cdot 2^2 + b_{3k+1}\cdot 2^1 + b_{3k}\cdot 2^0,
\end{aligned}
\right.
$$

we can write, since $0 \leq d_i \leq 7$ for $i = 0, 1, 2, \ldots, k$,

3.19
$$
\begin{aligned}
N &= d_0 8^0 + d_1 8^1 + \cdots + d_k 8^k, \\
&= \sum_{i=0}^{k} d_i 8^i,
\end{aligned}
$$

which is the octal representation of N. Therefore, to convert a binary integer to octal, we group the bits* by threes and convert each group of three bits to an octal digit. This grouping starts *at the binary point*.

---

* A bit is a binary digit, that is, either 0 or 1.

3.20 PROBLEM. If $\phi_2(x) = 10111001$, find $\phi_8(x)$.

Solution. By inspection,

$$\phi_8(x) = 271.$$

To convert an octal integer to binary, the octal digits are converted individually into binary triples.

3.21 PROBLEM. If $\phi_8(x) = 1074$, find $\phi_2(x)$.

Solution. By inspection,

$$\phi_2(x) = 1000111100.$$

Similar algorithms apply in the case of binary-to-octal and octal-to-binary conversion of fractions.

3.22 PROBLEM. If $\phi_2(x) = 0.10111001$, find $\phi_8(x)$.

Solution. By inspection,
$$\phi_8(x) = 0.562.$$

3.23 PROBLEM. If $\phi_8(x) = 0.3704$, find $\phi_2(x)$.

Solution. By inspection,

$$\phi_2(x) = 0.0111110001.$$

EXERCISES 2.3

1. Convert the number 12.85 to binary. Take the resulting number and convert back to decimal. Both procedures should be carried out first working in decimal. Then do the same thing working in binary. Find m and p such that $-12.85 = m2^p$ where p is an integer and $\frac{1}{2} \leqq |m| < 1$.

2. Convert 21.16 to base 7, working in decimal, and also working in base 7.

2.4 REPRESENTATION OF INTEGERS ON A BINARY COMPUTER

Let us consider the representation of integers using a fixed number (say k) of digits $b_0, b_1, \ldots, b_{k-1}$ each of which is one of the integers $0, 1, 2, \ldots, B - 1$. We wish to represent both positive and negative integers using the B^k different representations. The string of digits $b_{k-1}b_{k-2} \ldots b_1 b_0$ corresponds to a *word** on a fixed-word-length computer. This word can be considered to be the k-digit nonnegative integer $b_{k-1}B^{k-1} + b_{k-2}B^{k-2} + \cdots + b_0.$

---

* The term *word* has another meaning; it also means the memory location which contains the digits $b_0, b_1, \ldots, b_{k-1}$.

We now describe three of the methods which could be used. In each case we let $[x]$ denote the representation of x.

i) *Sign and absolute value representation*
Here we represent the absolute value of x by the $k - 1$ digits $b_{k-2}b_{k-3} \ldots b_0$. The sign is represented by b_{k-1}. If the integer is nonnegative we let $b_{k-1} = 0$; if it is nonpositive we let $b_{k-1} = 1$.
For example, if $B = 2, k = 3$ we have

$$
\begin{array}{ll}
[0] = 000 & [-0] = 100 \\
[1] = 001 & [-1] = 101 \\
[2] = 010 & [-2] = 110 \\
[3] = 011 & [-3] = 111.
\end{array}
$$

The representation of minus zero is somewhat questionable. One might consider this representation as nonadmissible.

ii) *Representation modulo B^k*
Given an integer x such that

$$- B^{k-1} \le x \le B^{k-1} - 1$$

we let

$$
[x] = \begin{cases} x & \text{if } x \ge 0 \\ x + B^k & \text{if } x < 0. \end{cases}
$$

For example, if $B = 2, k = 3$ we have

$$
\begin{array}{ll}
[0] = 000 & \\
[1] = 001 & [-1] = 111 \\
[2] = 010 & [-2] = 110 \\
[3] = 011 & [-3] = 101 \\
& [-4] = 100.
\end{array}
$$

Letting B^{k-1} represent $- B^{k-1}$ is somewhat questionable. One might consider B^{k-1} either as a nonadmissible representation or else interpret it as minus zero.

iii) *Representation modulo $B^k - 1$*
Given an integer x such that

$$- B^{k-1} + 1 \le x \le B^{k-1} - 1$$

we let

$$
[x] = \begin{cases} x & \text{if } x > 0 \\ x + B^k - 1 & \text{if } x < 0 \\ 0 \text{ or } B^k - 1 & \text{if } x = 0. \end{cases}
$$

For example, if $B = 2$, $k = 3$ we have

$$[0] = 000 \qquad [-0] = 111$$
$$[1] = 001 \qquad [-1] = 110$$
$$[2] = 010 \qquad [-2] = 101$$
$$[3] = 011 \qquad [-3] = 100.$$

The representation of minus zero by $B^k - 1$ is somewhat questionable. One might consider $B^k - 1$ as a nonadmissible representation.

The reader should verify that given a representation of a nonnegative integer x, one obtains a representation of $-x$ as follows:

a) With (ii) one replaces the least significant nonzero digit, say b_s, by $B - b_s$. Each digit b_t to the left of b_s is replaced by $B - 1 - b_t$. Thus, in the case $B = 2$, we leave the least significant digit of one alone and replace all zeros and ones to the left by ones and zeros, respectively. This is called the *twos complement representation* of negative integers even though only the least significant nonzero digit is complemented with respect to two and all more significant digits are complemented with respect to one.

b) With (iii) one replaces *each* digit b_s by $B - 1 - b_s$. Thus, in the case $B = 2$, we replace all zeros by ones and all ones by zeros. This is called the *ones complement representation* of negative integers.

Table 4.1 exhibits the representations of the integers under schemes (i), (ii), and (iii) for the case $k = 3$, $B = 2$. Table 4.2 gives the interpretation of the $2^3 = 8$ possible bit patterns for the same case.

We note that for all three representation schemes the first bit determines the sign. Thus, the number is nonnegative if the first digit is a zero and nonpositive if

4.1 Table
Representations of Binary Integers

Integer	(i) Signed Absolute Value	(ii) Modulo 2^3	(iii) Modulo $2^3 - 1$
0	000	000	000
1	001	001	001
2	010	010	010
3	011	011	011
-0	100	—	111
-1	101	111	110
-2	110	110	101
-3	111	101	100
-4	—	100	—

4.2 Table

Interpretations of Binary Bit Patterns

	Binary Bit Pattern	(i) Signed Absolute Value	(ii) Modulo 2^3	(iii) Modulo $2^3 - 1$
Positive Integers	000	0	0	0
	001	1	1	1
	010	2	2	2
	011	3	3	3
Negative Integers	100	-0	-4	-3
	101	-1	-3	-2
	110	-2	-2	-1
	111	-3	-1	-0

the first digit is a one. We also note that with two of the schemes, namely (i) and (iii) we have a representation for -0, whereas in scheme (ii) we have a representation for -4 even though there is no representation for 4.

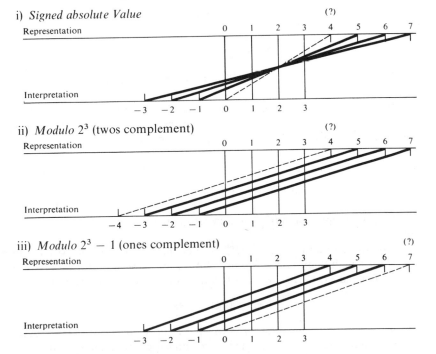

i) *Signed absolute Value*

ii) *Modulo 2^3* (twos complement)

iii) *Modulo $2^3 - 1$* (ones complement)

4.3 Fig. Mapping the Interval, $[0, 7]$ onto $[-3, 3]$.

Many mathematicians find the "minus zero" or other unusual interpretations to be a constant source of annoyance. However, many digital computers in use today have both the positive and the negative zero among their representable numbers and the reader should be aware of this fact.

Figure 4.3 shows that we can interpret these representation schemes as mappings of the integers in the interval $[0, 7]$ onto the integers in the interval $[-3, 3]$. In (iii), the ones complement case, we have a problem with seven, and in the other two cases we have a problem with four.

The source of the difficulty in (i) and (iii), of course, is that we are mapping an *even number* of nonnegative integers onto an *odd number* of integers, symmetric with respect to the origin. In (ii) the difficulty is due to the fact that B^{k-1} corresponds to $-B^{k-1}$.

Let us now examine number representation in a modern scientific computer with a word-length of sixty binary digits (bits). These blocks of sixty bits usually are used to represent either *fixed-point* or *floating-point* numbers.

We designate the sixty bits as $a_0, a_1, \ldots, a_{59}$ and write

4.4 $$a \sim a_{59}a_{58} \ldots a_2 a_1 a_0,$$

when a is a fixed-point integer, with a_{59} as our sign bit. Let $a > 0$ (then $a_{59} = 0$). In this case a can be written

4.5 $$a = a_{58}2^{58} + a_{57}2^{57} + \cdots + a_1 2 + a_0.$$

This allows us to represent the $(2^{59} - 1)$ positive integers, $1, 2, \ldots, (2^{59} - 1)$, uniquely.

Suppose the ones complement system is used to represent negative integers. Then, for $a > 0$, we write $-a$ by complementing each bit with respect to one. Complementation gives us $(1 - a_{59}) = a'_{59} = 1$, and so

4.6 $$-a = (1 - a_{58})2^{58} + (1 - a_{57})2^{57} + \cdots + (1 - a_1)2 + (1 - a_0)$$
$$= a'_{58}2^{58} + a'_{57}2^{57} + \cdots + a'_1 2 + a'_0.$$

This allows us to represent the $(2^{59} - 1)$ negative integers $-1, -2, \ldots, -(2^{59} - 1)$, uniquely.

So far, we have accounted for $(2^{60} - 2)$ of the 2^{60} possible bit patterns in a 60-bit word-length computer. The only bit patterns we have not discussed are $0000 \ldots 0000$ and $1111 \ldots 1111$. An obvious choice is the representation

4.7 $$\text{zero} \sim 0000 \ldots 0000,$$

and this leaves us with the question as to what to do with $1111 \ldots 1111$. It could represent "minus zero" or be treated as an unacceptable number. (Recall the case (iii) in Table 4.2.) The computer designer makes this decision for us, of course, and whether we like his decision or not, we must be aware of it and be prepared to live with it.

An important advantage of the use of scheme (ii) or scheme (iii) over scheme (i), from the point of view of the machine designer, is that *subtraction* can be performed by *adding* the complement of the subtrahend to the minuend, and we may use either the ones complement scheme or the twos complement scheme when we complement. Thus, special electronic circuits for subtraction are avoided in designing the arithmetic unit. We are assuming, of course, that the circuitry for complementing a number is simpler than the circuitry for subtraction.

The rules for addition are as follows.

Scheme (ii) (representation modulo B^k)
If $0 \leq \alpha \leq B^k - 1, 0 \leq \beta \leq B^k - 1$, then the result $\psi(\alpha + \beta)$ of adding α and β is an integer between 0 and $B^k - 1$ such that

$$\psi(\alpha + \beta) = \begin{cases} \alpha + \beta & \text{if } 0 \leq \alpha + \beta \leq B^k - 1 \\ \alpha + \beta - B^k & \text{if } B^k \leq \alpha + \beta \leq 2B^k - 2. \end{cases}$$

For example, if we wish to compute $2 - 3$ we write $2 + (-3)$, and if $k = 3$, $B = 2$ we have

$$[2] = 010$$
$$[-3] = 101.$$

Hence,

$$\psi([2] + [-3]) = \psi(010 + 101)$$
$$= \psi(111)$$
$$= 111.$$

Note that the answer is correct, since

$$[-1] = 111.$$

On the other hand, if we wish to compute $(-3) - 3$ we write $(-3) + (-3)$, and obtain

$$[-3] + [-3] = 101 + 101$$
$$= 1010.$$

Since $1010 \geq 2^3 = 1000$ we have

$$\psi([-3] + [-3]) = 1010 - 1000$$
$$= 010$$

and this is not the correct answer. This answer, however, is congruent to the correct answer, modulo 2^3.

Scheme (iii) (representation modulo $B^k - 1$)

If $0 \leq \alpha \leq B^k - 1, 0 \leq \beta \leq B^k - 1$, then the result of adding α and β is an integer between 0 and $B^k - 2$ such that

$$\psi(\alpha + \beta) = \begin{cases} \alpha + \beta & \text{if } 0 \leq \alpha + \beta \leq B^k - 2 \\ \alpha + \beta - (B^k - 1) & \text{if } B^k - 1 \leq \alpha + \beta \leq 2B^k - 2. \end{cases}$$

We now state a theorem concerning the representation scheme (iii).

4.8 Theorem. If $0 \leq |x| \leq B^{k-1} - 1$, $0 \leq |y| \leq B^{k-1} - 1$, and $0 \leq |x + y| \leq B^{k-1} - 1$, then

$$\psi([x] + [y]) = [x + y].$$

Proof. A proof can be given using the fact that $[x] \equiv x(\text{mod}(B^k - 1))$, $[y] \equiv y(\text{mod}(B^k - 1))$ and hence $[x] + [y] \equiv x + y(\text{mod}(B^k - 1))$. We also note that $\psi(\alpha + \beta) \equiv (\alpha + \beta)(\text{mod}(B^k - 1))$.

However, it is, perhaps, instructive to prove the theorem for each of the various special cases. First let us consider the case $x > 0$ and $y > 0$. Evidently $[x] = x$, $[y] = y$, and $[x + y] = x + y$. Consequently,

$$\begin{aligned} \psi([x] + [y]) &= \psi(x + y) \\ &= x + y \\ &= [x + y], \end{aligned}$$

since $x + y \leq B^{k-1} - 1$.

If $x > 0$ and $y < 0$ we have $[x] = x$, $[y] = y + B^k - 1$ and $[x] + [y] = x + y + B^k - 1$. We consider separately the cases where $x + y \geq 0$ and $x + y < 0$. If $x + y \geq 0$, then $B^k - 1 \leq [x] + [y] \leq 2B^k - 2$ and

$$\begin{aligned} \psi([x] + [y]) &= x + y \\ &= [x + y]. \end{aligned}$$

If $x + y < 0$ then $[x] + [y] \leq B^k - 2$ and

$$\begin{aligned} \psi([x] + [y]) &= [x] + [y] \\ &= x + y + B^k - 1 \\ &= [x + y]. \end{aligned}$$

The reader should verify the cases

$$\begin{array}{cc} x < 0, & y < 0 \\ x = 0, & y = 0 \\ x = 0, & y > 0 \\ x = 0, & y < 0. \end{array}$$

EXERCISES 2.4

1. Verify the complementation procedures described in the text for representing negative numbers.

2. Work out the details of the proof of Theorem 4.8 for the cases: $x = y = 0$; $x = 0, y > 0$; $x = 0, y < 0$; $x < 0, y < 0$.

3. State and prove a theorem analogous to Theorem 4.8 for the representation scheme (ii).

4. Consider the representation scheme (iii) of integers using three binary digits. Compute all possible sums of pairs of three binary digit positive integers according to the rules indicated and verify Theorem 4.8. Do the same thing for the representation scheme (ii) and verify the analogue of Theorem 4.8 considered in the preceding exercise.

2.5 FLOATING-POINT REPRESENTATIONS

We now discuss the representation of floating-point numbers (not necessarily integers) of the form

5.1 $$x = a \cdot 2^b,$$

where a and b are integers and where a satisfies an additional property so that the representation is unique. On some digital computers a is chosen so that $\frac{1}{2} \leq |a| < 1$. However, we shall insist here that a be an integer.

Suppose we assign forty-eight of the sixty bits to a. This leaves us ten bits for b since two bits must be reserved as sign bits (one for x and one for b). A convenient notation, then, is the following. Instead of (4.4) we write

5.2 $$x \sim s_1 s_2 b_9 b_8 b_7 \ldots b_2 b_1 b_0 a_{47} a_{46} a_{45} \ldots a_2 a_1 a_0.$$

In other words, the integers a and b correspond to the bit patterns

5.3 $$a \sim s_1 a_{47} a_{46} a_{45} \ldots a_2 a_1 a_0,$$

and

5.4 $$b \sim s_2 b_9 b_8 b_7 \ldots b_2 b_1 b_0.$$

We should remark at this point that, when the ones complement scheme (iii) is used to represent negative numbers, we can represent $-x$ by complementing each bit of (5.2) with respect to one. *This makes the procedure the same for both fixed-point and floating-point numbers.* We assume scheme (iii) in what follows.

First we look at a. Everything we stated earlier about fixed-point integer representation applies to (5.3). For example, if $x > 0$, then $s_1 = 0$ and

5.5 $$a = a_{47}2^{47} + a_{46}2^{46} + \cdots + a_1 2 + a_0.$$

This means that, for $-x$, $(1 - s_1) = s_1' = 1$, and

5.6
$$-a = (1 - a_{47})2^{47} + (1 - a_{46})2^{46} + \cdots + (1 - a_1)2 + (1 - a_0)$$
$$= a_{47}'2^{47} + a_{46}'2^{46} + \cdots + a_1'2 + a_0'.$$

For the moment we postpone discussion of the case $x = 0$.

For b there is an advantage* if we reverse the sign convention so that $b \geqq 0$ means $s_2 = 1$. What this really means is that we "bias" the exponent by a translation along the real axis. We can illustrate this by going back to the simple case where we have only three bits

5.7
$$b \sim s_2 b_1 b_0.$$

The eight possible bit patterns are shown in Table 5.8. The mapping of the integers in the interval $[0, 7]$ onto the integers of the interval $[-3, 3]$ is shown in Fig. 5.9.

5.8 Table
The "Reversed Sign" Convention or Biased System

	The Binary Bit Patterns	Biased System Using Ones Complement Representation (iii)
Positive Integers	111	3
	110	2
	101	1
	100	0
Negative Integers	011	-0
	010	-1
	001	-2
	000	-3

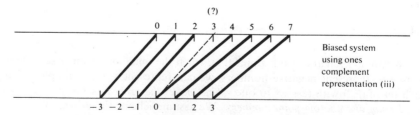

5.9 Fig. Mapping of the Interval $[0, 7]$ onto $[-3, 3]$.

---

* Floating-point numbers can easily be compared for size if the reverse sign convention is used for b.

As we saw in Fig. 4.3, one bit pattern always presents a problem. This time it is 011. As before, it could represent "minus zero" or be treated as an unacceptable number.

With this background we return to a discussion of (5.4). If $b \geqq 0$, then $s_2 = 1$ and

5.10 $$b = b_9 2^9 + b_8 2^8 + \cdots + b_1 2 + b_0.$$

This means that for $-b$, $(1 - s_2) = s_2' = 0$, and

5.11 $$\begin{cases} -b = (1 - b_9)2^9 + (1 - b_8)2^8 + \cdots + (1 - b_1)2 + (1 - b_0) \\ = b_9' 2^9 + b_8' 2^8 + \cdots + b_1' \cdot 2 + b_0'. \end{cases}$$

Suppose we separate the leftmost twelve bits in (5.2) from the remaining forty-eight bits. The term *mantissa* is used to describe the bit pattern

5.12 $$a_{47} a_{46} \ldots a_2 a_1 a_0$$

and the term *characteristic* or *exponent* is used to describe the bit pattern

5.13 $$s_1 s_2 b_9 b_8 \ldots b_2 b_1 b_0.$$

5.14 Table
Biased Characteristics

Corresponding Value of b Expressed in Decimal	Octal Representation of $s_1 s_2 b_9 b_8 \ldots b_1 b_0$ for $x > 0$	Octal Representation of $s_1 s_2 b_9 b_8 \ldots b_1 b_0$ for $x < 0$
1023	3777	4000
1022	3776	4001
⋮	⋮	⋮
1	2001	5776
0	2000	5777
−0	1777	6000
−1	1776	6001
−2	1775	6002
⋮	⋮	⋮
−1022	0001	7776
−1023	0000	7777

Table 5.14 exhibits all possible characteristics (5.13) where the twelve bits have

been converted to four octal digits. Notice that the bit pattern $[010\,000\,000\,000]_{two}$ $\sim [2000]_{eight}$ represents a zero exponent whereas $[001\,111\,111\,111]_{two} \sim [1777]_{eight}$ is the troublesome pattern corresponding to 011 in Table 5.14. In each case $s_2 = 0$ is followed by all ones. Most computer designers either declare 1777 to be an unacceptable characteristic or let it correspond to a "negative zero" exponent.

Until now we have not considered a representation for $x = 0$. Mathematically speaking, if $a = 0$ in (5.1), then b could be any acceptable exponent. However, we usually prefer a unique representation for zero (although we sometimes have to put up with the troublesome "minus zero" mentioned above) and the machine designer is faced with the decision as to which characteristic to use with $a = 0$ as a representation for $x = 0$.

One choice might be the characteristic 2000. However, many machine designers prefer 0000 since $x = 0$ would then be represented by a block of sixty zeros. Another reason is that a characteristic of 0000 corresponds to the exponent

5.15 $$b = -1023,$$

and small positive numbers have such exponents.

Another point needs to be discussed. We usually deal with *normalized* floating-point numbers, namely, numbers for which $a_{47} = 1$ when $x > 0$, and $a_{47} = 0$ when $x < 0$. Thus, for $x > 0$,

5.16 $$2^{47} \leqq a \leqq 2^{48} - 1$$

and since

5.17 $$-1023 \leqq b \leqq 1023,$$

the range of positive, representable, normalized, floating-point numbers is

5.18 $$2^{-1023} \cdot 2^{47} \leqq x \leqq 2^{1023}(2^{48} - 1).$$

This is approximately 2^{-976} to 2^{1071} or approximately 10^{-294} to 10^{322}. Since our representable numbers are symmetric about the origin we have a similar range of negative, representable, normalized, floating-point numbers.

In our subsequent discussion we shall refer to an idealized computer with the floating-point format described above as "Machine C." The Control Data 6600 has the same format as Machine C.

5.19 EXAMPLE. Determine the normalized floating-point representation of the numbers ± 10 and $\pm \frac{1}{10}$ on Machine C.

Solution. For the case $x = 10$ we have $\phi_2(x) = 1010$ and

$$x = a2^b = \underbrace{(1010000\ldots0)}_{48 \text{ bits}}2^{-44}.$$

If the exponent were 44 instead of -44 its binary representation would be

$$0\ 00001\ 01100.$$

The biased exponent would be

$$1\ 00001\ 01100.$$

Since the exponent is negative we take the one's complement obtaining

$$0\ 11110\ 10011.$$

The entire word is then

$$0\ 0\ \underbrace{1111010011}_{\text{10 bits}}\ \underbrace{1010000\ldots0}_{\text{48 bits}}$$

which in octal is

$$[10] = 17235\ 00000\ 00000\ 00000.$$

To represent -10 we simply complement each binary digit with respect to one. After converting to octal we get

$$[-10] = 60\ 542\ 77777\ 77777\ 77777.$$

(This is equivalent to complementing each octal digit with respect to 7.)

Let us now consider the case $x = \frac{1}{10}$. By performing a long division in binary of $\phi_2(10) = 1010$ into $\phi_2(1) = 1$ we get

$$\phi_2(\tfrac{1}{10}) = 0.000110011001100\ldots$$

where the digits are repeated. Thus we have

$$x \doteq \underbrace{(11001100\ldots)}_{\text{48 bits}}2^{-51}.$$

The biased exponent corresponding to 51 is

$$1\ 00001\ 10011$$

and that corresponding to -51 is therefore

$$0\ 11110\ 01100.$$

The entire word is then

$$0\,0\ \underbrace{1111\,0\,01100}_{\text{10 bits}}\ \underbrace{110011001100\ldots}_{\text{48 bits}}$$

which in octal is

$$[\tfrac{1}{10}] \doteq 17146\ 31463\ 14631\ 46314.$$

As before, we have

$$[-\tfrac{1}{10}] \doteq 60631\ 46314\ 63146\ 31463.$$

EXERCISES 2.5

1. For Machine C determine the normalized floating-point representations of ± 1, ± 2, ± 5, $\pm\tfrac{1}{2}$, $\pm\tfrac{1}{5}$, 0, -0, 2^{172}, 2^{-172} using the format described in the text. Express the results in each case as 20 octal digits.

2. What numbers are represented by the following (octal) machine representations on Machine C? Which are normalized?

71654	77777	77777	77777
71650	77777	77777	77777
21245	00000	00000	00000
21111	00000	00000	00000

2.6 COMPUTER-REPRESENTABLE NUMBERS

The totality of numbers which can be represented on a fixed-word-length computer forms a finite subset of the set of rational numbers. We call this finite subset of the rational numbers the *set of computer-representable numbers*. If we wish to represent a nonrepresentable number in the computer we must approximate it by some computer-representable number.

We have seen that the set of all positive computer-representable numbers x lies in the range

6.1
$$2^{-976} \leqq x \leqq (2^{48} - 1)2^{1023}.$$

Given a real number x (not necessarily computer-representable) in the above range, we let $\underline{x}$ and $\bar{x}$ denote, respectively, the largest computer-representable number not greater than x and the smallest computer-representable number not less than x. Thus, we have

6.2
$$\underline{x} \leqq x \leqq \bar{x}.$$

We assume that our computer-representable numbers are in normalized floating-point form (see Section 2.5). Now if

6.3
$$\underline{x} = a2^b,$$

where b is an integer and

6.4
$$2^{47} \leqq a \leqq 2^{48} - 1,$$

then we have

6.5
$$\bar{x} = (a + 1) 2^b.$$

We should point out that if $a = 2^{48} - 1$ then $\bar{x}$, when normalized, becomes $(2^{47})2^{b+1}$. If x is not computer-representable, then we have

6.6
$$x = (a + \delta)2^b$$

for some δ lying in the range

6.7
$$0 < \delta < 1.$$

CASE I.　　If $0 < \delta < \frac{1}{2}$, then we let x be represented by $\underline{x}$. Evidently we have

6.8
$$\underline{x} = x(1 + \underline{\varepsilon})$$

where the relative error $\underline{\varepsilon}$ is given by

6.9
$$\underline{\varepsilon} = \frac{\underline{x} - x}{x}$$
$$= \frac{-\delta 2^b}{(a + \delta)2^b}$$
$$= -\frac{\delta}{a + \delta}.$$

Thus,

6.10
$$|\underline{\varepsilon}| = \frac{\delta}{a + \delta}$$
$$= \frac{1}{a/\delta + 1}$$
$$< \frac{1}{2^{48} + 1}$$
$$< 2^{-48}$$

CASE II.   If $\frac{1}{2} \leqq \delta < 1$, then we let x be represented by $\bar{x}$. Evidently we have

6.11 $\bar{x} = x(1 + \bar{\varepsilon})$

where the relative error $\bar{\varepsilon}$ is given by

6.12 $\bar{\varepsilon} = \dfrac{\bar{x} - x}{x}$

$$= \frac{(1 - \delta)2^b}{(a + \delta)2^b}$$

$$= \frac{1 - \delta}{a + \delta}.$$

Since $(1 - \delta)/(a + \delta)$ decreases as δ increases in the interval $\frac{1}{2} \leqq \delta < 1$ we have

6.13 $|\bar{\varepsilon}| \leqq \dfrac{\frac{1}{2}}{a + \frac{1}{2}}$

$$= \frac{1}{2a + 1}$$

$$< 2^{-48}.$$

Thus we conclude that, given a number x in the range (6.1), there exists a computer-representable number, say $[x]$, such that

6.14 $[x] = x(1 + \varepsilon), \qquad \varepsilon = \dfrac{[x] - x}{x}$

where

6.15 $|\varepsilon| < 2^{-48}.$

Since there is a one-to-one correspondence between negative computer-representable numbers and positive computer-representable numbers, an analysis of the negative numbers produces similar bounds.

Frequently we desire to find a computer-representable number corresponding to a number x which can be represented exactly in double-precision arithmetic. Thus

6.16 $x = [a_U + a_L 2^{-48}]2^b$

where a_U and a_L have the same sign, with

6.17 $\begin{cases} 2^{47} \leqq |a_U| \leqq 2^{48} - 1 \\ 0 \ \ \ \leqq |a_L| \leqq 2^{48} - 1 \end{cases}$

and

6.18
$$|b| \leq 2^{10} - 1.$$

The representation of the number $a_U + a_L 2^{-48}$ requires a double-length accumulator, with 96 bits plus a sign bit. The most significant digit is 1 if the number is positive. The binary point is placed between the upper half and the lower half of the accumulator as shown.

The process of obtaining a single-precision number of the form

6.19
$$\begin{cases} (x)_R = a2^{b'} \\ 2^{47} \leq |a| \leq 2^{48} - 1 \end{cases}$$

to represent x, given x in double precision, is known as "rounding." A crude procedure for rounding is to simply ignore the lower half of the accumulator obtaining the rounded value $(x)_R$ given by

6.20
$$(x)_R = a_U 2^b.$$

This process is called *chopping*. Evidently we have the actual error

6.21
$$(x)_R - x = -2^{-48} a_L 2^b$$

and the relative error

6.22
$$\varepsilon_c = \frac{(x)_R - x}{x}$$
$$= -\frac{a_L 2^{-48}}{a_U + a_L 2^{-48}}$$

where, since $|a_U| \geq 2^{47}$, and $|a_L| < 2^{48}$,

6.23
$$|\varepsilon_c| < 2^{-47}.$$

A more accurate process of rounding corresponds to the procedure described at the beginning of this section for finding the representable number closest to a given number x. If $x > 0$ we add 2^{b-1} to x, renormalize if necessary, and then chop. Adding 2^{b-1} to x amounts to adding 1 to the leading bit of the lower half of the accumulator. If the leading bit of the lower half of the accumulator is 0, there is no "carry" bit and so we chop. If the leading bit is 1, a "carry" bit results and so the rounded number is

6.24
$$(x)_R = \begin{cases} (a_U + 1)2^b & \text{if } a_U \leq 2^{48} - 2 \\ (2^{47})2^{b+1} & \text{if } a_U = 2^{48} - 1. \end{cases}$$

With the more accurate process of rounding we obtain a rounded value $(x)_R$ where

6.25 $(x)_R = x(1 + \varepsilon_R)$

and

6.26 $|\varepsilon_R| < 2^{-48}.$

This follows from the analysis given earlier in this section.

Other processes* can be used for rounding a double-precision number. However, for any reasonable process we have, certainly,

6.27 $(x)_R = x(1 + \varepsilon_R)$

where

6.28 $|\varepsilon_R| < |\delta_R|2^{-47}$

and

6.29 $\frac{1}{2} \leq |\delta_R| \leq 1.$

EXERCISES 2.6

1. For Machine C represent the numbers 1/10 and 1/3 in normalized floating-point format with 96 bits in the mantissa. Round by chopping and also by the more accurate procedure described in the text. Verify in each case the bound given for the relative error.

2. Find the numbers representable on Machine C closest to 1/10, −1/10. Also find the representable number closest to $\sqrt{2}$. Show that the relative error is less than 2^{-48}.

3. Which of the following numbers are in the range of computer-representable numbers for Machine C?

$$10^{400}, 10^{300}, 10^{-300}, 10^{-400}.$$

---

* One such procedure is simply to place a one in the least significant bit in the upper half of the accumulator and then chop.

2.7 FLOATING-POINT ARITHMETIC OPERATIONS

We now describe how floating-point arithmetic operations are carried out on our mythical computer, which we have referred to as "Machine C." Our discussion is based on that of Wilkinson [1963, Chapter 1] (although the floating-point format is different) and on that given in the Control Data Computer Systems Reference Manual [Publication No. 60100000 (1969)].

We also assume that a double-length accumulator, as described in Section 2.6, is available for use in carrying out arithmetic operations. Let the two arguments of each arithmetic operation under discussion be normalized floating-point numbers of the form

7.1
$$\begin{cases} x_1 = a_1 2^{b_1} \\ x_2 = a_2 2^{b_2}, \end{cases}$$

where $a_1, a_2, b_1,$ and b_2 are integers such that

7.2
$$\begin{cases} 2^{47} \leq |a_1| \leq 2^{48} - 1 \\ 2^{47} \leq |a_2| \leq 2^{48} - 1, \end{cases}$$

and

7.3
$$\begin{cases} |b_1| \leq 2^{10} - 1 \\ |b_2| \leq 2^{10} - 1. \end{cases}$$

We assume, for the moment, that the exact result of each arithmetic operation lies in the range of computer-representable numbers. If this is not the case, then we have either *exponent overflow* or *exponent underflow* and these conditions are discussed later on in this section in (7.51).

We now discuss addition (which includes subtraction), multiplication, and division of floating-point numbers.

Addition (i). Let us assume, without loss of generality, that $b_2 \leq b_1$ and, for the moment, let us also assume that $b_1 - b_2 \leq 48$. We now write

7.4
$$\begin{aligned} x_1 + x_2 &= a_1 2^{b_1} + a_2 2^{b_2} \\ &= [a_1 + a_2 2^{-(b_1 - b_2)}] 2^{b_1} \\ &= \alpha_s 2^{b_1}, \end{aligned}$$

where

7.5
$$\alpha_s = a_1 + a_2 2^{-(b_1 - b_2)}.$$

To carry out the floating-point addition in Machine C the integer a_2 is placed in the upper half of the double-length accumulator and shifted right $b_1 - b_2$

places. Then a_1 is added to the upper half of the accumulator, forming α_s. If overflow occurs in the accumulator during this process, then the result is shifted right one place and b_1 is increased to $b_1 + 1$. Otherwise, the result is normalized, if necessary, by shifting left and adjusting b_1 accordingly.

Now the exact value of α_s (normalized) lies in the double-length accumulator. At this point it is customary to round the double-length result using, for example, the rounding procedure associated with either (6.20) or (6.24). Rounding could cause overflow in the accumulator, in which case a right shift would be required with its corresponding adjustment in the exponent.

We shall use Wilkinson's [1963, p. 9] notation $fl(x_1 + x_2)$ for the single-length number produced by the algorithm described above. If the accurate rounding procedure corresponding to (6.24) is used, then we can write, using (6.25),

7.6 $$fl(x_1 + x_2) = (x_1 + x_2)(1 + \varepsilon_s)$$

From the point of view of backward error analysis,* the *computed* sum can be interpreted as the *exact* sum of perturbed data if we write

7.6a $$fl(x_1 + x_2) = x_1(1 + \varepsilon_s) + x_2(1 + \varepsilon_s).$$

From either point of view, however,

7.7 $$|\varepsilon_s| < 2^{-48}.$$

For a more general rounding procedure we can use (6.27) and (6.28) to write

7.8 $$|\varepsilon_s| < |\delta_R|2^{-47}$$

where δ_R satisfies (6.29). Note that we can solve (7.6) for ε_s and obtain

7.9 $$\varepsilon_s = \frac{fl(x_1 + x_2) - (x_1 + x_2)}{(x_1 + x_2)}.$$

Hence, ε_s is the relative error in the computed sum.

These results are based on the assumption that $b_1 - b_2 \leqq 48$. We now show that (7.7) and (7.8) hold even when $b_1 - b_2 > 48$. In this case, if a_2 is shifted right $(b_1 - b_2)$ places, then $(b_1 - b_2 - 48)$ bits of a_2 may be lost. Let $\hat{a}_2$ represent a_2 minus the $(b_1 - b_2 - 48)$ least significant bits and let

7.10 $$\hat{x}_2 = \hat{a}_2 2^{b_2}.$$

---

* See (4.10) of the previous chapter.

CASE I.   If both x_1 and x_2 are positive and if

7.11 $$\text{fl}(x_1 + x_2) = \text{fl}(x_1 + \hat{x}_2)$$

$$= x_1,$$

then the actual error is x_2 and the relative error satisfies the inequality

7.12 $$|\varepsilon_s| = \frac{x_2}{x_1 + x_2}$$

$$= \frac{1}{x_1/x_2 + 1}$$

$$\leqq \frac{1}{2^{48} + 1}$$

$$< 2^{-48}.$$

On the other hand, if the computed result is not x_1 (see footnote, Section 2.6), then we must have

7.13 $$\text{fl}(x_1 + x_2) = \text{fl}(x_1 + \hat{x}_2)$$

$$= (a_1 + 1)2^{b_1}$$

$$= x_1 + 2^{b_1}.$$

In this case it is easily verified that

7.14 $$(x_1 + \hat{x}_2) \leqq (x_1 + x_2) \leqq \text{fl}(x_1 + x_2).$$

Hence,

7.15 $$|\text{fl}(x_1 + x_2) - (x_1 + x_2)| \leqq |\text{fl}(x_1 + x_2) - (x_1 + \hat{x}_2)| = |\delta_R|2^{b_1},$$

where the rounding error δ_R satisfies (6.29). Thus, the relative error ε_s satisfies the inequality

7.16 $$|\varepsilon_s| = \frac{|\text{fl}(x_1 + x_2) - (x_1 + x_2)|}{|x_1 + x_2|}$$

$$< \frac{|\delta_R|2^{b_1}}{x_1}$$

$$= \frac{|\delta_R|2^{b_1}}{a_1 2^{b_1}}$$

$$\leqq |\delta_R|2^{-47}.$$

CASE II.   If $b_1 - b_2 > 48$ and if x_1 and x_2 have opposite signs then, without loss of generality, we can assume* $x_2 < 0 < x_1$. If

7.17
$$\text{fl}(x_1 + x_2) = \text{fl}(x_1 + \hat{x}_2)$$
$$= x_1,$$

then the actual error is x_2 and the relative error can be written

7.18
$$|\varepsilon_s| = \left| \frac{x_2}{x_1 + x_2} \right|$$
$$= \frac{1}{|x_1/x_2| - 1}.$$

However, $x_1 \geq 2^{47} 2^{b_1}$ and $|x_2| \leq (2^{48} - 1)2^{b_2}$, and so

7.19
$$\left| \frac{x_1}{x_2} \right| \geq \frac{2^{47} 2^{b_1}}{[(2^{48} - 1)2^{-(b_1 - b_2)}]2^{b_1}}$$
$$\geq \left[\frac{2^{47}}{2^{48} - 1} \right] 2^{49}.$$

Hence

7.20
$$|\varepsilon_s| \leq \frac{1}{2^{96}/(2^{48} - 1) - 1}$$
$$= \frac{2^{48} - 1}{2^{96} - 2^{48} + 1}$$
$$< \frac{2^{48} - 1}{2^{48}(2^{48} - 1)}$$
$$= 2^{-48}.$$

On the other hand, if the computed result is not x_1 (with $b_1 - b_2 > 48$ and $x_2 < 0 < x_1$) then it must be

7.21
$$\text{fl}(x_1 + x_2) = (a_1 - \gamma)2^{b_1}$$
$$= x_1 - \gamma 2^{b_1}$$

---

* The case in which both x_1 and x_2 are negative is the same as Case I. If either one (or both) of the operands is zero we get an exact answer.

where

7.22
$$\gamma = \begin{cases} 1 & \text{if } a_1 > 2^{47} \\ \frac{1}{2} & \text{if } a_1 = 2^{47}. \end{cases}$$

It is easily verified that

7.23
$$\text{fl}(x_1 + x_2) \le (x_1 + x_2) \le (x_1 + \hat{x}_2).$$

Since the addition of a_1 and $\hat{a}_2 2^{-(b_1 - b_2)}$ can be carried out exactly in the double-length accumulator we can write, as in the case where $b_1 - b_2 \le 48$, by (7.6),

7.24
$$\text{fl}(x_1 + x_2) = \text{fl}(x_1 + \hat{x}_2)$$
$$= (x_1 + \hat{x}_2)(1 - \hat{\varepsilon}_s)$$

where

7.25
$$0 \le \hat{\varepsilon}_s \le |\delta_R| 2^{-47},$$

and where δ_R satisfies (6.29). The fact that $\hat{\varepsilon}_s \ge 0$ follows from (7.23).
Thus, from (7.24),

7.26
$$\text{fl}(x_1 + x_2) - (x_1 + \hat{x}_2) = -\hat{\varepsilon}_s(x_1 + \hat{x}_2)$$

and so

7.27
$$\text{fl}(x_1 + x_2) - (x_1 + x_2) = -\hat{\varepsilon}_s(x_1 + \hat{x}_2) + (\hat{x}_2 - x_2).$$

Hence,

7.28
$$\varepsilon_s = \frac{\text{fl}(x_1 + x_2) - (x_1 + x_2)}{x_1 + x_2}$$
$$= \frac{-\hat{\varepsilon}_s(x_1 + \hat{x}_2) + (\hat{x}_2 - x_2)}{(x_1 + \hat{x}_2) - (\hat{x}_2 - x_2)}$$
$$= -\frac{\hat{\varepsilon}_s - \theta}{1 - \theta}$$

where

7.29
$$\theta = \frac{\hat{x}_2 - x_2}{x_1 + \hat{x}_2}.$$

From (7.23) and the fact that $x_1 + \hat{x}_2$ is positive we note that $0 \le \theta < 1$.

Also, from (7.23) and the fact that $x_1 + x_2$ is positive, we note that ε_s is negative. Consequently,

7.30
$$|\varepsilon_s| = \frac{\hat{\varepsilon}_s - \theta}{1 - \theta}.$$

Since the denominator on the right is positive, this implies that the numerator is, also, and so

7.31
$$0 \leq \theta \leq \hat{\varepsilon}_s.$$

Now the expression for $|\varepsilon_s|$ above is a decreasing function of θ for θ in the interval (7.31). Hence,

7.32
$$|\varepsilon_s| = \frac{\hat{\varepsilon}_s - \theta}{1 - \theta}$$

$$\leq \hat{\varepsilon}_s$$

$$\leq |\delta_R| 2^{-47}.$$

Multiplication (ii). We may write

7.33
$$x_1 x_2 = [a_1 2^{b_1}][a_2 2^{b_2}]$$

$$= [a_1 a_2 2^{-48}] 2^{b_1 + b_2 + 48}$$

$$= \alpha_m 2^{b_1 + b_2 + 48}$$

where

7.34
$$\alpha_m = a_1 a_2 2^{-48}.$$

The *exact* value of α_m lies in the double-length accumulator. Before it is stored in the computer's memory it is normalized, if necessary, and then rounded (which could cause renormalization).

Again we adopt Wilkinson's notation for the computed result and write

7.35
$$\text{fl}(x_1 x_2) = x_1 x_2 (1 + \varepsilon_m).$$

Again, from the point of view of backward error analysis,* the *computed* product can be interpreted as the *exact* product of perturbed data, if we write

7.36
$$\text{fl}(x_1 x_2) = [x_1(1 + \varepsilon_m)] x_2$$

$$= x_1 [x_2(1 + \varepsilon_m)].$$

_____

* See (4.10) of the previous chapter.

From either point of view, however, the relative error ε_m satisfies the inequality

7.37 $$|\varepsilon_m| < |\delta_R|2^{-47}.$$

Division (iii). An approximation to the quotient x_1/x_2 can be computed as long as $x_2 \neq 0$. The integer a_1 is placed in the upper half of the double-length accumulator and shifted right one place if $|a_1| \geq |a_2|$ (see (7.38) below). Next, a finite number of bits of the quotient are computed, at least enough to fill the upper half of the double-length accumulator.

Suppose we define α_q to be

7.38 $$\alpha_q = \begin{cases} (a_1/a_2)2^{48} & \text{if } |a_1| < |a_2| \\ [a_1/(2a_2)]2^{48} & \text{if } |a_2| \leq |a_1|. \end{cases}$$

Then

7.39 $$2^{47} \leq \alpha_q < 2^{48},$$

and the double-length accumulator contains at least 48 bits of α_q.

If we compute *only* 48 bits of the quotient then the upper half of the double-length accumulator contains an approximation to α_q which can be considered to be a rounded approximation to α_q, with the rounding accomplished simply by chopping.

If we compute at least 49 bits of the quotient and add a one to the leading bit of the lower half of the accumulator, then the upper half contains a rounded approximation to α_q which corresponds to (6.25). In this case

7.40 $$\text{fl}\left(\frac{x_1}{x_2}\right) = \begin{cases} (\alpha_q)_R 2^{b_1 - b_2 - 48} & \text{if } |a_1| < |a_2| \\ (\alpha_q)_R 2^{b_1 - b_2 - 47} & \text{if } |a_2| \leq |a_1| \end{cases}$$
$$= \frac{x_1}{x_2}(1 + \varepsilon_q).$$

Again, from the point of view of backward error analysis, the *computed* quotient can be interpreted as the *exact* quotient of perturbed data, if we write

7.41 $$\text{fl}\left(\frac{x_1}{x_2}\right) = \frac{x_1(1 + \varepsilon_q)}{x_2}.$$

From either point of view, however, we see from (6.26) that the relative error ε_q satisfies the inequality

7.42 $$|\varepsilon_q| < 2^{-48}.$$

If we compute only 48 bits and, in effect, round by chopping, then

7.43 $|\varepsilon_q| < 2^{-47}$,

from (6.23). Other rounding schemes would give us

7.44 $|\varepsilon_q| < |\delta_R| 2^{-47}$,

where δ_R satisfies (6.29).

7.45 *Remark.* Frequently an attempt is made to compensate for rounding by adjusting the arguments before carrying out the arithmetic operation. Thus, for rounded floating-point addition on the Control Data 6600, a rounding bit of unity is attached to the right end of both operands (see Control Data [1969]). For rounded floating-point divide, the dividend is sometimes modified before division (Wilkinson [1963]).

For rounded floating-point multiplication on the Control Data 6600, a one is added in the second most significant bit of the lower half of the accumulator after the multiplication, but before normalization.

7.46 *Remark.* As indicated by Wilkinson [1963] the restriction to the use of a single-length accumulator is much less serious for multiplication and division than it is for addition and subtraction. For multiplication and division it seems reasonable to expect that when the result is in normalized form the maximum error in the mantissa will not exceed one in the least significant digit. This corresponds to the case $|\delta_R| \leq 1$ in each case.

For addition and subtraction the problem is more serious. One can no longer be sure that when the result is in normalized form the mantissa is correct to within one in the least significant digit. The relative error can be very large as can be seen by considering the case $x_1 = 2^{47}$ and $x_2 = -2^{-1}(2^{48} - 1)$.

On the other hand, Wilkinson shows that we can write

$$\mathrm{fl}(x_1 + x_2) = x_1(1 + \varepsilon_1) + x_2(1 + \varepsilon_2)$$

where ε_1 and ε_2 are of the order of ε_s but are in general not equal.* This change is due to the fact that if $|x_2| < |x_1|$ then x_2 will have to be rounded, following the right shift of $b_1 - b_2$ places, before the addition can take place in a single-length accumulator. The addition process itself may involve rounding so there may be *two* rounding operations, whereas in the case of the double-length accumulator there is only *one* rounding operation.

_____

* Compare this result with (7.6a).

7.47 *Remark.* Space does not permit a complete analysis of what Wilkinson [1963, p. 23], [1965, p. 117] calls $fl_2(\ )$ arithmetic. However, the basic idea is that computations such as floating-point sums

$$fl_2(a_1 + a_2 + \cdots + a_n)$$

and the floating-point accumulation of inner-products

$$fl_2(a_1 b_1 + a_2 b_2 + \cdots + a_n b_n)$$

can be carried out in a double-length accumulator with *no rounding until the complete sum* (or inner-product) *is formed.*

Guaranteed bounds on the errors in $fl_2(a_1 + a_2 + \cdots + a_n)$ and $fl_2(a_1 b_1 + a_2 b_2 + \cdots + a_n b_n)$ are considerably better than the guaranteed bounds on the errors in computing $fl(a_1 + a_2 + \cdots + a_n)$ and $fl(a_1 b_1 + a_2 b_2 + \cdots + a_n b_n)$. For this reason $fl_2(\ )$ arithmetic is often recommended at critical stages in an algorithm which requires that sums or inner-products be computed with high accuracy. See 12-(19.23) and 12-(21.7), for example.

7.48 EXAMPLE
As an example of the use of some of the above formulas, let us consider the accuracy of the evaluation of the quadratic polynomial

$$P(x) = a_0 x^2 + a_1 x + a_2$$

using floating-point operations. We evaluate $P(x)$ by computing z_0, z_1, z_2 where

$$z_0 = a_0$$

$$z_1 = x z_0 + a_1, \qquad (z_1 = a_0 x + a_1)$$

$$z_2 = x z_1 + a_2, \qquad (z_2 = a_0 x^2 + a_1 x + a_2).$$

We assume that a_0, a_1, a_2 and x are computer-representable numbers. In general, if a quantity y is computed by a sequence of floating-point operations, we let $\{y\}$ be the computed value of y. Thus we have

$$\{z_1\} = fl(fl(x a_0) + a_1)$$

$$\{z_2\} = fl(fl(x\{z_1\}) + a_2).$$

Therefore,

$$\{z_1\} = [fl(x a_0) + a_1](1 + \varepsilon_s^{(1)})$$

$$= [x a_0 (1 + \varepsilon_m^{(1)}) + a_1](1 + \varepsilon_s^{(1)})$$

$$= x a_0 (1 + \varepsilon_m^{(1)})(1 + \varepsilon_s^{(1)}) + a_1 (1 + \varepsilon_s^{(1)})$$

where

$$|\varepsilon_s^{(1)}| \leqq \varepsilon, \qquad |\varepsilon_m^{(1)}| \leqq \varepsilon$$

and

$$\varepsilon < 2^{-48}.$$

Similarly, we can show that

$$\{z_2\} = [\mathrm{fl}(x\{z_1\}) + a_2](1 + \varepsilon_s^{(2)})$$
$$= b_0 x^2 + b_1 x + b_2$$

where

$$\begin{cases} b_0 = (1 + \varepsilon_m^{(2)})(1 + \varepsilon_s^{(2)})(1 + \varepsilon_m^{(1)})(1 + \varepsilon_s^{(1)})a_0 \\ b_1 = (1 + \varepsilon_m^{(2)})(1 + \varepsilon_s^{(2)})(1 + \varepsilon_s^{(1)})a_1 \\ b_2 = (1 + \varepsilon_s^{(2)})a_2 \end{cases}$$

and

$$|\varepsilon_s^{(2)}| \leqq \varepsilon, \qquad |\varepsilon_m^{(2)}| \leqq \varepsilon.$$

Therefore

$$\begin{cases} a_0(1 - \varepsilon)^4 \leqq b_0 \leqq a_0(1 + \varepsilon)^4 \\ a_1(1 - \varepsilon)^3 \leqq b_1 \leqq a_1(1 + \varepsilon)^3 \\ a_2(1 - \varepsilon) \;\; \leqq b_2 \leqq a_2(1 + \varepsilon). \end{cases}$$

As we shall show later, in Section 5.6, for any positive integer k such that $k\varepsilon \leqq 0.1$ we have

7.49 $$1 - (1 - \varepsilon)^k \leqq (1 + \varepsilon)^k - 1 \leqq k\bar{\varepsilon},$$

where

7.50 $$\bar{\varepsilon} = 1.06\varepsilon.$$

Hence,

$$\begin{cases} |b_0 - a_0| \leqq 4|a_0|\bar{\varepsilon} \\ |b_1 - a_1| \leqq 3|a_1|\bar{\varepsilon} \\ |b_2 - a_2| \leqq \;\; |a_2|\bar{\varepsilon}. \end{cases}$$

Therefore

7.51 $|\{z_2\} - P(x)| \le (4|a_0|\,|x^2| + 3|a_1|\,|x| + |a_2|)\varepsilon.$

Exponent Overflow or Underflow

7.52 *Remark.* Exponent overflow or underflow can occur with any of the above operations when the result lies outside of the range of representable numbers. If exponent underflow occurs, the result is normally set equal to zero. The possibility of exponent overflow or underflow is much more likely in the case of multiplication and division than in the case of addition or subtraction. For addition, exponent overflow cannot occur unless both numbers are of like sign and at least one of them is greater than or equal to 2^{1070} in absolute value. Exponent underflow cannot occur unless the exact sum is zero or the numbers are of opposite sign and the largest in absolute value is less than 2^{-928} in absolute value. This follows since there can be a cancellation of at most 48 bits in the mantissa. Hence if $\max(|x_1|, |x_2|) \ge 2^{-928}$ then $|x_1 + x_2| \ge 2^{-976} - 2^{47} \cdot 2^{-1023}$ which is representable. On the other hand, if

$$\begin{cases} x_1 = (2^{48} - 1)2^{-977} \\ x_2 = -(2^{48} - 2)2^{-977} \end{cases}$$

then x_1 and x_2 are representable, but

$$x_1 + x_2 = 2^{-977} < 2^{-976} = 2^{47} \cdot 2^{-1023}.$$

Hence exponent underflow would occur.

We shall discuss the problem of controlling exponent overflow and underflow in Chapter 3.

EXERCISES 2.7

1. Let us consider a computer which we designate as "Machine D" with a 12-bit word length and with the same floating-point format as described in the text except that the mantissa has 6 bits plus sign and the exponent has 4 bits plus sign.

   a) What range of numbers can be represented?
   b) Represent the numbers ± 10, ± 2, $\pm\frac{1}{2}$, $\pm\frac{1}{10}$ on the machine. (Express each result as 4 octal digits.)
   c) Show the various steps involved in performing the following arithmetic operations

   $$2 + \tfrac{1}{10}, \qquad 10 \times \tfrac{1}{10}, \qquad 2 \times \tfrac{1}{10}, \qquad 2 \div 10, \qquad 40 - \tfrac{1}{10}.$$

   What members are represented by the results in each case? Find the relative error in each case and show that it is less than the theoretical error. (Assume the existence of a double-length accumulator and that in each case rounding is performed after normalization by adding 1 to the most significant digit of the lower half of the accumulator. For division assume that 12 significant digits of the quotient are obtained before normalization and rounding.)

2. For the floating-point addition/subtraction process described in the text for Machine C, show that if $x_1 + x_2 = 0$ then $fl(x_1 + x_2) = 0$.

3. With Machine D consider the addition of x_1 and x_2 where $x_1 = 1$ and $x_2 = -(2^5 + 2^2 + 1)2^{-12}$. Show that $fl(x_1 + x_2) \neq x_1$ but that nevertheless

$$\frac{fl(x_1 + x_2) - (x_1 + x_2)}{x_1 + x_2} \leqq 2^{-6}.$$

4. Consider a machine with only a single-length accumulator for addition/subtraction. Assume that to compute $x_1 + x_2$ where $|x_1| \geqq |x_2|$ and $x_1 = a_1 2^{b_1}$, $x_2 = a_2 2^{b_2}$ one first puts a_2 in the accumulator, then shifts right $b_1 - b_2$ places and adds in a_1. If overflow occurs, one shifts right one place and adds one to the exponent. Otherwise one shifts left if necessary to normalize. If the machine has the same word length and the same floating-point format as Machine C, show that there exist computer-representable numbers x_1 and x_2 with $x_1 + x_2 \neq 0$ such that

$$\left| \frac{fl(x_1 + x_2) - (x_1 + x_2)}{x_1 + x_2} \right| \geqq 1 = \frac{(2^{47})}{2^{47}}.$$

(Hint: Let $x_1 = 2^{47}$, $x_2 = -(2^{47} - \frac{1}{2})$.)

5. With Machine D find upper and lower bounds for $\{z\}$ where $z = x_1 + x_2 + x_3$ and

$$x_1 = 11.5, \quad x_2 = \tfrac{1}{8}, \quad x_3 = 1.5.$$

6. Give a bound on the relative error in the evaluation of the quadratic polynomial $11x^2 - 21x + 7$ on Machine C.

7. Derive a formula analogous to (7.5i) for the case of the cubic polynomial $P(x) = a_0 x^3 + a_1 x^2 + a_2 x + a_3$.

8. With Machine D verify that exponent overflow or underflow occur in the following cases: $2^{20} + 2^{20}$; $(2^6 - 1)2^{15} - (2^6 - 2)2^{15}$; $2^8 \div 2^8$; $2^8 \div 2^{-8}$.

9. Consider the use of a k-bit accumulator with $48 \leqq k \leqq 96$ instead of the 96-bit double-length accumulator normally used. What is the smallest value of k such that the accuracy obtained by the procedures described above will be as good as that obtained when $k = 48$.

2.8 FORTRAN ANALYSIS OF A FLOATING-POINT NUMBER

Probably the most effective and efficient way to develop a good library of programs for the computation of elementary functions would be to program in machine language, or an assembly language. By doing so one could minimize the number of machine instructions necessary to carry out a given computation and one could also take advantage of the fact that the machine operates in binary arithmetic. Thus, for instance, loss of accuracy can be minimized by using shift operations rather than multiplying or dividing by powers of two. If one were to write such programs using an algorithmic language, such as Fortran, one might not be able to take direct advantage of the properties of the computer since many versions of

Fortran do not allow complete facility for binary operations. It should be noted that a library of elementary functions is usually supplied to the computer purchaser along with the compiler, but frequently the performance of such routines is far from satisfactory.

Our object here is to describe how one can, using an algorithmic language such as Fortran, develop a satisfactory set of elementary function subroutines. Because of the loss of speed, it would not be practical to actually use such programs over any extended period of time. However, it is intended that the student can get an idea of some of the mathematical aspects involved in the preparation of elementary function routines and can actually test such routines without getting involved at this stage in machine language programming.

What is needed is the ability to obtain certain information about a number stored in the machine using the Fortran compiler. Such information is provided by the following two programs:

a) *Subroutine ANALYZE*: given a number x stored in the machine, find an integer p and the number m such that

8.1
$$x = m2^p$$

where $m = 0$ or else

8.2
$$\tfrac{1}{2} \leq |m| < 1.$$

We remark that if $x = a2^b$ where $2^{47} \leq |a| \leq 2^{48} - 1$ and b is an integer, then

8.3
$$m = \frac{a}{2^{48}}, \qquad p = b + 48.$$

b) *Subroutine INT*: Given a number x stored in the machine, find an integer n and the nonnegative fraction f such that

8.4
$$x = n + f.$$

Subroutine ANALYZE

We assume that the Fortran compiler is such that the instructions $X = 1.0$ and $Y = 1.0 + 1.0$ will produce the exact floating-point computer representations of unity and two, respectively, and that then we can generate the integers $z_k = 2^{2^k}$, $k = 1, 2, \ldots, 10$ by repeated squarings, all of which give exact results. To get the exact representation of $1/2$ we let $Z = 1.0/2.0$, an operation which is again assumed exact. By repeated squarings we can get exact values of $v_k = 2^{-2^k}, k = 0, 1, 2, \ldots, 9$. We can also get exact values of $u_k = 2^{-k}, k = 1, 2, \ldots, 48$ by repeated multiplications.

We also assume the existence of a completely reliable conditional transfer statement in the Fortran compiler (see Section 3.1).

We seek to obtain the integer p and the binary digits $e_0, e_1, \ldots, e_{10}$ such that

$$\phi_2(p) = e_{10}e_9 \ldots e_0,$$

if $|x| \geq 1$ and

$$\phi_2(p) = -e_9 e_8 \ldots e_0,$$

if $|x| < 1$. We also seek the fraction m and the binary digits $b_2, b_3, \ldots, b_{48}$ such that

$$\phi_2(|m|) = \cdot b_1 b_2 \ldots b_{48}, \qquad (b_1 = 1)$$

unless $x = 0$ in which case $p = 0$, $m = 0$, and all of the e_i and b_i vanish.

We first test whether $x = 0$. If so, then $p = m = 0$. Otherwise, we consider separately the cases where $|x| \geq 1$ and $|x| < 1$. In the case $|x| \geq 1$, we determine successively the digits $e_{10}, e_9, \ldots, e_0$ and then the digits $b_2, b_3, \ldots, b_{48}$. To get e_{10} we test whether $|x| \geq \frac{1}{2} 2^{2^{10}}$. If so, then $e_{10} = 1$; otherwise $e_{10} = 0$. Next, we test whether $|x| \geq \frac{1}{2}(2^{e_{10}2^{10}+2^9})$. If so, then $e_9 = 1$; otherwise $e_9 = 0$. Continuing in this way we find $e_8, e_7, \ldots, e_0$. Evidently

$$\phi_2(p) = e_{10}e_9 \ldots e_0.$$

Before discussing the determination of the b_k let us consider the case $|x| < 1$. To get e_9 we test whether $|x| < 2^{-2^9}$. If so, then $e_9 = 1$; otherwise $e_9 = 0$. Next we test whether $|x| < 2^{-2^9 e_9 - 2^8}$. If so, then $e_8 = 1$; otherwise $e_8 = 0$. Continuing in this way we find $e_7, e_6, \ldots, e_0$. Evidently

$$\phi_2(p) = -e_9 e_8 \ldots e_0.$$

In each of the cases $|x| \geq 1$ and $|x| < 1$ we compute $w = 2^p$. We then test whether $(\frac{1}{2} + \frac{1}{4})w \leq |x|$. If so, then $b_2 = 1$; otherwise $b_2 = 0$. Next we test whether $(\frac{1}{2} + \frac{1}{4}b_2 + \frac{1}{8})w \leq |x|$. If so, then $b_3 = 1$; otherwise $b_3 = 0$. Continuing in this way we determine $b_4, b_5, \ldots, b_{48}$. The sign of m is determined by the sign of x.

A flow diagram for Subroutine ANALYZE is given in Fig. 8.5. It should be noted that for brevity we have introduced the auxiliary variables y, y', m, and m'. In both the case $|x| \geq 1$ and the case $|x| < 1$, we assign an initial value of y as indicated. The y' are trial values of y. At the final stage $y = \frac{1}{2}2^p$ or 2^p in the cases $|x| \geq 1$ and $|x| < 1$, respectively. Thus $w = 2^p$ is given by $2y$ and y, respectively. The variable m is set initially to $\frac{1}{2}$. Using various tentative values m' for m we are able to successively determine the binary digits of m. At the end of the computation the required value of m is obtained.

We remark that for many practical purposes where the binary representation of m is not required it would be sufficient for the computation of m to simply compute $x/2^p$ using the divide instruction, which we assume will yield an exact result. One can then obtain n and f for INT. (See Exercise 3.)

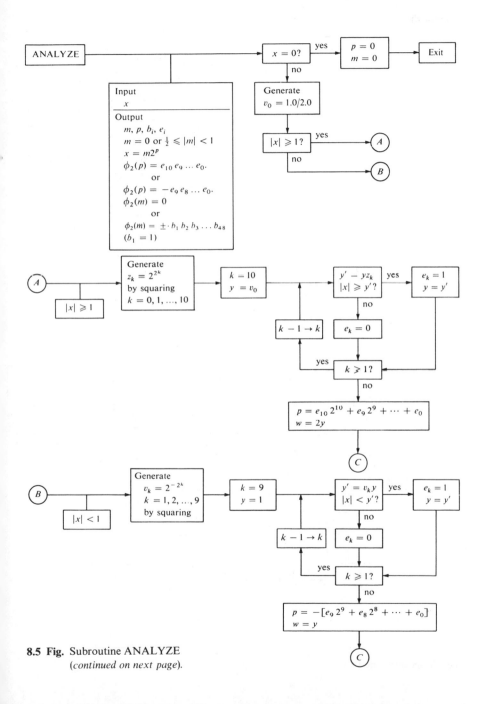

8.5 Fig. Subroutine ANALYZE
(continued on next page).

(*continued*).

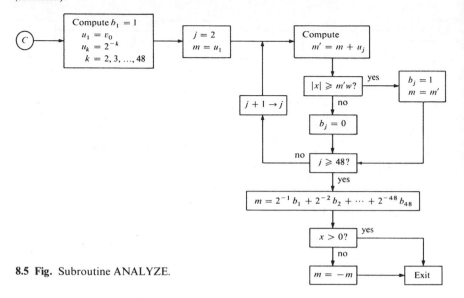

8.5 Fig. Subroutine ANALYZE.

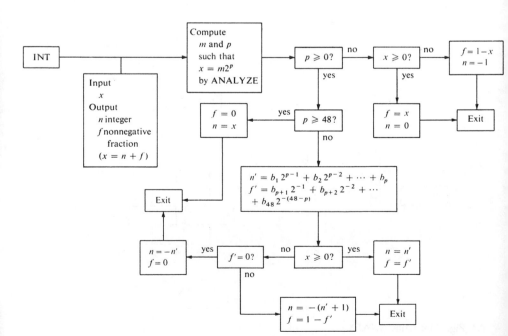

8.6 Fig. Subroutine INT.

Subroutine INT

Given x we first use ANALYZE to determine p and m such that $x = m2^p$. If $p < 0$ then $f = x$ and $n = 0$ when $x \geq 0$, and $f = 1 - x$ and $n = -1$, otherwise. If $p \geq 48$, then x is an integer and $f = 0$, $n = x$. If $0 \leq p$ and $p < 48$, then we, in effect, shift the binary point p places from the leftmost position in the binary representation of

$$|m| = \cdot b_1 b_2 \ldots b_{48}.$$

This is done as indicated in the flow diagram in Fig. 8.6.

EXERCISES 2.8

1. Write a program for Subroutine ANALYZE and apply it to the following cases.

   $27.125, \quad 10^{200}, \quad -1.0125, \quad 4715.815, \quad 10^{-157}, \quad 2^{85}, \quad 2^{-85}, \quad 1.25 \times 10^{27}, \quad 1, \quad 2, \quad \frac{1}{2},$
   $0.00010105, \quad 85, \quad 612, \quad 781.00012.$

   In each case, the program should print out the binary digits for p and m (or their octal equivalents) as well as the decimal representations. In each case, compute and print out $m2^p$ and x (in decimal) and compare.

2. Write a program for Subroutine INT and apply it to the numbers in the preceding example.

3. Write a modified version of ANALYZE where, when p is obtained as in ANALYZE, but where m is determined by dividing x by 2^p. Apply the modified version to the cases of Exercise 1.

4. Suppose that m and p are given in floating-point and fixed-point format, respectively, such that $x = m2^p$ where $x > 0$, $\frac{1}{2} \leq m < 1$ and p is an integer. If $|x| \leq 2^{48} - 1$ one can use the Fortran statement $N = X$ to yield the largest integer not greater than x if $x \geq 0$, and not less than x if $x < 0$. Show how this can be used to perform the functions of INT for all x. Write a modified version of INT based on this procedure.

5. Apply the modified version of INT to the cases of Exercise 1.

2.9 CALCULATION OF ELEMENTARY FUNCTIONS

In this section we describe some procedures for evaluating certain elementary functions such as $\sqrt{x}$, polynomials, e^x, $\log x$ and x^y. The methods given are very elementary and by no means optimal. Better methods based on the theory of polynomial approximation and rational function approximation are available and will be discussed later in Chapter 6.

Square Root of a Nonnegative Number

If $x = 0$, then $\sqrt{x} = 0$. If $x > 0$, then we use subroutine ANALYZE of the preceding section to obtain m and p such that p is an integer, $1/2 \leq m < 1$, and

9.1 $x = m2^p.$

We define m' and p' by

9.2 $$x = m'2^{p'}$$

where

9.3 $$p' = \begin{cases} p, & \text{if } p \text{ is even} \\ p + 1, & \text{if } p \text{ is odd} \end{cases}$$

and

9.4 $$m' = m2^{p-p'}.$$

Evidently p' is even and

9.5 $$1/4 \leqq |m'| < 1.$$

Therefore, we have

9.6 $$\sqrt{x} = \sqrt{m'}2^{p'/2}.$$

Since $2^{p'/2}$ can be computed easily and exactly, the problem is reduced to finding $\sqrt{m'}$ where m' satisfies (9.5). This can be done using the elementary procedure of Example 1-(6.3). As shown in Chapter 1, if we compute the initial approximation to $\sqrt{m'}$, namely, $(1 + m')/2$, followed by four iterations, we obtain $\sqrt{m'}$ correct to 2^{-48}.

By the use of a more sophisticated initial approximation for $\sqrt{m'}$ we can obtain $\sqrt{m'}$ to the desired accuracy in fewer than 4 iterations. See Exercise 2.

Evaluation of Polynomials

Let us consider the evaluation of the polynomial

9.7 $$P(x) = a_0 x^n + a_1 x^{n-1} + \cdots + a_n.$$

Polynomials are important not only in their own right but also for an approximate evaluation of more complicated functions. Many such functions can be represented in terms of infinite series. For example, the exponential function e^x can be represented by the power series

9.8 $$e^x = 1 + x + \frac{x^2}{2!} + \frac{x^3}{3!} + \cdots + \frac{x^n}{n!} + \cdots .$$

While this series converges for all x, the convergence is unacceptably slow for large $|x|$. However, for small $|x|$, say for $|x| < 1$, the convergence is quite rapid. Similarly, for small $|x|$ the series

9.9
$$\begin{cases} \sin x = x - \dfrac{x^3}{3!} + \dfrac{x^5}{5!} - \dfrac{x^7}{7!} + \cdots \\[2ex] \cos x = 1 - \dfrac{x^2}{2!} + \dfrac{x^4}{4!} - \dfrac{x^6}{6!} + \cdots \end{cases}$$

are satisfactory. In each case, given x we can obtain any desired accuracy, subject to roundoff limitations, by taking sufficiently many terms of the series, that is, by evaluating a polynomial of sufficiently high order.

A polynomial (9.7) can be conveniently evaluated by the following "nesting procedure."

9.10 $P(x) = ((\cdots(((a_0)x + a_1)x + a_2) + a_3)x + \cdots + a_{n-1})x + a_n)$

⓪   ①   ②   ③ $(n-1)$ ⓝ

where numbers under the right parentheses represent the partial sums

9.11
$$\begin{cases} z_0 = a_0 \\ z_1 = xz_0 + a_1 \\ z_2 = xz_1 + a_2 \\ \quad \cdots \\ z_n = xz_{n-1} + a_n. \end{cases}$$

A flow chart for the evaluation of a polynomial is as follows:*

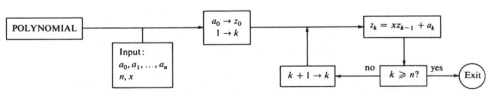

Thus a polynomial of degree n can be evaluated by carrying out n additions and n multiplications. This is much more efficient than computing the individual powers of x, multiplying by the appropriate coefficients and adding.

Of course, if one wishes to use the nesting procedure to evaluate a partial sum of a power series, one needs to know in advance how many terms of the series to use. However, given the series $\alpha_0 + \alpha_1 x + \alpha_2 x^2 + \cdots$ one could compute the partial sums $w_0 = \alpha_0$, $w_1 = x^{-1}w_0 + \alpha_1$, $w_2 = x^{-1}w_1 + \alpha_2, \ldots$. The numbers $w_0' = w_0$, $w_1' = xw_1$, $w_2' = x^2w_2$, $w_3' = x^3w_3$, etc. represent the partial sums. This procedure might be effective in some cases.

---

* In Chapter 5, we will give a bound on the roundoff error involved in computing the value of a polynomial.

Evaluation of Logarithms

Let us first consider the evaluation of the natural logarithm of x, $\log x$, which is the logarithm to the base e where

$$e = 2 + \frac{1}{2!} + \frac{1}{3!} + \frac{1}{4!} + \cdots + \frac{1}{n!} + \cdots \doteq 2.71828\ 18284\ 59045.$$

The Taylor's series expansion for $\log x$ for values of x near 1 is given by

9.12
$$\log x = t - \frac{t^2}{2} + \frac{t^3}{3} - \frac{t^4}{4} + \cdots$$

where

$$t = x - 1.$$

The series converges for all x in the range $0 < x < 2$. However, it is rather slowly converging, especially for values of x near 0 and 2. We can accelerate the convergence somewhat by noting that

9.13
$$\log(1 - z) = -z - \frac{z^2}{2} - \frac{z^3}{3} - \cdots$$

and hence

9.14
$$\log \frac{1 + z}{1 - z} = 2\left(z + \frac{z^3}{3} + \frac{z^5}{5} + \cdots\right).$$

Thus we have

9.15
$$\log x = 2\left[\left(\frac{x - 1}{x + 1}\right) + \frac{1}{3}\left(\frac{x - 1}{x + 1}\right)^3 + \frac{1}{5}\left(\frac{x - 1}{x + 1}\right)^5 + \cdots\right]$$

which converges provided $|(x - 1)(x + 1)^{-1}| < 1$, that is, provided $x > 0$. Moreover, the series converges more rapidly than (9.12). Indeed, the error in (9.12) due to neglecting all terms after the term $(-1)^{n+1}t^n/n$ in (9.12) is approximately

9.16
$$R_n = \frac{|x - 1|^{n+1}}{n + 1}.$$

This follows since (9.12) is an alternating series with terms of decreasing absolute values.* The error S_n due to neglecting all terms after the term

$$\frac{2}{2n + 1}\left(\frac{x - 1}{x + 1}\right)^{2n+1}$$

---

* See Theorem A-2.16.

in (9.15) is bounded by

9.17
$$S_n \leqq \left(\frac{2}{2n+3}\right)\left|\frac{x-1}{x+1}\right|^{2n+3}\frac{1}{1-|(x-1)/(x+1)|^2}$$

$$= \frac{2}{2n+3}\left|\frac{x-1}{x+1}\right|^{2n+3}\frac{|x+1|^2}{4x}$$

$$= \frac{1}{2(2n+3)}\frac{|x-1|^{2n+3}}{x|x+1|^{2n+1}}.$$

Clearly, for $x > 0$ and for sufficiently large n we have $S_n < R_n$. Moreover, for $x \geqq \frac{1}{4}$, it is evident that S_n is considerably less than R_n.

Thus (9.15) gives a satisfactory, but by no means optimum, procedure for computing $\log x$ provided x is reasonably close to unity. For example, if $\frac{1}{2} \leqq x \leqq 2$, then* we can get $\log x$ to within an error of 2^{-48} by choosing $n = 15$. We can get a relative accuracy of 2^{-48} by choosing $n = 20$. (See Exercise 12.)

We now describe an alternative procedure for computing $\log x$ which is based on the use of Subroutine ANALYZE and on the determination of $\log_2 x$. First we find m and p such that $x = m2^p$ using ANALYZE. We then compute $\log m$ as above, since $\frac{1}{2} \leqq m \leqq 1$, and then $\log_2 m$ by†

9.18
$$\log_2 m = \frac{\log m}{\log 2}$$

$$\doteq \frac{\log m}{0.693147180559945}.$$

Finally, we compute

9.19
$$\log_2 x = p + \log_2 m$$

and

9.20
$$\log x = \log_2 x \cdot \log 2.$$

Evaluation of x^y

---

*If $\frac{1}{2} \leqq x \leqq 2$, we have
$$S_n \leqq \frac{1}{2n+3}\left(\frac{1}{3}\right)^{2n+1}$$

†Actually, rather than reading in and storing $\log 2$, we can compute $\log 2$ using the procedure described above for $-\log\frac{1}{2}$.

Let us first consider the evaluation of the exponential function

9.21 $$e^x = 1 + x + \frac{x^2}{2!} + \frac{x^3}{3!} + \cdots + \frac{x^n}{n!} + \cdots .$$

Evidently the error R_n caused by neglecting all terms after $x^n/n!$ is bounded by

9.22 $$R_n = \frac{x^{n+1}}{(n+1)!} e^c$$

where c lies between 0 and x. This follows from Theorem A-3.64. One can easily verify that if $|x| \leq \frac{1}{2}$, the convergence is very rapid. As a matter of fact if $n = 15$ then

9.23 $$R_n \leq 2^{-48}$$

for all x in the specified range.

To evaluate e^x for values of x outside of the range $-\frac{1}{2} \leq x \leq \frac{1}{2}$ we again use logarithms to the base 2. Evidently the function e^x is a special case of x^y for $x > 0$, since

9.24 $$x^y = e^{y \log x} = 2^{y \log_2 x}.$$

Our procedure is as follows:

1. Determine m and p such that $x = m2^p$ using Subroutine ANALYZE.
2. Compute $\log m$, $\log_2 m = \log m/\log 2$ and $\log_2 x = p + \log_2 m$.
3. Compute $y \log_2 x$ and use Subroutine INT to determine the integer n and the nonnegative fraction f such that $y \log_2 x = n + f$. If $f > \frac{1}{2}$, then let $n' = n + 1$, $f' = f - 1$; otherwise $n' = n$, $f' = f$.
4. Compute $f' \log 2$ and $e^{f' \log 2}$ using the series expansion for the exponential. Since $|f' \log 2| \leq 0.35$, we require fewer than 15 terms to obtain an accuracy of 2^{-48}.
5. Finally, we have

9.25 $$x^y = 2^{n'} e^{f' \log 2}.$$

A flow diagram for Subroutine POWER which is based on the above procedure is given in Fig. 9.26, page 69.

EXERCISES 2.9

1. Write a program for evaluating the polynomial

$$P(x) = a_0 x^n + a_1 x^{n-1} + \cdots + a_n$$

where the coefficients $a_0, a_1, \ldots, a_n$ and x are real. Use the nesting procedure. Use the program to evaluate

$$P(x) = x^3 + x^2 - 4x + 6$$

for $x = 0, 1, 2$.

2. Develop the first two terms of the Taylor series for the function $f(x) = \sqrt{x}$ around $x = \frac{1}{2}$. Show that if $\frac{1}{4} \leq x \leq 1$ then the error in neglecting all terms beyond the second term (involving $x - \frac{1}{2}$) does not exceed $3\sqrt{2}/4 - 1 \doteq 0.06$.

3. Give an upper bound to the error in using the polynomial

$$1 + x + \frac{x^2}{2!} + \frac{x^3}{3!} + \frac{x^4}{4!}$$

to evaluate e^x for $x = 0.6$.

4. Develop a program for finding $\tan x$ based on the use of the Taylor's series expansion about $x = 0$ and the identity

$$\tan\left(\frac{\pi}{2} - x\right) = (\tan x)^{-1}.$$

5. Develop a program for finding $\sin x$ and $\cos x$ using the series

$$\sin x = x - \frac{x^3}{3!} + \frac{x^5}{5!} - \frac{x^7}{7!} + \cdots$$

$$\cos x = 1 - \frac{x^2}{2!} + \frac{x^4}{4!} - \frac{x^6}{6!} + \cdots$$

for $|x| < \pi/4$. For values of x such that $|x| > \pi/4$ use the following scheme. Compute $y = (2/\pi)x$. Let n be the closest integer to y and let

$$z = y - n.$$

Then compute $\sin x$ and $\cos x$ in terms of $\cos(\pi/2)z$ and $\sin(\pi/2)z$ using identities such as

$$\sin\left(\frac{\pi}{2} \pm x\right) = \mp \cos x$$

$$\cos\left(\frac{\pi}{2} \pm x\right) = \mp \sin x$$

$$\sin(\pi \pm x) = \pm \sin x$$

$$\cos(\pi \pm x) = -\cos x.$$

Use as many terms of the series so that the truncation error does not exceed 2^{-48}.

6. Prepare a program for finding the square root of a real number and apply it to the following cases

$$0.001012578$$

$$385.12057$$

$$2.8158 \times 10^{175}.$$

7. If $|x| \leqq 0.35$ how many terms of the exponential series will give an accuracy of 2^{-48}?

8. Verify that

$$\max_{\frac{1}{2} \leqq x \leqq 2} \frac{|x - 1|^{2n+3}}{x|x + 1|^{2n+1}} \leqq 2(\tfrac{1}{3})^{2n+1}.$$

9. Find the smallest integer n such that

$$\frac{1}{2n + 3} (\tfrac{1}{3})^{2n+1} \leqq 2^{-48}.$$

10. Find the smallest integer n such that

$$\frac{(\tfrac{1}{2})^{n+1}}{(n + 1)!} e^{\frac{1}{2}} \leqq 2^{-48}.$$

11. Write a program for computing x^y based on the procedure described in the text. Apply the program to the cases

$$27.25^{1.25}, 0.0015^{0.2718}.$$

12. Show that the relative error in the use of 20 terms of (9.15) is less, in absolute value, than 2^{-48}.

SUPPLEMENTARY DISCUSSION

There are many good references on the computer evaluation of mathematical functions. See Fike [1968], for example. For a recent paper on finding improved starting values for the determination of $\sqrt{x}$ see Moursund [1967].

Many attempts have been made to control or to estimate the errors in automatic computation by a variety of schemes. For example, in Chapter 13 we discuss the exact solution of a system of linear algebraic equations using *residue arithmetic*. Workers in this field include Takahasi and Ishibashi [1961], Lindamood [1964], Borosh and Fraenkel [1966], Newman [1967], Szabó and Tanaka [1967], Howell and Gregory [1970], and McClellan [1971].

Moore [1966], Hansen [1969], Griffith [1971] and others have explored *interval arithmetic* as a means of carrying out computations for which error bounds can be guaranteed. Ashenhurst and Metropolis and their collaborators (see Fraser and Metropolis [1968], for example) have investigated computational algorithms in *unnormalized floating-point arithmetic* as a means of estimating errors at each stage of the computation. Chartres [1966] advocates *variable-precision arithmetic* for the same purpose.

An abstract mathematical theory of floating-point arithmetic has been described by Matula [1969], who is a strong advocate of computer arithmetic as a component of a computer science curriculum.

Cody [1967] has studied the influence of the design of the arithmetic units of various digital computers (now in existence) on numerical algorithms, and many active numerical analysts have attempted to influence the design of the arithmetic units of future computers so that it will be easier to construct computational algorithms that are accurate. See, for example, Gregory [1966].

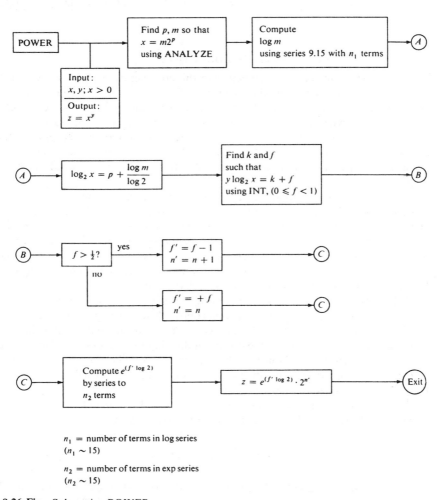

n_1 = number of terms in log series
$(n_1 \sim 15)$

n_2 = number of terms in exp series
$(n_2 \sim 15)$

9.26 Fig. Subroutine POWER.

CHAPTER 3

SURVEILLANCE OF NUMBER RANGES

3.1 INTRODUCTION

As we pointed out in Section 2.7, when an arithmetic operation is performed on the computer there is a possibility that the result of the operation may be too large to be represented on the machine. In such a case exponent overflow will occur. On most machines, when exponent overflow occurs there is an error message and the calculation is interrupted. If the result of the operation is too small to be represented on the machine, exponent underflow occurs. On most machines, in such a case the result is automatically replaced by zero and the calculation proceeds without interruption.

It is often possible to organize the computational procedure for a given problem in such a way that the possibility of exponent overflow can be minimized or eliminated. In some cases, however, it is desirable to perform a surveillance of the process in order to prevent the attempted execution of an operation which would result in an overflow. Such a surveillance can often be carried out using overflow and underflow testing features built into the computer hardware. However, because of the variations in computers and the differences in the way various compilers such as Fortran make use of the hardware, in order to achieve complete control of the surveillance process it seems desirable to develop a software system which is as independent as possible of the computer hardware and is applicable to a wide class of machines. One of the objects of this chapter is to describe some of the ingredients which would go into such a system. While the actual use of such a system would appear to be justified only in special cases, nevertheless, it is hoped that a study of the techniques used will be instructive.

The system we describe is based on the use of the Fortran algebraic language. It is assumed that the following conditional transfer statement

$$IF(X - Y)A, B, C$$

is exact in the following sense. If* $(X) = \pm 0$ and $(Y) = \pm 0$ or if $(X) = (Y)$, control passes to Statement B. If $(X) > (Y)$, control passes to C; and if $(X) < (Y)$, control passes to A. The correct programming for such a statement within the compiler would involve a bit-by-bit analysis of the floating-point representations of (X) and

---

* Here we let (X) denote the contents of the memory location designated by X.

(Y). It should be noted that a floating-point subtraction might not be adequate. For example, if $(X) = 2^{-(2^{10}-1)+47}m_x$ and $(Y) = 2^{-(2^{10}-1)+47}m_y$ where $\phi_2(m_x) = 0.1111$ and $\phi_2(m_y) = 0.1110$ we would get exponent underflow if the machine normalizes after subtraction. Thus, the result $(X) - (Y)$ would be interpreted as zero and control would pass to Statement B rather than to C.

With a Fortran compiler which does not have an exact conditional transfer statement, the use of the system described below would require an appropriate change in the compiler.

For the proper use of the surveillance programs, the user should be familiar with the characteristics of the input and output programs for his particular machine. The surveillance programs described in this section are concerned only with numbers which are stored in the machine. They are not concerned with the machine-dependent questions of what number is actually placed in a particular memory location, given a number on an input card, and of what number is printed out, given a number in a memory location. Some light on these questions can be shed by testing various cases using the programs described in Section 2.8, which enable one to determine the exact binary representation of a number which is stored in the machine.

3.2 ALLOWABLE NUMBER RANGES

Corresponding to a given machine, let α, β, $-\gamma$, $-\omega$ denote, respectively, the smallest positive, the largest positive, the largest negative, and the smallest negative numbers which can be represented (exactly) in the machine. Let

2.1 $$\mathscr{R} = \mathscr{Z} + \mathscr{P} + \mathscr{N}$$

where

2.2 $$\begin{cases} \mathscr{Z} = \{0\} \\ \mathscr{P} = \{x : \alpha \le x \le \beta\} \\ \mathscr{N} = \{x : -\omega \le x \le -\gamma\} \end{cases}$$

and let $\circledR$, $\circledP$, and $\circledN$ denote, respectively, the set of numbers in $\mathscr{R}$, $\mathscr{P}$, and $\mathscr{N}$ which can be represented exactly on the machine. We assume that for any binary operation, $\circ$, if $x \in \circledR$ and $y \in \circledR$ and if the exact result $x \circ y$ is in $\mathscr{R}$ then the machine will give a result, denoted by $\text{fl}(x \circ y)$, which is in $\mathscr{R}$. Of course, $\text{fl}(x \circ y)$ will not in general equal $x \circ y$. We also assume certain obvious monotonicity properties. For instance, if x, y, x', and y' are in $\circledP$, if $x \le x'$ and $y \le y'$, and if $x' + y' \le \beta$, then

2.3 $$\text{fl}(x + y) \le \text{fl}(x' + y').$$

Similarly, if x and y are in $\circledP$, if $x \le y$, and if x^{-1} and y^{-1} are in $\circledP$, then

2.4 $$\text{fl}(1 \div y) \le \text{fl}(1 \div x).$$

The reader should note that the numbers in $\circledR$ are the *computer-representable numbers* of Section 2.6. We shall refer to the numbers in $\mathscr{R}$, on the other hand, as *representable numbers*. Obviously, any number in $\mathscr{R}$ can be represented approximately, but not necessarily exactly, on the computer.

As an example, let us consider Machine C. The set $\circledR$ consists of the number zero and certain numbers which can be represented in normalized floating-point form. Thus $x \in \circledR$ if $x = 0$ or if x can be written in the form

2.5
$$x = a2^b$$

where a and b are any integers such that

2.6
$$\begin{cases} 2^{47} \leqq |a| \leqq 2^{48} - 1 \\ \phantom{2^{47} \leqq} |b| \leqq 2^{10} - 1. \end{cases}$$

Thus we have

2.7
$$\begin{cases} \alpha = \gamma = 2^{47}2^{-(2^{10}-1)} = 2^{-976} \\ \beta = \omega = (2^{48} - 1)2^{2^{10}-1} = (2^{48} - 1)2^{1023}. \end{cases}$$

We now seek to define the set $\mathscr{A}$ of *allowable numbers* and the set $\circledA$ of *computer-allowable numbers*. We first choose $\bar{a}$ and $\bar{b}$ such that

2.8
$$\bar{b} = \frac{1}{\bar{a}}$$

such that $\bar{a}$ and $\bar{b}$ are in $\circledR$, and such that (even when $\alpha \neq \gamma$ and $\beta \neq \omega$)

2.9
$$\begin{cases} \bar{a} \geqq \max(\alpha, \gamma) \\ \bar{b} \leqq \min(\beta, \omega). \end{cases}$$

We assume that $\mathrm{fl}(1 \div \bar{a}) = \bar{b}$ and that $\mathrm{fl}(1 \div \bar{b}) = \bar{a}$.

We define the sets $\mathscr{S}$, $\mathscr{M}$, and $\mathscr{L}$ by

2.10
$$\begin{cases} \mathscr{S} = \{x : 0 < |x| < \bar{a}\} \\ \mathscr{M} = \{x : \bar{a} \leqq |x| \leqq \bar{b}\} \\ \mathscr{L} = \{x : |x| > \bar{b}\}. \end{cases}$$

We let

2.11
$$\mathscr{A} = \mathscr{M} + \mathscr{L}$$

and

2.12
$$\circledA = \circledM + \mathscr{L}$$

where

2.13
$$\textcircled{$\mathscr{M}$}= \mathscr{M} \cap \textcircled{$\mathscr{R}$}.$$

We desire that the following property holds:

If $x \in \textcircled{$\mathscr{A}$}$ and $y \in \textcircled{$\mathscr{A}$}$ then

2.14
$$x + y \in \mathscr{R}.$$

In order to achieve this property, we choose $\bar{a}$ so that, even if x and y have opposite signs, we have

2.15
$$|x + y| \geqq \max(\alpha, -\gamma)$$

provided $|x| \geq \bar{a}$ and $|y| \geq \bar{a}$. It is easy to verify that for Machine C a satisfactory choice of $\bar{a}$ and $\bar{b}$ is

2.16
$$\begin{cases} \bar{a} = 2^{-928} \sim 10^{-279} \\ \bar{b} = 2^{928} \sim 10^{279}. \end{cases}$$

To summarize, we have the following properties:

1. If $x \in \mathscr{A}$ and $x \neq 0$, then $x^{-1} \in \mathscr{A}$.
2. If $x \in \mathscr{A}$ then $-x \in \mathscr{A}$.
3. If $x \in \textcircled{$\mathscr{A}$}$ then $-x \in \textcircled{$\mathscr{A}$}$. (We assume that $\mathrm{fl}(-x) = -x$ if $x \in \textcircled{$\mathscr{A}$}$.)
4. If $x \in \textcircled{$\mathscr{A}$}$ and $y \in \textcircled{$\mathscr{A}$}$, then $x + y \in \mathscr{R}$.

A basic ingredient in a package of surveillance subroutines is a program which we shall call "Subroutine REDIT." Subroutine REDIT accepts as input a number $x \in \textcircled{$\mathscr{R}$}$ and produces as output the real number $\rho \in \textcircled{$\mathscr{A}$}$ and the integer θ where:

1. if $x \in \textcircled{$\mathscr{A}$}$ then $\rho = x$. In this case

$$\theta = 1 \text{ if } x \in \mathscr{M}$$
$$\theta = 0 \text{ if } x \in \mathscr{Z}$$

2. if $x \in \mathscr{S}$ then $\rho = 0$ and $\theta = -1$.
3. if $x \in \mathscr{L}$ then $\rho = \bar{b}$ and $\theta = 2$.

A flow diagram for REDIT is given in Fig. 2.17.

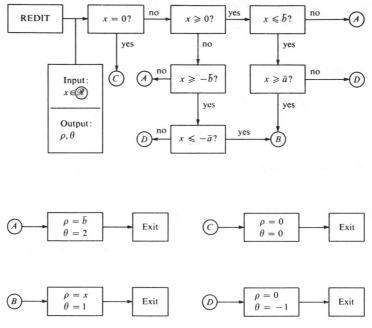

2.17 Fig. Subroutine REDIT.

EXERCISES 3.2

1. Verify that the choices of $\bar{a}$ and $\bar{b}$ given by (2.16) for Machine C are satisfactory.

2. Verify that if the range of representable numbers is $-\omega \leqq x \leqq -\gamma$, $x = 0$, $\alpha \leqq x \leqq \beta$, then a satisfactory choice of $\bar{b}$ is 2^p for some integer p such that $2^p \leqq \min(\beta, \omega, \alpha^{-1}, \gamma^{-1})$.

3. Prepare a Fortran program for REDIT.

4. What would be the result of applying REDIT to the following computer-representable numbers with Machine C? Give both ρ and θ in each case:

$$2^{-970}, 2^{970}, 2^{900}, 2^{-900}, 0$$
$$-2^{-970}, -2^{970}, -2^{900}, -2^{-900}.$$

5. Find a satisfactory choice of $\bar{a}$ and $\bar{b}$ for a machine similar to Machine C except that the word-length is only 36 bits with $t = 27$ and with only eight bits for the exponent including the sign bit.

3.3 BASIC REAL ARITHMETIC OPERATIONS

We now describe the three arithmetic subroutines of addition/subtraction, multiplication, and division. In each case the input arguments x and y are in $\mathcal{R}$.

The output is the number ρ which belongs to $\mathcal{A}$ and an indicator integer θ.

   For addition and subtraction (see Fig. 3.1, Subroutine RADD) we are assured that since x and y are in $\mathcal{A}$ the calculation of $x + y$ will produce a result in $\mathcal{R}$ and hence will not produce exponent overflow or underflow. Hence, we may compute $x + y$ and use Subroutine REDIT to edit the result.

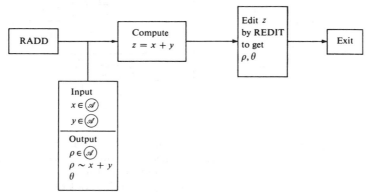

3.1 Fig. Subroutine RADD.

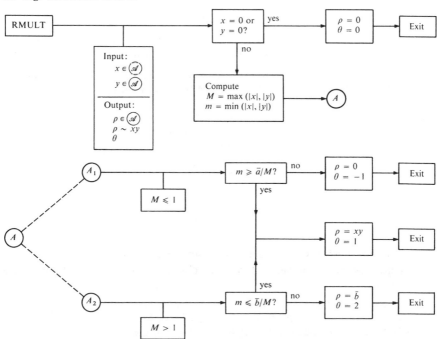

3.2 Fig. Subroutine RMULT.

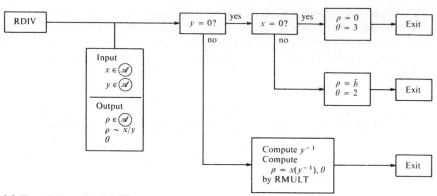

3.3 Fig. Subroutine RDIV.

Flow diagrams for multiplication and division are given in Figs. 3.2 and 3.3 respectively. The reader should verify that neither exponent overflow nor exponent underflow can occur at any stage of the computation. For the case of division, if both arguments are zero the result is indeterminant and we let $\theta = 3$.

EXERCISES 3.3

1. Prepare Fortran programs for RADD, RMULT, and RDIV. Give the programs based on these routines and on REDIT to determine AB, $A + B$, A/B where A and B are in $\mathcal{R}$.

2. For each of the following cases what would be the result of applying REDIT to x and to y and then computing $x + y$ by RADD. Do the same for xy and x/y.

x	y
2^{970}	2^{970}
2^{970}	2^{-970}
2^{-970}	2^{-970}
2^{900}	2^{900}
2^{900}	2^{-900}
2^{-900}	2^{-900}
0	1
1	0
0	0

Assume that Machine C is used and that $\bar{a} = (\bar{b})^{-1} = 2^{-928}$.

3. Suppose that we are dealing with Machine C with $\bar{a} = 2^{-928}$, $\bar{b} = 2^{928}$. Prove the following

a) If $A \in \mathcal{A}$, $B \in \mathcal{A}$, and $AB \in \mathcal{L}$ then $|A| \geq 1$ and $|B| \geq 1$.
b) If $A \in \mathcal{A}$, $B \in \mathcal{A}$, and $AB \in \mathcal{S}$ then $|A| \leq 1$ and $|B| \leq 1$.
c) If $A \in \mathcal{A}$, $B \in \mathcal{A}$, and $|A| \geq 1$ or $|B| \geq 1$ then $|A| - |B| \in \mathcal{A}$.
d) If $A \in \mathcal{A}$, $B \in \mathcal{A}$, and $|A| \leq 1$ or $|B| \leq 1$ then $|A| + |B| \notin \mathcal{L}$.

4. Show how the surveillance routines could be used to determine $ABC, (A + B)/C, A/(BC)$, $AB/C, A/(B + C)$ where A, B, and C are in $\mathcal{A}$.

5. Show how one could use the surveillance subroutines to handle the following expressions. Assume in every case that each argument is in $\mathcal{A}$.

a) ABC.   d) $A_1 A_2 \ldots A_n$.

b) $A + B + C$.   e) $A + B + C + D$.

c) $A(B + C)$.   f) $A_1 + A_2 + \cdots + A_n$.

6. Prepare a flow diagram and a Fortran program for the computation of $m2^p$ where $m = 0$ or $\frac{1}{2} \leq |m| < 1$ and where p is an integer such that $-(2^{10} - 1) \leq p \leq 2^{10} - 1$. (This program will be referred to as RSYNTH.)

7. Prepare a flow diagram and a Fortran program for the computation of $a_1 a_2 + b_1 b_2$ where a_1, a_2, b_1, and b_2 are in $\mathcal{A}$. Also determine m and the integer p such that $a_1 a_2 + b_1 b_2 = m2^p$ where $\frac{1}{2} \leq |m| < 1$ or $m = 0$ and $-(2^{10} - 1) \leq p \leq 2^{10} - 1$ (m and p should be found if possible even if $a_1 a_2 + b_1 b_2$ is in $\mathcal{L}$). The program may use ANALYZE and RSYNTH as well as the surveillance programs for real arithmetic. (This program will be referred to as RSCAL2.)

3.4 THE QUADRATIC EQUATION

Let us seek to develop an accurate and reliable procedure for solving the quadratic equation

4.1 $Ax^2 + Bx + C = 0$

where A, B, and C are in $\mathcal{R}$.

The first step in the procedure is to edit the coefficients using REDIT. If any coefficient is in $\mathcal{L}$, we do not attempt to solve the problem. Any coefficient which is in $\mathcal{S}$ is replaced by zero. Our objective is to find an accurate value of each root of the edited equation which is in $\mathcal{A}$. If a root is in $\mathcal{L}$ or $\mathcal{S}$ we wish to determine that fact also.

Elementary Special Cases

Having "edited" the numbers A, B, and C, we now examine the elementary special cases. If $A = 0$, then (4.1) reduces to

4.2 $Bx + C$.

If $B \neq 0$, then the solution is

4.3 $x = -C/B$.

We determine $-C/B$ by the Subroutine RDIV. As output for the problem we indicate that $x_1 \in \mathcal{L}$ and $x_2 = -C/B$ provided $C/B \in \mathcal{M}$. Otherwise, $x_2 \in \mathcal{S}$, $x_2 = 0$, or $x_2 \in \mathcal{L}$, as appropriate.

If $B = 0$ and if $A \neq 0$, then we have*

4.4
$$x_1 = \sqrt{-C/A}, \qquad x_2 = -\sqrt{-C/A}.$$

If $AC < 0$, then x_1 and x_2 are real. Otherwise, both are purely imaginary. We remark that $\sqrt{|C|/|A|} = 0$ if $C = 0$; otherwise, it is in $\mathcal{M}$.

If $B = 0$ and $A = 0$, then there is no solution if $C \neq 0$; on the other hand, every value of x is a solution if $C = 0$.

If $C = 0$ and $A \neq 0$, then the roots are

4.5
$$x_1 = 0, \qquad x_2 = -B/A.$$

If $C = 0$, $B = 0$, and $A \neq 0$, then the roots are

4.6
$$x_1 = 0, \qquad x_2 = 0.$$

Solution When All Coefficients Are in Ⓜ

If A and C are in Ⓜ, then we can compute $\sqrt{|A|}$ and $\sqrt{|C|}$. Moreover, $\sqrt{|C/A|} = \sqrt{|C|}/\sqrt{|A|}$ is in $\mathcal{M}$. We now proceed to determine which of the sets $\mathcal{S}$, $\mathcal{M}$, and $\mathcal{L}$ contain the following three critical quantities.

4.7
$$\frac{B}{2A}, \quad \frac{B}{2C}, \quad \delta = \frac{|B|}{2\sqrt{|A||C|}} = \sqrt{\frac{B}{2A}}\,\sqrt{\frac{B}{2C}}.$$

We also wish to evaluate each quantity when it is in $\mathcal{M}$. We remark that δ may be in $\mathcal{M}$ even if $B/(2A)$ and/or $B/(2C)$ are not.

Let us consider first the quantity $B/(2A)$. If $|A| \leqq 1$, then we can compute $2A$ without exponent overflow. We then use RDIV for $B/(2A)$. If $|A| > 1$ we compute B/A, which must be in $\mathcal{S}$ or $\mathcal{M}$. If $B/A \in \mathcal{S}$ then $B/(2A)$ is in $\mathcal{S}$. Otherwise, we compute $(B/A) \div 2$ by RDIV.

A similar procedure can be used for $B/(2C)$. We note that $\sqrt{|A||C|} = \sqrt{|A|}\,\sqrt{|C|} \in \mathcal{M}$. Thus we can handle δ in the same way.

If $B/(2A)$ is in $\mathcal{L}$, then at least one root of (4.1) is in $\mathcal{L}$. Let x_1 denote the root of (4.1) of largest modulus (or either root if both have the same modulus) and let x_2 denote the other. Then we have

4.8 Theorem. If $B/(2A) \in \mathcal{L}$, then $x_1 \in \mathcal{L}$.

Proof. The sum of the roots of (4.1) is $-B/A$. Hence we have

$$x_1 + x_2 = -\frac{B}{A}$$

---

* We note that if $x \in \mathcal{M}$ then $\sqrt{x} \in \mathcal{M}$ and $\sqrt{x}$ can be computed using the algorithm described in Section 2.9.

and

$$\left|\frac{B}{A}\right| = |x_1 + x_2| \leq |x_1| + |x_2| \leq 2|x_1|.$$

Hence,

$$|x_1| \geq \left|\frac{B}{2A}\right|.$$

Therefore, $x_1 \in \mathscr{L}$.

Similarly, we have

4.9 Theorem. If $B/(2C) \in \mathscr{L}$, then $x_2 \in \mathscr{S}$.

Proof. Consider the quadratic equation

4.10 $$C(1/x)^2 + B(1/x) + A = 0.$$

By Theorem 4.8 since $B/(2C) \in \mathscr{L}$, one of the roots of (4.10) is in $\mathscr{L}$. Hence, one of the roots of (4.1) is in $\mathscr{S}$, since the roots of (4.1) are the reciprocals of the roots of (4.10).

Consider now the case where all three of the basic quantities $B/(2A)$, $B/(2C)$, and δ are in $\mathscr{M}$. (We note that if $B/(2A) \in \mathscr{M}$ and $B/(2C) \in \mathscr{M}$, then $\delta \in \mathscr{M}$.) We write the solution of (4.1) in the form

4.11 $$x_1, x_2 = \frac{-B \pm \sqrt{B^2 - 4AC}}{2A} = -\frac{B}{2A} \pm \sqrt{\left(\frac{B}{2A}\right)^2 - \frac{C}{A}}.$$

If $\delta \geq 1$, then the roots are real. We have

$$x_1, x_2 = -\frac{B}{2A}\left(1 \pm \sqrt{1 - \frac{4AC}{B^2}}\right).$$

We compute x_1 using the plus sign before the radical, and we compute x_2 by $C/(Ax_1)$. Thus we obtain

4.12
$$\begin{cases} x_1 = -\frac{B}{2A}\left(1 + \sqrt{1 - \frac{4AC}{B^2}}\right) \\ x_2 = -\frac{2C}{B}\left(1 + \sqrt{1 - \frac{4AC}{B^2}}\right)^{-1} \end{cases}$$

If $\delta < 1$, then the roots will be real or complex depending on whether $AC \leq 0$ or $AC > 0$. If $AC > 0$, then we have

4.13 $$x_1, x_2 = -\frac{B}{2A} \pm i\sqrt{\frac{C}{A}}\sqrt{1 - \delta^2}.$$

On the other hand, if $AC \leqq 0$, then from (4.11) we have (see (5.21), page 88)

4.14
$$x_1 = -\left(\frac{B}{2A}\right) - \text{sgn}\left(\frac{B}{2A}\right)\sqrt{\left(\frac{B}{2A}\right)^2 + \left|\frac{C}{A}\right|}$$

$$= -\text{sgn}\left(\frac{B}{2A}\right)\left\{\frac{|B|}{2|A|} + \sqrt{\left(\frac{B}{2A}\right)^2 + \left|\frac{C}{A}\right|}\right\}$$

$$= -\text{sgn}\left(\frac{B}{2A}\right)\sqrt{\left|\frac{C}{A}\right|}\left(\delta + \sqrt{\delta^2 + 1}\right).$$

Moreover, $x_2 = C/(Ax_1)$ is given by

4.15
$$x_2 = \text{sgn}\left(\frac{B}{2A}\right)\sqrt{\left|\frac{C}{A}\right|}\left(\delta + \sqrt{\delta^2 + 1}\right)^{-1}.$$

We remark that in each case the formula has been chosen so that subtraction of two quantities of like sign is avoided, wherever possible. However, there does not appear any simple way to avoid subtraction in $\sqrt{1 - \delta^2}$ and $\sqrt{1 - 4AC/B^2}$ when $AC > 0$. Thus, in order to achieve maximum accuracy, double-precision arithmetic should be used to calculate these quantities.

Special Cases

If $B/(2A) \in \mathscr{S}$ and $B/(2C) \in \mathscr{S}$, then $\delta \in \mathscr{S}$. Therefore, by (4.13), (4.14), and (4.15) we have

4.16
$$x_1, x_2 = \pm\sqrt{-\frac{C}{A}} = \begin{cases} \pm i\sqrt{\dfrac{C}{A}} \text{ if } AC > 0 \\ \\ \pm\sqrt{\left|\dfrac{C}{A}\right|} \text{if } AC < 0. \end{cases}$$

If $B/(2A) \in \mathscr{M}$ and $B/(2C) \in \mathscr{S}$, then $\delta < 1$. If $AC > 0$, then the roots are complex and x_1 and x_2 are given by (4.13). If $AC \leqq 0$, then x_1 and x_2 are given by (4.14) and (4.15).

It is not possible for $B/(2A) \in \mathscr{L}$ and $B/(2C) \in \mathscr{S}$, since we would have $|B/(2A)| \geq 1/\bar{a}$, $|B/(2C)| \leq \bar{a}$; hence $|C/A| \geq 1/\bar{a}^2$, which is a contradiction since $\sqrt{|C/A|} \in \mathscr{M}$. Similarly, it is not possible for $|B/(2A)| \in \mathscr{S}$ and $|B/(2C)| \in \mathscr{L}$.

If $B/(2A) \in \mathscr{S}$ and $B/(2C) \in \mathscr{M}$, then $\delta < 1$. If $AC > 0$, then from (4.13) we have

4.17
$$x_1, x_2 = \pm i\sqrt{\frac{C}{A}}\sqrt{1 - \delta^2}.$$

If $AC \leqq 0$, then x_1 and x_2 are given by (4.14) and (4.15).

4.21 Table   Roots of the quadratic equation.

	$\frac{B}{2A} \in \mathscr{S}$	$\frac{B}{2A} \in \mathscr{M}$	$\frac{B}{2A} \in \mathscr{L}$
$\frac{B}{2C} \in \mathscr{S}$	$(\delta \in \mathscr{S})$ $x_1 = \sqrt{-\dfrac{C}{A}}$ $x_2 = -\sqrt{-\dfrac{C}{A}}$	$(\delta < 1)$ If $AC > 0$, $x_1, x_2 = -\dfrac{B}{2A} \pm i\sqrt{\dfrac{C}{A}}\sqrt{1-\delta^2}$ If $AC \leqq 0$, $x_1 = -\operatorname{sgn}\left(\dfrac{B}{2A}\right)\sqrt{\left|\dfrac{C}{A}\right|}\,[\delta + \sqrt{1+\delta^2}]$ $x_2 = \operatorname{sgn}\left(\dfrac{B}{2A}\right)\sqrt{\left|\dfrac{C}{A}\right|}\,[\delta + \sqrt{1+\delta^2}]^{-1}$	Not possible
$\frac{B}{2C} \in \mathscr{M}$	$(\delta < 1)$ If $AC > 0$, $x_1, x_2 = \pm i\sqrt{\dfrac{C}{A}}\sqrt{1-\delta^2}$ If $AC \leqq 0$, $x_1 = -\operatorname{sgn}\left(\dfrac{B}{2A}\right)\sqrt{\left|\dfrac{C}{A}\right|}\,[\delta + \sqrt{1+\delta^2}]$ $x_2 = \operatorname{sgn}\left(\dfrac{B}{2A}\right)\sqrt{\left|\dfrac{C}{A}\right|}\,[\delta + \sqrt{1+\delta^2}]^{-1}$	If $\delta \leqq 1$, *and* $AC > 0$, $x_1, x_2 = -\dfrac{B}{2A} \pm i\sqrt{\dfrac{C}{A}}\sqrt{1-\delta^2}$ If $\delta \leqq 1$ *and* $AC \leqq 0$, $x_1 = -\operatorname{sgn}\left(\dfrac{B}{2A}\right)\sqrt{\left|\dfrac{C}{A}\right|}\,[\delta + \sqrt{1+\delta^2}]$ $x_2 = \operatorname{sgn}\left(\dfrac{B}{2A}\right)\sqrt{\left|\dfrac{C}{A}\right|}\,[\delta + \sqrt{1+\delta^2}]^{-1}$ If $\delta > 1$, $x_1 = -\dfrac{B}{2A}\left(1 + \sqrt{1 - \dfrac{4AC}{B^2}}\right)$ $x_2 = -\dfrac{2C}{B}\left(1 + \sqrt{1 - \dfrac{4AC}{B^2}}\right)^{-1}$	$(\delta > 1)$ $x_1 \in \mathscr{L}$ $x_2 = -\dfrac{2C}{B}\left(1 + \sqrt{1 - \dfrac{4AC}{B^2}}\right)^{-1}$
$\frac{B}{2C} \in \mathscr{L}$	Not possible	$(\delta > 1)$ $x_1 = -\dfrac{B}{2A}\left(1 + \sqrt{1 - \dfrac{4AC}{B^2}}\right)$ $x_2 \in \mathscr{L}$	$(\delta \in \mathscr{L})$ $x_1 \in \mathscr{L}$ $x_2 \in \mathscr{S}$

$$\delta = \frac{|B|}{2\sqrt{|AC|}}$$

If $B/(2A) \in \mathcal{L}$ and $B/(2C) \in \mathcal{M}$, then $\delta > 1$. Hence, by (4.12) we have

4.18
$$x_1 \in \mathcal{L}, \quad x_2 = -\frac{2C}{B}\left(1 + \sqrt{1 - \frac{4AC}{B^2}}\right)^{-1}.$$

If $B/(2A) \in \mathcal{M}$ and $B/(2C) \in \mathcal{L}$, then again $\delta > 1$ and by (4.12) we have

4.19
$$x_2 \in \mathcal{S}, \quad x_1 = -\frac{B}{2A}\left(1 + \sqrt{1 - \frac{4AC}{B^2}}\right).$$

Finally, if $B/(2A) \in \mathcal{L}$, $B/(2C) \in \mathcal{L}$, then by Theorems 4.8 and 4.9 we have

4.20
$$x_1 \in \mathcal{L}, \quad x_2 \in \mathcal{S}.$$

A summary of the various cases is given in Table 4.21.

EXERCISES 3.4

1. Carry out the procedure outlined in the text, simulating the computer as closely as possible, to solve the quadratic equation (4.1) in the following cases. (Note that $(1 \pm x)^{1/2}$ would be computed as $1 \pm (x/2)$ for very small x.)

A	B	C
2^{950}	1	1
1	2^{950}	1
1	1	2^{950}
1	0	2^{-950}
2^{-950}	1	1
2^{-950}	1	2^{-950}
2^{-950}	2^{-950}	2^{-950}
2^{-950}	2^{-950}	1
1	2^{-950}	1
1	10^{-5}	1
4	-2×10^{200}	1
1	-10^5	1
10^{-30}	10^{-30}	10^{-30}
10^{-120}	-10^{120}	10^{120}
10^{-250}	10	10^{250}
10^{-200}	10^{200}	10^{200}
10^{200}	10^{-200}	10^{-200}
1	-18	1
6	5	4
6×10^{30}	5×10^{30}	-4×10^{30}
6×10^{30}	5×10^{30}	4×10^{30}
10^{300}	10^{-300}	10^{300}
1	-4	3.9999999
0.999999992	3.999999968	3.999999968

2. For the quadratic equations of Exercise 1, what would one obtain in each case if one were to use the formula

$$x_1, x_2 = -\frac{B}{2A} \pm \sqrt{\left(\frac{B}{2A}\right)^2 - \frac{C}{A}}.$$

3. Show that if $B/(2A) \in \mathcal{M}$ and $B/(2C) \in \mathcal{M}$ then $\delta = |B|/(2\sqrt{|A|}\sqrt{|C|})$ is also in $\mathcal{M}$.

4. Verify Theorem 4.8 for the case where $B = 2^{900}$, $A = 2^{-900}$, $C = 1$.

5. Verify Theorem 4.9 for the case $B = 2^{900}$, $A = 1$, $C = 2^{-900}$.

6. Prepare a flow diagram and write a computer program in Fortran for solving the quadratic equation (4.1) where A, B, and C are in $\circledR$. Follow the procedure outlined in the text.

3.5 COMPLEX ARITHMETIC OPERATIONS

We now describe some surveillance programs for complex arithmetic. Let us first define the sets $\mathcal{R}^*$ and $\circledR^*$, the sets of *representable complex numbers* and *computer-representable complex numbers*, respectively, as follows:

We let $z = x + iy$, where x and y are real, be in $\mathcal{R}^*$ if

5.1 $x \in \mathcal{R}$ and $y \in \mathcal{R}$.

We say that $z \in \circledR^*$ if

5.2 $x \in \circledR$ and $y \in \circledR$.

Let us now define the additional sets (see Fig. 5.9):

5.3
$$\begin{cases} \mathcal{M}_C^* = \{(x, y): x \in \mathcal{M} \text{ and } y \in \mathcal{M}\} \\ \mathcal{M}_R^* = \{(x, y): x \in \mathcal{M} \text{ and } y = 0\} \\ \mathcal{M}_I^* = \{(x, y): x = 0 \text{ and } y \in \mathcal{M}\} \end{cases}$$

5.4
$$\begin{cases} \mathcal{S}_C^* = \{(x, y): x \in \mathcal{S} \text{ and } y \in \mathcal{S}\} \\ \mathcal{S}_R^* = \{(x, y): x \in \mathcal{M} \text{ and } y \in \mathcal{S}\} \\ \mathcal{S}_I^* = \{(x, y): x \in \mathcal{S} \text{ and } y \in \mathcal{M}\} \end{cases}$$

5.5 $\mathcal{Z}^* = \{(x, y): x = y = 0\}$

5.6 $\mathcal{L}^* = \{(x, y): x \in \mathcal{L} \text{ or } y \in \mathcal{L}\}$

We define the set $\mathcal{A}^*$ of *allowable complex numbers* by

5.7 $\mathcal{A}^* = \mathcal{M}^* + \mathcal{Z}^*$

where

5.8 $\mathcal{M}^* = \mathcal{M}_C^* + \mathcal{M}_R^* + \mathcal{M}_I^*.$

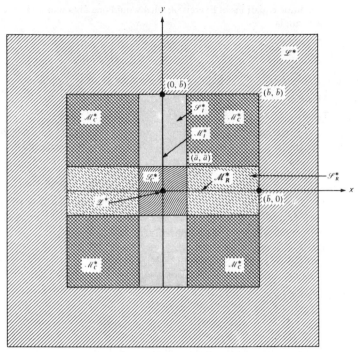

5.9 Fig. Sets of Complex Numbers.

In a similar way we define the set of *computer-allowable complex numbers* $\left(\mathscr{A}^*\right)$ by

5.10
$$\left(\mathscr{A}^*\right) = \mathscr{A}^* \cap \left(\mathscr{R}^*\right),$$

Similarly, we let

5.11
$$\left\{\begin{array}{l} \left(\mathscr{M}_C^*\right) = \mathscr{M}_C^* \cap \left(\mathscr{R}^*\right), \\[2ex] \left(\mathscr{M}_R^*\right) = \mathscr{M}_R^* \cap \left(\mathscr{R}^*\right), \text{ etc.} \end{array}\right.$$

Complex Editing Subroutines

We shall consider two complex editing subroutines. One, Subroutine CEDIT, is described in Fig. 5.12 and is primarily used in editing complex numbers in $\left(\mathscr{R}^*\right)$ prior to applying the complex surveillance subroutines. The other, Subroutine CPOSTED, is described in Fig. 5.13 and is primarily used in editing the output of complex arithmetic operations.

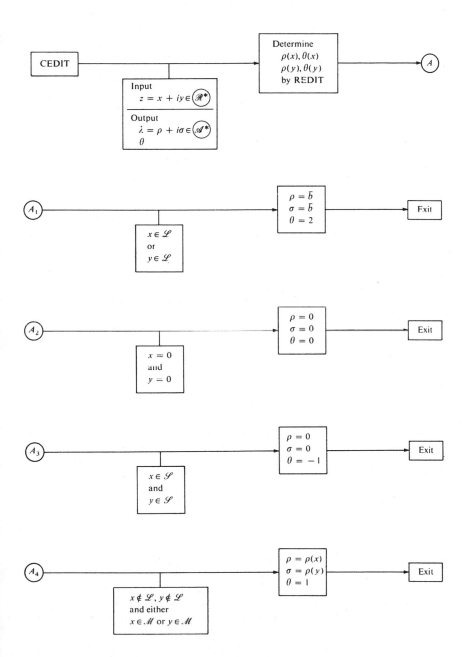

5.12 Fig. Subroutine CEDIT.

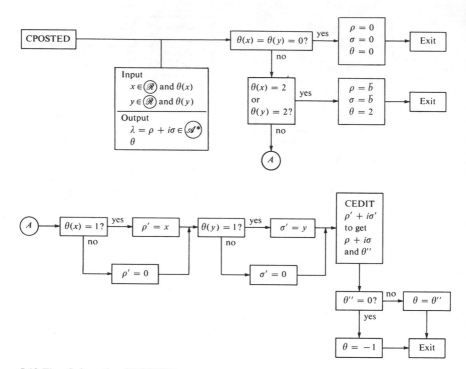

5.13 Fig. Subroutine CPOSTED.

The input to CEDIT consists of the real numbers $x \in \mathscr{R}$ and $y \in \mathscr{R}$. These numbers are edited separately by REDIT to give $(\rho(x), \theta(x))$ and $(\rho(y), \theta(y))$. The subroutine then determines the indicator integer θ as $-1, 0, 1, 2$ as follows:

$$\begin{aligned}
\theta &= -1, && \text{if } z = x + iy \in \mathscr{S}^* \\
\theta &= 0, && \text{if } z = 0 \\
\theta &= 1, && \text{if } z \in \mathscr{M}^* + \mathscr{S}_R^* + \mathscr{S}_I^* \\
\theta &= 2, && \text{if } z \in \mathscr{L}^*.
\end{aligned}$$

In a more sophisticated program one could have separate indicator digits for the cases $z \in \mathscr{M}_R^*$, $z \in \mathscr{M}_I^*$, $z \in \mathscr{S}_R^*$, and $z \in \mathscr{S}_I^*$. The output of CEDIT is $(\rho(x), \rho(y), 1)$ if $\theta = 1$, $(\bar{b}, \bar{b}, 2)$ if $\theta = 2$, $(0, 0, 0)$ if $\theta = 0$, and $(0, 0, -1)$ if $\theta = -1$.

The input to CPOSTED consists of two numbers x and y which are in $\mathscr{R}$ as well as $\theta(x)$ and $\theta(y)$. Here $\theta(x)$ and $\theta(y)$ may be 2, 1, 0, or -1. If $\theta(x) = \theta(y) = 0$, then we have the output $(0, 0, 0)$. If $\theta(x)$ or $\theta(y) = 2$, then we have the output $(\bar{b}, \bar{b}, 2)$. If $\theta(x)$ and $\theta(y)$ do not vanish and if neither equals 2, then we let $\rho' = 0$ if $\theta(x) = -1$ and $\sigma' = 0$ if $\theta(y) = -1$. If $\theta(x) = 1$ we let $\rho' = x$, and if $\theta(y) = 1$ we let $\sigma' = y$. We then enter CEDIT with ρ' and σ' obtaining (ρ, σ, θ''). The output of CPOSTED in this case is (ρ, σ, θ) where $\theta = \theta''$ if $\theta'' \neq 0$ and $\theta = -1$ if $\theta'' = 0$.

The addition of two complex numbers $z_1 = x_1 + iy_1$ and $z_2 = x_2 + iy_2$ is easily carried out using RADD and CPOSTED since

5.14 $z_1 + z_2 = (x_1 + iy_1) + (x_2 + iy_2) = (x_1 + x_2) + i(y_1 + y_2).$

The procedure is indicated by the flow chart given in Fig. 5.15.

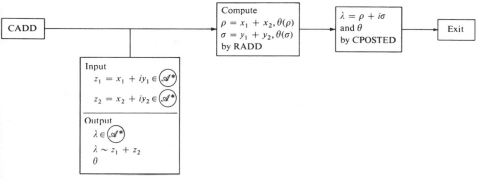

5.15 Fig Subroutine CADD.

The subroutines for complex multiplication and complex division are somewhat more complicated and are left as exercises. (See Young and McDonald [1969].) Instead of giving the details for these programs, we shall give two simpler examples of the use of surveillance routines for complex arithmetic.

Let us consider the computation of the modulus $|z|$ of a complex number $z = x + iy$ which is given by

5.16 $$|z| = \sqrt{x^2 + y^2}.$$

A direct computation by squaring x, squaring y, adding and taking the square root might lead to overflow even if $x \in \mathcal{A}$, $y \in \mathcal{A}$, and $|z| \in \mathcal{A}$. Instead of the direct computation we write (5.16) in the form

5.17 $$|z| = M\sqrt{1 + (m/M)^2}$$

where we let

5.18 $$M = \max(|x|, |y|), m = \min(|x|, |y|).$$

The calculation can be carried out as indicated in the following flow diagram.

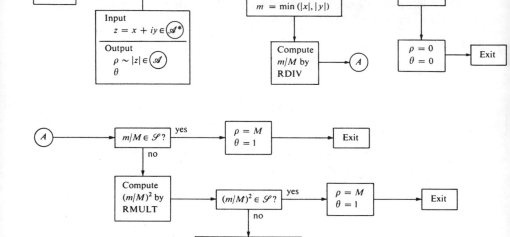

5.19 Fig. Subroutine CMOD.

As another example, let us consider the problem of finding the square root of a complex number $z = x + iy$. Evidently we have

5.20 $$\sqrt{z} = \sqrt{x + iy} = \pm \left\{ \sqrt{\tfrac{1}{2}(\sqrt{x^2 + y^2} + x)} + i \operatorname{sgn} y \sqrt{\tfrac{1}{2}(\sqrt{x^2 + y^2} - x)} \right\}$$

where

5.21 $$\operatorname{sgn} y = \begin{cases} 1 & \text{if } y \geqq 0 \\ -1 & \text{if } y < 0. \end{cases}$$

The above formula is easily verified since

5.22 $$(u + iv)^2 = (u^2 - v^2) + i(2uv)$$

and hence, if

5.23 $$u = \alpha \sqrt{\tfrac{1}{2}(\sqrt{x^2 + y^2} + x)}, \qquad v = \alpha \operatorname{sgn} y \sqrt{\tfrac{1}{2}(\sqrt{x^2 + y^2} - x)}$$

where $\alpha = \pm 1$, we have

5.24
$$\begin{cases} u^2 - v^2 = x \\ 2uv = 2 \operatorname{sgn} y \sqrt{\dfrac{y^2}{4}} = |y| \operatorname{sgn} y = y. \end{cases}$$

Our procedure is as follows. If $x \geqq 0$ we compute

5.25
$$u = \sqrt{\tfrac{1}{2}(\sqrt{x^2 + y^2} + x)}, \qquad v = \frac{y}{2u}$$

while, if $x < 0$, we compute

5.26
$$v = \operatorname{sgn} y \sqrt{\tfrac{1}{2}(\sqrt{x^2 + y^2} - x)}, \qquad u = \frac{y}{2v}.$$

Having obtained the solution $u + iv$ we also have the alternative solution $-(u + iv)$. The flow diagram of Fig. 5.27 shows how $\sqrt{z}$ would actually be found.

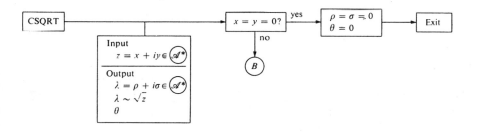

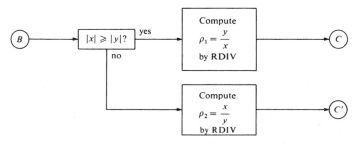

5.27 Fig. Subroutine CSQRT (*continued on next page*).

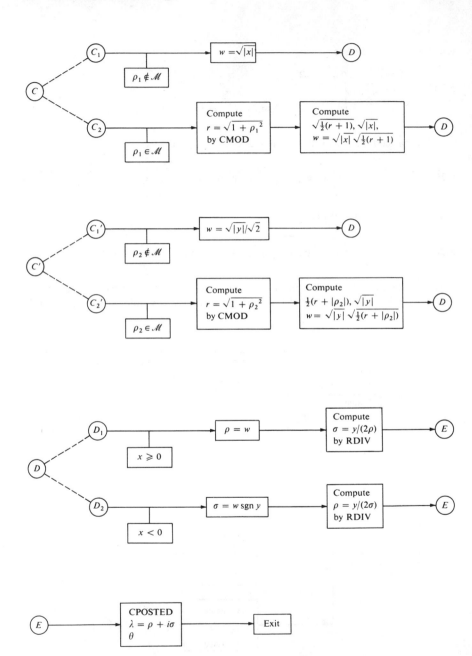

5.27 Fig. Subroutine CSQRT.

EXERCISES 3.5

1. Prepare Fortran programs for CEDIT and CPOSTED.
2. What would be the output of CPOSTED with Machine C in the following cases

x	$\theta(x)$	y	$\theta(y)$
b	1	1	1
b	2	1	1
b	1	b	1
$\bar{a}$	0	1	1
$\bar{a}$	-1	1	1
$\bar{a}$	1	1	1
2^{950}	1	1	1
2^{950}	2	1	1
1	1	$\bar{a}$	-1
0	0	0	0
0	-1	0	0

What would be the output of CEDIT where in each case the input data includes only x and y?

3. Compute the following: $(2 + 3i)(5 - 6i)$; $(2 + 3i)/(5 - 6i)$; $(3 - 4i)/(12 + 5i)$; $\sqrt{5 - 12i}$; $\sqrt{1 + i}$; $\sqrt{2 + 3i}$; $|5 + 12i|$ obtaining answers in the form $a + ib$ where a and b are real.

4. Prepare Fortran programs for CMOD and CSQRT. What would be the result of applying these programs to the following cases?

x	y	x	y	x	y
b	b	10^{30}	10^{20}	2^{900}	2^{900}
b	$\bar{a}$	1	1	2^{900}	2^{-900}
$\bar{a}$	b	5	-12	2^{-900}	2^{-900}
$\bar{a}$	$\bar{a}$	2	3		
3	4	10^{30}	10^{-20}		

5. Show that if $z \in \mathcal{M}^*$ then $\sqrt{z} \in \mathcal{M}^*$. Also show that Subroutine CSQRT when applied to a complex number $z \in \mathcal{M}^*$ produces a result in $\mathcal{M}^*$.

6. Prepare flow diagrams and Fortran programs for complex multiplication and complex division. The subroutines RSYNTH, and RSCAL2 should be used. (See Exercises 6 and 7, Section 3.3.) In the case of complex division, if both arguments are zero the result should be indicated as indeterminant.

7. Prepare a flow diagram and a Fortran program for finding r, c, and s such that $z = x + iy = r(c + is)$, where
$$c = \frac{x}{\sqrt{x^2 + y^2}}, \qquad s = \frac{y}{\sqrt{x^2 + y^2}}, \qquad r = \sqrt{x^2 + y^2}.$$
What would be the result of applying this program to the cases of Exercise 4?

8. Describe how one would solve the quadratic equation $Ax^2 + Bx + C = 0$ where the coefficients A, B, and C are in $\mathcal{R}^*$.
   (If A, B, or C is in $\mathcal{L}^*$, do not attempt to solve the problem.)

SUPPLEMENTARY DISCUSSION

The development of a set of surveillance routines for real and complex arithmetic is described by Young and McDonald [1969]. The use of the routines for solving the general quadratic equation with complex coefficients is described. Previous work on the development of a reliable scheme for solving the quadratic equation was done by W. Kahan (see Forsythe [1966], [1967], and [1969]).

An interesting paper describing some of the computer-dependent characteristics of a computer is given by Redish and Ward [1971].

CHAPTER 4

SOLUTION OF EQUATIONS

4.1 INTRODUCTION

In this chapter we shall be concerned with the numerical solution of equations of the form

1.1 $$f(x) = 0$$

where $f(x)$ is a function of the real or complex variable x. The function $f(x)$ may be complex valued but in most cases we shall assume that $f(x)$ is real for real values of x. This is the case, for example, if $f(x)$ is a polynomial with real coefficients. Normally, we shall assume that $f(x)$ is continuous and we shall frequently add appropriate differentiability assumptions.

Examples of some of the equations we shall consider are

1.2 $$f(x) = \sin x - x + 2 = 0$$

1.3 $$f(x) = x^3 - 2 = 0$$

1.4 $$f(x) = e^x + \log x - 3 = 0.$$

1.5 Definition. If $f(\bar{x}) = 0$, then $\bar{x}$ is a *root* of (1.1) and $\bar{x}$ is a *zero* of $f(x)$.
Thus $\sqrt[3]{2}$ is a root of (1.3) and a zero of $f(x) = x^3 - 2$.
Let us now define the multiplicity of a zero of a function $f(x)$ as follows.

1.6 Definition. If $\bar{x}$ is a zero of $f(x)$, then the *multiplicity*, m, of $\bar{x}$ as a zero of $f(x)$ is the least upper bound of all numbers k such that

1.7 $$\overline{\lim_{x \to \bar{x}}} \frac{|f(x)|}{|x - \bar{x}|^k} < \infty.$$

Thus, for example, the function $f(x) = x^{1/2}$ has a zero of multiplicity $\frac{1}{2}$ at $x = 0$. Similarly, the function $f(x) = x^{1/2} \log x$ has a zero of multiplicity $\frac{1}{2}$ at $x = 0$, since for any $\varepsilon > 0$ we have

1.8 $$\lim_{x \to 0} \frac{|f(x)|}{|x|^{1/2 - \varepsilon}} = 0.$$

We now prove a result in which m is an integer.

1.9 Theorem. If $\bar{x}$ is a zero of $f(x)$ and if for some integer m, $f(x)$ is m times continuously differentiable at $\bar{x}$, then the multiplicity of $\bar{x}$ is at least m if and only if

1.10 $$f(\bar{x}) = f'(\bar{x}) = \cdots = f^{(m-1)}(\bar{x}) = 0.$$

The multiplicity is exactly m if (1.10) holds and if

1.11 $$f^{(m)}(\bar{x}) \neq 0.$$

If $f(x)$ is $m + 1$ times continuously differentiable at $\bar{x}$ and if the multiplicity of $\bar{x}$ is m then (1.11) holds.

Proof. We consider first the case where $f(x)$ is m times continuously differentiable at $\bar{x}$. By Taylor's theorem we have

1.12 $$f(x) = f(\bar{x}) + (x - \bar{x})f'(\bar{x}) + \cdots + \frac{(x - \bar{x})^{m-1}}{(m-1)!} f^{(m-1)}(\bar{x}) + \frac{(x - \bar{x})^{m}}{m!} f^{(m)}(c(x))$$

where $c(x)$ lies between $\bar{x}$ and x. If (1.10) holds, then

1.13 $$\frac{f(x)}{(x - \bar{x})^{m}} = \frac{1}{m!} f^{(m)}(c(x))$$

and

1.14 $$\lim_{x \to \bar{x}} \frac{f(x)}{(x - \bar{x})^{m}} = \frac{1}{m!} \lim_{x \to \bar{x}} f^{(m)}(c(x)) = \frac{1}{m!} f^{(m)}(\bar{x})$$

by the continuity of $f^{(m)}(x)$. Hence the multiplicity of $\bar{x}$ is at least m. If (1.11) also holds, then the multiplicity is m since for any $\varepsilon > 0$ we have

$$\lim_{x \to \bar{x}} \frac{|f(x)|}{|x - \bar{x}|^{m+\varepsilon}} = \frac{1}{m!} \lim_{x \to \bar{x}} \frac{|f^{(m)}(c(x))|}{|x - \bar{x}|^{\varepsilon}} = \infty.$$

If, on the other hand, the multiplicity of $\bar{x}$ is at least m, then by (1.12) we have, since $f(\bar{x}) = 0$,

1.15 $$\frac{f(x)}{x - \bar{x}} = f'(\bar{x}) + (x - \bar{x})\frac{f''(\bar{x})}{2} + \cdots + \frac{(x - \bar{x})^{m-2}}{(m-1)!} f^{(m-1)}(\bar{x})$$

$$+ \frac{(x - \bar{x})^{m-1}}{m!} f^{(m)}(c(x)).$$

By (1.7), if $m \geq 2$, we have

1.16
$$\overline{\lim_{x \to \bar{x}}} \frac{|f(x)|}{|x - \bar{x}|} = 0$$

so that, taking limits of both sides of (1.15) as $x \to \bar{x}$, we have

1.17
$$f'(\bar{x}) = 0.$$

In the same way we can show that $f^{(2)}(\bar{x}) = f^{(3)}(\bar{x}) = \cdots = f^{(m-1)}(\bar{x}) = 0$ and hence (1.10) holds.

Suppose now that $f(x)$ is $m + 1$ times continuously differentiable at $\bar{x}$ and that $\bar{x}$ is a zero of $f(x)$ of multiplicity m. As we have seen, (1.10) holds. If (1.11) does not hold, then the multiplicity is $m + 1$ and not m. Hence (1.11) holds. This completes the proof of Theorem 1.9.

An explicit closed-form solution of (1.1) is available only in special cases such as the linear equation $Ax + B = 0$ and the quadratic equation $Ax^2 + Bx + C = 0$. Closed-form solutions are also available for the cubic and for the quartic equations, but these are relatively complicated and are seldom used in practice. For polynomial equations of degree higher than four it is well known that there exist cases where no closed-form solution is possible. As one might expect, closed-form solutions are seldom available for equations such as (1.2) and (1.4) which involve transcendental functions.

Because explicit closed-form solutions are seldom available, we shall be primarily concerned with approximate numerical methods such as iterative methods, which we have considered in Chapter 1 in connection with the equation $x^2 - N = 0$. In Section 4.2 we study the limitations on the attainable accuracy caused by the fact that $f(x)$ cannot be computed exactly on the computer. Graphical methods for obtaining approximate values of the roots are described in Section 4.3. Various iterative methods are discussed in Sections 4.4–4.10 including the method of bisection, the method of false position, the secant method, the Newton method, and Muller's method. Estimates of the convergence rates of three of the methods are derived in Section 4.11. Methods for accelerating the convergence are given in Section 4.12. Finally, Section 4.13 describes methods for solving systems of equations.

In an introductory course it may be appropriate to omit Sections 4.11, 4.12, and 4.13.

In this and subsequent chapters we assume the availability of subroutines for performing basic arithmetic operations in complex arithmetic and for the computation of elementary functions such as e^z, log z, sin z, etc. It is assumed that in the construction of such routines some attempt has been made to minimize the possibility of the occurrence of unnecessary overflow and underflow as in Chapter 3.

4.2 ATTAINABLE ACCURACY

We have already indicated that the accuracy attainable for a root of (1.1) is limited by the fact that $f(x)$ cannot be computed exactly. In the first place, unless $f(x)$ is a very simple function, it is necessary to replace $f(x)$ by a simpler function $F(x)$. Thus, for instance, if $f(x) = \sin x$ we might let $F(x)$ be a polynomial corresponding to a given number of terms of the Taylor's series for $f(x)$. In general, the replacement of $f(x)$ by $F(x)$ will result in some error. Another source of error is the fact that the computation of $F(x)$ involves rounding errors. A third source of error lies in the fact that $F(x)$ can only be evaluated for values of x which are *computer-representable numbers* as defined in Section 2.6.

Given a function $f(x)$ and an approximating function $F(x)$, let us define the function $\tilde{f}(x)$ as follows:

a) If x is a computer-representable number, $\tilde{f}(x)$ is the value of $F(x)$ produced by the computer.

b) If x is not a computer-representable number, let $\tilde{f}(x)$ be determined by linear interpolation* in $\tilde{f}(x_1)$ and $\tilde{f}(x_2)$. Here x_1 and x_2 are, respectively, the largest computer-representable number less than x and the smallest computer-representable number greater than x.

Evidently, between any pair of consecutive computer-representable numbers the function $\tilde{f}(x)$ is a linear function of x. An example is given in Fig. 2.1 which has a greatly expanded scale in x. Thus, for example, x_1 differs from x_2 by one unit in the least significant binary digit in the floating-point binary representations. The computed values of some approximating function $F(x)$ are used to define $\tilde{f}(x)$ at the computer-representable values $x_1, x_2, \ldots, x_6$. Linear interpolation is used to define $\tilde{f}(x)$ at other values of x.

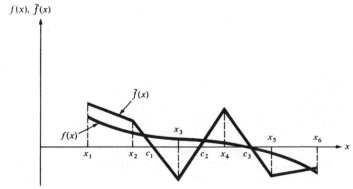

2.1 Fig. The Functions $f(x)$ and $\tilde{f}(x)$.

---

* See Section 6.2.

Joseph Krauskopf Memorial Library

Since $\tilde{f}(x)$ is continuous, it follows by Theorem A-3.24 that if $\tilde{f}(a)\tilde{f}(b) < 0$ then for some c lying between a and b we have $\tilde{f}(c) = 0$. Thus, for example, since in the example of Fig. 2.1 we have $\tilde{f}(x_3)\tilde{f}(x_2) < 0$, there is a value, say c_1, between x_2 and x_3 such that $\tilde{f}(c_1) = 0$.

Even if $\tilde{f}(x)$ is a good representation of $f(x)$ in the interval under consideration, the zeros of $\tilde{f}(x)$ may differ considerably from those of $f(x)$. In fact, $\tilde{f}(x)$ may have one or more zeros when $f(x)$ has none and vice versa. Figure 2.3 shows graphs of $f(x)$, $f(x) + \varepsilon$, $f(x) - \varepsilon$, and $\tilde{f}(x)$ in a hypothetical example where

2.2
$$|\tilde{f}(x) - f(x)| \leqq \varepsilon$$

through the range of interest. For negative x there are several zeros of $\tilde{f}(x)$ but none of $f(x)$, while the positive zero of $f(x)$ is quite different from either positive zero of $\tilde{f}(x)$.

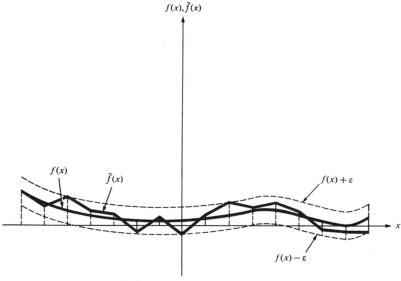

2.3 Fig. Graphs of $f(x)$ and $\tilde{f}(x)$.

Such pathological situations cannot occur if $f(x)$ has a continuous derivative which is bounded sufficiently far away from zero in an interval containing a zero $\bar{x}$ of $f(x)$. We prove

2.4 Theorem. Let $\bar{x}$ be a zero of $f(x)$, and for some $\alpha > 0$ and $\varepsilon > 0$ let $f(x)$ be continuous and continuously differentiable in I:

2.5
$$\bar{x} - \frac{\varepsilon}{\alpha} \leqq x \leqq \bar{x} + \frac{\varepsilon}{\alpha}.$$

If $\tilde{f}(x)$ is a continuous function in I such that (2.2) holds and if

2.6 $$|f'(x)| \geq \alpha$$

in I, then there exists a zero of $\tilde{f}(x)$ in I.

Proof. By the mean-value theorem we have

$$f(x) = f(\bar{x}) + (x - \bar{x})f'(c) = (x - \bar{x})f'(c)$$

for some c between $\bar{x}$ and x. Hence for some c_1 and c_2 in I we have

$$f\left(\bar{x} + \frac{\varepsilon}{\alpha}\right) = \frac{\varepsilon}{\alpha}f'(c_1)$$

$$f\left(\bar{x} - \frac{\varepsilon}{\alpha}\right) = -\frac{\varepsilon}{\alpha}f'(c_2).$$

If $f'(x) > 0$ in I then we have

$$f\left(\bar{x} + \frac{\varepsilon}{\alpha}\right) \geq \varepsilon, \qquad f\left(\bar{x} - \frac{\varepsilon}{\alpha}\right) \leq -\varepsilon.$$

From (2.2) we have

$$\tilde{f}\left(\bar{x} + \frac{\varepsilon}{\alpha}\right) \geq 0, \qquad \tilde{f}\left(\bar{x} - \frac{\varepsilon}{\alpha}\right) \leq 0,$$

and hence there is a zero of $\tilde{f}(x)$ in I. A similar argument holds for the case where $f'(x) < 0$ and the proof is complete.

As an example consider the problem of solving the equation

2.7 $$f(x) = \sin x - 0.750 = 0.$$

One root is

$$\bar{x} = \sin^{-1}(0.750) \doteq 0.8481.$$

Let us assume that the truncated Taylor's series

2.8 $$F(x) = x - \frac{x^3}{6} - 0.750.$$

is used to represent $f(x)$. We shall neglect the effect of rounding errors in the computation of $F(x)$ so that $\tilde{f}(x) = F(x)$ for computer-representable values of x. We seek to determine α and ε by trial and error. As a first approximation we let

$$\alpha_0 = f'(\bar{x}) = \cos \bar{x} \doteq \cos(0.8481) \doteq 0.6614,$$

and we let

$$\varepsilon_0 = |\tilde{f}(\bar{x}) - f(\bar{x})| = |\tilde{f}(\bar{x})| = |\bar{x} - \frac{\bar{x}^3}{6} - 0.750| \doteq 0.0036.$$

Thus we have

$$\frac{\varepsilon_0}{\alpha_0} \doteq \frac{0.0036}{0.6614} \doteq 0.0054.$$

We consider the interval $I_0 : |x - \bar{x}| \leq 0.0054$, that is, the interval

$$0.8427 \leq x \leq 0.8535.$$

Evidently, in I_0, our approximation α_0 is too large, since

$$f'(x) = \cos x \geq \cos(0.8535) \doteq 0.6573.$$

Since the series

2.9
$$\sin x = x - \frac{x^3}{3!} + \frac{x^5}{5!} - \frac{x^7}{7!} + \cdots$$

is an alternating series with terms of decreasing magnitude it follows that

2.10
$$|\tilde{f}(x) - f(x)| \leq \frac{x^5}{120}.$$

Hence in I_0 we have

2.11
$$|\tilde{f}(x) - f(x)| \leq \frac{(0.8535)^5}{120} \doteq 0.0038.$$

Let $\alpha_1 = 0.6573$ and $\varepsilon_1 = 0.0038$. Then $\varepsilon_1/\alpha_1 \doteq 0.0058 > 0.0054$. In order for Theorem 2.4 to be applicable with $\alpha = \alpha_1$ and $\varepsilon = \varepsilon_1$ it is necessary that $f'(x) \geq \alpha_1$ and $|\tilde{f}(x) - f(x)| \leq \varepsilon_1$ in the interval $I_1 : |x - \bar{x}| \leq 0.0058$. However, $f'(0.8539) = \cos(0.8539) \doteq 0.6570 = \alpha_2$ and $0.8539 \in I_1$. Hence Theorem 2.4 is not applicable with $\alpha = \alpha_1$ and $\varepsilon = \varepsilon_1$. However, in I_1 we have $f'(x) \geq \alpha_2$ and $|\tilde{f}(x) - f(x)| \leq (0.8539)^5/120 \doteq 0.0038$, so that $\varepsilon_2/\alpha_2 \doteq 0.0058$. Therefore Theorem 2.4 is applicable with $\alpha = \alpha_2$ and $\varepsilon = \varepsilon_2$, and there is a zero of $\tilde{f}(x)$ in the interval I_1:

2.12
$$|x - 0.8481| \leq 0.0058.$$

As a matter of fact there is a zero of $F(x)$ at 0.8537 which is within 0.0056 of $\bar{x}$. Thus the bound (2.12) is quite good. It is easy to show that for Machine C the additional error caused by the fact that $\tilde{f}(x)$ is defined by linear interpolation, if x is not computer-representable, is negligible. (See Section 6.3.)

The above analysis shows that under normal circumstances there is a zero of $\tilde{f}(x)$ near a zero of $f(x)$. In practice, of course, one seeks a zero of $\tilde{f}(x)$. Since one cannot expect to find a computer-representable value of x such that $\tilde{f}(x) = 0$ one usually accepts* as a zero of $\tilde{f}(x)$ a computer-representable value $\hat{x}$ such that

2.13 $$|\tilde{f}(\hat{x})| < \varepsilon_1$$

for some prescribed ε_1. We now seek to show that if $f(x)$ has a continuous derivative which is bounded sufficiently far away from zero in an interval containing $\hat{x}$ then there is a zero of $f(x)$ in that interval. We prove

2.14 Theorem. For some $\alpha > 0$ let $f(x)$ be continuous and continuously differentiable in I:

2.15 $$\hat{x} - \frac{|f(\hat{x})|}{\alpha} \leqq x \leqq \hat{x} + \frac{|f(\hat{x})|}{\alpha}.$$

If

2.16 $$|f'(x)| \geqq \alpha$$

in I then there is a zero of $f(x)$ in I.

Proof. By the mean-value theorem we have

$$f(x) = f(\hat{x}) + (x - \hat{x})f'(c)$$

where c lies between x and $\hat{x}$. If $f'(x) > 0$ in I, we have

$$f\left(\hat{x} + \frac{|f(\hat{x})|}{\alpha}\right) = f(\hat{x}) + \frac{|f(\hat{x})|}{\alpha} f'(c_1) \geqq 0$$

and

$$f\left(\hat{x} - \frac{|f(\hat{x})|}{\alpha}\right) = f(\hat{x}) - \frac{|f(\hat{x})|}{\alpha} f'(c_2) \leqq 0.$$

Hence $f(x)$ changes sign in I and has a zero in I. A similar argument holds if $f'(x) < 0$ in I, and the proof is complete.

As an example, let us consider the case

2.17 $$f(x) = x^2 - 2 = 0$$

---

* Alternatively, one might accept $\hat{x}$ as a zero of $\tilde{f}(x)$ if $\hat{x}$ lies in the interval $[a, b]$ where $\tilde{f}(a)\tilde{f}(b) < 0$ (see Exercise 5).

and $\hat{x} = 1.4$. In this case

2.18 $\tilde{f}(\hat{x}) = f(\hat{x}) = -0.04.$

In the interval $I : \hat{x} - \delta \leq x \leq \hat{x} + \delta$, where $\delta = |f(\hat{x})|/\alpha > 0$, we have

2.19 $f'(x) \geq 2(\hat{x} - \delta).$

Thus, by Theorem 2.14, there is a zero of $f(x)$ in I if

2.20 $2(\hat{x} - \delta) \geq \alpha$

and

$$\frac{0.04}{\alpha} = \delta.$$

This leads to the condition

2.21 $2(\hat{x} - \delta) \geq \dfrac{0.04}{\delta}.$

The smallest value of δ for which the above condition is satisfied is $\delta \doteq 0.01443$. Thus, there is a zero of $f(x)$ in the interval

$$1.38557 \leq x \leq 1.41443.$$

Actually, the zero $\bar{x} = \sqrt{2} \doteq 1.41421$ lies within this interval.

In practice, one seldom has an explicit expression for the minimum of $|f'(x)|$ in I. Frequently one chooses a fairly large interval and finds a lower bound for $|f'(x)|$ in that interval. If one applies Theorem 2.14, one frequently obtains a considerably smaller interval. This leads to a smaller value of α which can be used to find a still smaller interval, etc. Thus, in the example just considered we know that $f'(x) \geq 2$ in the interval $1 \leq \hat{x} \leq 2$. Hence, by Theorem 2.14, there is a root in the interval $1.38 \leq x \leq 1.42$. In this smaller interval $f'(x) \geq 2.76$. This leads to the interval $|x - 1.4| \leq 0.0145$. In this smaller interval $f'(x) \geq 2.7710$ which leads to a still smaller interval. Eventually this process converges to the interval $|x - 1.4| \leq 0.01443$.

Frequently, in order to apply Theorem 2.14, it is necessary to use the inequalities

2.22 $|f(\hat{x})| \leq |\tilde{f}(\hat{x}) - f(\hat{x})| + |\tilde{f}(\hat{x})|$

and

2.23 $|\tilde{f}(\hat{x}) - f(\hat{x})| \leq |\tilde{f}(\hat{x}) - F(\hat{x})| + |F(\hat{x}) - f(\hat{x})|.$

Here we assume as before that $F(\hat{x})$ is an approximation to $f(\hat{x})$ and that $\tilde{f}(\hat{x})$ is the computed value of $F(\hat{x})$. Thus having found $\hat{x}$ such that $|\tilde{f}(\hat{x})|$ is small, we need a bound on the rounding error $|\tilde{f}(\hat{x}) - F(\hat{x})|$ in the computation of $F(\hat{x})$, as well as a bound on $|F(\hat{x}) - f(\hat{x})|$.

Let us now consider the case where $f(x)$ is given by (2.7) and where we represent $f(x)$ by $F(x)$ given by (2.8). As before, we neglect the effect of rounding errors in the computation of $F(x)$. If we let

$$\hat{x} = 0.850$$

then we have

$$F(\hat{x}) = \hat{x} - \frac{(\hat{x})^3}{6} - 0.750 \doteq -0.0024.$$

Moreover,

$$|F(\hat{x}) - f(\hat{x})| \leqq \frac{(\hat{x})^5}{120} \doteq 0.0037.$$

Therefore, since $\tilde{f}(\hat{x}) = F(\hat{x})$, we have from (2.22)

$$|f(\hat{x})| \leqq |\tilde{f}(\hat{x})| + |\tilde{f}(\hat{x}) - f(\hat{x})| \leqq 0.0061.$$

In the interval $0.840 \leqq x \leqq 0.860$ we have

$$f'(x) = \cos x \geqq \cos(0.860) \doteq 0.652.$$

Since $|f(\hat{x})|/0.652 \leqq 0.00936 < 0.010$, it follows that Theorem 2.14 is applicable with $\alpha = 0.652$, and that there is a zero of $f(x)$ in the interval

$$|x - 0.850| \leqq 0.00936.$$

This interval contains the zero $\bar{x} \doteq 0.8481$ of $f(x)$.

EXERCISES 4.2

1. How accurately can one expect to get the positive zero of $f(x) = x^3 - 6x + 4.5$ (which is approximately 0.854) if one can compute $f(x)$ correctly to within 10^{-6}?

2. For the equation $f(x) = x^3 - 2 = 0$ find, using the analysis given in the text, the smallest positive δ so that there is a zero of $f(x)$ in the interval $|x - 1.2| \leqq \delta$. Compare δ with $|\sqrt[3]{2} - 1.2|$.

3. Consider the equation $f(x) = \cos x - 0.600 = 0$. If one uses the approximation

$$F(x) = 1 - \frac{x^2}{2} - 0.600$$

to represent $f(x)$, how close to $\bar{x}$ will there be a zero of $F(x)$? Here $\bar{x}$ is the zero of $f(x)$ in the range $0 < x < \pi/2$. Verify this bound by actually finding the zero of $F(x)$. Also, using Theorem 2.4, assess the accuracy of 0.940 as a zero of $f(x)$.

4. Let $f(x)$ be continuous and continuously differentiable in the interval $I: \hat{x} - \delta \leq x \leq \hat{x} + \delta$ where $\delta > 0$. Let $F(x)$ be an approximation to $f(x)$ and let $\tilde{f}(x)$ be the computed value of $F(x)$ for each computer-representable value of x. Show that if $|f'(x)| \geq \alpha$ in I and if $|F(\hat{x}) - \tilde{f}(\hat{x})| \leq \varepsilon_1$, $|f(\hat{x}) - F(\hat{x})| \leq \varepsilon_2$, and $|\tilde{f}(\hat{x})| \leq \varepsilon$, then there is a zero of $f(x)$ in the interval

$$|x - \hat{x}| \leq \frac{\varepsilon + \varepsilon_1 + \varepsilon_2}{\alpha}$$

provided

$$\delta \geq \frac{\varepsilon_1 + \varepsilon_2 + \varepsilon}{\alpha}$$

5. For some $\alpha > 0$ let $f(x)$ be continuous and continuously differentiable in I:

$$a - \frac{\varepsilon}{\alpha} \leq x \leq b + \frac{\varepsilon}{\alpha}$$

where $a < b$. Let $\tilde{f}(x)$ be continuous in $I' = [a, b]$, let $\tilde{f}(a)\tilde{f}(b) < 0$ and let

$$|\tilde{f}(x) - f(x)| \leq \varepsilon$$

in I'. Show that if $|f'(x)| \geq \alpha$ in I then there is a zero of $f(x)$ in I.

4.3 GRAPHICAL METHODS

It is sometimes convenient to determine approximate values of real roots of (1.1) by using a graphical method. We shall consider two such methods here. While these methods are primarily useful for hand computation rather than for use on digital computers, nevertheless, they do provide some insight into the problem and to the more practical methods which will be considered later.

The first graphical method which we shall consider consists simply of plotting the variable $y = f(x)$ against x. One computes $y = f(x)$ for several values of x and then plots the points (x, y). A curve is then drawn through the points, introducing additional points if necessary. Wherever the curve crosses the x-axis one has an approximate value of a real root.

As an example, let us consider the equation

3.1 $$f(x) = x^3 - 2 = 0.$$

Letting $x = 0, 0.5, 1.0, 1.5$, and 2.0 we get $y = f(x) = -2, -1.875, -1, 1.375$, and 6, respectively. Plotting the corresponding points on the graph (see Fig. 3.2) we construct the curve as indicated. Evidently, the curve crosses the x-axis at approximately $x = 1.25$. Since $f(x)$ is non-decreasing for all x, it is clear that there are no other real roots; hence no additional points on the graph need be plotted.

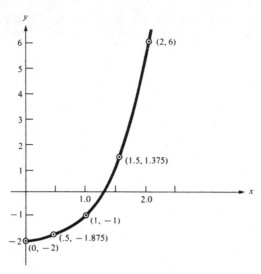

3.2 Fig. Graph of $y = x^3 - 2$.

We remark that it is frequently convenient to have different scales for x and y. This creates no problem as long as one takes into account the difference in scale when one uses derivatives.

It is sometimes convenient to use a second graphical method based on the plotting of two dependent variables y_1 and y_2. To illustrate, consider the case

3.3 $f(x) = \sin x - x + 2 = 0.$

We split the function $f(x)$ into the difference of two functions $g(x)$ and $h(x)$ such that

3.4
$$\begin{cases} f(x) = g(x) - h(x) \\ g(x) = \sin x = y_1 \\ h(x) = x - 2 = y_2. \end{cases}$$

Evidently $f(x) = 0$ if and only if $y_1 = y_2$. Our method is to plot both $y_1 = g(x)$ and $y_2 = h(x)$ and to determine points (x, y) where the curves intersect. The values of x corresponding to such intersection points will be the real roots of (1.1). The advantage of this method over the simpler method lies primarily in the fact that in many cases it is easier to plot y_1 and y_2 separately than it is to plot the function $f(x)$. In (3.4), we calculate y_1 and y_2 for the following values:

x	0	$\pi/2 \doteq 1.57$	$\pi \doteq 3.14$	$3\pi/2 \doteq 4.71$	$2\pi \doteq 6.28$
$y_1 = \sin x$	0	1	0	-1	0

x	0	1	2	4	5
$y_2 = x - 2$	-2	-1	0	2	3

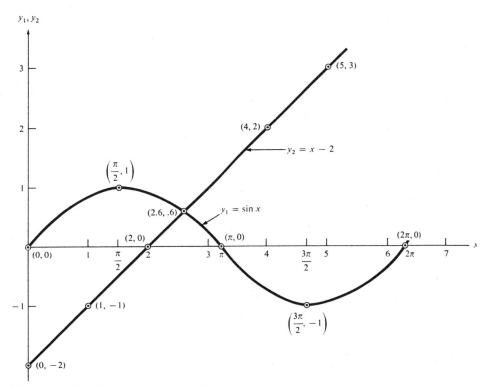

3.5 Fig. Graphs of $y_1 = \sin x$, $y_2 = x - 2$.

The graphs of y_1 and y_2 for our example are given in Fig. 3.5. The curves intersect at about (2.6, 0.6). Thus we obtain $x \doteq 2.6$ as an approximation to the root.

The approximate values of the roots obtained by the graphical procedures can frequently be used as starting values from which more accurate values can be obtained through the use of iterative methods, as will be described later in this chapter.

EXERCISES 4.3

1. Obtain by graphical methods approximate values of the real roots of $x = \log x + 3$.

4.4 THE METHOD OF BISECTION

We now describe a method known as the *method of bisection* for finding a zero of a continuous function $f(x)$, given a and b such that

4.1 $$f(a)f(b) < 0.$$

By Theorem A-3.24 we know that there is a number, say $\bar{x}$, in (a, b) such that

4.2
$$f(\bar{x}) = 0.$$

We seek to find $\bar{x}$ by determining a sequence of intervals, each half as large as the previous one and each containing at least one zero of $f(x)$.

In order to carry out the method of bisection we let

4.3
$$r_0 = a, \qquad s_0 = b,$$

and compute an initial estimate

4.4
$$z_0 = \frac{r_0 + s_0}{2} = \frac{a + b}{2}.$$

If $f(z_0) = 0$, the process is terminated. Otherwise, we let

4.5
$$\left. \begin{cases} r_1 = r_0 \\ s_1 = z_0 \end{cases} \right\} \qquad \text{if } f(z_0)f(r_0) < 0$$

and

4.6
$$\left. \begin{cases} r_1 = z_0 \\ s_1 = s_0 \end{cases} \right\} \qquad \text{if } f(z_0)f(r_0) > 0.$$

In general, given r_n and s_n such that

4.7
$$f(r_n)f(s_n) < 0$$

we let

4.8
$$z_n = \frac{r_n + s_n}{2}.$$

If $f(z_n) = 0$ the process is terminated. Otherwise we let

4.9
$$\left. \begin{cases} r_{n+1} = r_n \\ s_{n+1} = z_n \end{cases} \right\} \qquad \text{if } f(z_n)f(r_n) < 0$$

and

4.10
$$\left. \begin{cases} r_{n+1} = z_n \\ s_{n+1} = s_n \end{cases} \right\} \qquad \text{if } f(z_n)f(r_n) > 0.$$

As an example, let us consider the case

4.11 $$f(x) = x^3 - 2$$

with

4.12 $$a = 1, \quad b = 2.$$

Since

4.13 $$f(a) = -1, \quad f(b) = 6$$

we have

4.14 $$f(a)f(b) < 0$$

and hence there is a zero of $f(x)$ between 1 and 2. We let

4.15 $$r_0 = 1, \quad s_0 = 2$$

and compute an initial estimate

4.16 $$z_0 = \frac{r_0 + s_0}{2} = 1.5.$$

Since $f(z_0) = 1.375$, we have

4.17 $$f(z_0)f(r_0) < 0$$

and hence

4.18 $$r_1 = r_0 = 1, \quad s_1 = z_0 = 1.5.$$

Continuing, we have

4.19 $\quad z_1 = 1.25, \quad f(z_1) \doteq -0.047, \quad r_2 = 1.25, \quad s_2 = 1.5$

and

4.20 $\quad z_2 = 1.375, \quad f(z_2) \doteq 0.5996, \quad r_3 = 1.25, \quad s_3 = 1.375.$

Eventually, this process will converge to the zero

4.21 $$\sqrt[3]{2} \doteq 1.2599.$$

However, if we were to terminate after three iterations, our final value would be

4.22 $$z_3 = \frac{r_3 + s_3}{2} = 1.3125,$$

and our final error would be

4.23 $|\sqrt[3]{2} - z_3| \doteq 1.3125 - 1.2599 = 0.0526.$

  The convergence of this process is easily proved. For the sequence $r_0, r_1, \ldots$ is a bounded nondecreasing sequence while the sequence $s_0, s_1, \ldots$ is a bounded nonincreasing sequence. Hence both sequences converge. Let r and s be the limits of the sequences $r_0, r_1, \ldots$ and $s_0, s_1, \ldots$, respectively. Since

4.24 $r_n < z_n < s_n$

and since

4.25 $\lim_{n \to \infty} |r_n - s_n| = 0$

it follows that $r = s$ and

4.26 $r = \lim_{n \to \infty} s_n = \lim_{n \to \infty} z_n = \lim_{n \to \infty} r_n.$

Moreover, since

4.27 $f(r_n)f(s_n) < 0$

for all n it follows that

4.28 $0 \geqq \lim_{n \to \infty} [f(r_n)f(s_n)] = \left(\lim_{n \to \infty} f(r_n) \right)\left(\lim_{n \to \infty} f(s_n) \right)$

$$= [f(r)]^2$$

by (4.26) and the continuity of $f(x)$. Therefore, since $[f(r)]^2 \geqq 0$, it follows that

4.29 $f(r) = 0$

and

4.30 $\lim_{n \to \infty} f(r_n) = 0.$

Hence by (4.26) it follows that the sequence $z_0, z_1, \ldots$ converges to a zero of $f(x)$.

  Let us now estimate the number of iterations necessary to obtain a zero of $f(x)$ by the bisection method. In this analysis we do not consider the effect of rounding errors in the computation of $f(x)$ or of z_n. Since after computing z_n there is a zero, say $\bar{x}$, of $f(x)$ between r_n and s_n, it follows that

4.31 $|z_n - \bar{x}| \leqq \dfrac{b - a}{2^{n+1}}.$

For a given $\varepsilon > 0$, in order to make

4.32 $|z_n - \bar{x}| \leqq \varepsilon$

we choose n such that

4.33
$$2^{n+1} \geqq \frac{b-a}{\varepsilon},$$

i.e., such that

4.34
$$n \geqq \frac{\log\left(\dfrac{b-a}{\varepsilon}\right)}{\log 2} - 1.$$

In the example (4.11), the value of n needed to obtain the zero $\sqrt[3]{2}$ to within $\varepsilon = 10^{-6}$ satisfies

4.35
$$n \geqq \frac{\log 10^6}{\log 2} - 1 \doteq 18.93$$

so that it would be necessary to compute z_{19}.

If one stops after 3 iterations, then by (4.31) the error satisfies

4.36
$$|z_3 - \bar{x}| \leqq 2^{-4} = 0.0625,$$

which is slightly larger than the observed error, namely 0.0526, given by (4.23).

In practice, of course, one must work with the computed function $\tilde{f}(x)$ as defined in Section 4.2 rather than with $f(x)$. We prescribe two positive quantities ε_1 and ε_2 and will agree to accept α as a root if either

4.37
$$|\tilde{f}(\alpha)| \leqq \varepsilon_1$$

or if α lies in an interval $\beta \leqq x \leqq \gamma$ such that

4.38
$$\tilde{f}(\beta)\tilde{f}(\gamma) \leqq 0$$

and such that

4.39
$$\gamma - \beta \leqq \varepsilon_2.$$

Our computational procedure is as follows. Given a and b we first test whether

4.40
$$|\tilde{f}(a)| \leqq \varepsilon_1, \qquad \text{or} \qquad |\tilde{f}(b)| \leqq \varepsilon_1.$$

If so, we accept a or b as a zero and terminate the process. We also test whether

4.41
$$\tilde{f}(a)\tilde{f}(b) < 0.$$

If not, then the method of bisection is not guaranteed to converge and we terminate the process. Otherwise we test whether

4.42 $b - a \leqq \varepsilon_2.$

If so, then we accept

4.43 $z_0 = \dfrac{b + a}{2}$

as a zero. Otherwise we let

4.44 $r_0 = a, \qquad s_0 = b,$

and determine r_1 and s_1 based on (4.5) and (4.6). The subsequent process is as indicated in the flow diagram of Fig. 4.45.

If ε_1 and ε_2 are chosen with reasonable care, the above process will terminate. However, the process may continue indefinitely if ε_1 and ε_2 are chosen too small. This may happen with certain rounding procedures if ε_1 is chosen less than the expected rounding error in $f(x)$ in the interval $[a, b]$, and if ε_2 is chosen less than the minimum distance between two consecutive computer-representable numbers in $[a, b]$. In such a case we may reach a situation where the computed value of z_{n+1} may be the same as r_n or s_n and where

4.46 $|f(z_n)| > \varepsilon_1.$

In this case we have

4.47 $z_{n+1} = z_{n+2} = z_{n+3} = \cdots$

and hence

4.48 $\varepsilon_2 < |r_{n+1} - s_{n+1}| = |r_{n+2} - s_{n+2}| = |r_{n+3} - s_{n+3}| = \cdots .$

Thus the process will not terminate. To prevent this possibility one should either test at each step to be sure that $z_n \neq r_n$ and $z_n \neq s_n$ or else one should specify a maximum number of iterations which can be performed before the process is to be arbitrarily terminated.

EXERCISES 4.4

1. Find correct to three decimal places a solution of $\sin x - 0.750 = 0$ by the method of bisection with $a = 0.80$ and $b = 0.90$.

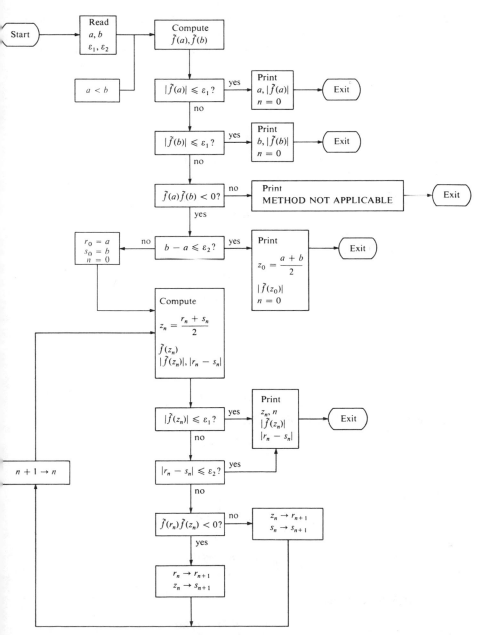

4.45 Fig. Flow Diagram for the Method of Bisection.

2. Verify Theorem A-3.24 for the case where $f(x) = Ax + B$ where A and B are real and for the case where $f(x) = Ax^2 + Bx + C$ where A, B, and C are real.

3. Write a computer program for solving the equation $f(x) = 0$ by the method of bisection given a and b such that $f(a)f(b) < 0$. Apply the program to the case $f(x) = x^3 - \frac{1}{2}$ where $a = 0, b = 1$. Estimate the number of bisections needed to obtain a root correct to 10^{-10} and compare with the observed results.

4.5 THE METHOD OF FALSE POSITION

The method of false position is very similar to the method of bisection except that at each step one determines z_n by

5.1
$$z_n = \frac{r_n f(s_n) - s_n f(r_n)}{f(s_n) - f(r_n)}.$$

The derivation of (5.1) is based on the following idea. Given a and b such that $f(a)f(b) < 0$ we determine the linear function $F(x)$ such that

5.2
$$\begin{cases} F(a) = f(a) \\ F(b) = f(b) \end{cases}$$

and then find c such that

5.3
$$F(c) = 0.$$

Since $F(x)$ is linear we have

5.4
$$F(x) = Ax + B$$

for some constants A and B. From (5.2) we have

5.5
$$\begin{cases} Aa + B = f(a) \\ Ab + B = f(b) \end{cases}$$

and hence

5.6
$$\begin{cases} A = \dfrac{f(b) - f(a)}{b - a} \\ B = \dfrac{bf(a) - af(b)}{b - a}. \end{cases}$$

Solving (5.3) for c we get

5.7
$$c = -\frac{B}{A} = \frac{af(b) - bf(a)}{f(b) - f(a)}.$$

We note that the denominator does not vanish because of (4.1). If we let

5.8 $r_n = a, \qquad s_n = b, \qquad z_n = c$

then we get (5.1).

Figure 5.10 illustrates the basic step in the method of false position. The graph of the function $F(x)$ is a straight line joining the points $(a, f(a))$ and $(b, f(b))$. From analytic geometry, using the equation for a straight line joining two points, we have

5.9 $$\frac{y - f(a)}{x - a} = \frac{f(b) - f(a)}{b - a}.$$

Letting $y = F(x)$ we verify that $F(x)$ has the form (5.4) with A and B given by (5.6).

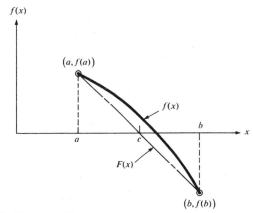

5.10 Fig. Method of False Position.

The procedure used for the method of false position is the same as that given for the method of bisection (see (4.3)–(4.10)) except that we replace (4.4) and (4.8) by (5.1). In the example (4.11) with a and b given by (4.12) we get

5.11 $$z_0 = \frac{(1)(6) - (2)(-1)}{6 - (-1)} = \frac{6 + 2}{6 + 1} = \frac{8}{7} \doteq 1.143$$

which is closer to the true value of the root, namely 1.2599, than is the value $z_0 = 1.5$ obtained by the method of bisection.

Since

5.12 $$f(z_0) = -\tfrac{174}{343} < 0$$

we let

5.13 $r_1 = 1.143, \quad s_1 = 2.0$

and proceed to determine $z_1, r_2, s_2, z_2, r_3, s_3$, etc. The reader should carry out the calculations and show that the convergence is much faster than for the method of bisection (in this example).

We now show that the method of false position converges provided $f(x) \in C[a, b]$ and $f(a)f(b) < 0$. In the first place, we observe that the sequence $r_0, r_1, \ldots$ is nondecreasing and bounded above by b. Hence the sequence converges to a limit, say r. Similarly, the sequence $s_0, s_1, \ldots$ is a nonincreasing sequence which converges to a limit $s \geq r$.

Suppose that neither r nor s is a zero of $f(x)$. Since $f(r) \neq 0$ and $f(s) \neq 0$, it follows by the continuity of $f(x)$ that for some $\varepsilon > 0$ and for some N

5.14 $|f(r_n)| \geq \varepsilon, \qquad |f(s_n)| \geq \varepsilon$

for all $n \geq N$. But by (5.1) we have

5.15 $z_n - r_n = \dfrac{(r_n - s_n)f(r_n)}{f(s_n) - f(r_n)}$

5.16 $z_n - s_n = \dfrac{(r_n - s_n)f(s_n)}{f(s_n) - f(r_n)}.$

If M is a bound on $|f(x)|$ in the interval $[a, b]$, then we have for all n sufficiently large

5.17 $|r_{n+1} - s_{n+1}| \leq \max(|z_n - r_n|, |z_n - s_n|)$

$$\leq \frac{|r_n - s_n|M}{M + \varepsilon}.$$

Therefore

5.18 $|r_{N+k} - s_{N+k}| \leq \left(\dfrac{M}{M + \varepsilon}\right)^k |r_N - s_N|.$

Hence

5.19 $\lim_{n \to \infty} r_n = \lim_{n \to \infty} s_n = r.$

But since

5.20 $f(r_n)f(s_n) < 0$

for all n, it follows as in the method of bisection that

5.21 $f(r) = 0$.

This contradiction proves that either r or s is a zero of $f(x)$.
  Suppose now that r is a zero of $f(x)$ but that s is not. By (5.15) we have

5.22 $$|z_n - r_n| \leqq \frac{|r_n - s_n| \, |f(r_n)|}{|f(s_n)|}.$$

Since the limit of the right-hand side exists and is equal to zero, it follows that

5.23 $$\lim_{n \to \infty} z_n = r.$$

Similarly, if s is a zero of $f(x)$ but r is not, then we have

5.24 $$\lim_{n \to \infty} z_n = s.$$

  Let us now consider the case where r and s are both zeros of $f(x)$. If $r \neq s$, then it is conceivable that the sequence $z_0, z_1, \ldots$ could have some subsequences converging to r and other subsequences converging to s. On the other hand, if $r = s$, then we clearly have

5.25 $$\lim_{n \to \infty} z_n = r.$$

We now show that unless r_0 or s_0 is a zero of $f(x)$ or unless z_n is a zero of $f(x)$ for some n, then $r = s$. It is easy to see that unless r_0 or s_0 is a zero of $f(x)$ or unless z_n is a zero of $f(x)$ for some n, neither r_n nor s_n can ever be a zero of $f(x)$. Suppose that $r \neq s$. For any $\delta > 0$ there exists N such that for all $n > N$ we have

5.26 $r - r_n < \delta, \qquad s_n - s < \delta$.

Evidently $z_n < r$ or $z_n > s$ for all n. Suppose that for some n we have $z_n < r$. Then

5.27 $$\frac{r_n f(s_n) - s_n f(r_n)}{f(s_n) - f(r_n)} < r$$

and

5.28 $$f(s_n) \geqq \frac{s_n - r}{r - r_n}[-f(r_n)] \geqq \frac{s - r}{\delta}[-f(r_n)].$$

Here we are assuming that $f(r_n) < 0$, $f(s_n) > 0$. In order that s_n converge to s where $s < s_n$ we must have $z_m > s$ for some $m > n$. But this implies that

5.29 $$[-f(r_m)] \geqq \frac{s - r_m}{s_m - s} f(s_m) \geqq \left(\frac{s - r}{\delta}\right)^2 [-f(r_m)].$$

Similarly, for some $p > n$ we must have

5.30
$$[-f(r_p)] \geq \left(\frac{s - r}{\delta}\right)^4 [-f(r_n)],$$

etc. Continuing this process we show that given any $F > 0$ there are values of x arbitrarily close to r for which $|f(x)| \geq F$. This certainly precludes the possibility that

5.31
$$\lim_{n \to \infty} f(r_n) = 0$$

and hence r is not a zero of $f(x)$. This contradiction proves that $r = s$. To summarize, we have the following possibilities:

1. Both r_0 and s_0 are zeros of $f(x)$ in which case we terminate the process and accept r_0 or s_0 as the desired zero of $f(x)$.

2. If r_0 is a zero of $f(x)$ but s_0 is not, the procedure gives $z_0 = z_1 = \cdots = r_0$.

3. If s_0 is a zero of $f(x)$ but r_0 is not, the procedure gives $z_0 = z_1 = \cdots = s_0$.

4. z_n is a zero of $f(x)$ for some n.

5. If $r = \lim_{n \to \infty} r_n$ is a zero of $f(x)$ but $s = \lim_{n \to \infty} s_n$ is not, then the sequence $z_0, z_1, \ldots$, converges to r. The reader should show that in this case for some N we have $s_N = s_{N+1} = \cdots$.

6. If s is a zero of $f(x)$ but r is not, then the sequence $z_0, z_1, \ldots$ converges to s. In this case for some N we have $r_N = r_{N+1} = \cdots$.

7. $r = s$ and r is a zero of $f(x)$. In this case $z_0, z_1, \ldots$ converges to r.

EXAMPLE

To show that the convergence may be very slow, consider the function $f(x)$ defined by

5.32
$$f(x) = \begin{cases} \delta & , \quad 0 \leq x \leq \frac{1}{2} \\ 4(1 + \delta)(x - x^2) - 1, & \frac{1}{2} \leq x \leq 1, \end{cases}$$

where $0 < \delta < 1$.

A graph of $f(x)$ is shown in Fig. 5.36. Evidently $f(x)$ and $f'(x)$ are continuous in the interval $0 \leq x \leq 1$. Consider the application of the method of false position with $a = 0, b = 1$. Since $f(a) = \delta$, $f(b) = -1$ we have

5.33
$$z_0 = \frac{(0)(-1) - (1)\delta}{-1 - \delta} = \frac{\delta}{1 + \delta} = 1 - \frac{1}{1 + \delta}.$$

In general we have

5.34
$$z_n = 1 - \frac{1}{(1 + \delta)^n}$$

provided that n is such that $z_{n-1} \leqq \frac{1}{2}$.

Suppose that δ is very small, but not so small that we would accept $x = 0$ as a root based on the condition $|f(0)| < \varepsilon$. The number of iterations so that $z_n \geqq \frac{1}{2}$ is given by

5.35
$$n \geqq \frac{\log 2}{\log(1 + \delta)} \doteq \frac{\log 2}{\delta} \doteq \frac{0.693}{\delta}$$

for δ very small. Hence if δ is 10^{-6} then about 693,000 iterations would be needed to reach $x = \frac{1}{2}$. On the other hand, the method of bisection will still converge to a relative accuracy of 2^{-48} in 48 iterations.

A flow diagram for the method of false position would be very similar to that of Fig. 4.45 for the method of bisection except for the different formula used to compute $z_0, z_1, \ldots$. Here it is essential to have a procedure for stopping the process after a fixed number of iterations in view of the fact that, as we have seen, the convergence can be arbitrarily slow. The procedure could be designed to switch over to the method of bisection if the method of false position fails.

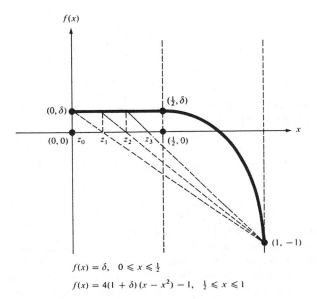

$$f(x) = \delta, \quad 0 \leqslant x \leqslant \tfrac{1}{2}$$
$$f(x) = 4(1 + \delta)(x - x^2) - 1, \quad \tfrac{1}{2} \leqslant x \leqslant 1$$

5.36 Fig. Application of the Method of False Position.

EXERCISES 4.5

1. Find correct to three decimal places a solution of $\sin x - 0.750 = 0$ by the method of false position with $a = 0.80$, $b = 0.90$.

2. Carry out five iterations with the method of false position for the example (5.32) where $\delta = 0.1$.

3. Write a computer program for solving the equation $f(x) = 0$ by the method of false position given x_0 and x_1 such that $f(x_0)f(x_1) < 0$. Apply the program to the case $f(x) = x^3 - \frac{1}{2} = 0$ where $x_0 = 0$, $x_1 = 1$. Also apply it to the following cases:

   a) $f(x) = \sin x - x^3 - 1 = 0$, $x_0 = -2$, $x_1 = -1$
   b) $f(x) = x - \log x - 3 = 0$, $x_0 = 2$, $x_1 = 6$

4.6 THE SECANT METHOD

The secant method is closely related to the method of false position except that it can be applied even if $f(a)f(b)$ is positive, though it may not always converge. We let $x_0 = a$, $x_1 = b$, and generate a sequence $x_2, x_3, \ldots$, determined by

6.1
$$x_{n+1} = \frac{x_{n-1}f(x_n) - x_n f(x_{n-1})}{f(x_n) - f(x_{n-1})} \qquad n = 1, 2, \ldots \; .$$

This formula is obtained from (5.7) by letting $a = x_{n-1}$, $b = x_n$, $c = x_{n+1}$.

Evidently the secant method will fail if at any stage we have $f(x_n) = f(x_{n-1})$. On the other hand, if the procedure is to converge, then $f(x_n)$ and $f(x_{n-1})$ must both approach zero; hence $f(x_n) - f(x_{n-1})$ approaches zero. To partially avoid this difficulty, one can rewrite (6.1) in the form

6.2
$$x_{n+1} = x_n - \frac{(x_n - x_{n-1})f(x_n)}{f(x_n) - f(x_{n-1})} \; .$$

Even this formula may not prove satisfactory in all cases, especially if $f'(x)$ is small near a root of (1.1).

In the secant method we are not sure that $f(x_n)f(x_{n-1}) < 0$ for all n even if $f(x_0)f(x_1) < 0$. This can be seen in the example (5.32). As a matter of fact, if $\delta < \sqrt{2} - 1$, the secant method does not converge if we let $x_0 = 0$, $x_1 = 1$. For we have

6.3
$$x_2 = 1 - (1 + \delta)^{-1}, \qquad x_3 = 1 - (1 + \delta)^{-2} < \tfrac{1}{2}.$$

Hence $f(x_2) = f(x_3) = \delta$ and the process fails.

We shall study the convergence properties of the secant method in more detail in Section 4.11. In essence, the secant method will normally converge faster than the method of false position and the method of bisection, but in some cases may not converge even when the other two methods do converge. If one suspects that the secant method will not converge, and if $f(a)f(b) < 0$, one can go ahead and apply the method. If convergence is not achieved, one can go back and start over with the method of false position or the method of bisection.

EXERCISES 4.6

1. Find correct to three decimal places a solution of $\sin x - 0.750 = 0$ by the secant method with $x_0 = 0.80$, $x_1 = 0.90$.

2. Attempt to find a zero of the function $f(x)$ given by (5.32), where $\delta = 0.1$ by the secant method with $x_0 = 0$, $x_1 = 1$.

3. Modify the program of Exercise 3, Section 4.5, to use the secant method. The program should not require that $f(a)f(b) < 0$. However, if $f(a)f(b) < 0$, and if convergence is not obtained after a specified number of iterations, the method of false position should be used. Compare the convergence rate with that of the method of false position in the special case $f(x) = x^3 - \frac{1}{2}$, with $x_0 = 0$, $x_1 = 1$, and to the other cases listed in Exercise 3, Section 4.5. Also apply the program to the example (5.32).

4.7 GENERAL PROPERTIES OF ITERATIVE METHODS

For each $n \geq 1$ let $\phi_n(t_0, t_1, t_2, \ldots, t_{n-1})$ be a function of n variables. Given such functions and given $k + 1$ starting values $x_0, x_1, \ldots, x_k$ we can define $x_{k+1}, x_{k+2}, \ldots$ by

7.1 $x_{n+1} = \phi_{n+1}(x_0, x_1, \ldots, x_n).$

Such a method is a *nonstationary iterative method*. If, for each n, the function ϕ_{n+1} depends at most on the variables $x_{n-s+1}, x_{n-s+2}, \ldots, x_{n-1}, x_n$, we say that the method is an *s-step* method. If, moreover, ϕ_{n+1} is independent of n, then the method is *stationary*. Evidently, if the method is a stationary s-step method, then we must have at least s starting values.

The method of bisection and the method of false position are examples of nonstationary methods. In each case the number of starting values, $k + 1$, is equal to two. In the method of bisection, the functions $\phi_{n+1}(x_0, x_1, \ldots, x_n)$ have the form

7.2 $\phi_{n+1}(x_0, x_1, \ldots, x_n) = \alpha_{n,0}x_0 + \alpha_{n,1}x_1 + \cdots + \alpha_{n,n-1}x_{n-1} + \alpha_{n,n}x_n$

where $\alpha_{n,n}$ and one of the other $\alpha_{n,i}$ equal $\frac{1}{2}$ and the rest vanish. Thus, for instance, $\alpha_{1,0} = \frac{1}{2}, \alpha_{1,1} = \frac{1}{2}$. If $f(x_0)f(x_2) < 0$, then $\alpha_{2,0} = \frac{1}{2}, \alpha_{2,1} = 0, \alpha_{2,2} = \frac{1}{2}$; otherwise $\alpha_{2,0} = 0, \alpha_{2,1} = \frac{1}{2}, \alpha_{2,2} = \frac{1}{2}$.

The secant method is an example of a stationary two-step method. In this case $k + 1 = 2$ and, for each n, $\phi(x_{n-1}, x_n)$ is given by

7.3 $\phi(x_{n-1}, x_n) = \dfrac{x_{n-1}f(x_n) - x_n f(x_{n-1})}{f(x_n) - f(x_{n-1})} = x_n - \dfrac{(x_n - x_{n-1})f(x_n)}{f(x_n) - f(x_{n-1})}.$

An important class of methods, and one which we shall be primarily concerned with in this section, is the class of stationary one-step methods. Here we choose a single starting value x_0 and a function $\phi(x)$. The sequence $x_1, x_2, \ldots$ is computed by

7.4
$$x_{n+1} = \phi(x_n), \qquad n = 0, 1, 2, \ldots .$$

The function $\phi(x)$ is sometimes known as an *iteration function* (see Traub [1964]). As an example of a stationary one-step method, consider the method

7.5
$$x_{n+1} = \frac{1}{2}\left(x_n + \frac{N}{x_n}\right)$$

for solving the equation

7.6
$$f(x) = x^2 - N = 0$$

where $N > 0$. Here $\phi(x) = \frac{1}{2}(x + N/x)$ which is continuous except when $x = 0$. Another iteration function would be $\phi(x) = N/x$ which corresponds to the iterative method

7.7
$$x_{n+1} = N/x_n.$$

Here again $\phi(x)$ is continuous except for $x = 0$.

Given an iteration function $\phi(x)$, there are many questions which should be considered in evaluating the associated iterative method.

1. Does the sequence defined by (7.4) converge to a (unique) limit?
2. If $\bar{x}$ is a root of (1.1) and if x_0 is sufficiently close to $\bar{x}$, does the sequence defined by (7.4) converge, and if so, does it converge to $\bar{x}$?
3. If the sequence defined by (7.4) converges, does it converge to a zero of $f(x)$?

The answer to the first question may depend upon the choice of x_0. For example, the sequence defined by (7.7) converges to $\sqrt{N}$ if $x_0 = \sqrt{N}$ and, provided $x_0 \neq 0$, oscillates between x_0 and N/x_0 otherwise. Ideally, the method should converge for all x_0 or, at least, for all x_0 sufficiently close to $\sqrt{N}$. In the latter case, a procedure for choosing x_0 close enough to $\sqrt{N}$ should be available. In Chapter 1 we saw that the sequence defined by (7.5) converges for any positive x_0. Actually, it converges to $\sqrt{N}$ if $x_0 > 0$ and to $-\sqrt{N}$ if $x_0 < 0$.

On the other hand, the sequence defined by

7.8
$$x_{n+1} = x_n(2 - Nx_n),$$

where $N > 0$, converges to $1/N$ if x_0 lies in the range $0 < x_0 < 2/N$ and diverges otherwise.* If $N < 1$, and we choose $x_0 = 1$, for example, we are sure that the method will converge. If $N > 1$, then one could find an integer p such that $2^p \geq N$ and then let $x_0 = 2^{-p}$.

---

* This method has actually been used on some of the earlier high-speed computing machines which did not have built-in division. It is given here for illustrative purposes only and should not be regarded as a method to be used in practice.

If the sequence $x_0, x_1, x_2, \ldots$ defined by (7.4) converges to a limit $\bar{x}$ and if $\phi(x)$ is continuous in a neighborhood of $\bar{x}$, then $\bar{x}$ satisfies

7.9 $$x = \phi(x).$$

To show this, we have from (7.4)

7.10 $$\lim_{n \to \infty} x_{n+1} = \lim_{n \to \infty} \phi(x_n).$$

But the left member of the above equation equals $\bar{x}$, while the right member equals $\phi(\bar{x})$ by the continuity of $\phi(x)$.

In our subsequent discussion we shall refer to (7.9) as the *related equation* corresponding to the iterative method (7.4).

Suppose now that the answer to Question 2 is affirmative. Then every root $\bar{x}$ of (1.1) must be the limit of some sequence $x_0, x_1, \ldots$ defined by (7.4) and hence, as we have just seen, $\bar{x}$ must be a root of (7.9). We are thus led to

7.11 Definiton. The iterative method (7.4) for solving (1.1) is *consistent* if every solution of (1.1) is a solution of the related equation (7.9).

Suppose the answer to Question 3 is affirmative, and let x^* be any root of (7.9). Then the sequence $x^*, x^*, x^*, \ldots$ which is generated by (7.4) with $x_0 = x^*$ clearly converges to x^*. Hence x^* is a root of (1.1). Thus every root of the related equation (7.9) is a root of (1.1). We are thus led to

7.12 Definition. The iterative method (7.4) for solving (1.1) is *reciprocally consistent* if every solution of (7.9) is a solution of (1.1).

7.13 Definition. The iterative method (7.4) for solving (1.1) is *completely consistent* if it is both consistent and reciprocally consistent.

EXAMPLES

7.14a $$f(x) = x^2 - N, \qquad N \geqq 0$$

$$\phi(x) = \frac{1}{2}\left(x + \frac{N}{x}\right).$$

The roots of $f(x) = 0$ are $x = \pm\sqrt{N}$, while the roots of $\phi(x) = x$ are also $\pm\sqrt{N}$ if $N > 0$. If $N = 0$, then zero is a double root of $f(x) = 0$, but a single root of $x = \phi(x)$. In any case, the method is *completely consistent*.

7.14b $$f(x) = N - \frac{1}{x}, \qquad N > 0$$

$$\phi(x) = x(2 - Nx).$$

The root of $f(x) = 0$ is $x = 1/N$. The roots of $x = \phi(x)$ are 0 and $1/N$. Hence the method is consistent but not reciprocally consistent.

7.14c
$$f(x) = N - \frac{1}{x}, \qquad N > 0$$

$$\phi(x) = -2.$$

Since the root of $x = \phi(x)$ is $x = -2$, the method is neither consistent nor reciprocally consistent.

7.14d
$$f(x) = x^2 - N, \qquad N \geqq 0$$

$$\phi(x) = x.$$

Here the equation $x = \phi(x)$ is satisfied by every x. Hence the method is consistent, but not reciprocally consistent.

7.14e
$$f(x) = x^2 - N, \qquad N > 0$$

$$\phi(x) = \sqrt{N}.$$

Here the root of $x = \phi(x)$ is $\sqrt{N}$ which is a root of $f(x) = 0$. Hence the method is reciprocally consistent but not consistent since $-\sqrt{N}$ is a root of $f(x) = 0$ but is not a root of $\phi(x) = \sqrt{N}$.

7.14f
$$f(x) = (x - 1)(x^2 - N), \qquad N > 0$$

$$\phi(x) = \frac{1}{2}\left(x + \frac{N}{x}\right).$$

Here the method is reciprocally consistent but not consistent unless $N = 1$.

If a method is consistent, then under certain conditions it will converge to a root $\bar{x}$ of (1.1) provided x_0 is sufficiently close to $\bar{x}$. Indeed, we have

7.15 Theorem. Let $\bar{x}$ be a root of (1.1) and let (7.4) define a consistent iterative method. If x_0 and $\phi(x)$ satisfy *either* of the following conditions, then the sequence $x_0, x_1, \ldots$, defined by (7.4) converges to $\bar{x}$.

  i) $\phi(x)$ is continuous and differentiable in the interval

$$I : \bar{x} - |x_0 - \bar{x}| \leqq x \leqq \bar{x} + |x_0 - \bar{x}|, \quad \text{and} \quad |\phi'(x)| \leqq M < 1 \quad \text{for all } x \text{ in } I.$$

 ii) $\phi(x)$ is continuous and differentiable in the closed interval between $\bar{x}$ and x_0 and $0 \leqq \phi'(x) \leqq M < 1$ in that same interval.

Proof. Let us first give a proof assuming (i). If $\bar{x}$ is a root of (1.1) then since (7.4) is consistent, we have $\bar{x} = \phi(\bar{x})$. Moreover, by the mean-value theorem

7.16 $$x_{n+1} - \bar{x} = \phi(x_n) - \phi(\bar{x}) = \phi'(c)(x_n - \bar{x})$$

where c lies between x_n and $\bar{x}$. If $n = 0$, we have $|x_1 - \bar{x}| \leq M|x_0 - \bar{x}| < |x_0 - \bar{x}|$ since $|\phi'(c)| \leq M < 1$. Similarly, $|x_2 - \bar{x}| \leq M^2|x_0 - \bar{x}|$, etc. Hence $|x_n - \bar{x}| \leq M^n|x_0 - \bar{x}|$ and $x_n \to \bar{x}$ as $n \to \infty$.

If (ii) holds, we have, as before,

7.17 $$x_{n+1} - \bar{x} = \phi'(c)(x_n - \bar{x})$$

where c lies between x_n and $\bar{x}$. Let us first assume $x_0 > \bar{x}$ and let $n = 0$. We obtain

7.18 $$x_1 - \bar{x} = \phi'(c)(x_0 - \bar{x}).$$

Therefore,

7.19 $$0 \leq x_1 - \bar{x} \leq M(x_0 - \bar{x}).$$

Similarly, we can show

7.20 $$0 \leq x_2 - \bar{x} \leq M^2(x_0 - \bar{x}).$$

In general, $0 \leq x_n - \bar{x} \leq M^n(x_0 - \bar{x})$ and hence $x_n \to \bar{x}$ as $n \to \infty$. A similar argument can be given if $x_0 < \bar{x}$.

Even if $\phi(x)$ is not differentiable, the conclusions of Theorem 7.15 are valid if $\phi(x)$ satisfies a Lipschitz condition in the domain D of interest, i.e., if there exists a constant M such that for any x', x'' in D we have

7.21 $$|\phi(x') - \phi(x'')| \leq M|x' - x''|.$$

We also require that the Lipschitz constant M be less than unity. Specifically, we have

7.22 Theorem. The conclusions of Theorem 7.15 hold if we replace (i) and (ii) by

i') $\phi(x)$ is continuous in $I: \bar{x} - |x_0 - \bar{x}| \leq x \leq \bar{x} + |x_0 - \bar{x}|$, and there exists a constant $M < 1$ such that (7.21) holds for all x', x'' in I.

ii') $\phi(x)$ is continuous in the closed interval J between $\bar{x}$ and x_0, and there exists a constant $M < 1$ such that if $x' \leq x''$ and if $x', x'' \in J$, then

7.23 $$0 \leq \phi(x'') - \phi(x') \leq M(x'' - x').$$

The proof of Theorem 7.22 is left as an exercise.

Even if the existence of a solution of the related equation $x = \phi(x)$ is not known in advance, if (7.21) holds for all x' and x'' for some $M < 1$, then we can prove the existence of a unique solution $\bar{x}$. We have

7.24 Theorem. If $\phi(x)$ is continuous for all x and if (7.21) holds for all x' and x'' and for some $M < 1$, then there exists a unique $\bar{x}$ such that $\bar{x} = \phi(\bar{x})$. Moreover, for any x_0, the sequence $x_0, x_1, x_2, \ldots$ where $x_1 = \phi(x_0)$, $x_2 = \phi(x_1)$, etc., converges to $\bar{x}$.

Proof. We first show that for any x_0 the sequence $x_0, x_1, x_2, \ldots$ is a Cauchy sequence.   Evidently $|x_2 - x_1| = |\phi(x_1) - \phi(x_0)| \leq M|x_1 - x_0|$, $|x_3 - x_2| \leq M^2|x_1 - x_0|$, and in general,

7.25
$$|x_{k+1} - x_k| \leq M^k|x_1 - x_0|.$$

Since $x_{n+p} - x_n = (x_{n+p} - x_{n+p-1}) + (x_{n+p-1} - x_{n+p-2}) + \cdots + (x_{n+1} - x_n)$, we have

7.26
$$|x_{n+p} - x_n| \leq (M^{n+p-1} + M^{n+p-2} + \cdots + M^n)|x_1 - x_0|$$
$$\leq \frac{M^n}{1 - M}|x_1 - x_0|.$$

Thus given $\varepsilon > 0$, there exists N such that if $m \geq N$, $n \geq N$, then $|x_m - x_n| < \varepsilon$. Thus the sequence $\{x_n\}$ is a Cauchy sequence and by Theorem A-2.12 has a limit, say $\bar{x}$. Moreover, by the continuity of $\phi(x)$, we have

7.27
$$\bar{x} = \phi(\bar{x}).$$

We now show that $\bar{x}$ is unique. Suppose y satisfies $y = \phi(y)$. If $x_0 = y$, then we have $x_1 = x_2 = \cdots = y$. But

7.28
$$|y - \bar{x}| = |\phi(y) - \phi(\bar{x})| \leq M|y - \bar{x}|.$$

Since $M < 1$ it follows that $y = \bar{x}$.

We now show that every sequence $x_0', x_1' = \phi(x_0')$, $x_2' = \phi(x_1'), \ldots$ converges to $\bar{x}$. Evidently

7.29
$$|x_1' - \bar{x}| = |\phi(x_0') - \phi(\bar{x})| \leq M|x_0' - \bar{x}|$$

and in general

7.30
$$|x_k' - \bar{x}| \leq M^k|x_0' - \bar{x}|.$$

Hence $x_k' \to \bar{x}$. This completes the proof of Theorem 7.24.

In a similar way we can prove

7.31 Corollary.   Let $\phi(x)$ be continuous in the interval $I = [a, b]$ and suppose that, for any x', x'' in I, (7.21) holds for some $M < 1$. If for each $x \in I$ we have $\phi(x) \in I$, then there exists a unique $\bar{x} \in I$ such that $\bar{x} = \phi(\bar{x})$. Moreover, for any

$x_0 \in I$ the sequence $x_0, x_1, \ldots$ where $x_1 = \phi(x_0)$, $x_2 = \phi(x_1)$, etc. converges to $\bar{x}$.
The function $\phi(x)$ is said to define a *contraction mapping* on the interval I.

EXAMPLES. Let us apply the above results to the iterative method

7.32
$$\phi_{n+1} = \frac{1}{2}\left(x_n + \frac{N}{x_n}\right)$$

for solving $x^2 - N = 0$ where $N > 0$. Evidently,

7.33
$$\phi'(x) = \frac{1}{2}\left(1 - \frac{N}{x^2}\right)$$

and $|\phi'(x)| \leqq 1$ for $x \geqq \sqrt{N}/\sqrt{3}$.
Consequently, if x_0 is chosen in the interval

7.34
$$\frac{\sqrt{N}}{\sqrt{3}} < x_0 < \left(2 - \frac{1}{\sqrt{3}}\right)\sqrt{N},$$

then $|x_0 - \sqrt{N}| \leqq (1 - 1/\sqrt{3})\sqrt{N}$ and $|\phi'(x)| < 1$ in the interval $\sqrt{N} - |x_0 - \bar{x}| \leqq x \leqq \sqrt{N} + |x_0 - \bar{x}|$. Thus by (i) of Theorem 7.15 we are assured of convergence to the solution $\sqrt{N}$.

Actually, of course, we have seen that the method converges to $\sqrt{N}$ for all $x_0 > 0$. For, we have shown that $x_1 \geqq \sqrt{N}$, $x_1 \geqq x_2 \geqq \sqrt{N}$, etc. Using Theorem 7.15 (ii) we can show that we have convergence provided $x_0 > \sqrt{N}$, for in that case we have

7.35
$$0 \leqq \phi'(x) < \tfrac{1}{2}.$$

While convergence holds for any $x_0 > 0$, this fact cannot be shown by Theorem 7.15 (ii); we can show that we have convergence provided $x_0 > \sqrt{N}$, for in that that $x_1 \geqq \sqrt{N}$; hence convergence follows.

We now give a slight generalization of Theorem 7.15.

*7.36 Theorem. Let $f(x)$ be a continuous function in $I = [\bar{x} - \delta, \bar{x} + \delta]$ where $\delta > 0$, and let $\bar{x}$ be the only root of (1.1) in I. If the iterative method (7.4) is completely consistent with (1.1), if $\phi(x)$ is continuous and differentiable in I and if $|\phi'(x)| < 1$ in I except possibly at $\bar{x}$, then the sequence $x_0, x_1, x_2, \ldots$ defined by (7.4) converges to $\bar{x}$ provided $x_0 \in I$.

Proof. Since the iterative method is consistent we have

7.37
$$\bar{x} = \phi(\bar{x})$$

---

*The material between the asterisk on this page and that on page 126 has been superseded by the text in Appendix B.

and, by the mean-value theorem,

7.38
$$x_{n+1} - \bar{x} = (x_n - \bar{x})\phi'(\xi)$$

where ξ lies between x_n and $\bar{x}$. Hence we have

7.39
$$|x_{n+1} - \bar{x}| < |x_n - \bar{x}| < \cdots < |x_0 - \bar{x}|.$$

Thus all of the x_n lie in the interval I. It follows by Theorem A-2.13 that there exists a subsequence $x_{k_0}, x_{k_1}, \ldots$ of the sequence $x_0, x_1, x_2, \ldots$ converging to a limit, say α, in I. By (7.4) and the continuity of $\phi(x)$ it follows that

7.40
$$\alpha = \phi(\alpha).$$

By the complete consistency this implies that α is a root of (1.1). But this contradicts the assumption that $\bar{x}$ is the only root (1.1) in I, and the proof of Theorem 7.36 is complete*

The result may be extended to the complex case if we assume that $f(z)$ and $\phi(z)$ are analytic in a circle C containing a root $\hat{z}$ and that $\hat{z}$ is the only zero of $f(z)$ in C. We also assume that $|\phi'(z)| < 1$ in C except possibly at $\hat{z}$. For we have

7.41
$$\phi(z) - \phi(\hat{z}) = \int_{\hat{z}}^{z} \phi'(t)dt,$$

the integral being taken over the straight line joining $\hat{z}$ to z. Thus $|\phi(z) - \phi(\hat{z})| < |z - \hat{z}|$. Thus, we can prove that if $z_0 \in C$ then all z_i lie in the circle $|z - \hat{z}| < |z_0 - \hat{z}|$. By Theorem A-2.13 (extended to sequence of complex numbers), there is a subsequence $\{z_{k_i}\}$ which converges to $\alpha \in C$. But by continuity this implies that $\alpha = \phi(\alpha)$ and, by the complete consistency α is a zero of $f(z)$. Thus $\alpha = \hat{z}$.

EXERCISES 4.7

1. Verify (7.2). Also show that for the method of false position ϕ_{n+1} has the form (7.2) where each $\alpha_{n,k}$ lies between zero and unity. Are the methods stationary or nonstationary? Is either method an s-step method for some s?

2. Consider the equation

$$f(x) = x^2 - \frac{1}{N} = 0$$

where $N > 0$. Indicate which of the following methods are consistent, which are reciprocally consistent, and which are completely consistent.

a) $x_{n+1} = \frac{1}{2}x_n(3 - x_n^2 N)$

b) $x_{n+1} = x_n$

c) $x_{n+1} = 1$

d) $x_{n+1} = \frac{1}{2}\left(x_n + \frac{1}{Nx_n}\right)$

e) $x_{n+1} = 1/\sqrt{N}$

f) $x_{n+1} = \left|\frac{1}{2}\left(x_n + \frac{1}{Nx_n}\right)\right|$

g) $x_{n+1} = x_n + 2.$

---

*The material between the asterisk on page 125 and that on this page has been superseded by the text in Appendix B.

Show that the first method will converge to $1/\sqrt{N}$ if x_0 is chosen in the interval

$$\left(2 - \sqrt{\frac{5}{3}}\right)\left(\frac{1}{\sqrt{N}}\right) < x_0 < \sqrt{\frac{5}{3}}\left(\frac{1}{\sqrt{N}}\right).$$

What range of starting values is acceptable if $N = 4$?

3. Show that the equation $xe^{-x} = e^{-3}$ has precisely two real roots.

4. Prove Theorem 7.22.

5. Prove Corollary 7.31.

6. Let S be a complete metric space with a distance function $d(x, y)$ and let S_1 be a closed subset of S. Let $\phi(x)$ be a function with domain S and range S which is a *contraction mapping* on S_1 in the sense that, for some $M < 1$, if x and y are in S_1 then $d(\phi(x), \phi(y)) \leq Md(x, y)$. Prove that if $S_1 = S$ then there exists a unique $\bar{x} \in S$ such that $\bar{x} = \phi(\bar{x})$ and, moreover, for any $x_0 \in S$ the sequence $x_1 = \phi(x_0)$, $x_2 = \phi(x_1), \ldots$ converges to $\bar{x}$. More generally, prove that, even if $S_1 \neq S$, if $\phi(x) \in S_1$ whenever $x \in S_1$ then there is a unique $\bar{x} \in S_1$ such that $\bar{x} = \phi(\bar{x})$ and, moreover, for any $x_0 \in S_1$ the sequence $x_1 = \phi(x_0)$, $x_2 = \phi(x_1), \ldots$ converges to $\bar{x}$.

4.8 GENERATION OF ITERATIVE METHODS

Given an equation $f(x) = 0$, one would, of course, like to choose an iteration function $\phi(x)$ so that the convergence of the corresponding iteration method would be as rapid as possible. In this section we shall consider a scheme for choosing $\phi(x)$ which is suggested by the second graphical method of Section 4.3. Thus we write $f(x)$ as the difference of two functions $g(x)$ and $h(x)$, i.e.,

8.1 $$f(x) = g(x) - h(x).$$

It is essential that we can easily solve either $g(x) = y$ or $h(x) = y$ for any given y. Thus, we can consider the iterative method defined by

8.2 $$g(x_{n+1}) = h(x_n)$$

and/or the method defined by

8.3 $$h(x_{n+1}) = g(x_n).$$

Symbolically, we can write (8.2) and (8.3) in the forms

8.4 $$x_{n+1} = g^{-1}[h(x_n)]$$

or

8.5 $$x_{n+1} = h^{-1}[g(x_n)],$$

respectively. These forms are useful in studying the convergence; however, in practice one would probably use the implicit forms (8.2) and (8.3).

As an example, consider the equation

8.6 $f(x) = \sin x - x + 2 = 0.$

Letting $g(x) = \sin x$ and $h(x) = x - 2$ we have

8.7 $f(x) = g(x) - h(x) = \sin x - (x - 2).$

In this case we can use either the iterative method

8.8 $x_{n+1} = \sin^{-1}[x_n - 2] = \phi_1(x_n)$
or
8.9 $x_{n+1} = 2 + \sin x_n = \phi_2(x_n).$

It is easy to verify that both methods are completely consistent. No extra roots are introduced because of the multiple-valued inverse sine function. To investigate the convergence we examine the derivatives. We have

8.10
$$\begin{cases} \phi_1'(x) = \dfrac{1}{\sqrt{1 - (x - 2)^2}} \\ \phi_2'(x) = \cos x. \end{cases}$$

In Section 4.3 we have found by graphical methods that there is a single root which is near $x = 2.6$. Evidently $|\phi_1'(x)| > 1$ near this value; hence we reject (8.8). On the other hand, $|\phi_2'(x)| < 1$ in an interval containing the root but not containing 0 or π. Thus if x_0 lies in any symmetric interval around the root, which does not contain 0 or π, the method will converge, by Theorem 7.15.

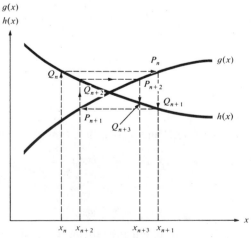

8.11 Fig. Graphical Interpretation of $x_{n+1} = g^{-1}[h(x_n)]$.

If we carry out the method (8.9) starting with $x_0 = 2.5$ we obtain the following results: $x_0 = 2.5, x_1 = 2.5985, x_2 = 2.5168, x_3 = 2.5849, x_4 = 2.5285, x_5 = 2.5754$, $x_6 = 2.5364$, etc.

There is an interesting graphical interpretation of these methods. For example, in Fig. 8.11 we have shown graphs of both $g(x)$ and $h(x)$ for a hypothetical case. Suppose we are considering the method defined by (8.2) and (8.4). Given x_n, one evaluates $h(x_n)$. This is like moving vertically from $(x_n, 0)$ up to the curve of $h(x)$ and to the point $Q_n: (x_n, h(x_n))$. Next, one solves the equation $g(x_{n+1}) = h(x_n)$ for x_{n+1}. This is like moving horizontally on the line $y = h(x_n)$ until one reaches the curve of $g(x)$ at the point $P_n:(g^{-1}(h(x_n)), h(x_n)) = (x_{n+1}, h(x_n))$. Next, we move to $Q_{n+1}:(x_{n+1}, h(x_{n+1}))$ and then to $P_{n+1}:(x_{n+2}, h(x_{n+1}))$, etc.

It is intuitively clear that the larger the slope of the g-curve relative to that of the h-curve, the greater the chance of convergence. For if the slope of the g-curve is large, then there will be a relatively small change in x in going from $(x_n, h(x_n))$ to $(x_{n+1}, h(x_n))$ whereas if $g'(x)$ is small, then $|x_{n+1} - x_n|$ may be quite large. Indeed, we have the following:

*8.12 Theorem Let $\bar{x}$ be a root of $g(x) - h(x) = 0$ and let $|g'(x)| \geqq \alpha$ and $|h'(x)| \leqq \beta$ in the interval $I = [\bar{x} - r, \bar{x} + r]$ for some $r > 0$, where $\alpha > 0$, $\beta \geqq 0$, and $\beta/\alpha < 1$. If x_0 is in I, then the method defined by $g(x_{n+1}) = h(x_n)$ converges to $\bar{x}$.

Proof. Since $g(\bar{x}) = h(\bar{x})$ and $g(x_{n+1}) - h(x_n)$ we have by the mean-value theorem*

8.13
$$g(x_{n+1}) - g(\bar{x}) = g'(c_1)(x_{n+1} - \bar{x})$$
$$= h(x_n) - h(\bar{x}) = h'(c_2)(x_n - \bar{x})$$

where c_1 lies between x_{n+1} and $\bar{x}$ and c_2 lies between x_n and $\bar{x}$. Therefore,

8.14
$$\left| \frac{x_{n+1} - \bar{x}}{x_n - \bar{x}} \right| = \left| \frac{h'(c_2)}{g'(c_1)} \right| \leqq \frac{\beta}{\alpha}.$$

In general we have

8.15
$$|x_n - \bar{x}| \leqq \left(\frac{\beta}{\alpha} \right)^n |x_0 - \bar{x}|$$

and, since $\beta/\alpha < 1$, convergence follows.

In the example where $g(x) = x - 2$ and $h(x) = \sin x$ we have $g'(x) = 1$ and $h'(x) = \cos x$. In any symmetric interval I containing the root but not containing 0 or π we have $|h'(x)| < 1$. Consequently if x_0 belongs to I the method converges.

---

*The material between asterisks has been superseded by the text in Appendix C.

If the derivative of $g(x)$ is larger in magnitude than the magnitude of the derivative of $h(x)$ near the root, then the method $x_{n+1} = g^{-1}(h(x_n))$ will converge, for

8.16
$$\phi'(x) = \frac{d}{dx}[g^{-1}(h(x))] = \frac{h'(x)}{g'[g^{-1}[h(x)]]}.$$

Evidently, near a root $\bar{x}$ of $f(x) = 0$ we have $g^{-1}[h(x)] \sim g^{-1}[h(\bar{x})] = \bar{x}$ and

8.17
$$\phi'(x) = \frac{d}{dx}[g^{-1}(h(x))] \sim \frac{h'(\bar{x})}{g'(\bar{x})}.$$

Hence if $|g'(\bar{x})| > |h'(\bar{x})|$ we have $|\phi'(x)| < 1$ and the method converges.

Let us now consider the case where $g(x) = x$, $h(x) = x + cf(x)$ where c is a constant. Evidently, by (8.2) we have

8.18
$$x_{n+1} = x_n + cf(x_n),$$

and

8.19
$$\phi(x) = x + cf(x).$$

Ideally, we would like to minimize $\max|\phi'(x)|$ in the interval $\bar{x} - |\bar{x} - x_n| \leq x \leq \bar{x} + |\bar{x} - x_n|$. However, since we do not know $\phi'(x)$ except for certain values of x, let us select a value, a, of x and require that $\phi'(a) = 0$. This gives

8.20
$$0 = \phi'(a) = 1 + cf'(a)$$

i.e.,

8.21
$$c = -\frac{1}{f'(a)}.$$

The iteration formula becomes

8.22
$$x_{n+1} = x_n - \frac{1}{f'(a)} f(x_n).$$

We now extend the method as follows. We seek a function $g(x)$ such that $\phi'(\bar{x}) = 0$, where

8.23
$$\phi(x) = x + g(x)f(x).$$

Differentiating, we have

8.24
$$\phi'(x) = 1 + g'(x)f(x) + g(x)f'(x).$$

If $\phi'(\bar{x}) = 0$ and $f'(\bar{x}) \neq 0$ we have

8.25
$$g(\bar{x}) = -\frac{1}{f'(\bar{x})}.$$

For our choice of $g(x)$ we can let $g(x) \equiv -1/f'(\bar{x})$. This, however, has the disadvantage that we do not know $\bar{x}$. A better choice is to let

8.26
$$g(x) = -\frac{1}{f'(x)}$$

and

8.27
$$\phi(x) = x - \frac{f(x)}{f'(x)}.$$

Thus we have the iterative method

8.28
$$x_{n+1} = x_n - \frac{f(x_n)}{f'(x_n)}.$$

This is the Newton method. Evidently the Newton method is a stationary one-step method. Moreover, if $f(\bar{x}) = 0$ and $f'(\bar{x}) \neq 0$, then if $x_0 = \bar{x}$, we have $x_n = \bar{x}$ for all n. We remark that (8.28) can be applied even if $f'(\bar{x}) = 0$, provided $f'(x_n) \neq 0$. As a matter of fact, we show in Section 4.11 that under fairly general conditions the method converges to a root $\bar{x}$ provided x_0 is sufficiently close to $\bar{x}$, regardless of the multiplicity of the root.

EXERCISES 4.8

1. Consider the equation $f(x) = \sin x - x^3 - 1 = g(x) - h(x)$ where $g(x) = \sin x$, $h(x) = x^3 + 1$. Which of the following methods will converge to the root near -1?
   a) $h(x_{n+1}) = g(x_n)$
   b) $g(x_{n+1}) = h(x_n)$.
   Carry out the convergent method with $x_0 = -1$. Draw a graph similar to Fig. 8.11.

2. Consider the equation $f(x) = x^3 - x - 1 = 0$. If we write $f(x) = g(x) - h(x)$ where $g(x) = x^3, h(x) = x + 1$, which method will converge to the positive root if x_0 is close enough to α?
   a) $h(x_{n+1}) = g(x_n)$
   b) $g(x_{n+1}) = h(x_n)$.
   Carry out the convergent iterative method with $x_0 = 1.5$.

3. Consider the equation of Exercise 3, Section 4.7. Show that the iterative method $x_{n+1} = 3 + \log x_n$ converges to the larger root, β, if $x_0 > \alpha$ and diverges if $x_0 < \alpha$ where α is the smaller root. Show that the method $x_{n+1} = e^{x_n - 3}$ converges to α if $x_0 < \beta$ and

diverges if $x_0 > \beta$. Find the larger root using the first method with $x_0 = 2$. Find the smaller root using the second method with $x_0 = 2$.

4.9 THE NEWTON METHOD

In the preceding section we derived (8.28) for the Newton method. We now give an alternative derivation based on the use of Taylor's theorem. If $\bar{x}$ is a solution of (1.1), then we have, by Taylor's theorem,

9.1
$$f(\bar{x}) = f(x_n) + (\bar{x} - x_n)f'(x_n) + \cdots$$
$$\doteq f(x_n) + (\bar{x} - x_n)f'(x_n).$$

Letting $f(\bar{x}) = 0$, and solving for $\bar{x}$ we obtain

9.2
$$\bar{x} \doteq x_n - \frac{f(x_n)}{f'(x_n)}.$$

We define the Newton method by

9.3
$$x_{n+1} = x_n - \frac{f(x_n)}{f'(x_n)}.$$

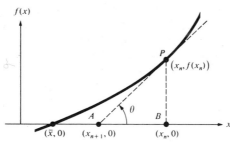

9.4 Fig. The Newton Method.

The Newton method has a simple geometric interpretation. Figure 9.4 shows the graph of a function $f(x)$. From the point $(x_n, f(x_n))$ one constructs a tangent to the curve and lets $(x_{n+1}, 0)$ be the point where this tangent crosses the x-axis. By geometry, we have

9.5
$$\tan \theta = f'(x_n) = \frac{BP}{AB} = \frac{f(x_n)}{x_n - x_{n+1}}.$$

Solving for x_{n+1} we again get (9.3).

As an example of the Newton method, consider the equation

9.6
$$f(x) = x^2 - N = 0$$

where $N \geq 0$. Since $f'(x) = 2x$, we have

9.7
$$x_{n+1} = x_n - \frac{x_n^2 - N}{2x_n} = \frac{1}{2}\left(x_n + \frac{N}{x_n}\right).$$

This is the formula which we have been using for the square root of N. We have already proved that the method converges if $x_0 > 0$.

Strictly speaking, the function $\phi(x)$ is not defined if $\bar{x}$ is a zero of $f'(x)$. However, if $\bar{x}$ is also a zero of $f(x)$, we can extend the definition of $\phi(x)$ to obtain a function $\hat{\phi}(x)$ as follows:

9.8
$$\hat{\phi}(x) = \begin{cases} \phi(x) & \text{if } x \neq \bar{x} \\ \bar{x} & \text{if } x = \bar{x}. \end{cases}$$

The continuity of $\hat{\phi}(x)$ near a k-fold root $\bar{x}$ can be seen as follows. Since $f(\bar{x}) = f'(\bar{x}) = \cdots = f^{(k-1)}(\bar{x}) = 0, f^{(k)}(\bar{x}) \neq 0$ we have by L'Hospital's rule

9.9
$$\lim_{x \to \bar{x}} \frac{f(x)}{f'(x)} = \lim_{x \to \bar{x}} \frac{f'(x)}{f''(x)} = \cdots = \lim_{x \to \bar{x}} \frac{f^{(k-1)}(x)}{f^{(k)}(x)} = 0;$$

hence $\hat{\phi}(\bar{x}) = \lim_{x \to \bar{x}} \phi(x)$.

Evidently then, the Newton method is consistent if we define it as follows:

9.10
$$x_{n+1} = \begin{cases} x_n - \dfrac{f(x_n)}{f'(x_n)}, & \text{if } f'(x_n) \neq 0 \\ x_n, & \text{if } f(x_n) = f'(x_n) = 0. \end{cases}$$

We remark that x_{n+1} is still not defined if $f'(x_n) = 0$ and $f(x_n) \neq 0$.

To examine the convergence in general we assume that $f(x)$ is continuously twice differentiable near a root $\bar{x}$. Differentiating the iteration function

9.11
$$\phi(x) = x - \frac{f(x)}{f'(x)}$$

we get

9.12
$$\phi'(x) = 1 - \frac{[f'(x)]^2 - f(x)f''(x)}{[f'(x)]^2} = \frac{f(x)f''(x)}{[f'(x)]^2}.$$

If $\bar{x}$ is a simple root of $f(x) = 0$, then $f'(\bar{x}) \neq 0$. Moreover, by the continuity of $f'(x)$, $|f'(x)| \geq \varepsilon$, for some $\varepsilon > 0$, in a suitable neighborhood of $\bar{x}$. Within this neighborhood we select a subneighborhood so that $|f(x)f''(x)| < \varepsilon^2$, which is possible since $f(\bar{x}) = 0$ and since $f(x)$ and $f''(x)$ are continuous. Therefore, in this subneighborhood we have

9.13
$$|\phi'(x)| < 1.$$

Therefore by Theorem 7.15 the method converges, provided we start close enough to $\bar{x}$.

Actually, as will be shown in Section 4.11, the Newton method converges to a multiple root $\bar{x}$ provided one starts sufficiently close to $\bar{x}$. However, in this case the convergence is somewhat slower. Also, as will be shown in Section 4.11 if $\bar{x}$ is a root of multiplicity p, then faster convergence can be achieved by using

9.14
$$x_{n+1} = x_n - p\frac{f(x_n)}{f'(x_n)}.$$

The Newton method can also be defined for the equation

9.15
$$f(z) = g(z) + ih(z) = 0$$

where $f(z)$ is an analytic function of the complex variable $z = x + iy$ and where $g(z)$ and $h(z)$ are real for all z. Thus, for example, if

9.16
$$f(z) = z^2 + 1$$

then

9.17
$$g(z) = x^2 - y^2 + 1, \qquad h(z) = 2xy.$$

The derivative $f'(z)$ is given by

9.18
$$f'(z) = g_x + ih_x = h_y - ig_y$$

and by the Cauchy-Riemann equations (see Theorem A-5.20) we have

9.19
$$g_x = h_y, \qquad h_x = -g_y.$$

The Newton method is defined by

9.20
$$z_{n+1} = z_n - \frac{f(z_n)}{f'(z_n)}.$$

Evidently this reduces to the ordinary Newton method if z_n is real and if $h(z) = 0$ for real z. From (9.18) and (9.19) we have

9.21
$$z_{n+1} = z_n - \frac{g(z_n) + ih(z_n)}{g_x(z_n) + ih_x(z_n)}$$

$$= z_n - \frac{(gg_x + hh_x) + i(hg_x - gh_x)}{g_x^2 + h_x^2}$$

i.e.,

9.22
$$\begin{cases} x_{n+1} = x_n - \dfrac{gg_x + hh_x}{g_x^2 + g_y^2} \\[3mm] y_{n+1} = y_n - \dfrac{hg_x - gh_x}{g_x^2 + g_y^2}. \end{cases}$$

Whether (9.22) or (9.20) is more convenient depends on the given equation.

As in the real case, to assure convergence it is necessary that the starting value z_0 be chosen sufficiently close to the desired root z^*. Moreover, if $f(z)$ is real for real z, as is the case for the example $f(z) = 1 + z^2$, in order to hope to get complex roots we must let z_0 be complex. Otherwise all iterants will be real. Thus in the example $f(z) = 1 + z^2$ the Newton method is given by

9.23
$$z_{n+1} = z_n - \frac{1 + z_n^2}{2z_n} = \frac{z_n^2 - 1}{2z_n}.$$

If z_0 is real, then $z_1, z_2, z_3, \ldots$ will all be real and one will never get either of the roots $\bar{z}^{(1)} = i, \bar{z}^{(2)} = -i$. On the other hand, if we let $z_0 = i/2$, then we get

9.24
$$z_1 = \frac{5}{4}i, \qquad z_2 = \frac{41}{40}i, \qquad z_3 = \frac{3281}{3280}i.$$

Obviously, the sequence is converging to $\bar{z}^{(1)} = i$.

Another drawback of the Newton method is that even though x_0 is closer to a root, say r_1, than to any other root, the method may not converge to r_1. In fact, it may not even converge at all. For example, consider the case

9.25
$$f(x) = (x - 1)(x + 1)^5.$$

Evidently $f'(\frac{2}{3}) = 0$; hence, if we let $x_0 = \frac{2}{3}$, the method will not converge. For x_0 in the range $-1 \leq x_0 < \frac{2}{3}$, the method will converge to -1 even though x_0 may be closer to 1 than to -1.

EXERCISES 4.9

1. Derive an iterative method based on the Newton method for finding $N^{1/5}$. Apply the method to obtain $2^{1/5}$ correct to three decimals.

2. Apply the Newton method to solve the following equations:

equation	starting value
a) $\sin x = x^3 + 1$	$x_0 = -1$
b) $x = \log x + 3$	$x_0 = 2$
c) $x^3 - x - 1 = 0$	$x_0 = 1.5$
d) $\sin x = x - 2$	$x_0 = 2.5$

3. For the equation $f(x) = \sin x - x + 2 = 0$, find an interval about the positive root so that $|\phi'(x)| \leq \frac{1}{2}$ where $\phi(x) = x - f(x)/f'(x)$. Does the method $x_{n+1} = \phi(x_n)$ converge for the starting value 2.4? 2.2? 2.0?

4. Use the Newton method to solve $f(x) = x^5 - 6x^4 + 15x^3 - 20x^2 + 14x - 4 = 0$ with $x_0 = 0.8$. Also apply the modified method (9.14) with $p = 2$.

5. Apply the Newton method to get a solution of $f(z) = \sin z - z + 2 = 0$ using $z_0 = -1 + i$.   (Note that $\sin(x + iy) = \sin x \cosh y + i \cos x \sinh y$ and $\cos(x + iy) = \cos x \cosh y - i \sin x \sinh y$.)

6. Carry out the Newton method for the equation $f(x) = (x - 1)(x + 1)^5 = 0$ with $x_0 = \frac{1}{2}$. Comment on the rapidity of convergence.

7. In the preceding exercise use (9.14) with $p = 5$.

8. Carry out three iterations of the Newton method for solving $1 + z^2 = 0$ with the starting value $z_0 = (1 + i)/2$. Perform the same three iterations using (9.22).

4.10 MULLER'S METHOD

Muller's method can be considered as a direct extension of the secant method. Given three distinct values a, b, and c, one constructs a second-degree polynomial $P(x)$ such that

10.1 $$P(a) = f(a), \qquad P(b) = f(b), \qquad P(c) = f(c).$$

(See Fig. 10.2.) One then finds the zeros of $P(x)$ and chooses one of these roots, d, to be the new approximation to the root. The process is then repeated using b, c, and d to find e, and so on.

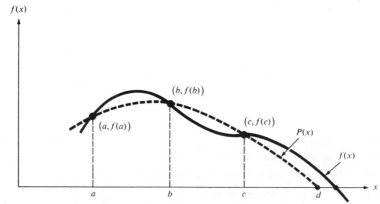

10.2 Fig. Muller's Method.

To construct $P(x)$ we seek coefficients A, B, and C so that

10.3 $$P(x) = Ax^2 + Bx + C$$

and so that the conditions (10.1) are satisfied. It is convenient to work with

10.4 $$P(x) = \hat{A}(x - c)^2 + \hat{B}(x - c) + \hat{C}$$

where

10.5 $$\hat{A} = A, \qquad \hat{B} = B + 2cA, \qquad \hat{C} = C + cB + c^2A.$$

Formulas (10.4) and (10.5) can be obtained by replacing x by $(c + (x - c))$ in (10.3). Actually, we shall not need A, B, and C in any of our calculations and hence we shall concentrate on determining $\hat{A}$, $\hat{B}$, and $\hat{C}$. We remark, however, that A, B, and C could be determined from $\hat{A}$, $\hat{B}$, $\hat{C}$ by

10.6 $$A = \hat{A}, \qquad B = \hat{B} - 2c\hat{A}, \qquad C = \hat{C} - c\hat{B} + c^2\hat{A}.$$

From (10.1) and (10.4) we have

10.7 $$\begin{cases} f(a) = \hat{A}\hat{a}^2 + \hat{B}\hat{a} + \hat{C} \\ f(b) = \hat{A}\hat{b}^2 + \hat{B}\hat{b} + \hat{C} \\ f(c) = \qquad\qquad\quad \hat{C} \end{cases}$$

where

10.8 $$\hat{a} = a - c, \qquad \hat{b} = b - c.$$

Solving (10.7) for $\hat{A}$, $\hat{B}$, $\hat{C}$ we have

10.9 $$\hat{A} = \frac{\hat{b}\Delta(a) - \hat{a}\Delta(b)}{\hat{a}\hat{b}(\hat{a} - \hat{b})}, \qquad \hat{B} = \frac{\hat{a}^2\Delta(b) - \hat{b}^2\Delta(a)}{\hat{a}\hat{b}(\hat{a} - \hat{b})}, \qquad \hat{C} = f(c)$$

where, for convenience, we let

10.10 $$\Delta(a) = f(a) - f(c), \qquad \Delta(b) = f(b) - f(c).$$

Having determined $\hat{A}$, $\hat{B}$, and $\hat{C}$ we now determine d so that $P(d) = 0$. This leads to the quadratic equation

10.11 $$\hat{A}\hat{d}^2 + \hat{B}\hat{d} + \hat{C} = 0$$

where

10.12 $$\hat{d} = d - c.$$

In general, we seek the root of (10.11) of smallest modulus. For solving this quadratic equation with complex coefficients, a procedure similar to that given in Section 3.4 for handling a quadratic with real coefficients should be used. In the normal case where $\hat{A}$, $\hat{B}$, $\hat{C}$ are in $\mathscr{M}^*$ and where the ratios $\hat{B}/(2\hat{A})$ and $\hat{B}/(2\hat{C})$ are in $\mathscr{M}^*$ we would use the formula

10.13
$$d - c = -\frac{2\hat{C}}{\hat{B}}\left(1 + \sqrt{1 - \frac{4\hat{A}\hat{C}}{\hat{B}^2}}\right)^{-1},$$

if $|\hat{B}^2/4\hat{A}\hat{C}| \geqq 1$ and,

10.14
$$d - c = \sqrt{\frac{\hat{C}}{\hat{A}}}\left[-\Delta + K\sqrt{\Delta^2 - 1}\right]^{-1}$$

if $|\hat{B}^2/4\hat{A}\hat{C}| < 1$, where $K = \pm 1$ is chosen so that $d - c$ has the smaller modulus and where

10.15
$$\Delta = \frac{\hat{B}/(2\hat{A})}{\sqrt{\hat{C}/\hat{A}}}.$$

Evidently, if a, b, and c are distinct, we can compute $\hat{A}$, $\hat{B}$, and $\hat{C}$ since $\hat{a} = a - c \neq 0$, $\hat{b} = b - c \neq 0$, and $\hat{a} - \hat{b} = a - b \neq 0$. We now show that unless $f(a) = f(b) = f(c)$, the quadratic (10.11) has at least one root. Clearly, the quadratic will have at least one root if $\hat{A} \neq 0$ or if $\hat{B} \neq 0$. If both $\hat{A}$ and $\hat{B}$ vanish we have, by (10.9),

10.16
$$\begin{cases} \hat{b}\Delta(a) - \hat{a}\Delta(b) = 0 \\ \hat{a}^2\Delta(b) - \hat{b}^2\Delta(a) = 0. \end{cases}$$

Multiplying the first equation by $\hat{b}$ and adding to the second, we get

10.17
$$\hat{a}(\hat{a} - \hat{b})\Delta(b) = 0$$

i.e., since $\hat{a} \neq 0$, $\hat{b} \neq 0$, and $\hat{a} - \hat{b} \neq 0$, we have $\Delta(b) = 0$. Similarly $\Delta(a) = 0$. Thus if $\hat{A} = \hat{B} = 0$ we have

10.18
$$f(a) = f(b) = f(c).$$

If this happens, then the quadratic through the points $(a, f(a)), (b, f(b)), (c, f(c))$ is a straight line parallel to the x-axis. Evidently there are no roots unless $f(a) = f(b) = f(c) = 0$.

If $\hat{A} = 0$, we have $\hat{b}\Delta(a) - \hat{a}\Delta(b) = 0$, or

10.19
$$\frac{f(b) - f(c)}{b - c} = \frac{f(a) - f(c)}{a - c}.$$

Therefore, all three of the points $(a, f(a)), (b, f(b)), (c, f(c))$ lie in a straight line. This presents no difficulty unless $f(a) = f(b) = f(c)$; unless this happens we have

10.20
$$\hat{d} = -\frac{\hat{C}}{\hat{B}} = -\frac{f(c)\hat{a}\hat{b}(\hat{a} - \hat{b})}{\hat{a}^2\Delta(b) - \hat{b}^2\Delta(a)}$$

$$= -\frac{f(c)\hat{a}\hat{b}(\hat{a} - \hat{b})}{\hat{a}(\hat{a} - \hat{b})\Delta(b)} = \frac{-f(c)\hat{b}}{\Delta(b)}$$

and

10.21
$$d - c = \frac{-f(c)(b - c)}{f(b) - f(c)}.$$

Hence,

10.22
$$d = c - \frac{f(c)(b - c)}{f(b) - f(c)}$$

which agrees with formula (6.2) for the secant method if we let $x_{n-1} = b$, $x_n = c$, and $x_{n+1} = d$. Thus, if the three points $(a, f(a))$, $(b, f(b))$, $(c, f(c))$ lie in a straight line, or equivalently, if $\hat{A} = 0$, the method reduces to the secant method. An illustration of this case is given in Fig. 10.23.

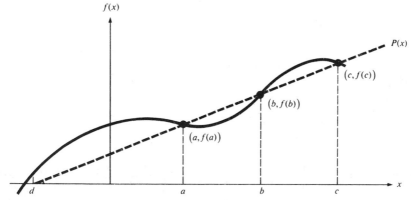

10.23 Fig. Muller's Method—Case $\hat{A} = 0$.

It is obvious that $d \neq c$ since in this case we would have $\hat{d} = 0$ and, by (10.11) we would have $\hat{C} = f(c) = 0$. Thus, unless $f(c) = 0$, in which case c is a root, it is not possible for d to equal c. (Of course, the computer representation of d may turn out to be indistinguishable from that of c, but this is another matter.) We now show that $d \neq a$ if $f(a) \neq 0$. If $d = a$, then by (10.11), $\hat{A}\hat{a}^2 + \hat{B}\hat{a} + \hat{C} = 0$. But by (10.7) it follows that $\hat{A}\hat{a}^2 + \hat{B}\hat{a} + \hat{C} = f(a) \neq 0$, and we have a contradiction. Similarly, if $f(b) \neq 0$, then $d \neq b$.

As an example, let us consider the equation

10.24 $$f(z) = z^3 - 2.$$

Let $a = 0$, $b = 1$, $c = 2$. Evidently $f(a) = -2$, $f(b) = -1$, $f(c) = 6$. Therefore, $\Delta(a) = -8$, $\Delta(b) = -7$, $\hat{a} = -2$, $\hat{b} = -1$. Substituting in (10.9) we get

10.25
$$\begin{cases} \hat{A} = \dfrac{(-1)(-8) - (-2)(-7)}{(-2)(-1)(-1)} = 3 \\[3mm] \hat{B} = \dfrac{4(-7) - (1)(-8)}{(-2)(-1)(-1)} = 10 \\[3mm] \hat{C} = 6. \end{cases}$$

Since $|\hat{B}^2/(4\hat{A}\hat{C})| = |100/72| > 1$, we use (10.13) and obtain

10.26 $$d - 2 = -\frac{12}{10}\left[1 + \sqrt{1 - \frac{72}{100}}\right]^{-1} = -\frac{6}{5}\left[1 + \frac{\sqrt{7}}{5}\right]^{-1}$$

$$= -\frac{6}{5 + \sqrt{7}} \doteq -0.785.$$

Hence we have

10.27 $$d \doteq 2 + (-0.785) = 1.215.$$

Since the roots of the given equation are

10.28
$$\begin{cases} r_1 = \sqrt[3]{2} \doteq 1.260 \\[3mm] r_2 = \sqrt[3]{2}\left(\cos\dfrac{2\pi}{3} + i\sin\dfrac{2\pi}{3}\right) = \sqrt[3]{2}\left(-\dfrac{1}{2} + \dfrac{\sqrt{3}}{2}i\right) \doteq -0.630 + 1.091i \\[3mm] r_3 = \sqrt[3]{2}\left(\cos\dfrac{4\pi}{3} + i\sin\dfrac{4\pi}{3}\right) = \sqrt[3]{2}\left(-\dfrac{1}{2} - \dfrac{\sqrt{3}}{2}i\right) \doteq -0.630 - 1.091i \end{cases}$$

we see that the value d is very close to the real root. One could reasonably expect that further iterations would lead to convergence to this root.

If we had chosen $a = -2$, $b = -1$, $c = 0$, we might expect to approach one of the complex roots. Indeed, we have $\hat{a} = -2$, $\hat{b} = -1$, $f(a) = -10$, $f(b) = -3$, $f(c) = -2$, $\Delta(a) = -8$, $\Delta(b) = -1$, $\hat{A} = -3$, $\hat{B} = -2$, $\hat{C} = -2$. Since $|\hat{B}^2/(4\hat{A}\hat{C})| = \frac{1}{6} < 1$, we use (10.14). Thus we obtain

10.29

$$d = \sqrt{\frac{2}{3}}\left[\frac{-\frac{1}{3}}{\sqrt{\frac{2}{3}}} \pm \sqrt{\frac{1}{6} - 1}\right]^{-1}$$

$$= \sqrt{\frac{2}{3}}\left[\frac{-1}{\sqrt{6}} \pm \frac{\sqrt{5}}{\sqrt{6}}i\right]^{-1} = -\frac{1}{3} \pm \frac{\sqrt{5}}{3}i$$

$$\doteq -0.333 \pm 0.745i.$$

By choosing the $+$ sign, since both values of d have the same modulus, we obtain a value close to the complex root r_2. Again, one would expect that further iterations would lead to convergence to r_2.

Figure 10.30 indicates the roots of the equation and the first four iterants with each set of starting values.

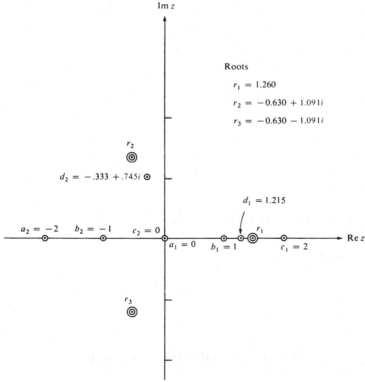

10.30 Fig. Application of Muller's Method to $z^3 - 2 = 0$.

Muller's method is a stationary three-step iterative method which requires three starting values. In the notation of Section 4.7 we have

10.31
$$x_{n+1} = \phi(x_{n-2}, x_{n-1}, x_n)$$

where

10.32
$$\phi(a, b, c) = c + (d - c)$$

and where $d - c$ is given by (10.13) or (10.14) as appropriate.

Muller's method has several advantages over the Newton method and the method of false position. First, it can be carried out using only the values of the function itself; derivatives are not required. This is especially important if the function $f(x)$ is not given by an explicit formula, but rather is found by a prescribed algorithm, e.g., the evaluation of a determinant of a matrix whose elements are functions of x. Another advantage of Muller's method is that even if the function $f(x)$ is real-valued for real x and even if the starting values are real, one may still be able to find complex roots. Such is not the case for the method of false position or the Newton method. Of course the latter methods can often be adapted to get around this difficulty.

One disadvantage of Muller's method as compared with the Newton method lies in the fact that in certain cases, no matter how close one is to a root $\bar{x}$, the case $f(a) = f(b) = f(c)$ may arise if $\bar{x}$ is a root of multiplicity greater than two, and hence convergence may fail. (See the discussion in Champagne [1964].)

One possibility of avoiding this difficulty which may in some cases be used if one is near a triple root is to use a method suggested by Cauchy, which we shall refer to as *Cauchy's method*. (See Traub [1964].) Cauchy's method is a direct extension of the Newton method where we consider $f(x)$, $f'(x)$, and $f''(x)$. From Taylor's series we have

10.33
$$f(\bar{x}) \doteq f(x_n) + (\bar{x} - x_n)f'(x_n) + \frac{(\bar{x} - x_n)^2}{2} f''(x_n).$$

We solve the quadratic equation

10.34
$$\hat{A}(x_{n+1} - x_n)^2 + \hat{B}(x_{n+1} - x_n) + \hat{C} = 0$$

where

10.35
$$\hat{A} = \tfrac{1}{2}f''(x_n), \qquad \hat{B} = f'(x_n), \qquad \hat{C} = f(x_n)$$

for x_{n+1}. If we let $x_{n+1} = d$ and $x_n = c$, we can use the procedures described above for solving (10.11). While Cauchy's method has the disadvantage of requiring two derivatives, nevertheless, this difficulty is less serious in the case of polynomial equations (see Section 5.4).

Having found a root α_1 of (1.1) by Muller's method, we may attempt to find another root by considering the function

10.36 $$f_1(x) = \frac{f(x)}{x - \alpha_1}.$$

In the case of a polynomial equation, $f_1(x)$ may be computed explicitly, as we shall see in Chapter 5. Otherwise, for each x one computes $f(x)$ and then divides by $x - \alpha_1$. In general, having found roots $\alpha_1, \alpha_2, \ldots, \alpha_k$ one works with the function

10.37 $$f_k(x) = \frac{f(x)}{(x - \alpha_1)(x - \alpha_2) \ldots (x - \alpha_k)}.$$

Let us now describe a simple program based on Muller's method to find a single root of (1.1). It is assumed that we have available a program which we shall refer to as Subroutine CQUAD which is analogous to the program for solving a quadratic equation with real coefficients described in Section 3.4 and which can handle complex coefficients. As input, one specifies three distinct values x_0, x_1, x_2 and a subroutine for computing $f(x)$ for any given value of x, real or complex. In addition, one specifies a root acceptance procedure; for example, a root is accepted if $|f(x)| < \varepsilon$ for some specified $\varepsilon > 0$. As shown in the flow diagram of Fig. 10.38, one first computes $f(x_0), f(x_1)$, and $f(x_2)$ to see if any of the values x_0, x_1, or x_2 is a root. If not, one begins the iteration process. Letting $x_0 = a, x_1 = b, x_2 = c$ one then determines $\hat{A}, \hat{B}$, and $\hat{C}$ from (10.9). While one would not normally use the surveillance routines described in Chapter 3 to compute f, it is recommended that they be used to determine $\hat{A}, \hat{B}$, and $\hat{C}$. If any of the three coefficients is in $\mathscr{L}$, then the process has failed. Otherwise CQUAD is used to solve the complex quadratic (10.11). We accept the root x_S which has smallest modulus if $x_S \in \mathscr{M}$. Then $x_3 = c + x_S$. If $x_S \in \mathscr{S}$, then we cannot iterate further since $x_3 = c$. Hence we terminate the process, printing $c + x_S$, our final iterant and $|f(c + x_S)|$. If $x_S \in \mathscr{L}$, then the process is considered to have failed. The entire procedure is continued until it fails or converges or until the number of iterations exceeds a certain specified number.

The program just described will find only one root, at most. A more complete program would provide for the use of Cauchy's method and/or a modification of the iterants in case Muller's method fails. There would also be the facility to find more than one root. It would be desirable to incorporate some method for selecting the starting values not only for the first root but for subsequent roots as well.

EXERCISES 4.10

1. Find $\hat{A}, \hat{B}$, and $\hat{C}$ so that

$$P(x) = 2x^2 + 3x - 1 = \hat{A}(x - 1)^2 + \hat{B}(x - 1) + \hat{C}.$$

Also find A, B, and C so that

$$P(x) = 7(x - 1)^2 - 2(x - 1) + 3 = Ax^2 + Bx + C.$$

2. Find d such that $P(d) = 0$ where $P(x) = Ax^2 + Bx + C$ and

$$P(1) = 6$$
$$P(2) = 3$$
$$P(3) = 1.$$

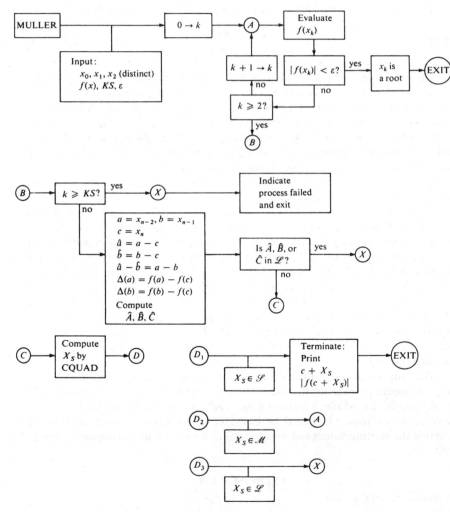

10.38 Fig. Flow Diagram for Muller's Method.

3. Carry out one iteration using Muller's method for the equation $f(z) = z^3 - z - 1 = 0$ with $z_0 = 0, z_1 = 1, z_2 = 2$. Also use the starting values $z_0 = 0, z_1 = -1, z_2 = -2$. What happens if we use the starting values $z_0 = -1, z_1 = 0, z_2 = 1$?

4. With the starting values $x_0 = \frac{1}{4}, x_1 = \frac{1}{2}, x_2 = 1$, perform 2 iterations with Muller's method for the equation $\log x - x + 3 = 0$.

5. Obtain two roots of the equation $f(z) = \sin z - z + 2 = 0$ using Muller's method as follows. Apply Muller's method with $z_0 = -1, z_1 = 0, z_2 = 1$, obtaining a root α. Then with the same starting values, apply the method to the function

$$f_1(z) = \frac{f(z)}{z - \alpha}.$$

Use built-in complex arithmetic subroutines, if available, otherwise write all formulas in terms of real operations.

6. Carry out two iterations with Cauchy's method and $z_0 = 1$ for the equation of Exercise 3.

7. Solve $z^3 - 2 = 0$ by Cauchy's method using $z_0 = 2$.

8. Develop a flow diagram to combine Cauchy's method and Muller's method. Assume that subroutines are available to evaluate $f'(x)$ and $f''(x)$ for any x.

4.11 ORDERS OF CONVERGENCE OF ITERATIVE METHODS

In this section we consider the convergence and rates of convergence of three of the methods which have been considered above, namely, the Newton method, the secant method, and Muller's method. We derive sufficient conditions for convergence under certain assumptions on the behavior of the function $f(x)$ in the neighborhood of a root. We also study the rate at which the method converges if each starting value is chosen sufficiently close to a root.

Order of Convergence

Let us first discuss the order of convergence of a sequence $x_0, x_1, x_2, \ldots$, to a limit $\bar{x}$. If $0 < C < 1$, and

11.1
$$\lim_{n \to \infty} \frac{|e_{n+1}|}{|e_n|} = C$$

where

11.2
$$e_n = x_n - \bar{x},$$

then the convergence is of *first order*. If, moreover, we have

11.3
$$\lim_{n \to \infty} \frac{e_{n+1}}{e_n} = C$$

where $0 < |C| < 1$, then the convergence is *linear*. If

11.4 $e_{n+1} = Ce_n, \qquad n = 0, 1, 2, \ldots$

where $0 < |C| < 1$, then the convergence is *geometric*.

If there exists $p > 1$ such that for some $K > 0$ we have

11.5 $$\lim_{n \to \infty} \frac{|e_{n+1}|}{|e_n|^p} = K$$

then we say that the *order of convergence* of the sequence is p. Evidently, if the order of convergence is p, then we have

11.6 $$|e_n| \leqq \frac{1}{L^{1/(p-1)}} \left[L^{1/(p-1)} |e_0| \right]^{p^n}$$

where

11.7 $$L = \max\left(\frac{|e_{n+1}|}{|e_n|^p} \right), \qquad n = 0, 1, 2, \ldots .$$

Thus, if $|e_0|$ is sufficiently small so that

11.8 $$L^{1/(p-1)} |e_0| < 1$$

then the convergence will be very rapid indeed.

The order of convergence of a sequence generated by an iterative method clearly may depend on the function $f(x)$, the multiplicity of the root $\bar{x}$ under consideration, and on the starting value x_0. In our subsequent discussion we shall assume that x_0 is chosen close to a simple root $\bar{x}$. It will be shown that the order of convergence of any such sequence generated by the Newton method is two. For the secant method and for Muller's method it can be shown (see Traub [1964]) that the orders of convergence are approximately 1.618 and 1.839, respectively. Rather than prove this, we shall be content to show that for the secant method, if the starting values are sufficiently close to a simple root, then for some nonnegative constants C_1 and C_2, with $C_2 < 1$, we have

11.9 $$|e_n| \leqq C_1 C_2^{p^n}$$

for $p \doteq 1.618$. Such a relation clearly holds, by (11.6), if the order of convergence is p. We give a nonrigorous argument for Muller's method.

The Newton Method

Let us first consider the case where $\bar{x}$ is a simple root of (1.1) and where $f'(x)$ and $f''(x)$ exist and are continuous in a suitable interval around $\bar{x}$. By (9.3) it follows that

11.10
$$e_{n+1} = e_n - \frac{f(x_n)}{f'(x_n)}$$

where

$$e_n = x_n - \bar{x}.$$

Moreover,

11.11
$$e_{n+1} = \frac{e_n f'(x_n) - f(x_n)}{f'(x_n)}.$$

By Taylor's theorem we have

11.12
$$f(\bar{x}) = f(x_n) + (\bar{x} - x_n)f'(x_n) + \frac{(\bar{x} - x_n)^2}{2} f''(c)$$

where c lies between $\bar{x}$ and x_n. Hence since $f(\bar{x}) = 0$ it follows that

11.13
$$f(x_n) - e_n f'(x_n) = -\frac{e_n^2}{2} f''(c)$$

and

11.14
$$e_{n+1} = \frac{f''(c)}{2f'(x_n)} e_n^2.$$

Suppose now that $|f'(x)| \geqq m_1 > 0$ and $|f''(x)| \leqq M_2$ in $I : \bar{x} - |x_0 - \bar{x}| \leqq x \leqq \bar{x} + |x_0 - \bar{x}|$ and

11.15
$$\frac{M_2}{2m_1} |e_0| < 1.$$

Then $x_n \in I$ for all n and

11.16
$$|e_n| \leqq \frac{2m_1}{M_2} \left(\frac{M_2}{2m_1} |e_0| \right)^{2^n}$$

Moreover, the quantity

$$\frac{M_2}{2m_1} |e_0|$$

is *squared* on each iteration. The method is said to be an "error squaring" method, or a second-order method. Actually, we have from (11.14)

11.17
$$\frac{|e_{n+1}|}{|e_n|^2} \leqq \frac{M_2}{2m_1}.$$

Hence the order of convergence is at least two. That the convergence is precisely two follows from (11.14) and the continuity of $f'(x)$ and $f''(x)$.

A more precise assessment of the behavior of e_n can be given for the case where

11.18 $f(x) = x^2 - N, \qquad N > 0.$

We have $\bar{x} = \sqrt{N}, f'(x) = 2x, f'(\bar{x}) = 2\sqrt{N} \neq 0$. Suppose that x_0 is chosen in the interval $I = [\sqrt{N}/2, 3\sqrt{N}/2]$. Then $m_1 \geqq \sqrt{N}$ and $M_2 = 2$ in I. Hence,

11.19 $\dfrac{M_2}{2m_1}|e_0| \leqq \dfrac{2}{2\sqrt{N}}\dfrac{\sqrt{N}}{2} = \dfrac{1}{2}.$

Therefore, by (11.16),

11.20 $|e_n| \leqq \dfrac{2\sqrt{N}}{2}\left(\dfrac{1}{2}\right)^{2^n} = \sqrt{N}\left(\dfrac{1}{2}\right)^{2^n}$

and

11.21 $\dfrac{|e_n|}{\sqrt{N}} \leqq \left(\dfrac{1}{2}\right)^{2^n}.$

Thus the relative error, $|e_n|/\sqrt{N}$, decreases very rapidly as n increases.

Near a multiple root the Newton method converges, but the rate of convergence is much slower than in the case of a simple root. Let us assume that $\bar{x}$ is a double root of (1.1), i.e., that $f(\bar{x}) = f'(\bar{x}) = 0$ but $f''(\bar{x}) \neq 0$. By (11.14) we have

11.22 $e_{n+1} = \dfrac{f''(c)}{2f'(x_n)}e_n^2 = -\dfrac{f''(c)e_n^2}{2e_n f''(\xi)} = -\dfrac{1}{2}e_n\dfrac{f''(c)}{f''(\xi)}$

where ξ lies between $\bar{x}$ and x_n. If $f''(x)$ is continuous near $\bar{x}$ we have

11.23 $|e_{n+1}| \sim \tfrac{1}{2}|e_n|$

provided x_n is close enough to $\bar{x}$.

As an example, let $f(x) = x^2$. Evidently $\bar{x} = 0$ and $f'(\bar{x}) = 0$, but $f''(\bar{x}) = 2$. In this case $x_1 = \tfrac{1}{2}x_0, x_2 = \tfrac{1}{2}x_1, \ldots$, hence $x_n = (\tfrac{1}{2})^n x_0$. Clearly $|e_{n+1}| = \tfrac{1}{2}|e_n|$.

Now let us assume that $\bar{x}$ is a p-fold root of (1.1); i.e., $f(\bar{x}) = f'(\bar{x}) = \cdots = f^{(p-1)}(\bar{x}) = 0$, but that $f^{(p)}(\bar{x}) \neq 0$. We also assume that f and all derivatives up to and including the pth are continuous in a suitable neighborhood of $\bar{x}$. By Taylor's theorem using the integral form of the remainder (see Theorem A-3.66) we have

11.24 $$f(x) = f(\bar{x}) + (x - \bar{x})f'(\bar{x}) + \cdots + \frac{(x - \bar{x})^{p-1}}{(p-1)!} f^{(p-1)}(\bar{x})$$
$$+ \int_{\bar{x}}^{x} \frac{(x - t)^{p-1}}{(p-1)!} f^{(p)}(t)\, dt$$

and

11.25 $$f'(x) = f'(\bar{x}) + (x - \bar{x})f''(\bar{x}) + \cdots + \frac{(x - \bar{x})^{p-2}}{(p-2)!} f^{(p-1)}(\bar{x})$$
$$+ \int_{\bar{x}}^{x} \frac{(x - t)^{p-2}}{(p-2)!} f^{(p)}(t)\, dt.$$

Consequently, we have

11.26 $(x - \bar{x})f'(x) - f(x) = \int_{\bar{x}}^{x} \frac{f^{(p)}(t)}{(p-1)!}(x - t)^{p-2}[(p-1)(x - \bar{x}) - (x - t)]\, dt.$

But since $(x - t)^{p-2}((p-1)(x - \bar{x}) - (x - t)) = (x - t)^{p-2}[(p-2)x + t - (p-1)\bar{x}]$ is one-signed for t between x and $\bar{x}$ we have, by the mean-value theorem for integrals (Theorem A-3.36)

11.27 $(x - \bar{x})f'(x) - f(x) = \dfrac{f^{(p)}(c)}{(p-1)!} \displaystyle\int_{\bar{x}}^{x} (x - t)^{p-2}[(p-1)(x - \bar{x}) - (x - t)]\, dt$
$$= \frac{f^{(p)}(c)}{(p-1)!}\left(\frac{1-p}{p}\right)(x - \bar{x})^p$$

where c lies between x and $\bar{x}$. Moreover,

11.28 $$f'(x) = \frac{(x - \bar{x})^{p-1}}{(p-1)!} f^{(p)}(\xi)$$

where ξ lies between $\bar{x}$ and x. Therefore,

11.29 $$e_{n+1} = \frac{e_n f'(x_n) - f(x_n)}{f'(x_n)} = \frac{1-p}{p} e_n \frac{f^{(p)}(c)}{f^{(p)}(\xi)}$$

Thus, as $e_n \to 0$ we have, by the continuity of $f^{(p)}(x)$

11.30 $$|e_{n+1}| \sim \frac{p-1}{p} |e_n|.$$

Evidently, for large p the convergence of the method is quite slow.

If it is known that the root which is being sought is a p-fold root, then the Newton method can be accelerated by using the formula

11.31
$$x_{n+1} = x_n - p\frac{f(x_n)}{f'(x_n)}.$$

If x_0 is chosen sufficiently near to $\bar{x}$, then the convergence will be of second order. To study the convergence, we have by (11.24) and (11.25) replacing p by $p + 1$

11.32 $(x - \bar{x})f'(x) - pf(x) = \displaystyle\int_{\bar{x}}^{x} \frac{f^{(p+1)}(t)}{(p-1)!}(x-t)^{p-1}((x-\bar{x})-(x-t))\,dt.$

Since $(x - \bar{x}) - (x - t) = t - \bar{x}$, we can again apply the mean-value theorem for integrals and obtain

11.33 $(x - \bar{x})f'(x) - pf(x) = \dfrac{f^{(p+1)}(c)}{(p-1)!}\displaystyle\int_{\bar{x}}^{x}(x-t)^{p-1}((x-\bar{x})-(x-t))\,dt$

$$= \frac{f^{(p+1)}(c)}{(p-1)!}\frac{(x-\bar{x})^{p+1}}{p(p+1)}$$

where c lies between $\bar{x}$ and x. From (11.28) we have

11.34
$$e_{n+1} = \frac{e_n^2}{p(p+1)}\frac{f^{(p+1)}(c)}{f^{(p)}(\xi)}.$$

Thus the procedure of second order.

The reader should note, however, that if (11.31) is used near a simple root, then the procedure may not converge. For example, if $f(x) = x$ and $x_0 = 1$, with $p = 2$ we have $f'(x) = 1$ and $x_1 = x_0 - 2x_0 = -x_0, x_2 = -x_1 = x_0$, etc. Even worse, divergence can occur if a larger p is chosen.

The Secant Method

We consider the case where $\bar{x}$ is a simple root and where $f'(x)$ and $f''(x)$ exist and are continuous in a suitable interval around $\bar{x}$. We also assume that in this interval $|f'(x)| \geq m_1 > 0$ and $|f''(x)| \leq M_2$. As in the case of the Newton method we have by (6.1)

11.35
$$e_{n+1} = \frac{e_{n-1}f(x_n) - e_n f(x_{n-1})}{f(x_n) - f(x_{n-1})}.$$

We now study the behavior of the expression

11.36
$$(y - \bar{x})f(x) - (x - \bar{x})f(y)$$

for all x and y near $\bar{x}$.

Using Taylor's theorem with the integral form of the remainder we have

11.37
$$\begin{cases} f(x) = f(\bar{x}) + (x - \bar{x})f'(\bar{x}) + \int_{\bar{x}}^{x} (x - t)f''(t)\,dt \\ f(y) = f(\bar{x}) + (y - \bar{x})f'(\bar{x}) + \int_{\bar{x}}^{y} (y - t)f''(t)\,dt. \end{cases}$$

First, let us assume that $\bar{x} \leq y \leq x$. We have, since $f(\bar{x}) = 0$,

11.38
$$(y - \bar{x})f(x) - (x - \bar{x})f(y) = \int_{\bar{x}}^{y} f''(t)(x - y)(t - \bar{x})\,dt$$
$$+ \int_{y}^{x} f''(t)(x - t)(y - \bar{x})\,dt.$$

Since $(x - y)(t - \bar{x}) \geq 0$ and $(x - t)(y - \bar{x}) \geq 0$ in the interval $\bar{x} \leq t \leq x$ the above integral equals

11.39
$$f''(c)\left[\int_{\bar{x}}^{y} (x - y)(t - \bar{x})\,dt + \int_{y}^{x} (x - t)(y - \bar{x})\,dt \right]$$
$$= f''(c)\left[(x - y)\frac{(y - \bar{x})^2}{2} + \frac{(x - y)^2}{2}(y - \bar{x}) \right]$$
$$= \tfrac{1}{2}f''(c)(x - y)(y - \bar{x})(x - \bar{x}).$$

Therefore,

11.40
$$(y - \bar{x})f(x) - (x - \bar{x})f(y) = \tfrac{1}{2}f''(c)(x - y)(y - \bar{x})(x - \bar{x}).$$

By the mean-value theorem we have

11.41
$$f(x_n) - f(x_{n-1}) = (x_n - x_{n-1})f'(\xi)$$

where ξ lies between x_n and x_{n-1}. Hence by (11.35) it follows that

11.42
$$e_{n+1} = \frac{1}{2}\frac{f''(c)}{f'(\xi)}e_n e_{n-1}$$

and

11.43
$$|e_{n+1}| \leq \frac{M_2}{2m_1}|e_n|\,|e_{n-1}|.$$

Let us now consider the case where $x \geq \bar{x}$, $y \leq \bar{x}$. By (11.37) we have

11.44
$$(y - \bar{x})f(x) - (x - \bar{x})f(y) = \int_{\bar{x}}^{x} (y - \bar{x})(x - t)f''(t)\, dt$$

$$- \int_{\bar{x}}^{y} (x - \bar{x})(y - t)f''(t)\, dt$$

$$= \int_{y}^{x} f''(t)g(t)\, dt$$

where

11.45
$$g(t) = \begin{cases} (y - \bar{x})(x - t) & \bar{x} \leq t \leq x \\ (x - \bar{x})(y - t) & y \leq t \leq \bar{x}. \end{cases}$$

Evidently $g(t) \leq 0$ for $y \leq t \leq \bar{x}$; hence by the mean-value theorem for integrals we have

11.46
$$(y - \bar{x})f(x) - (x - \bar{x})f(y) = f''(c) \int_{y}^{x} g(t)\, dt$$

$$= f''(c)\left[(y - \bar{x})\frac{(x - \bar{x})^2}{2} - (x - \bar{x})\frac{(y - \bar{x})^2}{2} \right]$$

$$= -f''(c)\frac{(\bar{x} - y)(x - \bar{x})(x - y)}{2},$$

Hence (11.40) holds.

Let us now solve the recurrence equation

11.47
$$q_{n+1} = q_n q_{n-1}$$

where q_0 and q_1 are arbitrary. If we let $z_n = \log q_n$ we have

11.48
$$z_{n+1} = z_n + z_{n-1}.$$

This is a linear recurrence relation with constant coefficients which we can solve by letting

11.49
$$z_n = c_1 \lambda_1^n + c_2 \lambda_2^n$$

where $\lambda_1 = (1 + \sqrt{5})/2$ and $\lambda_2 = (1 - \sqrt{5})/2$ are roots of the quadratic equation

11.50
$$\lambda^2 - \lambda - 1 = 0.$$

One can verify that (11.49) is a solution of (11.48) for any constants c_1 and c_2 by direct substitution. We determine c_1 and c_2 by letting $z_0 = \log q_0$, $z_1 = \log q_1$ obtaining

11.51
$$z_0 = c_1 + c_2, \qquad z_1 = c_1 \lambda_1 + c_2 \lambda_2;$$

hence

11.52
$$c_1 = \frac{\lambda_2 z_0 - z_1}{\lambda_2 - \lambda_1}, \qquad c_2 = \frac{z_1 - \lambda_1 z_0}{\lambda_2 - \lambda_1}.$$

Therefore,

11.53
$$z_n = \frac{1}{\lambda_2 - \lambda_1} \left\{ \lambda_1^n \log \frac{q_0^{\lambda_2}}{q_1} + \lambda_2^n \log \frac{q_1}{q_0^{\lambda_1}} \right\}$$

and, since $\lambda_2 - \lambda_1 = -\sqrt{5}$, and $\lambda_1 \lambda_2 = -1$,

11.54
$$q_n = \left[\left(\frac{q_0^{\lambda_2}}{q_1} \right)^{\lambda_1^n} \left(\frac{q_1}{q_0^{\lambda_1}} \right)^{\lambda_2^n} \right]^{-1/\sqrt{5}} = \left(\frac{q_0^{-\lambda_1^{n-1} + \lambda_2^{n-1}}}{q_1^{\lambda_1^n - \lambda_2^n}} \right)^{-1/\sqrt{5}}.$$

Therefore,

11.55
$$q_n = q_0^{(\lambda_1^{n-1} - \lambda_2^{n-1})/\sqrt{5}} q_1^{(\lambda_1^n - \lambda_2^n)/\sqrt{5}}.$$

Since $\lambda_1 \doteq 1.618$, $\lambda_2 \doteq -0.618$, we have for large n

11.56
$$q_n \sim [(q_0^{1/\lambda_1} q_1)^{1/\sqrt{5}}]^{\lambda_1^n}$$

and

11.57
$$q_{n+1} \sim q_n^{\lambda_1}.$$

Evidently, by (11.43) and (11.55) we have

11.58
$$|e_{n+1}| \leq \frac{2m_1}{M_2} \left\{ \left[\left(\frac{|e_0| M_2}{2m_1} \right)^{1/\lambda_1} \left(\frac{|e_1| M_2}{2m_1} \right) \right]^{1/\sqrt{5}} \right\}^{\lambda_1^{n+1}},$$

provided $|e_0|$ and $|e_1|$ are sufficiently small so that

11.59
$$\left| \frac{|e_i| M_2}{2m_1} \right| < 1 \qquad i = 0, 1.$$

Thus, if (11.59) holds, then the method will converge. Hence (11.9) holds with $p \doteq 1.618$.

We remark that if $\bar{x}$ is a double root, then $m_1 = 0$ and the secant method may not converge. Indeed there exist cases such that no matter how small a positive number δ we choose, we can always find x_0 and x_1 in the interval $[\bar{x} - \delta, \bar{x} + \delta]$ such that $x_0 \neq x_1$ and such that $f(x_1) = f(x_0)$. For example, let $f(x) = x^2$.

For any $\delta > 0$ there exist $x_0 \neq x_1$ in the interval $[-\delta, \delta]$ such that $f(x_0) = f(x_1)$. (For example, we can let $x_0 = \delta/2$, $x_1 = -\delta/2$.)

Muller's Method

In order to study the convergence of Muller's method it is convenient to write the polynomial $P(x)$ given by (10.3), and (10.4) in the form

11.60
$$P(x) = \tilde{A}(x - \bar{x})^2 + \tilde{B}(x - \bar{x}) + \tilde{C}$$

where

11.61
$$\begin{cases} \tilde{A} = \hat{A} \\ \tilde{B} = \hat{B} + 2(\bar{x} - c)\hat{A} \\ \tilde{C} = \hat{C} + \hat{B}(\bar{x} - c) + \hat{A}(\bar{x} - c)^2. \end{cases}$$

This form can be obtained by replacing $x - c$ by $(x - \bar{x}) + (\bar{x} - c)$ in (10.4). As before, the coefficients $\tilde{A}$, $\tilde{B}$, and $\tilde{C}$ can be determined from the conditions

11.62
$$P(a) = f(a), \qquad P(b) = f(b), \qquad P(c) = f(c).$$

To see that $\tilde{A}$, $\tilde{B}$, and $\tilde{C}$ can be uniquely determined if a, b, and c are distinct, consider finding $\hat{A}$, $\hat{B}$, and $\hat{C}$ by (10.9) and using (11.61). One can then determine d by solving

11.63
$$P(d) = 0$$

and taking the root nearest to c.

Now let $a = \bar{x} + e_{n-2}, b = \bar{x} + e_{n-1}, c = \bar{x} + e_n$, and $d = \bar{x} + e_{n+1}$. We have

11.64
$$f(\bar{x} + e_{n-i}) = a_1 e_{n-i} + a_2 e_{n-i}^2 + a_3 e_{n-i}^3 + \cdots, \qquad i = 0, 1, 2$$

where

11.65
$$a_1 = f'(\bar{x}), \qquad a_2 = \tfrac{1}{2}f''(\bar{x}), \qquad a_3 = \tfrac{1}{6}f'''(\bar{x}), \text{ etc.}$$

Neglecting terms of order greater than three in the e_n we have by (11.62) and (11.64)

11.67
$$\begin{cases} \tilde{A}e_{n-2}^2 + \tilde{B}e_{n-2} + \tilde{C} = a_1 e_{n-2} + a_2 e_{n-2}^2 + a_3 e_{n-2}^3 \\ \tilde{A}e_{n-1}^2 + \tilde{B}e_{n-1} + \tilde{C} = a_1 e_{n-1} + a_2 e_{n-1}^2 + a_3 e_{n-1}^3 \\ \tilde{A}e_n^2 + \tilde{B}e_n + \tilde{C} = a_1 e_n + a_2 e_n^2 + a_3 e_n^3. \end{cases}$$

These equations can be written in the form

11.68
$$\begin{cases} a_3 e_{n-2}^3 + (a_2 - \tilde{A})e_{n-2}^2 + (a_1 - \tilde{B})e_{n-2} - \tilde{C} = 0 \\ a_3 e_{n-1}^3 + (a_2 - \tilde{A})e_{n-1}^2 + (a_1 - \tilde{B})e_{n-1} - \tilde{C} = 0 \\ a_3 e_n^3 \quad + (a_2 - \tilde{A})e_n^2 \quad + (a_1 - \tilde{B})e_n \quad - \tilde{C} = 0. \end{cases}$$

Therefore e_{n-2}, e_{n-1}, and e_n are distinct roots of the cubic equation

11.69
$$a_3 \lambda^3 + (a_2 - \tilde{A})\lambda^2 + (a_1 - \tilde{B})\lambda - \tilde{C} = 0;$$

hence we have

11.70
$$\begin{cases} -\dfrac{a_2 - \tilde{A}}{a_3} = e_{n-2} + e_{n-1} + e_n \\ \dfrac{a_1 - \tilde{B}}{a_3} = e_{n-2}e_{n-1} + e_{n-2}e_n + e_{n-1}e_n \\ \dfrac{\tilde{C}}{a_3} = e_{n-2}e_{n-1}e_n, \end{cases}$$

and

11.71
$$\begin{cases} \tilde{A} = a_2 + a_3(e_{n-2} + e_{n-1} + e_n) \\ \tilde{B} = a_1 - a_3(e_{n-2}e_{n-1} + e_{n-2}e_n + e_{n-1}e_n) \\ \tilde{C} = a_3 e_{n-2}e_{n-1}e_n. \end{cases}$$

But by (11.63) we have, since $d = \bar{x} + e_{n+1}$,

11.72
$$P(\bar{x} + e_{n+1}) = \tilde{A}e_{n+1}^2 + \tilde{B}e_{n+1} + \tilde{C} = 0;$$

hence

11.73
$$[a_2 + a_3(e_{n-2} + e_{n-1} + e_n)]e_{n+1}^2$$
$$+ [a_1 - a_3(e_{n-2}e_{n-1} + e_{n-2}e_n + e_{n-1}e_n)]e_{n+1} + a_3 e_{n-2}e_{n-1}e_n = 0.$$

Let us assume now that $a_1 \neq 0$ and that $|e_{n-2}|$, $|e_{n-1}|$, and $|e_n|$ are small relative to $|a_3/a_1|$. Writing the solution of (11.72) in the form

11.74
$$e_{n+1} = \frac{-2\tilde{C}}{\tilde{B} \pm \sqrt{\tilde{B}^2 - 4\tilde{A}\tilde{C}}}$$

it can be seen that to within terms of higher order in e_n, e_{n-1}, and e_{n-2} we have

11.75
$$e_{n+1} \doteq \frac{a_3}{a_1} e_n e_{n-1} e_{n-2}.$$

We are thus led to seek a bound on the solution of the equation

11.76
$$|e_{n+1}| = \frac{M_3}{6m_1} |e_n e_{n-1} e_{n-2}|$$

where M_3 is an upper bound for $|f'''(x)|$ and m_1 is a lower bound for $|f'(x)|$ in the interval under consideration. (We assume $m_1 > 0$.)

Let us solve the recurrence equation

11.77
$$q_{n+1} = q_n q_{n-1} q_{n-2}.$$

As in the case of the secant method, we let $z_n = \log q_n$ and solve the linear recurrence equation

11.78
$$z_{n+1} = z_n + z_{n-1} + z_{n-2}.$$

The general solution is

11.79
$$z_n = c_1 \lambda_1^n + c_2 \lambda_2^n + c_3 \lambda_3^n$$

where $\lambda_1, \lambda_2, \lambda_3$ are roots of

11.80
$$\lambda^3 - \lambda^2 - \lambda - 1 = 0.$$

The solution of (11.80) is

11.81
$$\begin{cases} \lambda_1 \doteq 1.8392868 \\ \lambda_2, \lambda_3 \doteq -0.4196434 \pm 0.6062907i \end{cases}$$

so that $|\lambda_2| = |\lambda_3| < 1$. The constants c_1, c_2, and c_3 are given by

11.82
$$c_1 = \frac{\lambda_2 \lambda_3 z_0 - (\lambda_2 + \lambda_3)z_1 + z_2}{(\lambda_3 - \lambda_1)(\lambda_2 - \lambda_1)}$$

and similar equations for c_2 and c_3. Using the fact $\lambda_1 + \lambda_2 + \lambda_3 = 1$, $\lambda_1 \lambda_2 + \lambda_1 \lambda_3 + \lambda_2 \lambda_3 = -1$, $\lambda_1 \lambda_2 \lambda_3 = 1$ we have

11.83
$$\begin{cases} c_1 = \dfrac{z_0 + \lambda_1(\lambda_1 - 1)z_1 + \lambda_1 z_2}{\lambda_1^2 + 2\lambda_1 + 3} \\[3mm] c_2 = \dfrac{z_0 + \lambda_2(\lambda_2 - 1)z_1 + \lambda_2 z_2}{\lambda_2^2 + 2\lambda_2 + 3} \\[3mm] c_3 = \dfrac{z_0 + \lambda_3(\lambda_3 - 1)z_1 + \lambda_3 z_2}{\lambda_3^2 + 2\lambda_3 + 3} . \end{cases}$$

Therefore

11.84
$$q_n = \prod_{i=1}^{3} [q_0 q_1^{\lambda_i(\lambda_i - 1)} q_2^{\lambda_i}]^{\lambda_i^n/(\lambda_i^2 + 2\lambda_i + 3)}.$$

For large n we have $|\lambda_2^n| \to 0$ and $|\lambda_3^n| \to 0$ since $|\lambda_2| = |\lambda_3| < 1$. Therefore, for large n

11.85
$$q_n \sim [q_0 q_1^{\lambda_1(\lambda_1 - 1)} q_2^{\lambda_1}]^{\lambda_1^n/(\lambda_1^2 + 2\lambda_1 + 3)}.$$

Consequently,

11.86
$$q_{n+1} \sim q_n^{\lambda_1}.$$

Evidently, by (11.85) we have, letting $q_n = M_3 |e_n|/(6m_1)$,

11.87
$$|e_{n+1}| \sim \frac{6m_1}{M_3} \left[\left\{ \left(\frac{M_3 |e_0|}{6m_1} \right) \left(\frac{M_3 |e_1|}{6m_1} \right)^{\lambda_1(\lambda_1 - 1)} \left(\frac{M_3 |e_2|}{6m_1} \right)^{\lambda_1} \right\}^{1/\theta} \right]^{\lambda_1^n}$$

where

11.88
$$\theta = \lambda_1^2 + 2\lambda_1 + 3 \doteq 10.06.$$

Based on the above nonrigorous argument, we would expect that Muller's method would converge if x_0, x_1, and x_2 are sufficiently close to $\bar{x}$ so that

11.89
$$\left| \frac{M_3 |e_i|}{6m_1} \right| < 1, \qquad i = 0, 1, 2.$$

Moreover, we would expect that (11.9) would hold with $p \doteq 1.839$.

We remark that if $\bar{x}$ is a triple root, then $m_1 = 0$ and Muller's method may not converge. Indeed, there exist cases such that no matter how small a positive number δ we choose we can always find distinct values x_0, x_1, and x_2 in the circle in the complex plane

11.90 $|x - \bar{x}| \leq \delta$

such that $f(x_0) = f(x_1) = f(x_2)$. For example, let $f(x) = x^3$. For any $\delta > 0$ there exist distinct values x_0, x_1, x_2 in the circle $|x| \leq \delta$ such that $f(x_0) = f(x_1) = f(x_2)$. As a matter of fact we can let

11.91 $x_0 = \dfrac{\delta}{2}, \quad x_1 = \dfrac{\delta}{2}\left(-\dfrac{1}{2} + \dfrac{\sqrt{3}}{2}i\right), \quad x_2 = \dfrac{\delta}{2}\left(-\dfrac{1}{2} - \dfrac{\sqrt{3}}{2}i\right).$

Actually, in practice, Muller's method usually converges for triple roots and even for roots of higher multiplicities. The difficulty described above seems to occur less often than might be expected. Moreover, the program can be instructed to move one or more of the iterants slightly in order that the process can continue whenever the problem arises. This modification works well in many cases.

EXERCISES 4.11

1. Discuss the convergence of the following sequences

   a) $x_n = 1 - (\tfrac{1}{2})^n$

   b) $x_n = 1 - n(\tfrac{1}{2})^n$

   c) $x_0 = 1, x_1 = \tfrac{1}{2}, x_2 = \tfrac{3}{4}, x_3 = \tfrac{9}{8}, x_4 = \tfrac{17}{16}, x_5 = \tfrac{3}{32}, x_6 = \tfrac{65}{64}, \ldots$

   d) $x_0 = \dfrac{3}{4}, x_{n+1} = \dfrac{1}{2}\left(\dfrac{1}{2} + \dfrac{1}{2x_n}\right), n = 0, 1, 2, \ldots$.

2. Apply the Newton method to the equation $f(x) = x^3 - \tfrac{1}{2} = 0$ and verify that the order of convergence is approximately two. Let $x_0 = \tfrac{1}{2}$.

3. Apply the secant method to the equation $f(x) = x^3 - \tfrac{1}{2} = 0$ with $x_0 = 0$, $x_1 = 1$, and verify that the order of convergence is approximately 1.618.

4. Apply Muller's method to the equation $f(x) = x^3 - \tfrac{1}{2} = 0$ with $x_0 = 0$, $x_1 = 1$, $x_2 = \tfrac{1}{2}$, and verify that the order of convergence is approximately 1.839.

5. Apply the Newton method, with $x_0 = 0.8$, the secant method with $x_0 = 0.8, x_1 = 1.2$, and Muller's method with $x_0 = 0.6$, $x_1 = 0.8$, $x_2 = 1.2$ to the equation $f(x) = x^3 - x^2 - x + 1 = 0$ and verify that the convergence is only of first order in each case. Then apply the modified Newton method (9.14) with $p = 2$ and verify that the convergence is second order.

6. Verify that for any constants c_1 and c_2 the function $z_n = c_1\lambda_1^n + c_2\lambda_2^n$ is a solution of $z_{n+1} = z_n + z_{n-1}$ where λ_1 and λ_2 satisfy $\lambda^2 - \lambda - 1 = 0$. If $z_0 = 1$, $z_1 = 2$, find c_1, c_2, and z_5.

7. Show that Cauchy's method has third-order convergence near a zero $\bar{x}$ of $f(x)$ where $f'(\bar{x}) \neq 0, f''(\bar{x}) \neq 0$, and $|f'''(x)|$ is bounded in a suitable interval containing $\bar{x}$.

8. Derive a formula for the Newton method for the equation

   $$\frac{f(x)}{f'(x)} = 0.$$

   Show that the convergence of the method to any zero of $f(x)$ is of second order even if the zero is not simple.

9. Carry out the method of the preceding exercise for the equation of Exercise 5 with $x_0 = 0.8$.

4.12 ACCELERATION OF THE CONVERGENCE

The convergence of an iterative method with linear convergence can often be accelerated by the use of the Aitken δ^2 process. Suppose that the sequence $x_0, x_1, x_2, \ldots$ has *geometric convergence* to $\bar{x}$ (see (11.4)), and hence for some C with $0 < |C| < 1$ we have

12.1
$$\frac{x_{n+1} - \bar{x}}{x_n - \bar{x}} = C, \qquad n = 0, 1, 2, \ldots .$$

In this case we can determine $\bar{x}$ exactly given any three consecutive numbers in the sequence. Thus we have

12.2
$$\frac{x_n - \bar{x}}{x_{n-1} - \bar{x}} = \frac{x_{n+1} - \bar{x}}{x_n - \bar{x}}$$

and solving for $\bar{x}$ we get

12.3
$$\bar{x} = \frac{x_{n-1}x_{n+1} - x_n^2}{x_{n+1} - 2x_n + x_{n-1}} = x_{n+1} - \frac{(x_{n+1} - x_n)^2}{x_{n+1} - 2x_n + x_{n-1}}.$$

As an example, consider the sequence

12.4
$$x_n = 1 - (\tfrac{1}{2})^n, \qquad n = 0, 1, 2, \ldots$$

which converges geometrically to the limit unity since

12.5
$$\frac{x_{n+1} - 1}{x_n - 1} = \frac{1}{2}.$$

Substituting (12.4) in the right member of (12.3) we get unity.

Even if the convergence of the sequence is not geometric, the convergence can often be accelerated by using the sequence $y_0, y_1, \ldots$, given by

12.6
$$y_n = \frac{x_{n+2}x_n - x_{n+1}^2}{x_{n+2} - 2x_{n+1} + x_n} = x_{n+2} - \frac{(x_{n+2} - x_{n+1})^2}{x_{n+2} - 2x_{n+1} + x_n}.$$

This is the Aitken δ^2 process. It can be shown (see Traub [1964]) that

12.7
$$\lim_{n \to \infty} \frac{y_n - \bar{x}}{x_n - \bar{x}} = 0$$

provided

12.8
$$\lim_{n \to \infty} \frac{x_{n+1} - \bar{x}}{x_n - \bar{x}} = C < 1,$$

and that under additional assumptions the convergence of the new sequence is second order. It should be noted that there is no gain in using the Aitken δ^2 process with a sequence whose convergence is second order.

EXERCISES 4.12

1. Verify that substituting (12.4) in the right member of (12.3) yields the value unity.
2. Apply the Aitken δ^2 process to the sequence of partial sums of the series $\log 2 = 1 - \frac{1}{2} + \frac{1}{3} - \frac{1}{4} \cdots$.
3. Apply the Aitken δ^2 process to the sequence $x_0, x_1, x_2, \ldots$ generated by

$$x_{n+1} = \log x_n + 3$$

   where $x_0 = 4$. Carry five decimal places.
4. Apply the Aitken δ^2 process to the sequence

$$x_0 = \frac{3}{4}, x_{n+1} = \frac{1}{2}\left(x_n + \frac{1}{2x_n}\right), n \geq 0$$

   and verify that the convergence is not accelerated. Carry out the calculations to five decimal places.
5. Apply the Newton method to solve the equation $x^3 - x^2 - x + 1 = 0$, letting $x_0 = 0.5$. Attempt to accelerate the convergence by using the Aitken δ^2 process.

4.13 SYSTEMS OF NONLINEAR EQUATIONS

In this section we shall consider the solution of the system of nonlinear equations

13.1
$$\begin{cases} f_1(x_1, x_2, \ldots, x_n) = 0 \\ f_2(x_1, x_2, \ldots, x_n) = 0 \\ \vdots \\ f_n(x_1, x_2, \ldots, x_n) = 0. \end{cases}$$

For the sake of simplicity we shall frequently consider the system of two equations in two unknowns,

13.2
$$\begin{cases} g(x, y) = 0 \\ h(x, y) = 0. \end{cases}$$

An important application of (13.2) is in the determination of complex roots of a single equation of the form

13.3
$$f(z) = 0.$$

For, if $f(x + iy) = g(x, y) + ih(x, y)$ where $g(x, y)$ and $h(x, y)$ are real, for real x and y, we can solve (13.3) if we can solve the system (13.2). For example, the equation

13.4
$$\phi(z) = e^z + z + 2 = 0$$

is equivalent to the system

13.5
$$\begin{cases} g(x, y) = e^x \cos y + x + 2 = 0 \\ h(x, y) = e^x \sin y + y \quad\quad = 0 \end{cases}$$

since

13.6
$$e^{x+iy} = e^x(\cos y + i \sin y).$$

We shall first discuss graphical methods both for the case $n = 2$ and also for the case $n = 3$. Next, we consider the generalized Newton method. Gradient methods are then considered, followed by a brief discussion of the Jacobi, Gauss-Seidel, and successive-overrelaxation iterative methods.

Use of Graphs and Linear Interpolation

For the case $n = 2$ we can often obtain approximate real solutions graphically as follows. First, we evaluate $g(x, y)$ and $h(x, y)$ at the mesh points of a rectangular grid $x = a, a + \Delta x, a + 2\Delta x, \ldots, a + p\Delta x, y = b, b + \Delta y, b + 2\Delta y, \ldots, b + q\Delta y$. Approximate contours where $g(x, y) = 0$ and $h(x, y) = 0$ are constructed, and where these intersect we have an approximate value of the solution. The construction of the contours can be done as follows. At the center of each mesh rectangle let a value of $g(x, y)$ be assigned as the average of the values of the four vertices of the rectangle. For each rectangle, consider the four triangles formed by the diagonals of the rectangle. For any such triangle, if the sign of $g(x, y)$ is the same for all three vertices, then no points of the contour are assigned to the rectangle. If g changes sign between two vertices, or if it vanishes at a vertex, we let a point of the contour be determined by linear interpolation as described below. It may happen that the entire side of the triangle belongs to the contour if the values at both ends are zero. In fact the entire triangle may be part of the contour if f vanishes at all three vertices. By connecting all pairs of points thus obtained, we get for each triangle either no points, a single point, a single line, or the whole triangle.

We now describe the linear interpolation process. We replace $g(x, y)$ by a function $G(x, y)$ of the form $G(x, y) = A + Bx + Cy$ where A, B, C are constants to be determined and where $G(x_1, y_1) = g(x_1, y_1)$, $G(x_2, y_2) = g(x_2, y_2)$. We seek a point $(\bar{x}, \bar{y})$ on the line L joining (x_1, y_1) and (x_2, y_2) so that $G(\bar{x}, \bar{y}) = 0$. Evidently, if $(\bar{x}, \bar{y})$ is on the line L we have

13.7
$$\begin{cases} \bar{x} = x_1 + t(x_2 - x_1) \\ \bar{y} = y_1 + t(y_2 - y_1) \end{cases}$$

for some t in the range $0 \leq t \leq 1$. The requirement $G(\bar{x}, \bar{y}) = 0$ is equivalent to

13.8
$$A + Bx_1 + Cy_1 + t(B(x_2 - x_1) + C(y_2 - y_1)) = 0.$$

But since $g(x_1, y_1) = A + Bx_1 + Cy_1$ and $g(x_2, y_2) = A + Bx_2 + Cy_2$ we have $B(x_2 - x_1) + C(y_2 - y_1) = g(x_2, y_2) - g(x_1, y_1)$. Therefore the condition becomes

13.9
$$g(x_1, y_1) + t[g(x_2, y_2) - g(x_1, y_1)] = 0,$$

or

13.10
$$t = \frac{g(x_1, y_1)}{g(x_1, y_1) - g(x_2, y_2)}.$$

Hence

13.11
$$\begin{cases} \bar{x} = x_1 + \dfrac{(x_2 - x_1)g(x_1, y_1)}{g(x_1, y_1) - g(x_2, y_2)} = \dfrac{x_2 g(x_1, y_1) - x_1 g(x_2, y_2)}{g(x_1, y_1) - g(x_2, y_2)} \\[3mm] \bar{y} = y_1 + \dfrac{(y_2 - y_1)g(x_1, y_1)}{g(x_1, y_1) - g(x_2, y_2)} = \dfrac{y_2 g(x_1, y_1) - y_1 g(x_2, y_2)}{g(x_1, y_1) - g(x_2, y_2)}. \end{cases}$$

It should be noted that the above formulas are each similar to (5.8) for the method of false position. We also note that if $g(x_1, y_1) = 0$ and $g(x_2, y_2) \neq 0$, then we have $\bar{x} = x_1$, $\bar{y} = y_1$. Similarly, if $g(x_1, y_1) \neq 0$ and $g(x_2, y_2) = 0$, then we get $\bar{x} = x_2$, $\bar{y} = y_2$. In any case, unless both $g(x_1, y_1)$ and $g(x_2, y_2)$ vanish we get a unique point on the line L.

The procedure we have just described will yield points of the contours $G(x, y) = 0$ and $H(x, y) = 0$ for each basic triangle. (Here $H(x, y)$ is the replacement for $h(x, y)$.) One can then determine whether these contours intersect and if so where.

Programs are available for high-speed computers for automatically producing contours of a function $g(x, y)$. See, for instance, Downing [1966]. As input to such programs one needs only to provide the mesh constants a, b, Δx, and Δy as well as the instructions for a subroutine for computing $g(x, y)$ for any x and y. Alternatively, one can read in values of $g(x, y)$ for a rectangular grid of points.

It is not difficult to make the procedure fully automatic. A basic part of an automatic program is the determination of whether in a given triangle the linear functions $G(x, y)$ and $H(x, y)$ which represent $g(x, y)$ and $h(x, y)$, respectively, vanish at a common point. (See Ostrowski [1966], p. 239–240.)

It is also possible to generalize the procedure to the system

13.12
$$\begin{cases} g(x, y, z) = 0 \\ h(x, y, z) = 0 \\ k(x, y, z) = 0. \end{cases}$$

We cover a rectangular parallelepiped in the (x, y, z) space by a mesh formed by the planes

13.13
$$\begin{cases} x = a_1, a_1 + \Delta x, a_1 + 2\Delta x, \ldots, a_1 + N_1\Delta x = b_1 \\ y = a_2, a_2 + \Delta y, a_2 + 2\Delta y, \ldots, a_2 + N_2\Delta y = b_2 \\ z = a_3, a_3 + \Delta z, a_3 + 2\Delta z, \ldots, a_3 + N_3\Delta z = b_3. \end{cases}$$

We replace $g(x, y, z)$, $h(x, y, z)$, $k(x, y, z)$ by piecewise linear functions $G(x, y, z)$, $H(x, y, z)$, and $K(x, y, z)$, respectively.

Consider a basic rectangular parallelepiped of sides Δx, Δy, and Δz. At the center of the parallelepiped we assign to $G(x, y, z)$ the average of the values of $g(x, y, z)$ at the 8 vertices. At the center of each face we assign G the average of the values of g at the corners of the face. We develop a linear function of the form

13.14
$$G(x, y, z) = \alpha + \beta x + \gamma y + \delta z$$

in each tetrahedron formed by two adjacent vertices, the center of the solid and the center of one of the faces containing the two vertices. (See Fig. 13.15.) $G(x, y, z)$ is

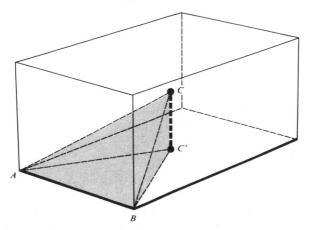

13.15 Fig.　Piecewise Linear Interpolation in Three Variables.

to agree with the given functions $g(x, y, z)$ at these four points. In each triangular face of the tetrahedron we can find the points where $G(x, y, z)$ vanishes as in the two-dimensional case. For the entire tetrahedron the set of all lines joining points on the triangular faces where $G(x, y, z)$ vanishes makes up the set of all points in the tetrahedron where $G(x, y, z)$ vanishes. We can then determine whether there are any points where $G(x, y, z)$, $H(x, y, z)$, and $K(x, y, z)$ vanish.

Generalized Newton Method

A method which can sometimes be used to solve (13.1) is the generalized Newton method. For simplicity we describe the method for the system (13.2). The basic step in the process is the following. For given x and y determine x' and y' so that to first-order terms in $\Delta x = x' - x$ and $\Delta y = y' - y$ we have

13.16 $g(x', y') = h(x', y') = 0.$

By Taylor's theorem we have (see Theorem A-4.6)

13.17 $\begin{cases} g(x', y') = g(x, y) + \Delta x g_x(x, y) + \Delta y g_y(x, y) + \cdots \\ h(x', y') = h(x, y) + \Delta x h_x(x, y) + \Delta y h_y(x, y) + \cdots \end{cases}$

Neglecting higher-order terms, solving for Δx and Δy, and using (13.16) we obtain

13.18 $\begin{cases} x' - x = \Delta x = \dfrac{-g(x, y)h_y(x, y) + h(x, y)g_y(x, y)}{J} \\[2ex] y' - y = \Delta y = \dfrac{-h(x, y)g_x(x, y) + g(x, y)h_x(x, y)}{J} \end{cases}$

where

13.19 $J = \det\begin{pmatrix} g_x & g_y \\ h_x & h_y \end{pmatrix} = g_x h_y - h_x g_y$

is the *Jacobian* of the functions f and g.

We note that if $g_y = 0$ then the first equation of (13.18) reduces to

13.20 $x' - x = -\dfrac{g(x, y)}{g_x(x, y)}$

which is simply the Newton method for solving $g(x, y) = 0$ for fixed y.

Suppose now that $f(x, y) = g(x, y) + ih(x, y)$ is an analytic function of $z = x + iy$. Then, by Theorem A-5.20, $g(x, y)$ and $h(x, y)$ satisfy the Cauchy-Riemann equations (9.19) and (13.18) becomes

13.21
$$\begin{cases} x' - x = -\dfrac{gg_x + hh_x}{g_x^2 + h_x^2} \\[3mm] y' - y = -\dfrac{hg_x - gh_x}{g_x^2 + h_x^2} \end{cases}$$

which agrees with (9.22).
    As an example, let

13.22
$$\begin{cases} g(x, y) = x^3 - 3xy^2 + 1 = 0 \\ h(x, y) = 3x^2y - y^3 \quad\;\; = 0. \end{cases}$$

Evidently,

13.23
$$\begin{cases} g_x = 3x^2 - 3y^2, \quad g_y = -6xy \\ h_x = 6xy \qquad\;\;, \quad h_y = 3x^2 - 3y^2. \end{cases}$$

Let $x = 1, y = 1$. Then $g = -1, h = 2$. Moreover, $g_x = 0, g_y = -6, h_x = 6,$ $h_y = 0$, and $J = 36$. Substituting in (13.18) we have

13.24
$$\begin{cases} x_1 = x' = 1 - \frac{1}{3} = \frac{2}{3} \doteq 0.66667 \\ y_1 = y' = 1 - \frac{1}{6} = \frac{5}{6} \doteq 0.83333. \end{cases}$$

Subsequent calculations are given in Table 13.27. After 4 iterations the exact solution $\bar{x} = 1/2, \bar{y} = \sqrt{3}/2 \doteq 0.86602$ is obtained to 5 decimals.
    We note that $g(x, y) + ih(x, y) = f(z) = z^3 + 1$. Hence the roots are easily seen to be

13.25
$$\begin{cases} z_1 = -1 \\[2mm] z_2 = \dfrac{1}{2} + \dfrac{\sqrt{3}}{2}i \doteq 0.5 + 0.86602\,i \\[3mm] z_3 = \dfrac{1}{2} - \dfrac{\sqrt{3}}{2}i \doteq 0.5 - 0.86602\,i \end{cases}$$

The Newton method applied to $f(z)$ gives

13.26
$$z_{n+1} = z_n - \frac{z_n^3 + 1}{3z_n^2} = \frac{2}{3}z_n - \frac{1}{3z_n^2}.$$

It can easily be verified that the use of the above formula with $z_0 = 1 + i$ gives the same results as in Table 13.27.
    The effectiveness of the Newton method is critically dependent, even more than in the one variable case, on the choice of starting values. In the case of two

equations in two unknowns, good starting values can often be obtained by the graphical method described previously.

13.27 Table

n	x_n	y_n	$g(x_n, y_n)$	$h(x_n, y_n)$	g_x	g_y
0	1.00000	1.00000	-1.00000	2.00000	0.00000	-6.00000
1	0.66667	0.83333	-0.09259	0.53241	-0.75000	-3.33333
2	0.50869	0.84110	0.05201	0.05791	-1.34604	-2.56716
3	0.49933	0.86627	0.00037	-0.00211	-1.50328	-2.59533
4	0.50000	0.86602	0.00000	0.00000	-1.50000	-2.59807

h_x	h_y	J	Δx	Δy
6.00000	0	36.00000	-0.33333	-0.16667
3.33333	-0.75000	11.67361	-0.15797	0.00777
2.56716	-1.34604	8.40217	-0.00936	0.02517
2.59533	-1.50328	8.99555	0.00067	-0.00024
2.59807	-1.50000	8.99998	0.00000	0.00000

Gradient Methods

Let us now consider the problem of minimizing a function $Q(x, y)$ in some region R. The solution of this problem would yield a solution to (13.2) if we let

13.28
$$Q(x, y) = \tfrac{1}{2}[g(x, y)]^2 + \tfrac{1}{2}[h(x, y)]^2$$

provided (13.2) has a solution.

The procedure is as follows. Given x and y, we seek x' and y' such that $Q(x', y') < Q(x, y)$. We search for (x', y') along a line L through (x, y) such that the directional derivative of $Q(x, y)$ along that line is minimized. If ϕ is the angle between the line L and the x-axis, then the directional derivative of Q along L is

13.29
$$\frac{\partial Q}{\partial v} = Q_x \cos \phi + Q_y \sin \phi.$$

Differentiating with respect to ϕ we get

13.30
$$\frac{d}{d\phi} \frac{\partial Q}{\partial v} = -Q_x \sin \phi + Q_y \cos \phi.$$

This derivative vanishes when

13.31
$$\tan \phi = Q_y/Q_x.$$

The corresponding values of $\cos \phi$ and $\sin \phi$ are

13.32 $$\cos \phi = \frac{\alpha Q_x}{\sqrt{Q_x^2 + Q_y^2}}, \quad \sin \phi = \frac{\alpha Q_y}{\sqrt{Q_x^2 + Q_y^2}}$$

where $\alpha = \pm 1$. If we choose $\alpha = -1$, then the directional derivative has a (negative) minimum given by

13.33 $$\frac{\partial Q}{\partial v} = -\sqrt{Q_x^2 + Q_y^2}.$$

The unit vector in the direction of L is

13.34 $$-\frac{1}{\sqrt{Q_x^2 + Q_y^2}} \nabla Q$$

where

13.35 $$\nabla Q = (Q_x, Q_y)$$

is the gradient of Q. Thus our line L is in the opposite direction from the gradient. A point (x', y') on L has the form

13.36 $$\begin{cases} x' = x - tQ_x \\ y' = y - tQ_y \end{cases}$$

where t is a real parameter.

Unless Q_x and Q_y both vanish, we can determine our direction uniquely. Moreover, by choosing t small enough we can certainly make $Q(x', y')$ less than $Q(x, y)$. On the other hand, if $Q_x = Q_y = 0$, then we have

13.37 $$\begin{cases} Q_x = g_x g + h_x h = 0 \\ Q_y = g_y g + h_y h = 0 \end{cases}$$

which imply that $g = h = 0$ unless the *Jacobian*

13.38 $$J = \frac{\partial(g, h)}{\partial(x, y)} = g_x h_y - g_y h_x$$

vanishes. Hence, if $Q_x = Q_y = 0$, then we either have a solution or else we are at a point where the Jacobian vanishes.

There are several ways of choosing t. If $Q(x, y)$ is given by (13.28), one possibility is to choose t so that Q would vanish if Q_x and Q_y were constant along the line L. Thus we have by Taylor's theorem, neglecting all terms beyond the third,

13.39 $$Q(x', y') \doteq Q(x, y) + (-tQ_x)Q_x + (-tQ_y)Q_y.$$

Letting $Q(x', y') = 0$ and solving for t we get

13.40
$$t = \frac{Q}{Q_x^2 + Q_y^2}.$$

Hence by (13.36) we have

13.41
$$x' = x - \frac{QQ_x}{Q_x^2 + Q_y^2}, \qquad y' = y - \frac{QQ_y}{Q_x^2 + Q_y^2}.$$

If $f(x, y)$ and $g(x, y)$ are the real and imaginary parts, respectively, of some analytic function, then by the Cauchy-Riemann equations (9.19) we have

13.42
$$\begin{cases} Q_x = gg_x + hh_x, \qquad Q_y = gg_y + hh_y \\ Q_x^2 + Q_y^2 = (h^2 + g^2)(g_x^2 + h_x^2). \end{cases}$$

Therefore,

13.43
$$\begin{cases} x' = x - \dfrac{1}{2} \dfrac{gg_x + hh_x}{g_x^2 + h_x^2} \\[2ex] y' = y - \dfrac{1}{2} \dfrac{gg_y + hh_y}{g_x^2 + h_x^2}. \end{cases}$$

This result should be compared with the formula (13.18) for the generalized Newton method where, by (9.19) we have

13.44
$$\begin{cases} x' = x - \dfrac{gg_x + hh_x}{g_x^2 + h_x^2} \\[2ex] y' = y - \dfrac{gg_y + hh_y}{g_x^2 + h_x^2}. \end{cases}$$

Thus this variant of the gradient method is equivalent to the generalized Newton method for the system (13.2) except that the "corrections" $x' - x$ and $y' - y$ are each multiplied by $\frac{1}{2}$. We remark that, as we have seen, the Newton method for the system (13.2) is equivalent to the Newton method applied to

13.45
$$f(x + iy) = 0$$

where

13.46
$$f(x + iy) = g(x, y) + ih(x, y).$$

Evidently the Jacobian $\partial(g, h)/\partial(x, y)$ is simply $g_x^2 + g_y^2 = |f'(z)|^2$; hence we expect the gradient method to converge either to a zero of $f(z)$ or to a zero of $f'(z)$.

Suppose that the method converges to a zero z_0 of $f'(z)$ such that $|f(z_0)| > 0$. Then, unless $f(z)$ is a constant function in every neighborhood of z_0, there is a value of z near z_0 such that $|f(z)| < |f(z_0)|$. This follows from the minimum-modulus principle* for analytic functions which states that if $f(z)$ is a nonconstant analytic function in a simply connected region R and if $f(z)$ does not vanish in R, then $|f(z)|$ cannot assume its minimum value in R.

In practice, one could consider a square around z_0 and search** for a value of z on the boundary of the square such that $|f(z)| < |f(z_0)|$. If no such z could be found, then there would be a zero of f in the square. One could then reduce the size of the square and search again until a suitable z has been found.

Having found z such that $|f(z)| < |f(z_0)|$ we can resume the basic gradient procedure designed to reduce $Q(z) = \frac{1}{2}|f(z)|^2$. Thus, we are sure that we will not return to z_0 since $Q(z)$ will always be less than $Q(z_0)$. The process is continued until we reach a zero of f or of f'. In the latter case we repeat the process.

If second partial derivatives of Q are conveniently available, we may use the following scheme. Letting (x', y') lie on L we may expand $Q(x', y')$ in a Taylor series

13.47 $$Q(x', y') = Q(x, y) - tQ_x^2 - tQ_y^2 + \frac{t^2 Q_x^2}{2} Q_{xx} + t^2 Q_x Q_y Q_{xy} + \frac{t^2 Q_y^2}{2} Q_{yy} + \cdots.$$

Neglecting all terms of order t^3 and higher, we seek to minimize $Q(x', y')$ as a function of t. Differentiating, we have

13.48 $$\frac{d}{dt} Q(x', y') \doteq -(Q_x^2 + Q_y^2) + 2t\left[\frac{Q_x^2}{2} Q_{xx} + Q_x Q_y Q_{xy} + \frac{Q_y^2}{2} Q_{yy}\right].$$

The above expression vanishes when

13.49 $$t = \frac{Q_x^2 + Q_y^2}{Q_x^2 Q_{xx} + 2Q_x Q_y Q_{xy} + Q_y^2 Q_{yy}}.$$

In our example (13.22) we have

13.50 $$Q = \frac{1}{2}(x^3 - 3xy^2 + 1)^2 + \frac{1}{2}(3x^2 y - y^3)^2.$$

---

* The minimum modulus principle for analytic functions follows from the maximum modulus principle (see, for instance, Ahlfors [1966] as applied to $[f(z)]^{-1}$).

** This could be done by evaluating $f(z)$ for a number of values of z on the boundary. In the case of a polynomial, one could use Lehmer's method (see Section 5.6).

If $x = y = 1$, we have $Q = \frac{1}{2} + \frac{1}{2}(2)^2 = 2.5$

13.51
$$\begin{cases} Q_x = gg_x + hh_x = 12 \\ Q_y = gg_y + hh_y = 6 \\ Q_{xx} = gg_{xx} + g_x^2 + hh_{xx} + h_x^2 = 42 \\ Q_{xy} = gg_{xy} + g_x g_y + hh_{xy} + h_x h_y = 18 \\ Q_{yy} = gg_{yy} + g_y^2 + hh_{yy} + h_y^2 = 30 \end{cases}$$

since

13.52
$$\begin{cases} g_{xx} = 6x = 6 \\ g_{xy} = -6y = -6 \\ g_{yy} = -6x = -6 \\ h_{xx} = 6y = 6 \\ h_{xy} = 6x = 6 \\ h_{yy} = -6y = -6. \end{cases}$$

Therefore by (13.49) we have

13.53 $t = \dfrac{180}{(12)(12)(42) + 2(12)(6)(18) + (6)(6)(30)} = \dfrac{180}{9720} = \dfrac{1}{54} \doteq 0.01852.$

Hence,

13.54
$$\begin{cases} x' \doteq 1 - (0.01852)(12) = 0.7778 \\ y' \doteq 1 - (0.01852)(6) = 0.8889. \end{cases}$$

Comparing the results obtained by the generalized Newton method, procedure (13.41), with the procedure (13.36) and (13.49), we have

	t	x'	y'	$Q(x', y')$
Newton	$\frac{1}{36} \doteq 0.02777$	0.66667	0.83333	0.146022
Procedure (13.41)	$\frac{1}{72} \doteq 0.01388$	0.83333	0.91667	0.785419
Procedure (13.36), (13.49)	$\frac{1}{54} \doteq 0.01852$	0.7778	0.8889	0.4846

In this case the generalized Newton method gives the best approximation to the true solution $x = 0.50000$, $y = 0.86602$. However, all three procedures give a substantial reduction in $Q(x', y')$ as compared with the value of 2.5 for $Q(1, 1)$. Figure 13.55 gives a graph of $Q(x', y')$, where $x' = 1 - 12t$, $y' = 1 - 6t$, as a function of t. The graph is based on the numerical values shown.

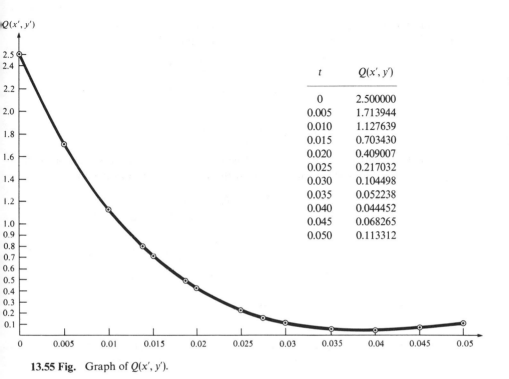

t	$Q(x', y')$
0	2.500000
0.005	1.713944
0.010	1.127639
0.015	0.703430
0.020	0.409007
0.025	0.217032
0.030	0.104498
0.035	0.052238
0.040	0.044452
0.045	0.068265
0.050	0.113312

13.55 Fig. Graph of $Q(x', y')$.

It can be seen that the optimum value of t is around 0.04 and is somewhat larger than the value 0.02777 given by the Newton method and the value 0.0186 given by (13.49). The latter value is not optimum since the function $Q(x', y')$ is not a quadratic function of t.

Another approach which appears somewhat safer for automatic computation is to select a value of δ and simply compute values of $Q(x', y')$ for $x' = x - k\delta Q_x$, $y' = y - k\delta Q_y$, $k = 1, 2, \ldots$. The value of k is chosen so that Q is minimized. In the above example, if we let $\delta = 0.05$, we would have $k = 8$. In the event that no value of k results in a smaller value of Q, then one can divide δ by 2. This process can be repeated until eventually one finds a value of δ and k such that $Q(x - k\delta Q_x, y - k\delta Q_y) < Q(x, y)$. This can fail to happen eventually only if Q_x and Q_y both vanish, in which case we are in the neighborhood of a local extremum.

A reasonable initial choice of δ might be some fraction, say $\frac{1}{4}$, of the value of t given by (13.40), by (13.49), or by the generalized Newton method.

Other Iterative Methods

We mention briefly another class of methods. Assume that in the system (13.1) we can solve the ith equation for x_i either explicitly in terms of the other x_j or numerically. The *Jacobi* method of iteration is defined by

13.56
$$\begin{cases} f_1(x_1^{(k+1)}, x_2^{(k)}, \ldots, x_n^{(k)}) = 0 \\ f_2(x_1^{(k)}, x_2^{(k+1)}, \ldots, x_n^{(k)}) = 0 \\ \vdots \\ f_n(x_1^{(k)}, x_2^{(k)}, \ldots, x_n^{(k+1)}) = 0. \end{cases}$$

This method will converge for a sufficiently good initial approximation $x_1^{(0)}$, $x_2^{(0)}, \ldots, x_n^{(0)}$ to the true solution provided the spectral radius of the matrix

13.57
$$M = \begin{pmatrix} \dfrac{\partial f_1}{\partial x_1} & \dfrac{\partial f_1}{\partial x_2} \cdots \dfrac{\partial f_1}{\partial x_n} \\ \dfrac{\partial f_2}{\partial x_1} & \dfrac{\partial f_2}{\partial x_2} \cdots \dfrac{\partial f_2}{\partial x_n} \\ \vdots \\ \dfrac{\partial f_n}{\partial x_1} & \dfrac{\partial f_n}{\partial x_2} \cdots \dfrac{\partial f_n}{\partial x_n} \end{pmatrix}$$

is less than unity. The spectral radius of a matrix is the maximum of the absolute values of its *eigenvalues*, which are the solutions λ of the polynomial equation

13.58
$$\det(M - \lambda I) = 0,$$

where I is the identity matrix. (See Chapter 11.)

In our example (13.22), we can solve the equation $h(x, y) = 0$ for x and $g(x, y) = 0$ for y obtaining

13.59
$$\begin{cases} x = \sqrt{\dfrac{y^2}{3}}, \quad y \neq 0 \\ y = \sqrt{\dfrac{x^3 + 1}{3x}}. \end{cases}$$

The iterative process is defined by

13.60
$$\begin{cases} x_{n+1} = \sqrt{\dfrac{y_n^2}{3}} \\ y_{n+1} = \sqrt{\dfrac{x_n^3 + 1}{3x_n}}. \end{cases}$$

Let $x_0 = y_0 = 1$. We get

13.61 $\begin{cases} x_1 = \sqrt{\frac{1}{3}} \doteq 0.57735, & x_2 = 0.47140, & x_3 = 0.47905 \\ y_1 = \sqrt{\frac{2}{3}} \doteq 0.81650, & y_2 = 0.82974, & y_3 = 0.88384 \end{cases}$

etc. The process appears to be converging (slowly) to the solution $\bar{x} = 0.50000$ and $\bar{y} \doteq 0.86602$.

In the Jacobi method one uses $x_1^{(k)}$ in the calculation of $x_2^{(k+1)}$ even though $x_1^{(k+1)}$ is available at that point in the calculation. In the *Gauss-Seidel method* one uses at each step the latest available iterants. Thus we have

13.62 $\begin{cases} f_1(x_1^{(k+1)}, x_2^{(k)}, \ldots, x_n^{(k)}) = 0, & \text{(solve for } x_1^{(k+1)}) \\ f_2(x_1^{(k+1)}, x_2^{(k+1)}, \ldots, x_n^{(k)}) = 0, & \text{(solve for } x_2^{(k+1)}) \\ \quad \vdots \\ f_n(x_1^{(k+1)}, x_2^{(k+1)}, \ldots, x_n^{(k+1)}) = 0, & \text{(solve for } x_n^{(k+1)}). \end{cases}$

The *successive overrelaxation method* involves the use of a "relaxation factor," ω, and the iterative procedure

13.63 $$x_i^{(k+1)} = x_i^{(k)} + \omega(\tilde{x}_i^{(k+1)} - x_i^{(k)})$$

where for $i = 1, 2, \ldots, n$ we determine $\tilde{x}_i^{(k+1)}$ by

13.64 $$f_i(x_1^{(k+1)}, x_2^{(k+1)}, \ldots, x_{i-1}^{(k+1)}, \tilde{x}_i^{(k+1)}, x_{i+1}^{(k)}, \ldots, x_n^{(k)}) = 0.$$

Frequently one can increase the rapidity of convergence over that of the Gauss-Seidel method by suitable choice of ω.

We shall discuss these methods further in Chapter 16 when we consider iterative methods for solving systems of linear algebraic equations.

EXERCISES 4.13

1. Show that the system

$$g(x, y) = \sin x \cosh y - x + 2 = 0$$
$$h(x, y) = \cos x \sinh y - y = 0$$

is equivalent to the single equation

$$f(z) = \sin z - z + 2 = 0$$

where $z = x + iy$. (Note that $\sin(x + iy) = \sin x \cosh y + i \cos x \sinh y$ and $\cos(x + iy) = \cos x \cosh y - i \sin x \sinh y$). Compute $g(x, y)$ and $h(x, y)$ for $x = -3.0(0.1)3.0$ and

$y = -3.0(0.1)3.0$ and determine approximate solutions by the graphical method.

2. Solve the system of the preceding exercise by the generalized Newton method
   a) if $x_0 = \quad 2, y_0 = 0$
   b) if $x_0 = -1, y_0 = 1$
   c) if $x_0 = -1, y_0 = 2$.

   Compare the results for the starting values (b) with those obtained in Exercise 5, Section 4.9.

3. For the system of Exercise 1, carry out two steps of the gradient method with $x_0 = -1$, $y_0 = 1$. Use both (13.40) and (13.49) to determine the length of the step to be taken.

4. Use the Jacobi method to solve the system

$$g(x, y) = 2x - y = 0$$
$$h(x, y) = -x + 2y - 3 = 0$$

with starting values $x_0 = y_0 = 0$. Also do the same thing with the Gauss-Seidel method and with the successive overrelaxation method with $\omega = 1.2$.

5. Attempt to find the complex roots of the equation $z^3 - 2z + 2 = 0$ by letting $z = x + iy$ and solving the system

$$g(x, y) = x^3 - 3xy^2 - 2x + 2 = 0$$
$$h(x, y) = 3x^2y - y^3 - 2y = 0$$

for x and y. Use the Gauss-Seidel method with $x_0 = y_0 = 1$ and

$$h(x_{k+1}, y_k) = 0$$
$$g(x_{k+1}, y_{k+1}) = 0$$

6. Carry out the example (13.22) using the Gauss-Seidel method and also the successive overrelaxation method with $\omega = 1.5$.

7. For the system of Exercise 1, carry out two iterations of the Gauss-Seidel method with $x_0 = -1$, $y_0 = 1$. Write the equations in reverse order and carry out the process with $h(x_{n+1}, y_n) = 0$, $g(x_{n+1}, y_{n+1}) = 0$. Will the method converge to the solution nearest to $(-1, 1)$?

SUPPLEMENTARY DISCUSSION

The reader is referred to Traub [1964] and to Householder [1970] for excellent treatments of methods for solving a single equation, to Ostrowski [1966] for methods for solving a single equation and systems of equations, and to Ortega and Rheinboldt [1970] for systems of equations.

Section 4.4

The error estimate (4.31) is also applicable to $\tilde{f}(x)$ if we replace $\bar{x}$ by $\hat{x}$, where $\tilde{f}(\hat{x}) = 0$, provided that z_n is computed exactly.

Section 4.6

Ostrowski [1966] considered the iterative method

$$x_{n+1} = \frac{x_0 f(x_n) - x_n f(x_0)}{f(x_n) - f(x_0)}$$

and gave conditions for its convergence.

Section 4.7
The usage of the terms "consistent," "reciprocally consistent," and "completely consistent" follows that of Young [1971, 1971a].

Section 4.11
Our definitions of order of convergence are based on those given by Traub [1964] and others. More general definitions are given by Ortega and Rheinboldt (1970, Chapter 9).

Section 4.12
See Traub [1964] for a discussion of methods for accelerating the convergence of iterative methods with order of convergence greater than one. See Shanks [1955] for a discussion of methods for accelerating the convergence of sequences.

Section 4.13
The use of graphical methods for solving a pair of nonlinear equations is discussed by Hartree [1952]. The procedure described in the text is based on a contour plotting procedure used by Downing [1966] for the two-dimensional case.

CHAPTER 5

ROOTS OF POLYNOMIAL EQUATIONS

5.1 INTRODUCTION

We now consider the problem of solving the polynomial equation

1.1 $$P(x) = a_0 x^n + a_1 x^{n-1} + a_2 x^{n-2} + \cdots + a_n = 0$$

where the coefficients $a_0, a_1, \ldots, a_n$ are given numbers, which may be real or complex. As we shall show, (1.1) has precisely n roots, not necessarily distinct. However, even in the most common case where all coefficients are real, some or all of the roots may be complex. On the other hand, if the coefficients are real, then the complex roots occur in complex conjugate pairs, $a \pm ib$.

Any of the methods described in Chapter 4 for solving the more general equation

1.2 $$f(x) = 0$$

can, of course, be used to solve polynomial equations. However, in many cases, such methods can be modified to take advantage of the special properties of polynomials. In addition, there are several other methods which are applicable primarily or exclusively to polynomial equations.

In the case of (1.2) we were usually content to find a few, or even one, real root. On the other hand, we frequently wish to find all roots, real and complex, of (1.1). This can be done, in principle at least, if we have a method capable of finding one root. For, as shown in Section 5.2, once a root has been found, one can determine a "reduced polynomial" which has the same zeros as the original polynomial except for the one zero already found.* One can then seek a zero of the reduced polynomial and, having found this, obtain another reduced polynomial. Continuing in this way, one can determine all of the roots of (1.1).

One of the most popular methods for solving polynomial equations is the Newton method. The relative ease in computing the derivative of a polynomial makes this method particularly convenient to apply. We discuss the Newton method and related methods, including the secant method, Lin's method, and the Lin-Bairstow method in Section 5.3.

In Section 5.4 we discuss Muller's method and Cauchy's method. The use of a procedure based on Muller's method, with Cauchy's method used when con-

---

* If the zero is of multiplicity greater than one, then it will also be a zero of the reduced polynomial but with its multiplicity reduced by one.

vergence difficulties appear, is recommended for the reader who wishes to develop a fairly reliable and accurate program with a minimum of effort. It is felt that the increased reliability of the procedure over the Newton method more than justifies the slight increase in computational effort.

For the effective application of most of the methods we shall consider, it is desirable to have means for locating roots. In Section 5.5 we give several procedures for determining the existence or nonexistence of real roots in certain intervals and of complex roots in certain circles. Such estimates are useful not only in finding suitable starting values for use with iterative processes but also in assessing the accuracy of roots which have been accepted according to the criteria of Section 5.6.

Section 5.6 is concerned with criteria for accepting a root. In general, no matter what α is found we cannot expect that $P(\alpha)$ will vanish. A crude procedure would be to assign a value of ε and to accept a root if $|P(\alpha)| \leq \varepsilon$. A more sophisticated procedure is to require that $|P(\alpha)| \leq q\delta(P(\alpha))$ where $\delta(P(\alpha))$ is a bound on the rounding error in the calculation of $P(\alpha)$ and q is a small integer (e.g., $q = 4$). Having accepted a root on this basis, one can then use the methods of Section 5.5 to assess its accuracy.

In some cases, even though a number α has been found which satisfies the convergence criterion, it may be desirable to find a number α^*, near α, which also satisfies the convergence criterion but has other properties as well. For example, α^* might satisfy the convergence criterion for the derivative polynomial and hence be a double root. Another possibility in the case where $P(x)$ has real coefficients would be for α^* to be real, in which case the reduced polynomial, $Q(x)$, will have real coefficients; alternatively, $\bar{\alpha}^*$ might satisfy the convergence criterion for $Q(x)$, in which case the reduced polynomial obtained by dividing out $\bar{\alpha}^*$ and α^* will have real coefficients.

Sections 5.7 and 5.8 are concerned with matrix-related methods such as Bernoulli's method and a method referred to as the "IP method," which is based on the inverse power method for finding eigenvalues of a matrix. These methods appear to have many advantages over other methods since they are less sensitive to the choice of starting values. The discussion of these methods presupposes a familiarity with many of the computational methods of linear algebra which are described in Chapters 11, 12, and 14. Hence, a study of Sections 5.7 and 5.8 should be deferred until such familiarity has been attained.

In Section 5.9 we describe some considerations involved in the construction of a "polyalgorithm" for solving polynomial equations. (A polyalgorithm is a combination of several algorithms designed to solve a class of problems.) The discussion is based on programs developed at the Computation Center of The University of Texas at Austin and described by Champagne [1964], B. Wilkinson [1969] and by Young, et al. [1971].

While it is only possible to consider in detail a few of the many available methods for solving polynomial equations, we give in Section 5.10 a brief discussion of certain other methods which have been found to be useful. These include

Laguerre's method, a procedure of Bareiss involving root squaring and a resultant procedure, a method based on the quotient difference algorithm, a three-stage variable shift procedure of Jenkins and Traub, and a family of globally convergent, multiplicity independent methods due to Shröder and McDonald.

5.2 GENERAL PROPERTIES OF POLYNOMIALS

In this section we review some of the properties of polynomials which we shall use later on. First, let us review the definition of a polynomial. A *polynomial* $P(x)$ is a function of the form

2.1 $$P(x) = a_0 x^n + a_1 x^{n-1} + \cdots + a_n = \sum_{i=0}^{n} a_i x^{n-i},$$

where the *coefficients* $a_0, a_1, \ldots, a_n$ are real or complex numbers. We assume that $a_0 \neq 0$ unless $n = 0$. The integer n is the *degree* of the polynomial.

Evidently $P(x) \equiv 0$ and $P(x) \equiv 2$ are polynomials of degree zero, while $P(x) = 2x^3 - 1$ is a polynomial of degree three.

An important property of polynomials is the following:

2.2 Theorem. If two polynomials $P(x)$ and $Q(x)$ are equal for all values of x, then their degrees and corresponding coefficients are equal.

Proof. Let $P(x) = a_0 x^n + a_1 x^{n-1} + \cdots + a_n$ and let $Q(x) = b_0 x^m + b_1 x^{m-1} + \cdots + b_m$. Evidently $R(x) \equiv P(x) - Q(x)$ is a polynomial of the form $c_0 x^r + c_1 x^{r-1} + \cdots + c_r$ where $r \leq \max(n, m)$. Since $R(x) \equiv 0$, it follows that R and all of its derivatives vanish everywhere, in particular for $x = 0$. Since $R(0) = 0$ we have $c_r = 0$. Since $R'(x) = rc_0 x^{r-1} + (r - 1)c_1 x^{r-2} + \cdots + c_{r-1}$ and $R'(0) = 0$, we have $c_{r-1} = 0$. Similarly, $c_{r-2} = c_{r-3} = \cdots = c_1 = c_0 = 0$.

Suppose now that $n \geq m$. Then if $n = 0$ we have $m = 0$ and $c_0 = a_0 - b_0 = 0$; hence $a_0 = b_0$. If $n > 0$, then $a_0 \neq 0$. Hence $a_0 = b_0$; otherwise the coefficient of x^n in $R(x)$ will not vanish. Hence $b_0 \neq 0$ and $m = n$. Similarly, if $m \geq n$, we have $m = n$. Thus $P(x)$ and $Q(x)$ have the same degree and their corresponding coefficients are equal.

It follows from Theorem 2.2 that if $P(x)$ is a polynomial which vanishes for all x, then all coefficients of $P(x)$ vanish and hence the degree of $P(x)$ is zero.

We now show that for any number α there exist constants $\hat{a}_0, \hat{a}_1, \ldots, \hat{a}_n$ such that

2.3 $$P(x) = \hat{a}_0 (x - \alpha)^n + \hat{a}_1 (x - \alpha)^{n-1} + \cdots + \hat{a}_n.$$

Moreover,

2.4 $$\hat{a}_p = \frac{P^{(n-p)}(\alpha)}{(n - p)!}, \qquad p = 0, 1, 2, \ldots, n.$$

This, of course, follows from Taylor's theorem (Theorem A-3.64) since the $(n + 1)$st and all higher derivatives of $P(x)$ vanish. The result can also be shown directly by noting that

2.5 $$x^p = ((x - \alpha) + \alpha)^p = (x - \alpha)^p + \binom{p}{1}\alpha(x - \alpha)^{p-1} + \cdots + \alpha^p.$$

Hence $P(x)$ is a polynomial in $x - \alpha$ of exact degree n. Letting $x = \alpha$ in (2.3), we get $\hat{a}_n = P(\alpha)$. The general result (2.4) is obtained by successively differentiating (2.3) and letting $x = \alpha$.

As an immediate consequence of (2.3) we have

2.6 Theorem. If $P(x)$ is a polynomial of degree n, then for any α there exists a unique polynomial $Q(x)$ such that

2.7 $$P(x) \equiv (x - \alpha)Q(x) + P(\alpha).$$

If $n \geq 1$, the degree of $Q(x)$ is $n - 1$; otherwise $Q(x) \equiv 0$.

Proof. Evidently (2.7) holds, by (2.3), with

2.8 $$Q(x) = \hat{a}_0(x - \alpha)^{n-1} + \hat{a}_1(x - \alpha)^{n-2} + \cdots + \hat{a}_{n-1}$$

since $\hat{a}_n = P(\alpha)$. To prove the uniqueness suppose that $n \geq 1$ and for some polynomial $Q^*(x)$ of degree $n - 1$ we have $P(x) \equiv (x - \alpha)Q^*(x) + P(\alpha)$. Then

2.9 $$(x - \alpha)R(x) \equiv 0$$

where

2.10 $$R(x) \equiv Q(x) - Q^*(x).$$

Evidently $R(x) = 0$ for all x except possibly when $x = \alpha$. But by (2.9) we have

2.11 $$\frac{d}{dx}[(x - \alpha)R(x)]_{x=\alpha} = [(x - \alpha)R'(x) + R(x)]_{x=\alpha} = R(\alpha) = 0$$

so that $Q(x) \equiv Q^*(x)$ and the uniqueness follows.

2.12 Corollary. If $P(x)$ is a polynomial of degree $n \geq 1$ and if $P(\alpha) = 0$ then there exists a unique polynomial of degree $n - 1$ such that

2.13 $$P(x) \equiv (x - \alpha)Q(x).$$

The polynomial $Q(x)$ is often referred to as the *reduced polynomial*.

By Theorem 4-1.9, since any polynomial has continuous derivatives of all orders for all values of x we have

2.14 Theorem. Let $P(x)$ be a polynomial of degree $n \geq 1$. The multiplicity of α as a zero of $P(x)$ is m if and only if

2.15 $P(\alpha) = P'(\alpha) = \cdots = P^{(m-1)}(\alpha) = 0$

and

2.16 $P^{(m)}(\alpha) \neq 0.$

We are now in a position to relate the zeros of the polynomial $P(x)$ which has a zero α and the zeros of the reduced polynomial.

2.17 Theorem. Let $P(x)$ be a polynomial of degree $n \geq 1$ and let α be a zero of $P(x)$. If $Q(x)$ is the polynomial of degree $n - 1$ such that (2.13) holds, then

a) the multiplicity of α as a zero of $P(x)$ is m if and only if the multiplicity of α as a zero of $Q(x)$ is $m - 1$. (If $m = 1$, then α is not a zero of $Q(x)$.)
b) if $\beta \neq \alpha$, then the multiplicity of β as a zero of $P(x)$ is the same as its multiplicity as a zero of $Q(x)$.

Proof. By successively differentiating (2.13) we obtain

2.18 $P^{(k)}(x) \equiv (x - \alpha)Q^{(k)}(x) + kQ^{(k-1)}(x), \qquad k = 1, 2, \ldots, n$

and hence

2.19 $P^{(k)}(\alpha) = kQ^{(k-1)}(\alpha), \qquad k = 1, 2, \ldots, n.$

By Theorem 2.14, α is a zero of $P(x)$ of multiplicity m if and only if (2.15) and (2.16) hold. But by (2.19) these conditions hold if and only if $Q(\alpha) = Q'(\alpha) = \cdots = Q^{(m-2)}(\alpha) = 0$ and $Q^{(m-1)}(\alpha) \neq 0$, i.e., if and only if α is a zero of $Q(x)$ of multiplicity $m - 1$.

Similarly, by (2.18) we have

2.20 $P^{(k)}(\beta) = (\beta - \alpha)Q^{(k)}(\beta) + kQ^{(k-1)}(\beta), \qquad k = 1, 2, \ldots, n.$

Evidently, by Theorem 2.14, to prove (b) it is sufficient to show that $P(\beta) = P'(\beta) = \cdots = P^{(r-1)}(\beta) = 0$ and $P^{(r)}(\beta) \neq 0$ if and only if $Q(\beta) = Q'(\beta) = \cdots = Q^{(r-1)}(\beta) = 0$ and $Q^{(r)}(\beta) \neq 0$. But this follows from (2.13) with $x = \beta$ and (2.20). Hence the proof of Theorem 2.17 is complete.

From Corollary 2.12 and Theorem 2.17 we have[*]

---

[*] An alternative proof can be given based on the use of (2.3) and Theorem 2.14. The uniqueness of $S(x)$ can be shown as in the proof of Theorem 2.6.

2.21 Corollary. If $P(x)$ is a polynomial of degree $n \geq 1$, then α is a zero of $P(x)$ of multiplicity m if and only if there exists a unique polynomial $S(x)$ of degree $n - m$ such that

2.22 $$P(x) \equiv (x - \alpha)^m S(x)$$

and such that

2.23 $$S(\alpha) \neq 0.$$

Moreover, if $\beta \neq \alpha$, then the multiplicity of β as a zero of $P(x)$ is the same as its multiplicity as a zero of $S(x)$.

2.24 Theorem. Let $P(x)$ be a polynomial which does not vanish identically and which has distinct zeros $\alpha_1, \alpha_2, \ldots, \alpha_s$ with multiplicities at least $v_1, v_2, \ldots, v_s$, respectively. Then there exists a unique polynomial $T(x)$ such that

2.25 $$P(x) \equiv (x - \alpha_1)^{v_1}(x - \alpha_2)^{v_2} \ldots (x - \alpha_s)^{v_s} T(x).$$

Proof. If $P(x)$ does not vanish identically, then since $P(\alpha_1) = 0$, the degree of $P(x)$ must be at least equal to one. Otherwise, $P(x) \equiv c$, a constant, and c would equal zero. The result (2.25) follows by repeated application of Corollary 2.21.

As a consequence of Theorem 2.24 we have

2.26 Corollary. Let $P(x)$ be a polynomial of degree n or less which has zeros of multiplicities at least $v_1, v_2, \ldots, v_s$ at the distinct points $\alpha_1, \alpha_2, \ldots, \alpha_s$, respectively. If

2.27 $$\sum_{j=1}^{s} v_j > n,$$

then

2.28 $$P(x) \equiv 0.$$

Moreover, if the degree of $P(x)$ is $n \geq 1$, and if the multiplicities of $\alpha_1, \alpha_2, \ldots, \alpha_s$ are $v_1, v_2, \ldots, v_s$, respectively, where

2.29 $$\sum_{j=1}^{s} v_j = n,$$

then for some unique constant c we have

2.30 $$P(x) \equiv c(x - \alpha_1)^{v_1}(x - \alpha_2)^{v_2} \ldots (x - \alpha_s)^{v_s}.$$

Proof. Evidently, by (2.25) the degree of $P(x)$ is at least $\sum_{j=1}^{s} v_j$ unless $T(x) \equiv 0$. Hence, by (2.27) we have $T(x) \equiv 0$ and (2.28) holds.

If the multiplicities of $\alpha_1, \alpha_2, \ldots, \alpha_s$ are $v_1, v_2, \ldots, v_s$, respectively, and if (2.29) holds, then the degree of $T(x)$ is zero. Since $P(x) \not\equiv 0$, it follows that $T(x) \not\equiv 0$; hence $T(x)$ is a constant and (2.30) holds. The uniqueness of the constant c follows from equating the coefficients of x^n in both sides of (2.30).

Given α, the determination of $Q(x)$ of Theorem 2.6 could be carried out using (2.8) and (2.4). However, this would involve computing various derivatives of $P(x)$. It is generally simpler to use the following procedure to determine $Q(x)$ and $P(\alpha)$. Let us define the numbers $z_0(\alpha), z_1(\alpha), \ldots, z_n(\alpha)$ by

2.31
$$\begin{cases} z_0(\alpha) = a_0 \\ z_k(\alpha) = \alpha z_{k-1}(\alpha) + a_k, \qquad k = 1, 2, \ldots, n. \end{cases}$$

We now show that if $n \geq 1$, then

2.32
$$Q(x) = z_0(\alpha)x^{n-1} + z_1(\alpha)x^{n-2} + \cdots + z_{n-1}(\alpha)$$

and

2.33
$$P(\alpha) = z_n(\alpha).$$

But by (2.31) we have

2.34
$$\begin{aligned}(x - \alpha)&[z_0(\alpha)x^{n-1} + z_1(\alpha)x^{n-2} + \cdots + z_{n-1}(\alpha)] \\ &= z_0(\alpha)x^n + (z_1(\alpha) - \alpha z_0(\alpha))x^{n-1} + (z_2(\alpha) - \alpha z_1(\alpha))x^{n-2} \\ &\quad + \cdots + (z_{n-1}(\alpha) - \alpha z_{n-2}(\alpha))x - \alpha z_{n-1}(\alpha) \\ &= a_0 x^n + a_1 x^{n-1} + \cdots + a_{n-1}x - \alpha z_{n-1}(\alpha) \\ &= P(x) - z_n(\alpha). \end{aligned}$$

Letting $x = \alpha$ we get (2.33). Evidently we have

2.35
$$P(x) \equiv (x - \alpha)Q^*(x) + P(\alpha)$$

where $Q^*(x)$ denotes the right member of (2.32). By the uniqueness of $Q(x)$ (see Theorem 2.6), it follows that $Q(x) \equiv Q^*(x)$ and hence (2.32) holds.

The method just described for evaluating $P(\alpha)$ can also be justified as follows. We note that if, as in Section 2.9, we write $P(x)$ in the form

2.36
$$P(x) = (\;(\;\cdots\;(\;(\,(a_0)x + a_1)x + a_2)x + \cdots)x + a_n)$$
$$\quad\;\; {\scriptstyle n \;\; n-1 \qquad\;\; 2\; 1\; 0\;\; 0 \qquad 1 \qquad\;\; 2 \qquad n-1 \qquad\;\; n}$$

the number between the parentheses labelled "0" is $z_0(\alpha)$, the number between the parentheses labelled "1" is $z_1(\alpha)$, etc. Moreover, the $z_k(\alpha)$ clearly satisfy the relations (2.31) and hence (2.33) holds.

As an example, consider the polynomial $P(x) = 2x^3 - 3x + 1$. If $\alpha = 2$, we have

2.37
$$\begin{cases} z_0(\alpha) = 2 \\ z_1(\alpha) = 2(2) + 0 = 4 \\ z_2(\alpha) = 2(4) - 3 = 5 \\ z_3(\alpha) = 2(5) + 1 = 11. \end{cases}$$

The value of $z_3(\alpha)$ with $\alpha = 2$ agrees with $P(2) = 2(2)^3 - 3(2) + 1 = 11$. Moreover, if we let $Q(x) = z_0(\alpha)x^2 + z_1(\alpha)x + z_2(\alpha) = 2x^2 + 4x + 5$, then we have

2.38 $P(x) = (x - 2)Q(x) + 11.$

Indeed, by direct multiplication we have

2.39 $(x - 2)(2x^2 + 4x + 5) + 11 = 2x^3 - 3x + 1 = P(x).$

It is convenient for hand computation to arrange the work in the form indicated below.

$$\begin{array}{cccc|c} a_0 & a_1 & a_2 & a_3 & \alpha \\ & \alpha z_0 & \alpha z_1 & \alpha z_2 & \\ \hline z_0 & z_1 & z_2 & z_3 & \end{array}$$

This is the familiar "synthetic division" process of elementary algebra. The numbers z_i are obtained by adding the numbers a_i and αz_{i-1}. The values of $P(\alpha) = z_3$ and the coefficients of $Q(x) = z_0 x^2 + z_1 x + z_2$ are obtained. Thus in the above example we would have

$$\begin{array}{cccc|c} 2 & 0 & -3 & 1 & 2 \\ & 4 & 8 & 10 & \\ \hline 2 & 4 & 5 & 11 & \end{array}$$

and hence $P(2) = 11$, $Q(x) = 2x^2 + 4x + 5$ as before.

On the other hand, if $\alpha = 1$, we have

$$\begin{array}{cccc|c} 2 & 0 & -3 & 1 & 1 \\ & 2 & 2 & -1 & \\ \hline 2 & 2 & -1 & 0 & \end{array}$$

Here we see that $P(1) = 0$. The reduced polynomial is $Q(x) = 2x^2 + 2x - 1$. One can verify directly that

$$(x - 1)Q(x) = (x - 1)(2x^2 + 2x - 1) = 2x^3 - 3x + 1 = P(x).$$

The problem of finding the zeros of (2.1) is solvable in principle. In fact we have

2.40 Theorem. If $P(x)$ is a polynomial of degree $n \geq 1$, then there exist distinct numbers $\alpha_1, \alpha_2, \ldots, \alpha_s$, possibly complex (which are unique except for order) and positive integers $v_1, v_2, \ldots, v_s$ such that for some unique constant c we have

2.41
$$P(x) \equiv c(x - \alpha_1)^{v_1}(x - \alpha_2)^{v_2} \cdots (x - \alpha_s)^{v_s}$$

where

2.42
$$\sum_{j=1}^{s} v_j = n.$$

Proof. We use induction. The result is true for $n = 1$ because in this case we have, since $a_0 \neq 0$,

2.43
$$P(x) \equiv a_0 x + a_1 = a_0\left(x + \frac{a_1}{a_0}\right)$$

which has the form (2.41) with $c = a_0$, $\alpha_1 = -a_1/a_0$. Let us assume the result to be true for n and let $P(x)$ be a polynomial of degree $n + 1$. By the fundamental theorem of algebra (Theorem A-8.6), there exists at least one zero, say α, of $P(x)$. Moreover, by Corollary 2.12 we have

2.44
$$P(x) \equiv (x - \alpha)Q(x)$$

where $Q(x)$ is a unique polynomial of degree n. The theorem now follows from the induction hypothesis and from Theorem 2.17.

As an example, let us consider the polynomial

2.45
$$P(x) = 2x^3 - 8x^2 + 10x - 4.$$

We note that $x = 1$ is a zero of $P(x)$. The reduced polynomial is $Q(x) = 2x^2 - 6x + 4$. Again $x = 1$ is a zero of the reduced polynomial and we have

2.46
$$P(x) = (x - 1)^2(2x - 4) = 2(x - 1)^2(x - 2)$$

which has the form (2.41).

By means of the representation (2.41) of the general polynomial $P(x)$ we have

2.47
$$P(x) = c\{x^n - s_1 x^{n-1} + s_2 x^{n-2} + \cdots + (-1)^n s_n\},$$

where the symmetric functions $s_1, s_2, \ldots, s_n$ are defined by

2.48

$$
\begin{cases}
s_1 = \sum_{i=1}^{n} \alpha_i \\[2ex]
s_2 = \sum_{\substack{i,j=1 \\ j<i}}^{n} \alpha_i \alpha_j \\[2ex]
\vdots \\[1ex]
s_n = \alpha_1 \alpha_2 \cdots \alpha_n.
\end{cases}
$$

Thus for $k = 1, 2, \ldots, n$, s_k is the sum of the products of the roots taken k at a time. Equating coefficients of (2.47) and (2.1) we have

2.49
$$
\frac{a_k}{a_0} = (-1)^k s_k, \qquad k = 1, 2, \ldots, n.
$$

Thus we have a simple relation between the symmetric functions and the coefficients of the polynomial.

The procedure described above for evaluating a polynomial can be extended to allow the evaluation of its derivatives. Let us define the polynomials $z_0(x)$, $z_1(x), \ldots, z_n(x)$ by the recurrence relation

2.50
$$
\begin{cases}
z_0(x) = a_0 \\
z_k(x) = x z_{k-1}(x) + a_k, \qquad k = 1, 2, \ldots, n.
\end{cases}
$$

This is the same as (2.31) with α replaced by x. Differentiating we have

2.51
$$
\begin{cases}
z_0'(x) = 0 \\
z_k'(x) = x z_{k-1}'(x) + z_{k-1}(x), \qquad k = 1, 2, \ldots, n.
\end{cases}
$$

We note that the determination of $z_0'(x), z_1'(x), \ldots, z_n'(x)$ by (2.51) is the same in terms of the quantities $0, z_0(x), z_1(x), \ldots, z_{n-1}(x)$ as the determination of $z_0(x)$, $z_1(x), \ldots, z_n(x)$ is in terms of $a_0, a_1, \ldots, a_n$. In other words, if in (2.50) we replaced a_0 by 0, a_1 by $z_0(x), \ldots, a_n$ by $z_{n-1}(x)$ then instead of $z_0(x), z_1(x), \ldots, z_n(x)$ we would get $z_0'(x), z_1'(x), \ldots, z_n'(x)$. Similarly, if we differentiate again we get

2.52
$$
\begin{cases}
z_0''(x) = 0 \\
z_k''(x) = x z_{k-1}''(x) + 2 z_{k-1}'(x), \qquad k = 1, 2, \ldots, n
\end{cases}
$$

or

2.53
$$
\begin{aligned}
&\tfrac{1}{2} z_0''(x) = 0 \\
&\tfrac{1}{2} z_k''(x) = x(\tfrac{1}{2} z_{k-1}''(x)) + z_{k-1}'(x), \qquad k = 1, 2, \ldots, n.
\end{aligned}
$$

In general, we have, for $p \geqq 1$,

$$
\textbf{2.54} \begin{cases} \dfrac{1}{p!} z_0^{(p)}(x) = 0 \\[2mm] \left(\dfrac{1}{p!} z_k^{(p)}(x)\right) = x\left(\dfrac{1}{p!} z_{k-1}^{(p)}(x)\right) + \left(\dfrac{1}{(p-1)!} z_{k-1}^{(p-1)}(x)\right), \qquad k = 1, 2, \ldots, n. \end{cases}
$$

Again the numbers

$$
\frac{1}{p!} z_0^{(p)}(x), \qquad \frac{1}{p!} z_1^{(p)}(x), \ldots, \frac{1}{p!} z_n^{(p)}(x)
$$

are determined from

$$
0, \qquad \frac{1}{(p-1)!} z_0^{(p-1)}, \ldots, \frac{1}{(p-1)!} z_{n-1}^{(p-1)}
$$

in the same way. As a matter of fact, the computation of $P(x)$ and all of its derivatives can be carried out using one table, as illustrated below for the case $n = 3$.

				a_0	a_1	a_2	a_3	x
					xz_0	xz_1	xz_2	
			0	z_0	z_1	z_2	$z_3 = P(x)$	
				xz'_0	xz'_1	xz'_2		
		0		z'_1	z'_2	$z'_3 = P'(x)$		
			$x\left(\frac{1}{2}z''_0\right)$	$x\left(\frac{1}{2}z''_1\right)$	$x\left(\frac{1}{2}z''_2\right)$			
	0	$\frac{1}{2}z''_0$	$\frac{1}{2}z''_1$	$\frac{1}{2}z''_2$	$\frac{1}{2}z''_3 = \frac{1}{2}P''(x)$			
	$x\left(\frac{1}{3!}z_0^{(3)}\right)$	$x\left(\frac{1}{3!}z_1^{(3)}\right)$	$x\left(\frac{1}{3!}z_2^{(3)}\right)$					
$\frac{1}{3!}z_0^{(3)}$	$\frac{1}{3!}z_1^{(3)}$	$\frac{1}{3!}z_2^{(3)}$	$\frac{1}{3!}z_3^{(3)} = \frac{1}{3!}P^{(3)}(x)$					

We remark that all numbers to the left of the vertical dotted line are zero and may be omitted in actual calculation. In the case of the polynomial $P(x) = 2x^3 - 8x^2 + 10x - 4$ we have, for $x = 1$,

$$
\begin{array}{rrrr|r}
2 & -8 & 10 & -4 & \,1 \\
  & 2 & -6 & 4 & \\
\hline
2 & -6 & 4 & 0 & \\
  & 2 & -4 & & \\
\hline
2 & -4 & 0 & & \\
  & 2 & & & \\
\hline
2 & -2 & & & \\
\hline
2 & & & & \\
\end{array}
$$

Thus we have

$$P(1) = 0, \qquad P'(1) = 0, \qquad \frac{1}{2}P''(1) = -2, \qquad \frac{1}{3!}P^{(3)}(1) = 2.$$

Moreover, since $x - 1$ is a root of $P(x) = 0$, the reduced polynomial is

$$Q(x) = 2x^2 - 6x + 4$$

which also vanishes at $x = 1$.

The procedure outlined above affords a convenient method for computing derivatives and also for "shifting the origin," i.e., expressing $P(x)$ in powers of $(x - \alpha)$ for any α. Thus, if $P(x)$ is a polynomial of degree n, then the coefficients of the powers of $(x - \alpha)^p$ given in the Taylor series expansion (2.3), namely the $(1/p!)P^{(p)}(\alpha) = (1/p!)z_n^{(p)}(\alpha)$ obtained in the above procedure are precisely the coefficients used in this expansion.

In the case $P(x) = 2x^3 - 8x^2 + 10x - 4$ we have

$$P(x) = 2(x - 1)^3 - 2(x - 1)^2.$$

Therefore, $x = 1$ is a double zero of $P(x)$.

Let us express $P(x)$ in powers of $x - 3$. We have

$$
\begin{array}{rrrr|r}
2 & -8 & 10 & -4 & \,3 \\
  & 6 & -6 & 12 & \\
\hline
2 & -2 & 4 & \,⑧ & \\
  & 6 & 12 & & \\
\hline
2 & 4 & ⑯ & & \\
  & 6 & & & \\
\hline
② & ⑩ & & & \\
\end{array}
$$

so that

$$P(x) = 2(x - 3)^3 + 10(x - 3)^2 + 16(x - 3) + 8.$$

In some cases, however, rather than using the procedure described above, it is preferable to compute the derivatives using polynomials obtained by analytically differentiating the polynomial $P(x)$. Thus we have

2.55
$$\begin{cases} P'(x) = na_0x^{n-1} + (n-1)a_1x^{n-2} + \cdots + a_{n-1} \\ P''(x) = n(n-1)a_0x^{n-2} + (n-1)(n-2)a_1x^{n-3} + \cdots + a_{n-2} \end{cases}$$

etc. By storing the values of the coefficients of the derivative polynomials we can in general calculate the derivatives for particular values of x with fewer operations and with fewer rounding errors.

In the example considered above we have

$$\begin{array}{ll} P(x) = 2x^3 - 8x^2 + 10x - 4; & P(3) = 8 \\ P'(x) = 6x^2 - 16x + 10; & P'(3) = 16 \\ P''(x) = 12x - 16; & P''(3) = 20 \\ P'''(x) = 12; & P'''(3) = 12. \end{array}$$

Hence

$$\begin{aligned} P(x) &= \tfrac{1}{6}P'''(3)(x-3)^3 + \tfrac{1}{2}P''(3)(x-3)^2 + P'(3)(x-3) + P(3) \\ &= 2(x-3)^3 + 10(x-3)^2 + 16(x-3) + 8 \end{aligned}$$

as before.

Let us now consider the problem of evaluating $P(x)$ for a complex value of the argument $\alpha = \beta + i\gamma$, where the coefficients $a_k = b_k + ic_k$ are complex. We can again use the recursion relation (2.31). Letting $z_k(\alpha) = Q_k + iR_k$ we have

2.56
$$\begin{cases} Q_0 = b_0, R_0 = c_0 \\ Q_k = Q_{k-1}\beta - R_{k-1}\gamma + b_k, & k = 1, 2, \ldots, n \\ R_k = R_{k-1}\beta + Q_{k-1}\gamma + c_k, & k = 1, 2, \ldots, n. \end{cases}$$

The work can be arranged as follows:

b_0 c_0	b_1	c_1	b_2	c_2	b_3	c_3	$\beta + i\gamma$
	$Q_0\beta - R_0\gamma$	$R_0\beta + Q_0\gamma$	$Q_1\beta - R_1\gamma$	$R_1\beta + Q_1\gamma$	$Q_2\beta - R_2\gamma$	$R_2\beta + Q_2\gamma$	
Q_0 R_0	Q_1	R_1	Q_2	R_2	Q_3	R_3	

Thus, for example, if $P(x) = (1+i)x^3 + 2$ we can evaluate $P(1-i)$ as follows:

1	1	0	0	0	0	2	0	$1-i$
		2	0	2	-2	0	-4	
1	1	2	0	2	-2	2	-4	

Thus $P(1-i) = 2 - 4i$, as can be verified from $(1+i)(1-i)^3 + 2 = 2(1-i)^2 + 2 = -4i + 2$. Derivatives can be computed in a similar manner.

It may be preferable to use the above procedure rather than complex arithmetic subroutines, especially if the a_k are real.

EXERCISES 5.2

1. Find a, b, and c so that $(a + b)x^3 + (a - b)x^2 + 1 \equiv 3x^3 + 5ax^2 + cx + 1$.

2. If $P(x) = x^4 - 5x^3 + 10x^2 - 10x + 4$ find the unique polynomial $Q(x)$ such that $P(x) \equiv (x - \alpha)Q(x) + P(\alpha)$ both for the case $\alpha = -1$ and for the case $\alpha = 1$.

3. Prove that if $P(x)$ and $Q(x)$ are polynomials such that $(x - \alpha)P(x) \equiv (x - \alpha)Q(x)$, then $P(x) \equiv Q(x)$.

4. Express $P(x) = 3x^4 - 7x^2 + 10x - 5$ in powers of $(x - 1)$ using Taylor's theorem and also by the shift of origin procedure described in the text. How many real zeros of $P(x)$ are greater than unity?

5. Determine the multiplicity of the root $\alpha = 1$ of the polynomial equation $P(x) = x^5 - 2x^4 + 4x^3 - x^2 - 7x + 5 = 0$.

6. Determine the unique fourth-degree polynomial which has zeros at ± 1 and $\pm i$ and has the value $1 + i$ at $z = 2$.

7. Evaluate the polynomial $P(z) = z^3 - (2 + i)z + 1$ for $z = 2 + 3i$ using (2.56) and the procedure described in the text. Check by direct computation.

8. Let $P(x)$ and $Q(x)$ be polynomials of degree n and m, respectively, where $1 \leq m \leq n$. Show that there exists a unique polynomial $S(x)$ of degree $n - m$ and a unique polynomial $T(x)$ of degree $m - 1$ or less such that

$$P(x) \equiv Q(x)S(x) + T(x).$$

5.3 THE NEWTON METHOD AND RELATED METHODS

In this section we consider the Newton method and related methods, including the secant method and the Lin and Lin-Bairstow methods, the latter two methods being designed to apply specifically to polynomial equations.

As in the case of the more general equation (1.2), the application of the secant method to the polynomial equation (1.1) involves only one functional evaluation per iteration. However, the method suffers from a number of disadvantages (some of which are shared by the Newton method). Thus, as already noted, the method may fail near a double root even though both of the previous iterants x_n and x_{n-1} are arbitrarily close to the root, since we may have $f(x_n) = f(x_{n-1})$. If the polynomial has real coefficients and if real starting values are used, one cannot hope to get complex roots. This is true also of the Newton method. It appears that the use of the secant method should be restricted primarily to the case where one is near a simple real root.

For a general nonlinear equation, the work required using the Newton method may be much greater than for the secant method if the computation of the derivative of $f(x)$ is complicated. In the case of a polynomial equation, however, the derivative of $f(x)$ is a polynomial and is no more difficult to compute than $f(x)$ itself. Moreover, as described in Section 5.2), the derivative may be computed by repeated synthetic division. In the general case, $f'(x)$ may be a considerably different kind of function than $f(x)$. The convergence of the Newton method is somewhat faster (of order 2 instead of 1.618 as in the secant method) but more important, the method will converge, in principle at least, to a root of any multiplicity provided we start out close enough to the root. Moreover, if one suspects a multiple root, then one can use the acceleration technique described in Section 4.12.

As an example of the application of the Newton method to polynomial equations, let us consider the problem of solving

3.1
$$P(z) = x^3 - 2 = 0.$$

With the starting value $x_0 = 1$, we have

$$
\begin{array}{rrrr|r}
1 & 0 & 0 & -2 & \underline{\;1\;} \\
  & 1 & 1 & 1 & \\
\hline
1 & 1 & 1 & -1 & \\
  & 1 & 2 & & \\
\hline
1 & 2 & 3 & &
\end{array}
$$

Hence $P(1) = -1$, $P'(1) = 3$, and $x_1 = 1 + 1/3 \doteq 1.333$

$$
\begin{array}{rrrr|r}
1 & 0 & 0 & -2 & \underline{\;1.333\;} \\
  & 1.333 & 1.777 & 2.369 & \\
\hline
1 & 1.333 & 1.777 & 0.369 & \\
  & 1.333 & 3.553 & & \\
\hline
1 & 2.666 & 5.330 & &
\end{array}
$$

3.2
$$x_2 \doteq 1.333 - (0.369/5.33) \doteq 1.333 - 0.069 = 1.264.$$

Thus after two iterations we have an approximate value of 1.264 as compared with the true value $\bar{x} = \sqrt[3]{2} \doteq 1.260$. Evidently the process is converging to the root $\bar{x} \doteq 1.260$. Clearly, for there to be any possibility of convergence to one of the complex roots $-0.630 \pm 1.091i$ we would have to use a complex starting value, for example $x_0 = i$. In this case we have, following the procedure of the previous section,

$$
\begin{array}{rr|rr|rr|rr|rr}
1 & 0 & 0 & 0 & 0 & 0 & -2 & 0 & \underline{0} & \underline{1} \\
  &   & 0 & 1 & -1 & 0 & 0 & -1 & & \\
\hline
1 & 0 & 0 & 1 & -1 & 0 & -2 & -1 & & \\
  &   & 0 & 1 & -2 & 0 & & & & \\
\hline
1 & 0 & 0 & 2 & -3 & 0 & & & &
\end{array}
$$

Thus we have $P(i) = -2 - i$ and $P'(i) = -3$ and

3.3 $$x_1 = x_0 - \frac{P(i)}{P'(i)} = i - \frac{-2 - i}{-3} = -\frac{2}{3} + \frac{2}{3}i.$$

Thus the method appears to be converging to the root $-0.630 + 1.091i$.

Having found a single root one can, of course, determine the reduced polynomial and proceed to find one of its zeros. However, the following procedure has been suggested by Wilkinson [1963]. Having found a root α, consider the function

3.4 $$T(x) = \frac{P(x)}{x - \alpha}.$$

The Newton method applied to $T(x)$ is

3.5 $$x_{n+1} = x_n - \frac{T(x_n)}{T'(x_n)} = x_n - \left[\frac{P'(x_n)}{P(x_n)} - \frac{1}{x_n - \alpha} \right]^{-1}.$$

Thus, one can work with the original polynomial instead of the reduced polynomial. In general, having found $\alpha_1, \alpha_2, \ldots, \alpha_s$ we use

3.6 $$x_{n+1} = x_n - \left[\frac{P'(x_n)}{P(x_n)} - \sum_{k=1}^{s} \frac{1}{x_n - \alpha_k} \right]^{-1}.$$

From (2.51) we have, letting $z_k(x) = Q_k(x) + iR_k(x)$ and $z'_k(x) = Q'_k(x) + iR'_k(x)$, $k = 0, 1, 2, \ldots, n$, where $Q_k(x)$, $R_k(x)$, $Q'_k(x)$, and $R'_k(x)$ are real,

3.7 $$\begin{cases} Q'_0(x) = 0, \\ R'_0(x) = 0 \\ Q'_k(x) = Q_{k-1}(x) + \beta Q'_{k-1}(x) - \gamma R'_{k-1}(x) \\ R'_k(x) = R_{k-1}(x) + \gamma Q'_{k-1}(x) + \beta R'_{k-1}(x), \qquad k = 1, 2, \ldots, n, \end{cases}$$

where $x = \beta + i\gamma$. Evidently

3.8 $$\begin{cases} P(x) = P(\beta + i\gamma) = Q_n(x) + iR_n(x) \\ P'(x) = P'(\beta + i\gamma) = Q'_n(x) + iR'_n(x). \end{cases}$$

Therefore, as in Section 4.9, we have

3.9 $$\hat{\beta} + i\hat{\gamma} = x - \frac{P(x)}{P'(x)} = \beta + i\gamma - \frac{Q_n + iR_n}{Q'_n + iR'_n}$$

$$= \left(\beta - \frac{Q_n Q'_n + R_n R'_n}{(Q'_n)^2 + (R'_n)^2} \right) + i\left(\gamma - \frac{R_n Q'_n - Q_n R'_n}{(Q'_n)^2 + (R'_n)^2} \right).$$

Hence we have

$$
3.10 \qquad
\begin{cases}
\hat{\beta} = \beta - \dfrac{Q_n Q_n' + R_n R_n'}{(Q_n')^2 + (R_n')^2} \\[4mm]
\hat{\gamma} = \gamma - \dfrac{R_n Q_n' - Q_n R_n'}{(Q_n')^2 + (R_n')^2} .
\end{cases}
$$

Thus if one wishes to work in real arithmetic one could use (3.10), while with complex arithmetic one could use

$$
3.11 \qquad x_{n+1} = x_n - \frac{P(x_n)}{P'(x_n)} ,
$$

or the modification (3.6), directly.

We mention briefly two methods which are somewhat similar to the Newton method. The first such is *Lin's method* where instead of using

$$
3.12 \qquad x_{n+1} = x_n - \frac{P(x_n)}{P'(x_n)}
$$

one uses

$$
3.13 \qquad x_{n+1} = x_n - \frac{P(x_n)}{z_{n-1}} .
$$

Here z_{n-1} is the next to last result obtained in the synthetic division process. Evidently, by (2.31) and (2.33), $a_n + x_n z_{n-1} = z_n = P(x_n)$; hence we have

$$
3.14 \qquad x_{n+1} = x_n - \frac{x_n P(x_n)}{P(x_n) - P(0)} .
$$

Thus we have the Newton method except that $P'(x_n)$ is replaced by the slope of the line joining $(0, P(0))$ with $(x_n, P(x_n))$.

Another related method is the Lin-Bairstow method which is applicable if the polynomial has real coefficients. Here one divides $P(x)$ by a quadratic polynomial $x^2 + px + q$ where p and q are real, obtaining a quotient plus a remainder which is of the form $Ax + B$. The division is carried out by a type of synthetic division procedure which we illustrate for the case $n = 4$.

a_0	a_1	a_2	a_3	a_4	$\lfloor x^2 + px + q$
	$-b_0 p$	$-b_0 q$			

$b_0 = a_0$ $b_1 = a_1 - b_0 p$ $c_1 = a_2 - b_0 q$

| | | $-b_1 p$ | $-b_1 q$ | | |

b_1 $b_2 = c_1 - b_1 p$ $c_2 = a_3 - b_1 q$

| | | | $-b_2 p$ | $-b_2 q$ | |

b_2 $b_3 = a_3 - b_1 q - b_2 p$ $b_4 = a_4 - b_2 q$

3.15
$$\frac{P(x)}{x^2 + px + q} = b_0 x^2 + b_1 x + b_2 + \frac{b_3 x + b_4}{x^2 + px + q}$$

where

3.16
$$\begin{cases} b_0 = a_0 \\ b_1 = a_1 - b_0 p \\ b_2 = a_2 - b_0 q - b_1 p \\ b_3 = a_3 - b_1 q - b_2 p \\ b_4 = a_4 - b_2 q. \end{cases}$$

The Lin-Bairstow method amounts to solving the system

3.17
$$\begin{cases} b_3(p, q) = 0 \\ b_4(p, q) = 0 \end{cases}$$

by the generalized Newton method. Thus, following the discussion of Section 4.13, given p and q, one iteration of the generalized Newton method would yield $\hat{p}$ and $\hat{q}$ where

3.18
$$\begin{cases} \hat{p} = p + \dfrac{-b_3 \dfrac{\partial b_4}{\partial q} + b_4 \dfrac{\partial b_3}{\partial q}}{J} \\[4mm] \hat{q} = q + \dfrac{-b_4 \dfrac{\partial b_3}{\partial p} + b_3 \dfrac{\partial b_4}{\partial p}}{J} \end{cases}$$

where

3.19
$$J = \frac{\partial b_3}{\partial p} \frac{\partial b_4}{\partial q} - \frac{\partial b_4}{\partial p} \frac{\partial b_3}{\partial q}.$$

The derivatives $\partial b_i / \partial p$ and $\partial b_i / \partial q$ can be evaluated using a recurrence formula similar to that used for the b_i themselves.

In the case in which the zeros of $x^2 + px + q$ are complex we can interpret the method as follows. We express $b_3(p, q)$ and $b_4(p, q)$ in terms of the real functions $U(\beta, \gamma)$ and $V(\beta, \gamma)$ where

3.20
$$P(\beta + i\gamma) = U(\beta, \gamma) + iV(\beta, \gamma),$$

and where $\alpha = \beta + i\gamma$ and $\bar{\alpha}$ are the roots of $x^2 + px + q = 0$. Since

3.21
$$P(x) = (x - \alpha)(x - \bar{\alpha})(b_0x^2 + b_1x + b_2) + b_3x + b_4$$

we have

3.22
$$\begin{cases} b_4 = P(\alpha) - b_3\alpha \\ b_4 = P(\bar{\alpha}) - b_3\bar{\alpha}. \end{cases}$$

Therefore

3.23
$$\frac{1}{2i}(P(\alpha) - P(\bar{\alpha})) = V(\beta, \gamma) = b_3\gamma$$

and

3.24
$$\begin{cases} b_3 = \dfrac{1}{\gamma} V(\beta, \gamma) \\[3mm] b_4 = U(\beta, \gamma) - \dfrac{\beta}{\gamma} V(\beta, \gamma). \end{cases}$$

Thus an iteration of the Lin-Bairstow method is equivalent to an iteration for the system

3.25
$$\begin{cases} U(\beta, \gamma) - \dfrac{\beta}{\gamma} V(\beta, \gamma) = 0 \\[3mm] \dfrac{1}{\gamma} V(\beta, \gamma) \qquad\quad = 0 \end{cases}$$

by the generalized Newton method considering p and q as independent variables. Note that the iterants we would obtain using the method for this system are not the same, in general, as for the system obtained by replacing the second equation by $V(\beta, \gamma) = 0$.

As noted in Section 4.13, the Newton method (3.11) for solving the single equation $P(z) = 0$ is equivalent to the generalized Newton method for the system

3.26
$$U(\beta, \gamma) = 0$$
$$V(\beta, \gamma) = 0$$

but with β and γ as independent variables instead of p and q. By virtue of the similarity between (3.25) and (3.26) it is not surprising that the Lin-Bairstow method has been observed to have convergence properties similar to the Newton method—rapid convergence in some cases, usually with good starting values, and non-convergence in other cases.

EXERCISES 5.3

1. Find the largest real root of the equation $3x^4 - 7x^2 + 10x - 5 = 0$ by the Newton method. Exhibit the reduced polynomial.

2. Find a root of the reduced polynomial of the preceding exercise both using the Newton method and also by using the procedure (3.5).

3. Carry out one more iteration of the Newton method for the example given in the text for the equation $x^3 - 2 = 0$.

4. Find the largest root of the equation $x^3 - x - 1 = 0$ by the Newton method and exhibit the reduced polynomial. Let $x_0 = 1$. Find the zeros of the reduced polynomial and verify that they are also zeros of the original polynomial.

5. Work out a recurrence formula for obtaining the derivatives $\partial b_3/\partial p$, $\partial b_4/\partial p$, $\partial b_3/\partial q$, $\partial b_4/\partial q$. Apply the formula to the case of $P(x) = x^3 - 24x + 1$ and the quadratic factor $x^2 - 4x + 13$.

6. For the equation $x^3 - 2 = 0$ and the starting value $x_0 = -\frac{2}{3} + \frac{2}{3}i$ carry out one iteration of the Lin-Bairstow method.

5.4 MULLER'S METHOD AND CAUCHY'S METHOD

The use of Muller's method, Muller [1956], for solving polynomial equations is the same as that described in Section 4.10, for the case of the more general equation (1.2) as far as finding a single root is concerned. The required values of the polynomial are found by synthetic division. However, the determination of subsequent roots can be carried out using reduced polynomials.

As mentioned earlier, in Section 4.10, no matter how close the iterants x_k, x_{k-1}, and x_{k-2} are to a root r, if r has multiplicity three or higher, the process may not converge. To avoid this difficulty, it is recommended that Cauchy's method be used. Cauchy's method bears the same relation to Muller's method as the Newton method does to the secant method. For the case of the general equation (1.2), we write

4.1 $$f(\bar{x}) \doteq f(x_k) + (\bar{x} - x_k)f'(x_k) + \frac{(\bar{x} - x_k)^2}{2}f''(x_k)$$

where $\bar{x}$ is a root of (1.2). Letting $f(\bar{x}) = 0$ and solving for $\bar{x}$, we obtain

4.2

$$\bar{x} \doteq x_{k+1} = x_k - \frac{2f(x_k)}{f'(x_k) \pm \sqrt{[f'(x_k)]^2 - 2f(x_k)f''(x_k)}}$$

where the sign of the radical is chosen to maximize the modulus of the denominator. We remark that both $f'(x_k)$ and $f''(x_k)$ can be found by repeated synthetic division if $f(x)$ is a polynomial.

If convergence difficulties are experienced with Muller's method it is probably best to arbitrarily modify the starting values x_0, x_1, and x_2 once or twice to see if convergence is achieved. If this fails, one should switch over to Cauchy's method.

As mentioned in the introduction, for the user who wants to develop a fairly reliable and accurate scheme without too much effort, the use of Muller's method is recommended. The program should have the capability of switching over to Cauchy's method if convergence difficulties arise. Such a program would give adequate results in most cases.

A first step in improving the program might be to use a more sophisticated root acceptance technique than that used in Section 4.10. Thus, instead of choosing an ε and accepting α as a root if $|P(\alpha)| < \varepsilon$, for some small ε, one would use a convergence criterion which involves the rounding errors in the computation of $P(x)$. Such a criterion is described in Section 5.6. The speed of convergence of the method can be improved by choosing the starting values based on estimates of the location of the roots as given in Section 5.5.

EXERCISES 5.4

1. For the polynomial $P(x) = 2x^3 - 8x^2 + 10x - 4$ perform one iteration with Muller's method with $x_0 = -\frac{1}{2}$, $x_1 = 0$, $x_2 = \frac{1}{2}$. Do the same with Cauchy's method with $x_0 = 0$.

2. Write a program for finding one root of the polynomial equation (1.1) where the coefficients $a_0, a_1, \ldots, a_n$ may be complex. The program should accept any three starting values (real or complex) and carry out Muller's method; if Muller's method fails, Cauchy's method should be attempted. A value of ε should be assigned so that a root α is accepted if $|P(\alpha)| \leqq \varepsilon$. Muller's method should be assumed to have failed if the difference between two successive iterants exceeds a given number, say δ, or if the coefficients of the auxiliary quadratic become too small. Cauchy's method should be assumed to fail if the difference between two iterants is too large or if $|P'(\alpha)|$ and $|P''(\alpha)|$ are too small. Also, the process should be terminated when the total number of iterations exceeds a prescribed number, say 50.

    The program should be tested for the polynomial $x^3 + x^2 - 4x + 6 = 0$ with $\varepsilon = 10^{-5}$, $\delta = 10$, and starting values -1, 0, 1. Also try the polynomial $x^5 - 6x^4 + 15x^3 - 20x^2 + 14x - 4$ with the starting values $-\frac{1}{2}, 0, \frac{1}{2}$.

5.5 LOCATION OF THE ROOTS

In this section we consider methods for approximately locating the roots of a polynomial equation. If the coefficients of the polynomial are real, then information on the number of real roots can be obtained by Descartes' rule of signs. More precise information can be found by the use of Sturm sequences. The question of whether or not there is a root in a given circle can be answered by the use of the Lehmer-Schur method. Also, for a given α we can give a family of upper bounds for the radius of a circle with α as a center and containing at least one root. Such bounds are useful not only in locating roots but also in determining the accuracy of a given root. Upper and lower bounds for the moduli of the roots are also given.

Descartes' Rule of Signs

Descartes' rule of signs is primarily useful for theoretical purposes in the study of methods for solving polynomial equations. It is seldom used in a computer program for actually solving polynomial equations.

Given a polynomial with real coefficients, we can obtain information on the number of positive real roots by considering the sign variations of the coefficients. We write down the signs of the nonvanishing coefficients and count the number of sign changes, i.e., a change from a " + " to a " − " or from a " − " to a " + ". For example, for the polynomial

5.1 $P(x) = 3x^5 - x^4 - x^3 + x + 1$

we have the signs

$$+ \quad - \quad - \quad + \quad +.$$

Hence there are *two* sign variations. The polynomial

5.2 $P(x) = x^5 + 1$

has no sign variations.

Descartes' rule of signs is stated as follows.

> The number of positive real roots, p, of a polynomial equation with real coefficients does not exceed v, the number of sign variations of the coefficients; moreover, $v - p$ is a nonnegative even integer.

Thus in the first example given above there are either two positive roots or none. In the second example there is none. For the polynomial

5.3 $P(x) = x^5 - x^3 - x^2 + x - 1$

there are 3 sign variations; hence there are either 3 or 1 positive real roots.

By considering $P(-x)$ we can estimate the number of negative real roots. Thus in (5.1) we have $P(-x) = -3x^5 - x^4 + x^3 - x + 1$ and there are 3 sign variations. Hence there are either 3 or 1 negative real roots. Thus, there may be a total of 5, 3, or 1 real roots. This is already known since complex roots occur in conjugate pairs. In the example (5.2) there are exactly one negative real root and four complex roots. In (5.3) there are 0 or 2 negative roots; hence there are either 0, 2, or 4 complex roots.

For any α we can estimate how many real roots of (5.1) are greater than α by shifting the origin. Thus we let

5.4 $$P(x) = \hat{a}_0(x - \alpha)^n + \hat{a}_1(x - \alpha)^{n-1} + \cdots + \hat{a}_n$$

and consider the sign variations of the $\hat{a}_i$. For example, if $\alpha = 1$, then the polynomial (5.1) becomes

5.5 $$P(x) = 3(x - 1)^5 + 14(x - 1)^4 + 25(x - 1)^3 + 21(x - 1)^2 + 9(x - 1) + 3$$

and there are no sign variations. Hence there are no real zeros of $P(x)$ greater than 1. Therefore, there are either two real zeros or none in the interval $0 \leq x \leq 1$.

If there are no sign variations, it is obvious that $P(x) \neq 0$ for any $x > 0$. We now give an intuitive proof of Descartes' rule for the case of one sign variation. Without loss of generality we assume that $a_0 > 0$. Thus for some k we have, $a_0, a_1, \ldots, a_k$ are nonnegative and $a_{k+1}, \ldots, a_n$ are nonpositive. Hence

5.6 $$P(x) = a_0 x^n + a_1 x^{n-1} + \cdots + a_k x^{n-k} - |a_{k+1}|x^{n-k-1} - \cdots - |a_n|$$

$$= Q(x) - R(x)$$

where $Q(x)$ and $R(x)$ are polynomials with nonnegative coefficients. Evidently, for positive x sufficiently large, $P(x) > 0$ since

$$P(x) = x^n\{a_0 + a_1 x^{-1} + \cdots + a_k x^{-k} - |a_{k+1}|x^{-k-1} - \cdots - |a_n|x^{-n}\}.$$

Let s be the largest integer such that $a_s < 0$. Then

5.7 $$P(x) = x^{n-s}\{a_0 x^s + a_1 x^{s-1} + \cdots + a_k x^{s-k} - |a_{k+1}|x^{s-k-1} - \cdots - |a_s|\}$$

which is clearly negative for x small enough, but positive. Since $P(x)$ changes sign in the range $0 < x < \infty$, it follows that there is at least one root $\bar{x}$ in that range. We now show that there is no more than one. But if t is the largest integer such that $a_t > 0$, we have

5.8 $$P(x) = x^{n-t}\{a_0 x^t + \cdots + a_t - |a_{k+1}|x^{t-k-1} - \cdots - |a_n|x^{-(n-t)}\}$$

$$= x^{n-t}\{T(x) - U(x)\}.$$

Evidently, $T(x)$ is an increasing function and $U(x)$ is a decreasing function of x. Since $P(\bar{x}) = 0$, we have $T(\bar{x}) = U(\bar{x})$. Hence $P(x) \neq 0$ for $x > \bar{x}$. Similarly, we can show that $P(x) \neq 0$ for $0 < x < \bar{x}$. Thus $P(x) = 0$ has exactly one positive root.

Suppose now that there are *two* sign variations. Then we may write

5.9 $$P(x) = Q(x) - R(x) + S(x)$$

where the polynomials $Q(x)$, $R(x)$, and $S(x)$ have nonnegative coefficients and at least one positive coefficient each. Let k be greater than the degree of $S(x)$ but not greater than the degree of the lowest-degree term of $R(x)$. Evidently $S^{(k)}(x) \equiv 0$ and $P^{(k)}(x)$ has one sign variation; hence the equation $P^{(k)}(x) = 0$ has exactly one positive real root as well as a possible root at $x = 0$ (see Fig. 5.10). If $P^{(k-1)}(0) = 0$, then $P^{(k-1)}(x)$ also has precisely one positive real zero. If $P^{(k-1)}(0) \neq 0$, then $P^{(k-1)}(0) > 0$ since the coefficients of $S(x)$ and hence those of $S^{(k-1)}(x)$ are non-negative. In this case, $P^{(k-1)}(x)$ has either two positive zeros, one double zero, or none (see Fig. 5.11).

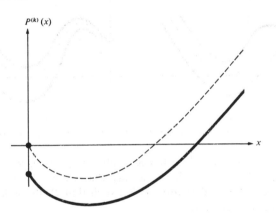

$P^{(k)}(x)$

x

5.10 Fig. Behavior of $P^{(k)}(x)$.

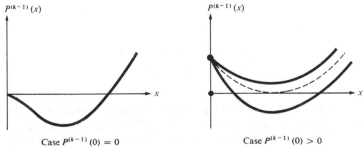

$P^{(k-1)}(x)$ $P^{(k-1)}(x)$

x x

Case $P^{(k-1)}(0) = 0$ Case $P^{(k-1)}(0) > 0$

5.11 Fig. Behavior of $P^{(k-1)}(x)$.

If $P^{(k-1)}(0) > 0$ and if $P^{(k-1)}(x)$ has no positive zeros or else has one double zero, then $P^{(k-1)}(x) > 0$ for all $x > 0$ except for at most one value of x. Since

$P^{(k-2)}(0) \geqq 0$ it follows that $P^{(k-2)}(x) > 0$ for all $x > 0$. (We remark that $P^{(k-2)}(0)$ may vanish.) Similarly, $P^{(k-3)}(x)$, $P^{(k-4)}(x)$, ..., $P(x)$ are positive for all $x > 0$ and hence have no positive zeros.

If $P^{(k-1)}(0) > 0$ and if $P^{(k-1)}(x)$ has two positive zeros, then since $P^{(k-2)}(0) \geqq 0$ it follows that $P^{(k-2)}(x)$ has either two positive zeros or none (see Fig. 5.12). If $P^{(k-2)}(x)$ has no positive zeros, then the same is true of $P^{(k-3)}(x)$, $P^{(k-4)}(x)$, ..., $P(x)$. If $P^{(k-2)}(x)$ has two positive zeros, then $P^{(k-3)}$ has either two positive zeros or none. Continuing this process we can show that $P(x)$ has either two positive zeros or none.

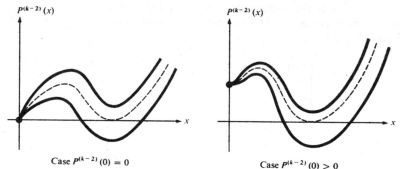

Case $P^{(k-2)}(0) = 0$ Case $P^{(k-2)}(0) > 0$

5.12 Fig. Behavior of $P^{(k-2)}(x)$ (assuming $P^{(k-1)}(0) > 0$ and $P^{(k-1)}(x)$ has two positive zeros).

If $P^{(k-1)}(0) = 0$, then $P^{(k-1)}(x)$ has precisely one positive zero (see Fig. 5.11). We then examine $P^{(k-2)}(x)$ in the same way that we examined $P^{(k-1)}(x)$. We continue until we obtain a derivative $P^{(q)}(x)$ such that $P^{(q)}(0) > 0$. We can then apply the argument given above.

As an example, consider the case

5.13 $P(x) = x^5 - x^4 + x^2 + 1.$

Evidently we have

5.14
$$\begin{cases} P'(x) = 5x^4 - 4x^3 + 2x \\ P''(x) = 20x^3 - 12x^2 + 2 \\ P^{(3)}(x) = 60x^2 - 24x \end{cases}$$

Therefore, $P^{(3)}(0) = 0$ and $P^{(3)}(x)$ has one positive zero. Hence $P^{(2)}(x)$ has either two positive zeros or none since $P^{(2)}(0) > 0$. If $P^{(2)}(x)$ has no positive zeros, then $P^{(2)}(x) > 0$ for all $x > 0$ and, since $P'(0) = 0$ it follows that $P'(x) > 0$ for all $x > 0$. Similarly, $P(x) > 0$ for all $x > 0$ and there are no positive zeros. If $P^{(2)}(x)$ has two positive zeros, then $P'(x)$ has either zero or two. If $P'(x)$ has no positive zeros, then neither does $P(x)$. If $P'(x)$ has two positive zeros, then $P(x)$ may have either zero or two. In any case, $P(x)$ has either no positive zeros or two positive zeros.

The above argument can be extended to the case of more sign variations. Thus, if there are three sign variations we write

5.15 $P(x) = Q(x) - R(x) + S(x) - T(x).$

We let k be greater than the degree of $T(x)$ but not greater than the degree of the lowest degree term of $S(x)$. Then $T^{(k)}(x) \equiv 0$ and $P^{(k)}(x)$ has two sign variations, hence, either two or no positive zeros. Therefore, $P^{(k-1)}(x)$ has either three or one positive zero if $P^{(k-1)}(0) < 0$. If $P^{(k-1)}(0) = 0$, then $P^{(k-1)}(x)$ has either two positive zeros or none. The argument can be continued to show that $P(x)$ has either one positive zero or three.

Sturm Sequences

Given a polynomial $P(x)$ with real coefficients and a real number a, we define the function $V(a)$ as follows:

$V(a)$ is the number of sign variations in the numbers $g_0(a), g_1(a), \ldots, g_n(a)$, ignoring zeros. Here the polynomials $g_0(x), g_1(x), \ldots$ are defined by

5.16 $$\begin{cases} g_0(x) = P(x) \\ g_1(x) = P'(x), \end{cases}$$

and for each $k \geq 2$, $g_k(x)$ is the unique polynomial of degree less than that of $g_{k-1}(x)$ such that

5.17 $$g_{k-2}(x) = Q_k(x)g_{k-1}(x) - g_k(x)$$

where $Q_k(x)$ is a polynomial. If $g_m(x)$ is a constant for some m, then we let $g_{m+1}(x) = g_{m+2}(x) = \cdots = g_n(x) = 0$. If $g_m(x)$ is not a constant but all subsequent g_k vanish, then $g_m(x)$ is a factor of all $g_i(x)$, $i = 0, 1, 2, \ldots, m - 1$. In this case $V(a)$ is the number of sign variations of

$$\frac{g_0(a)}{g_m(a)}, \frac{g_1(a)}{g_m(a)}, \ldots, \frac{g_{m-1}(a)}{g_m(a)}.$$

As an example, consider the polynomial

5.18 $P(x) = x^3 - x^2 + x + 1.$

Evidently,

5.19 $g_1(x) = P'(x) = 3x^2 - 2x + 1$

and

5.20 $g_0(x) = (\tfrac{1}{3}x - \tfrac{1}{9})g_1(x) - (-\tfrac{4}{9}x - \tfrac{10}{9})$

so that

5.21
$$g_2(x) = -\tfrac{4}{9}x - \tfrac{10}{9}.$$

Next, we have

5.22
$$g_1(x) = (-\tfrac{27}{4}x + \tfrac{171}{8})g_2(x) - (-\tfrac{99}{4})$$

and

5.23
$$g_3(x) = -\tfrac{99}{4}.$$

If we let $a = 1$, then we have

5.24 $g_0(1) = 2,$ $g_1(1) = 2,$ $g_2(1) = -\tfrac{14}{9},$ $g_3(1) = -\tfrac{99}{4};$

hence there is one sign variation and

5.25
$$V(1) = 1.$$

Sturm's theorem states that if $P(\alpha) \neq 0$ and $P(\beta) \neq 0$, then the number of distinct real roots of (1.1) in the interval $\alpha \leq x \leq \beta$ is exactly $V(\alpha) - V(\beta)$. A proof of this result can be found in Householder [1964, 1970] and in Ralston [1965].

In our example, if β is sufficiently large, we have

5.26 $g_0(\beta) > 0,$ $g_1(\beta) > 0,$ $g_2(\beta) < 0,$ $g_3(\beta) < 0.$

Thus, since $V(1) = V(\beta) = 1$, it follows that there is no real root greater than 1. On the other hand, if we let $\alpha = -3$, we have $g_0(\alpha) < 0$, $g_1(\alpha) > 0$, $g_2(\alpha) > 0$, and $g_3(\alpha) < 0$. Hence $V(-3) = 2$. Therefore, there is one real root between -3 and 1. Since $V(\alpha) = 2$ for α very small, it follows that there are no roots less than -3. Hence there is exactly one distinct real root. Actually, since $g_3(x)$ is a constant, there are no multiple roots; thus there is exactly one real root and hence two complex roots.

In applying Sturm's theorem, care must be taken to insure that rounding errors do not occur in the determination of the signs of the $g_i(\alpha)$ and $g_i(\beta)$. The difficulty of doing this often limits the usefulness of the procedure as a practical means of locating real roots.

The Lehmer-Schur Method

As shown by Lehmer [1961], the following procedure can be used to determine whether or not there exists a root of (1.1) inside of the unit circle, $|z| \leq 1$. Let $P^*(z)$ be defined by

5.27
$$P^*(z) = z^n \bar{P}(\bar{z}^{-1}) = \bar{a}_0 + \bar{a}_1 z + \bar{a}_2 z^2 + \cdots + \bar{a}_n z^n.$$

We compute $T[P(z)]$, $T^2[P(z)] = T[T[P(z)]]$, etc., where

5.28 $$T[P(z)] = \bar{a}_n P(z) - a_0 P^*(z),$$

which is a polynomial of degree less than n and with constant term

5.29 $$T[P(0)] = |a_n|^2 - |a_0|^2.$$

We let k be the smallest integer such that

5.30 $$T^k[P(0)] = 0.$$

The rules for determining whether or not there is a root in the unit circle are as follows:

a) If $P(0) \neq 0$ and if for some j with $1 \leq j < k$, we have $T^j[P(0)] < 0$ then $P(z)$ has at least one zero in the unit circle.
b) If $P(0) \neq 0$ and $T^j[P(0)] > 0$ for all j such that $1 \leq j < k$ and if $T^{k-1}[P(z)]$ is a constant, then there is no zero in the unit circle.
c) The test fails if $T^k[P(0)] = 0$ but $T^{k-1}[P(z)]$ is not a constant.

If the test fails, one can simply scale the polynomial by considering the polynomial $P_1(x) = P(\rho x) = (\rho^n a_0)x^n + (\rho^{n-1}a_1)x^{n-1} + \cdots + a_n$. We then test whether there is a zero of $P_1(x)$ in the unit circle, i.e., a zero of $P(z)$ in the circle $|z| \leq \rho$.

In practice, one usually wishes to find out whether or not there is a root in the circle $|z - \alpha| \leq r$. To this end we let

5.31 $$w = \frac{z - \alpha}{r}, \qquad z = \alpha + rw.$$

We first shift the origin, i.e., compute $\hat{a}_0, \hat{a}_1, \ldots, \hat{a}_n$ such that

5.32 $$P(z) = \hat{a}_0(z - \alpha)^n + \hat{a}_1(z - \alpha)^{n-1} + \cdots + \hat{a}_n.$$

We then consider the modified polynomial

5.33 $$P_1(w) = (\hat{a}_0 r^n)w^n + (\hat{a}_1 r^{n-1})w^{n-1} + \cdots + \hat{a}_n.$$

Evidently there will be a zero of $P_1(w)$ in the circle $|w| \leq 1$ if and only if there is a zero of $P(z)$ in the circle $|z - \alpha| \leq r$.

A Family of Bounds

For any complex number α we can give a family of bounds for the radius of a circle about α containing at least one root of (1.1). Thus, for each integer $k = 1, 2, \ldots, n$ we assert that there is a root in the circle C_{ρ_k}

5.34 $$|x - \alpha| \leq \rho_k$$

where

5.35 $\rho_k = \sqrt[k]{n(n-1)\ldots(n-k+1)\left|\dfrac{P(\alpha)}{P^{(k)}(\alpha)}\right|} = \sqrt[k]{\binom{n}{k}\left|\dfrac{\hat{a}_n}{\hat{a}_{n-k}}\right|}$, $k = 1, 2, \ldots, n$.

Here we define $\hat{a}_0, \hat{a}_1, \ldots, \hat{a}_n$ by

5.36 $P(x) = \hat{a}_0(x-\alpha)^n + \hat{a}_1(x-\alpha)^{n-1} + \cdots + \hat{a}_n$.

In particular, if $k = 1$ we have

5.37 $\rho_1 = n\left|\dfrac{P(\alpha)}{P'(\alpha)}\right| = n\left|\dfrac{\hat{a}_n}{\hat{a}_{n-1}}\right|$;

thus the bound is n times the "Newton correction." In the case $k = n$ we have

5.38 $\rho_n = \sqrt[n]{\left|\dfrac{P(\alpha)}{a_0}\right|} = \sqrt[n]{\left|\dfrac{\hat{a}_n}{a_0}\right|}$

since $a_0 = (1/n!)P^{(n)}(\alpha) = \hat{a}_0$.

To prove (5.34) we first note that the coefficients appearing in (5.36) are given by

5.39 $\hat{a}_{n-k} = \dfrac{P^{(k)}(\alpha)}{k!}$, $k = 0, 1, 2, \ldots, n$.

Evidently, the polynomial

5.40 $Q(y) = \hat{a}_n y^n + \hat{a}_{n-1} y^{n-1} + \cdots + \hat{a}_0$

has zeros $(\lambda_1 - \alpha)^{-1}, (\lambda_2 - \alpha)^{-1}, \ldots, (\lambda_n - \alpha)^{-1}$, where $\lambda_1, \lambda_2, \ldots, \lambda_n$ are the roots of $P(x) = 0$. Consequently, as shown in Section 5.2, for each k, $\hat{a}_{n-k}/\hat{a}_n$ is $(-1)^k$ times the sum of all possible products of the zeros of (5.40) taken k at a time. For example,

5.41 $\begin{cases} -\dfrac{\hat{a}_{n-1}}{\hat{a}_n} = \displaystyle\sum_{i=1}^{n} \dfrac{1}{\lambda_i - \alpha} \\[2ex] \dfrac{\hat{a}_{n-2}}{\hat{a}_n} = \displaystyle\sum_{i,j=1, i<j}^{n} \left(\dfrac{1}{\lambda_i - \alpha}\right)\left(\dfrac{1}{\lambda_j - \alpha}\right) \\[2ex] \quad\vdots \\[1ex] (-1)^n \dfrac{\hat{a}_0}{\hat{a}_n} = \left(\dfrac{1}{\lambda_1 - \alpha}\right)\left(\dfrac{1}{\lambda_2 - \alpha}\right)\cdots\left(\dfrac{1}{\lambda_n - \alpha}\right). \end{cases}$

Since there are $_nC_k$ products of the n roots taken k at a time, we have

5.42
$$\left|\frac{\hat{a}_{n-k}}{\hat{a}_n}\right| \leqq {}_nC_k \frac{1}{(\min_i|\lambda_i - \alpha|)^k}$$

and hence

5.43
$$\min_i|\lambda_i - \alpha| \leqq \sqrt[k]{{}_nC_k \left|\frac{\hat{a}_n}{\hat{a}_{n-k}}\right|}$$

and (5.35) follows from (5.39).

Given a knowledge of the location of some of the roots, we can improve the above bounds. Suppose, for example, that $\lambda_1, \lambda_2, \ldots, \lambda_s$ are known exactly. In the case $k = 1$ we have

5.44
$$\left|-\frac{\hat{a}_{n-1}}{\hat{a}_n} - \sum_{i=1}^s \frac{1}{\lambda_i - \alpha}\right| \leqq \sum_{i=s+1}^n \left|\frac{1}{\lambda_i - \alpha}\right| \leqq (n-s)\frac{1}{\min_i|\lambda_i - \alpha|}$$

so that

5.45
$$\min_i|\lambda_i - \alpha| \leqq (n-s)\left|\frac{P'(\alpha)}{P(\alpha)} + \sum_{i=1}^s \frac{1}{\lambda_i - \alpha}\right|^{-1}.$$

Similarly, in the case $k = n$ we have

5.46
$$\min_i|\lambda_i - \alpha| \leqq \sqrt[n-s]{\frac{|P(\alpha)|}{|a_0(\lambda_1 - \alpha)(\lambda_2 - \alpha)\cdots(\lambda_s - \alpha)|}}.$$

Frequently one does not have the roots $\lambda_1, \lambda_2, \ldots, \lambda_s$ exactly, but instead has bounds for them. Naturally, one would only "remove" known roots if better bounds would be obtained thereby.

Normally, if α is very near a simple root, the bound based on $k = 1$ will be the best. Near a double root the bound based on $k = 2$ will be best, etc.

The treatment of other values of k is more complicated. We illustrate by the case $n = 3, k = 2, s = 1$. We have

5.47
$$\frac{\hat{a}_1}{\hat{a}_3} = \frac{1}{(\lambda_1 - \alpha)(\lambda_2 - \alpha)} + \frac{1}{(\lambda_1 - \alpha)(\lambda_3 - \alpha)} + \frac{1}{(\lambda_2 - \alpha)(\lambda_3 - \alpha)}.$$

Let $q = \min_{i=1,2}|\lambda_i - \alpha|$. Then

5.48
$$\left|\frac{\hat{a}_1}{\hat{a}_3}\right| \leqq q^{-2} + q^{-1}\frac{2}{|\lambda_3 - \alpha|}.$$

The above inequality can be solved for q. In this case a direct solution is possible, but in general an iterative procedure might be used starting with a value of q determined by (5.35).

As an example, let us consider the polynomial

5.49 $$P(x) = x^3 - x^2 - 1.01x + 0.99$$

whose roots are 0.9, 1.1, and -1. Letting $\alpha = 1$ we have

5.50 $$P(x) = (x-1)^3 + 2(x-1)^2 - 0.01(x-1) - 0.02$$

so that

5.51 $$\hat{a}_0 = 1, \quad \hat{a}_1 = 2, \quad \hat{a}_2 = -0.01, \quad \hat{a}_3 = -0.02.$$

Using (5.35) we find that there is at least one root in the circle $|x - 1| \leqq \rho_k$, $k = 1, 2, 3$, where

5.52 $$\rho_1 = 3\left(\frac{0.02}{0.01}\right) = 6, \quad \rho_2 = \sqrt{3\left(\frac{0.02}{2}\right)} \doteq 0.173, \quad \rho_3 = \sqrt[3]{\frac{0.02}{1}} \doteq 0.271.$$

If we use the fact that there is a root at -1 we do not obtain an improved value of ρ_1, since, by (5.45),

5.53 $$\rho_1' = (3 - 1)\left|\frac{0.01}{0.02} + \frac{1}{-1 - 1}\right|^{-1} = \infty.$$

For ρ_2' we can solve (5.48) for q, where

5.54 $$\left|\frac{\hat{a}_1}{\hat{a}_3}\right| = 100 \leqq q^{-2} + q^{-1}.$$

Solving for q we get $q^{-1} \doteq 9.5125$ and hence $\rho_2' = q \doteq 0.105$. This bound is very good.

For ρ_3' we have by (5.46)

5.55 $$\rho_3' = \sqrt{\left|\frac{\hat{a}_3}{\hat{a}_0}\right|\left(\frac{1}{2}\right)} = \sqrt{0.01} = 0.1$$

which is an exact bound.

Another Family of Bounds

McDonald [1970] considered a family of bounds which are perhaps less simple to compute than those given above but which are better adapted to take advantage of a knowledge of some roots.

If $\lambda_1, \lambda_2, \ldots, \lambda_n$ are the (not necessarily distinct) zeros of $P(x)$ we have for some constant c

5.56 $$P(x) = c(x - \lambda_1)(x - \lambda_2)\cdots(x - \lambda_n).$$

Therefore

5.57
$$\frac{P'(x)}{P(x)} = \sum_{i=1}^{n} \frac{1}{x - \lambda_i}$$

and

5.58
$$\left(\frac{P'(x)}{P(x)}\right)' = -\sum_{i=1}^{n} \frac{1}{(x - \lambda_i)^2}.$$

In general,

5.59
$$\left(\frac{P'(x)}{P(x)}\right)^{[k-1]} = (-1)^{k-1}(k-1)! \sum_{i=1}^{n} \frac{1}{(x - \lambda_i)^k}.$$

Therefore,

5.60
$$\min_i |\lambda_i - \alpha| \leq \sqrt[k]{n \left|\left(\frac{P'(\alpha)}{P(\alpha)}\right)^{[k-1]}\right|^{-1} (k-1)!} = \sigma_k.$$

Evidently (5.60) and (5.43) agree if $k = 1$, since $P(\alpha) = \hat{a}_n$ and $P'(\alpha) = \hat{a}_{n-1}$. To compute the successive derivatives of $P'(x)/P(x)$ we let

5.61
$$q_k(x) = \left[\frac{P'(x)}{P(x)}\right]^{[k]},$$

and we use the recurrence relation (McDonald [1970])

5.62
$$q_k(x) = \frac{P^{(k+1)}(x) - \sum_{j=1}^{k} \binom{k}{j} q_{k-j}(x) P^{(j)}(x)}{P(x)}.$$

In particular we have

5.63
$$\begin{cases} q_0(x) = \dfrac{P'(x)}{P(x)} \\[3mm] q_1(x) = \dfrac{P''(x) - q_0(x)P'(x)}{P(x)} \\[3mm] q_2(x) = \dfrac{P'''(x) - 2q_1(x)P'(x) - q_0(x)P''(x)}{P(x)}. \end{cases}$$

If $\lambda_1, \lambda_2, \ldots, \lambda_s$ are known exactly, then by (5.59) we have

5.64
$$\left| \left(\frac{P'(\alpha)}{P(\alpha)} \right)^{[k-1]} - (-1)^{k-1}(k-1)! \sum_{i=1}^{s} \frac{1}{(\alpha - \lambda_i)^k} \right| \leqq \frac{n-s}{\min_i |\lambda_i - \alpha|^k}(k-1)!$$

and hence

5.65
$$\min_i |\lambda_i - \alpha| \leqq \sigma'_k = \sqrt[k]{(n-s)\left| \frac{(-1)^{k-1}}{(k-1)!}\left[\frac{P'(\alpha)}{P(\alpha)} \right]^{[k-1]} - \sum_{i=1}^{s} \frac{1}{(\alpha - \lambda_i)^k} \right|^{-1}}$$

EXAMPLE

Let us again consider the example (5.49). Evidently we have, for $\alpha = 1$,

5.66 $P(\alpha) = -0.02, \qquad P'(\alpha) = -0.01, \qquad P''(\alpha) = 4, \qquad P'''(\alpha) = 6$

and

5.67
$$\begin{cases} q_0(\alpha) = \dfrac{-0.01}{-0.02} = 0.5 \\[2mm] q_1(\alpha) \doteq \dfrac{4 - (0.5)(-0.01)}{-0.02} = -200.25 \\[2mm] q_2(\alpha) = \dfrac{6 - 2(-200.25)(-0.01) - 0.5(4)}{-0.02} = -0.25. \end{cases}$$

Therefore,

5.68
$$\begin{cases} \sigma_1 = \rho_1 = 6 \\[2mm] \sigma_2 = \sqrt{\dfrac{3}{200.25}} = 0.122 \\[2mm] \sigma_3 = \sqrt[3]{\dfrac{3}{0.25}} = 2.289. \end{cases}$$

The bound σ_2 is somewhat better than that given by the previous method.
Now, let us assume that the root at -1 is known. By (5.65) we have

5.69
$$\sigma'_2 = \sqrt[2]{(2)|200.25 - \tfrac{1}{4}|^{-1}}$$
$$= \sqrt{\tfrac{2}{200}}$$
$$= 0.1.$$

This bound is exact.

Cauchy Bounds

By the solution of the polynomial equation

5.70 $$T(x) = |a_0|x^n - |a_1|x^{n-1} - \cdots - |a_n| = 0$$

we can find an upper bound for the moduli of the roots of (1.1). Thus let R be the positive real root of (5.70), (a unique such root exists by Descartes' rule of signs). Then all roots of (1.1) lie in the circle

5.71 $$|x| \leqq R.$$

To prove this we have

5.72 $$P(x) = x^n \left\{ a_0 + \frac{a_1}{x} + \cdots + \frac{a_n}{x^n} \right\}.$$

If $|x| > R$, then, unless $a_1 = a_2 = \cdots = a_n = 0$ we have

5.73 $$\left| \frac{a_1}{x} + \cdots + \frac{a_n}{x^n} \right| < \frac{|a_1|}{R} + \frac{|a_2|}{R^2} + \cdots + \frac{|a_n|}{R^n} = |a_0|$$

since $T(R) = 0$. Hence since

5.74 $$|P(x)| \geqq |x|^n \left\{ |a_0| - \left| \sum_{i=1}^{n} \frac{a_i}{x^i} \right| \right\}$$

it follows that $P(x) \neq 0$ and x is not a root.

In a similar manner we can show that no roots lie inside the circle

5.75 $$|x| \leqq r$$

where r is the positive root of

5.76 $$|a_0|x^n + |a_1|x^{n-1} + \cdots + |a_{n-1}|x - |a_n| = 0.$$

For the case of (5.49) if one finds the positive root for each of the equations

5.77 $$x^3 - x^2 - 1.01x - 0.99 = 0$$

5.78 $$x^3 + x^2 + 1.01x - 0.99 = 0$$

one obtains 1.841 and 0.538, respectively. Thus for each zero λ_i of (5.49) we have

5.79 $$0.538 \leqq |\lambda_i| \leqq 1.841.$$

Miscellaneous Bounds

The following bounds on the roots of the polynomial equation (1.1) can be obtained using known bounds on the eigenvalues of the associated companion matrix (see Wilf [1960] and McDonald [1970]). Direct proofs of some of the results, not involving the use of matrix theory, are given as exercises.

a) Every root of (1.1) lies in one of the two circles

5.80
$$\begin{cases} |z| \leq 1 \\ \left| z + \dfrac{a_1}{a_0} \right| \leq \displaystyle\sum_{j=2}^{n} \left| \dfrac{a_j}{a_0} \right| \end{cases}.$$

b) Every root of (1.1) lies in one of the three circles

5.81
$$\begin{cases} \left| z + \dfrac{a_1}{a_0} \right| \leq 1 \\ |z| \leq 1 + \displaystyle\max_{2 \leq j \leq n-1} \left| \dfrac{a_j}{a_0} \right| \\ |z| \leq \left| \dfrac{a_n}{a_0} \right| \end{cases}.$$

c) Every root of (1.1) lies in the annulus

5.82
$$\frac{|a_n|}{\sqrt{\sigma}} \leq |z| \leq \max\left\{ 1, \sqrt{\frac{\sigma}{|a_0|^2} - \frac{|a_n|^2}{\sigma}} \right\}$$

where

5.83
$$\sigma = \sum_{j=0}^{n} |a_j|^2 .$$

d) There are no roots of (1.1) in the circle $|z| \leq \rho$ where $\rho = \max(\rho_1, \rho_2, \rho_3)$ and where

5.84
$$\rho_1^{-1} = \max\left\{ 1, \sum_{j=0}^{n-1} \left| \frac{a_j}{a_n} \right| \right\}$$

5.85
$$\rho_2^{-1} = \max\left\{ \left| \frac{a_0}{a_n} \right|, 1 + \max_{1 \leq j \leq n-1} \left| \frac{a_j}{a_n} \right| \right\}$$

5.86
$$\rho_3 = \frac{|a_n|}{\sqrt{\sigma}} .$$

EXERCISES 5.5

1. Give a proof, based on that given in the text for the case of two sign variations, of Descartes' rule for the case of three sign variations.

2. Apply Descartes' rule of signs to estimate the number of positive zeros of the polynomial $P(x) = x^4 - 5x^3 + 10x^2 - 10x + 4$ and the number of real zeros greater than $\frac{1}{2}$. Do the same thing using Sturm sequences.

3. Find the number of real roots between 0 and 3 of the equation

$$P(x) = x^4 - 4x^3 + 3x^2 + 4x - 4 = 0$$

using Sturm sequences.

4. In the preceding exercise determine whether or not there is a root in the circle $|z + 2| \leq 2$ using Lehmer's method.

5. Apply Lehmer's method to test whether there is a root of

$$P(x) = x^4 - 4x^3 + 3x^2 + 4x - 4 = 0$$

in the circle $|z| \leq \frac{1}{2}$; in the circle $|z| \leq \frac{3}{4}$.

6. Apply the Lehmer-Schur method to determine if there is a zero of $P(z) = z^4 - 5z^3 + 10z^2 - 10z + 4$ in the following circles: $|z| \leq \frac{1}{2}, |z| \leq \frac{3}{2}, |z - \frac{1}{2}| \leq \frac{1}{4}, |z - \frac{1}{2}| \leq 1$.

7. For the polynomial $P(x) = x^5 - 6x^4 + 15x^3 - 20x^2 + 14x - 4$ compute $\rho_1, \rho_2, \rho_3, \rho_4$, and ρ_5 by (5.35) with $\alpha = \frac{1}{2}$. Also obtain improved bounds if it is known that there is a root at $x = 2$. Do the same for the bounds (5.60).

8. Verify (5.62) for the cases $k = 1, 2, 3$. Give a general proof by mathematical induction.

9. Prove (5.75)–(5.76).

10. Apply the Cauchy bounds to determine upper and lower bounds for the moduli of the zeros of the polynomial $P(x) = x^4 - 5x^3 + 10x^2 - 10x + 4$. Also find the bounds given by (5.80)–(5.86).

11. Obtain the bounds (5.80)–(5.86) for the polynomial (5.49).

12. Prove (5.81) using the following argument, which is illustrated for the case $n = 5$. Let $P(x) = x^5 + b_1 x^4 + b_2 x^3 + b_3 x^2 + b_4 x + b_5$ where $b_i = a_i/a_0$, $i = 1, 2, 3, 4, 5$. If $P(x) = 0$ then

$$x + b_1 = \frac{1}{x}\left\{b_2 + \frac{1}{x}\left\{b_3 + \frac{1}{x}\left[b_4 + \frac{b_5}{x}\right]\right\}\right\}.$$

Then show that if $|x| \geq |b_5|$ and $|x| \geq \max_{2 \leq i \leq 4} |b_i| + 1$ then $|x + b_1| \leq 1$.

13. Prove (5.80) by writing (1.1) in the form

$$z + \frac{a_1}{a_0} = -\sum_{j=2}^{n} \frac{a_j}{a_0} z^{-j}.$$

14. Derive (5.84) and (5.85) from (5.80) and (5.81), respectively, by considering the backward polynomial

$$a_n x^n + a_{n-1} x^{n-1} + \cdots + a_0.$$

15. Apply the matrix bounds to determine inclusion and exclusion regions for the roots of the equation

$$4x^5 - 16x^4 + 17x^3 + 6x^2 - 21x + 10 = 0.$$

5.6 ROOT ACCEPTANCE AND REFINEMENT

In general, we cannot hope to find an exact root of (1.1). In the first place, in general, the desired root will not be a rational number, much less a computer-representable number. Even if the root, r, is a computer-representable number, the value of $P(r)$ produced by the machine will in general not vanish. As a matter of fact, it may well happen that the computed value of $P(x)$ will not vanish for *any* computer-representable number. Rather than attempting to find an α such that the computed value of $P(\alpha)$ vanishes, instead we seek an α such that $|P(\alpha)|$ is "small" in some sense, and then we use one of the procedures described below to determine how far α is from a root.

A simple criterion which is sometimes used for root acceptance is to choose a value of ε, say 10^{-10}, and to accept α as a root if $|P(\alpha)| \leq \varepsilon$. On the other hand, if ε is chosen too large, the accepted root may be unnecessarily inaccurate. Frequently a value of ε which is good for one polynomial may be bad for another. For example, if $\varepsilon = 10^{-10}$ is good for the polynomial $P(x)$, it will not be good for the polynomial $P_1(x)$ whose coefficients are those of $P(x)$ multiplied by 10^{-20}. Clearly, $P_1(x)$ has the same zeros as $P(x)$ but a value of α such that $|P_1(\alpha)| \leq \varepsilon$ will yield a value of $10^{20}P_1(\alpha)$ for $P(\alpha)$. Hence α will presumably not be close to a zero of $P_1(x)$.

We now describe a procedure based on that considered by Adams [1967] which is designed to overcome this difficulty. Each time $P(x)$ is computed, we also compute a bound for $\delta(P(x))$ where

6.1 $$\delta(P(x)) = |P(x) - \tilde{P}(x)|$$

and where $\tilde{P}(x)$ is the value of $P(x)$ produced by the computer. We then test whether or not for a given value α we have

6.2 $$|\tilde{P}(\alpha)| \leq qr(P(\alpha))$$

where $r(P(\alpha))$ is a positive number such that

6.3 $$\delta(P(\alpha)) \leq r(P(\alpha))$$

and where q is an appropriately chosen small integer. If (6.2) is satisfied, then α is accepted as a zero of $P(x)$. From considerations described below we let

6.4 $$q = \begin{cases} 3, & \text{if } \alpha \text{ and the coefficients of } P(x) \text{ are real} \\ 4, & \text{otherwise.} \end{cases}$$

Let us assume that we are working with a computer with t binary digits in the mantissa of a floating-point number. Let us define

6.5
$$\varepsilon = 2^{-t}$$

and, following J. Wilkinson [1963], let

6.6
$$\bar{\varepsilon} = 1.06\varepsilon.$$

Thus, for instance, on Machine C (see Chapter 2), we have $t = 48$ and

6.7
$$\varepsilon = 2^{-48}$$

and

6.8
$$\bar{\varepsilon} = (1.06)2^{-48}.$$

We seek to show that

6.9
$$\delta(P(\alpha)) \leq r(P(\alpha))$$

where

6.10 $r(P(\alpha)) = \{(2n + 1)|a_0| \, |\alpha|^n + (2n - 1)|a_1| \, |\alpha|^{n-1} + \cdots + |a_n|\}\bar{\sigma}$

and where

6.11 $\bar{\sigma} = \begin{cases} \bar{\varepsilon}, & \text{if } \alpha \text{ and the coefficients of } P(x) \text{ are real,} \\ 2.47\bar{\varepsilon}, & \text{otherwise.} \end{cases}$

Let us first consider the case where α and the coefficients of $P(x)$ are real. In Section 2.7 we showed that for any binary operation $\circ$ (which may be $+, -, \times, \div$) we have

6.12
$$\mathrm{fl}(x_1 \circ x_2) = (x_1 \circ x_2)(1 + \varepsilon_0)$$

where

6.13
$$|\varepsilon_0| \leq \varepsilon.$$

In order to prove (6.9) we seek to show that if $P(\alpha)$ is computed by the algorithm defined by

6.14 $\begin{cases} z_0 = a_0 \\ z_k = \alpha z_{k-1} + a_k, & k = 1, 2, \ldots, n \end{cases}$

then one obtains a number $\hat{z}_n$ which is the exact value of $\hat{P}(\alpha)$, where $\hat{P}(x)$ is defined by

6.15
$$\hat{P}(x) = \hat{a}_0 x^n + \hat{a}_1 x^{n-1} + \cdots + \hat{a}_n.$$

Here

6.16
$$\hat{a}_k = (1 + \varepsilon_k)a_k$$

and

6.17
$$(1 - \varepsilon)^{2(n-k)+1} \leqq 1 + \varepsilon_k \leqq (1 + \varepsilon)^{2(n-k)+1}.$$

Evidently, as in Section 2.7, we have

6.18
$$\hat{z}_1 = \text{fl}(\text{fl}(xz_0) + a_1)$$
$$= (1 + \varepsilon_a^{(1)})(xz_0(1 + \varepsilon_m^{(1)}) + a_1)$$

where $|\varepsilon_a^{(1)}| \leqq \varepsilon$, $|\varepsilon_m^{(1)}| \leqq \varepsilon$.   Hence

6.19
$$\hat{z}_1 = xa_0(1 + \varepsilon_a^{(1)})(1 + \varepsilon_m^{(1)}) + a_1(1 + \varepsilon_a^{(1)}).$$

Thus $\hat{z}_1$ is the exact value of the polynomial $\hat{a}_0^{(1)}x + \hat{a}_1^{(1)}$ where

6.20
$$\hat{a}_0^{(1)} = a_0(1 + \varepsilon_a^{(1)})(1 + \varepsilon_m^{(1)}), \quad \hat{a}_1^{(1)} = a_1(1 + \varepsilon_a^{(1)}).$$

Moreover,

6.21
$$\hat{a}_0^{(1)} = a_0(1 + \varepsilon_0)$$

where

6.22
$$(1 - \varepsilon)^3 \leqq (1 - \varepsilon)^2 \leqq 1 + \varepsilon_0 \leqq (1 + \varepsilon)^2 \leqq (1 + \varepsilon)^3$$

and

6.23
$$\hat{a}_1^{(1)} = a_1^{(1)}(1 + \varepsilon_1), \qquad \text{where} \qquad |\varepsilon_1| \leqq \varepsilon.$$

Next, we have

6.24
$$\hat{z}_2 = \text{fl}(\text{fl}(x\hat{z}_1) + a_2)$$
$$= (1 + \varepsilon_a^{(2)})((1 + \varepsilon_m^{(2)})x\hat{z}_1 + a_2)$$
$$= \hat{a}_0^{(2)}x^2 + \hat{a}_1^{(2)}x + \hat{a}_2^{(2)}$$

where

6.25
$$\begin{cases} \hat{a}_0^{(2)} = a_0(1 + \varepsilon_a^{(1)})(1 + \varepsilon_m^{(1)})(1 + \varepsilon_a^{(2)})(1 + \varepsilon_m^{(2)}) \\ \hat{a}_1^{(2)} = a_1(1 + \varepsilon_a^{(1)})(1 + \varepsilon_a^{(2)})(1 + \varepsilon_m^{(2)}) \\ \hat{a}_2^{(2)} = a_2(1 + \varepsilon_a^{(2)}). \end{cases}$$

Therefore,

6.26
$$\hat{a}_i^{(2)} = a_i(1 + \varepsilon_i)$$

where

6.27
$$(1 - \varepsilon)^5 \leqq (1 - \varepsilon)^4 \leqq 1 + \varepsilon_0 \leqq (1 + \varepsilon)^4 \leqq (1 + \varepsilon)^5$$
$$(1 - \varepsilon)^3 \leqq 1 + \varepsilon_1 \leqq (1 + \varepsilon)^3$$
$$1 - \varepsilon \leqq 1 + \varepsilon_2 \leqq 1 + \varepsilon.$$

The general result (6.17) follows by induction.

We now show that for any $\varepsilon > 0$ and for any positive integer k such that

6.28
$$k\varepsilon < 0.1$$

we have

6.29
$$1 - k\bar{\varepsilon} \leqq (1 - \varepsilon)^k,$$
$$(1 + \varepsilon)^k \leqq 1 + k\bar{\varepsilon}.$$

To show this we first observe that

6.30
$$1 - (1 - \varepsilon)^k = \binom{k}{1}\varepsilon - \binom{k}{2}\varepsilon^2 + \binom{k}{3}\varepsilon^3 + \cdots + (-1)^{k-1}\varepsilon^k$$
$$\leqq \binom{k}{1}\varepsilon + \binom{k}{2}\varepsilon^2 + \binom{k}{3}\varepsilon^3 + \cdots + \varepsilon^k$$
$$= (1 + \varepsilon)^k - 1.$$

Therefore, it is sufficient to show that $(1 + \varepsilon)^k - 1 \leqq k\bar{\varepsilon}$. But since $\log(1 + x) \leqq x$ for $x > 0$ we have $k \log(1 + \varepsilon) \leqq k\varepsilon$ and

6.31
$$(1 + \varepsilon)^k \leqq e^{k\varepsilon}.$$

Therefore, since $(e^x - 1)x^{-1} = 1 + x/2! + x^2/3! + \cdots$ is an increasing function of x we have

6.32
$$\frac{(1 + \varepsilon)^k - 1}{k\varepsilon} \leqq \frac{e^{k\varepsilon} - 1}{k\varepsilon} \leqq \frac{e^{0.1} - 1}{0.1} \doteq 1.052 \leqq 1.06.$$

From (6.29) and (6.17) we have

6.33 $$|\varepsilon_k| \leqq (2(n - k) + 1)\bar{\varepsilon}.$$

Therefore, by (6.33) we have

6.34 $$|\hat{P}(\alpha) - P(\alpha)| = |\hat{z}_n - z_n| \leqq |\hat{a}_0 - a_0||\alpha|^n + |\hat{a}_1 - a_1||\alpha|^{n-1} + \cdots + |\hat{a}_n - a_n|$$
$$\leqq r(P(\alpha))$$

and (6.9) holds when α and the coefficients of $P(x)$ are real.

Let us now consider the case where α is complex and/or some of the coefficients of $P(x)$ are complex. It can be shown using methods similar to those used by Champagne* [1964] that if x_1 and x_2 are complex, then (6.12) holds for addition, subtraction, and multiplication with

6.35 $$|\varepsilon_o| \leqq 2.47\varepsilon.$$

The rest of the proof of (6.9) is the same, except for obvious modifications, as the proof for the real case.

We now seek to show, given a real polynomial $P(x)$, that there exists a computer-representable number α such that $\hat{P}(\alpha)$, the computed value of $P(\alpha)$, satisfies the condition

6.36 $$|\hat{P}(\alpha)| \leqq 3r(P(\alpha))$$

where $r(P(\alpha))$ is given by (6.10). Let us first assume that $P(\alpha)$ has a real zero, say α^*, and let α be the representable number closest to α^* such that $|\alpha| \geqq |\alpha^*|$. By the mean-value theorem we have

6.37 $$P(\alpha) - P(\alpha^*) = (\alpha - \alpha^*)P'(\xi)$$

where ξ lies between α and α^*. Moreover, by the methods of Section 2.7, we can show that

6.38 $$|\alpha - \alpha^*| \leqq 2|\alpha|\varepsilon$$

where ε is given by (6.5). Therefore,

6.39 $$|P(\alpha) - P(\alpha^*)| \leqq 2|\alpha|\varepsilon|P'(\xi)|$$
$$\leqq 2|\alpha|\varepsilon\{n|a_0||\xi|^{n-1} + \cdots + |a_{n-1}|\}$$
$$\leqq 2|\alpha|\varepsilon\{n|a_0||\alpha|^{n-1} + \cdots + |a_{n-1}|\}$$
$$\leqq 2\bar{\varepsilon}\{(2n + 1)|a_0||\alpha|^n + \cdots + |a_{n-1}||\alpha|\}$$
$$\leqq 2r(P(\alpha)).$$

---

* Champagne's methods are based on the error analysis procedures described by J. Wilkinson [1963].

Hence,

6.40
$$|\hat{P}(\alpha)| \leq |\hat{P}(\alpha) - P(\alpha)| + |P(\alpha) - P(\alpha^*)|$$
$$\leq 3r(P(\alpha)).$$

Suppose now that $P(x)$ has complex coefficients and α^* is complex. If α^* is a zero of $P(x)$ let $\alpha = \alpha_1 + i\alpha_2$ where α_1 and α_2 are the smallest computer-representable real numbers such that $|\alpha_1| \geq |\mathrm{Re}\,\alpha^*|$, $|\alpha_2| \geq |\mathrm{Im}\,\alpha^*|$. Evidently

6.41
$$|\alpha - \alpha^*| \leq 2\sqrt{2}|\alpha|\varepsilon.$$

As before, we can show that

6.42
$$|P(\alpha) - P(\alpha^*)| \leq 2\sqrt{2}r(P(\alpha))$$

and hence

6.43
$$|\hat{P}(\alpha)| \leq (1 + 2\sqrt{2})r(P(\alpha))$$
$$\leq 4r(P(\alpha)).$$

Multiple Roots

Suppose we have found a number α such that $|\tilde{P}(\alpha)| \leq qr(P(\alpha))$. Before accepting α as a root we should first test whether α is a multiple root or else if there is a multiple root nearby. More precisely, we test whether for some α^* near α we have $|\tilde{P}'(\alpha^*)| \leq qr(P'(\alpha^*))$, $|\tilde{P}''(\alpha^*)| \leq qr(P''(\alpha^*))$, ..., $|\tilde{P}^{(k-1)}(\alpha^*)| \leq qr(P^{(k-1)}(\alpha^*))$ for some $k > 1$.

The following procedure is based on that used by Champagne and Kripke. (See Champagne [1964].) Iterate with $P'(x)$ starting with α and using the basic iterative procedure. If at any step the convergence condition (6.2) for $P(x)$ is violated, stop and accept α as a simple root. If at some iteration we obtain α_1 for which the convergence criteria are satisfied for both $P(x)$ and $P'(x)$, then start iterating on $P''(x)$, beginning with α_1, checking at each stage that the convergence criteria for both $P(x)$ and $P'(x)$ are satisfied. This process is continued as long as possible.

It appears to be more accurate to evaluate derivatives of $P(x)$ by using the polynomials for the derivatives obtained by algebraic differentiation, rather than by using repeated synthetic division. Thus, for $P'(x)$, for instance, one would use

6.44
$$P'(x) = (na_0)x^{n-1} + ((n-1)a_1)x^{n-2} + \cdots + a_{n-1}.$$

Real Coefficients—Complex Roots

If the polynomial has real coefficients, then we frequently desire to find either real roots or complex conjugate pairs. By so doing, we obtain reduced polynomials with real coefficients. Normally, if we have found α such that $|\tilde{P}(\alpha)| \leqq qr(P(\alpha))$ then $|\tilde{Q}(\bar{\alpha})| \leqq qr(Q(\bar{\alpha}))$ where $\tilde{Q}(x)$ is the computed reduced polynomial obtained by the synthetic division process, i.e.,

6.45 $P(x) = (x - \alpha)Q(x) + P(\alpha).$

However, if $\gamma = \text{Im}(\alpha)$ is very small, it may happen that $\bar{\alpha}$ does not satisfy the convergence criterion for $Q(x)$. In that case we seek to find either:

a) a real value α^* near α which satisfies the convergence criterion (in this case we can divide out the root and obtain a reduced polynomial with real coefficients), or

b) a complex value α^{**} such that $|\tilde{P}(\alpha^{**})| \leqq q(r(P(\alpha^{**})))$ and $|\tilde{Q}(\alpha^{**})| \leqq qr(Q(\alpha^{**}))$.

The procedure for doing this is to first find a circle C about α containing at least one root. One of the bounds (5.34)–(5.35) can be used. We then seek to determine whether or not there is a real root in C. An accurate procedure* for doing this based on the use of Hermite interpolation is described by B. Wilkinson [1969]. If there is a real root in C, we determine it and divide it out. Otherwise, we use a search procedure to find a value of α^{**} in C which satisfies the convergence criteria for $P(x)$ while $\bar{\alpha}^{**}$ satisfies the convergence criterion for $Q(x)$. In this connection we remark that from (6.45) it follows that $\beta = \text{Re } \alpha$ and $\gamma = \text{Im } \alpha$, then

6.46 $Q(\beta - i\gamma) = \dfrac{1}{\gamma} \text{Im}[P(\beta + i\gamma)].$

Our search procedure is based on a gradient method for minimizing

6.47 $J = |P(\alpha)|^2 + v\left[\dfrac{1}{\gamma}\text{Im}(P(\alpha))\right]^2$

where v is an approximate value of

$$r(P(\alpha))/r\left(\frac{1}{\gamma}\text{Im}(P(\alpha))\right).$$

Details of this procedure are also given by B. Wilkinson [1969].

When the above process has been carried out and a value of α^{**} obtained, we divide out the approximate roots α^{**} and $\bar{\alpha}^{**}$, obtaining a reduced polynomial with real coefficients.

In the case of multiple roots, for polynomials with real coefficients, it is recommended that priority be given to finding roots which are exact complex conjugates even at the expense of accepting a lower multiplicity.

---

* Of course Sturm sequences can be used, but care must be taken to insure that rounding errors do not give erroneous conclusions about the existence of a real root.

If there is a root of very small modulus, it may happen that $r(P(x))$ decreases more rapidly than $|\tilde{P}(x)|$ and an excessive amount of time may be required to satisfy (6.2). In that case it is suggested that one of the family of bounds given in Section 5.5 be used, for example those given by (5.34). Thus α would be accepted as a root if, for some k, ρ_k, given by (5.35), is less than some prescribed value. We remark that the results obtained by such a process are invariant under multiplication of all coefficients of $P(x)$ by a given constant.

EXERCISES 5.6

1. Compute a bound on the rounding error for the computer determination of the polynomials

$$P_1(x) = 8x^5 + 1.25x^4 - 3.17x^3 + 2.11$$

$$P_2(x) = x^4 - 5x^3 + 10x^2 - 10x + 4$$

for $x = 2.67$. Assume the computer carries 48 binary digits for the mantissa of a floating-point number.

2. Compute a bound on the rounding error for the polynomial

$$(1 + i)z^2 + (2 - i)z + 3 + 3i$$

for $z = \frac{1}{2} - \frac{1}{2}i$. Assume that there are 48 binary digits in the mantissa of a floating-point number.

3. Modify the program of Exercise 2, Section 5.4 as follows. Accept α as a root when $|P(\alpha)| \leq 4\delta(P(\alpha))$ where $\delta(P(\alpha))$ is a bound on the rounding error in the computation of $P(\alpha)$. Also compute the bounds (5.34)–(5.35) and accept α if there is a root within, say $\delta = 10^{-10}$, of α.

5.7 MATRIX RELATED METHODS: THE MODIFIED BERNOULLI METHOD

In this section and the next section we describe methods for solving polynomial equations which are based on methods for finding eigenvalues of matrices.* The methods for finding eigenvalues of matrices (which we shall consider) are the power method and the inverse power method, which are treated in Chapter 14. The reader who is not familiar with these methods is advised to defer a study of this section until he has studied the appropriate parts of Chapters 11, 12, and 14.

The problem of solving the polynomial equation (1.1) is equivalent to that of finding the eigenvalues of the companion matrix

7.1

$$C = \begin{pmatrix} 0 & 1 & 0 & 0 & \cdots & 0 & 0 \\ 0 & 0 & 1 & 0 & \cdots & 0 & 0 \\ 0 & 0 & 0 & 1 & \cdots & 0 & 0 \\ \cdot & \cdot & \cdot & \cdot & \cdots & \cdot & \cdot \\ 0 & 0 & 0 & 0 & \cdots & 0 & 1 \\ -b_n & -b_{n-1} & -b_{n-2} & -b_{n-3} & \cdots & -b_2 & -b_1 \end{pmatrix}$$

* See Section 11.21.

where

7.2
$$b_i = \frac{a_i}{a_0}, \qquad i = 1, 2, \ldots, n.$$

For example, in the case $n = 4$ we have

7.3
$$C = \begin{pmatrix} 0 & 1 & 0 & 0 \\ 0 & 0 & 1 & 0 \\ 0 & 0 & 0 & 1 \\ -b_4 & -b_3 & -b_2 & -b_1 \end{pmatrix}.$$

The characteristic polynomial* for C is

7.4
$$\det(C - \lambda I) = \det\begin{pmatrix} -\lambda & 1 & 0 & 0 \\ 0 & -\lambda & 1 & 0 \\ 0 & 0 & -\lambda & 1 \\ -b_4 & -b_3 & -b_2 & (-b_1 - \lambda) \end{pmatrix}$$

$$= b_4 + b_3\lambda + b_2\lambda^2 + (b_1 + \lambda)\lambda^3$$

$$= \frac{1}{a_0}\{a_4 + a_3\lambda + a_2\lambda^2 + a_1\lambda^3 + a_0\lambda^4\}$$

$$= \frac{1}{a_0} P(\lambda).$$

In general, we have, for an $n \times n$ matrix C,

7.5
$$\det(C - \lambda I) = (-1)^n \frac{P(\lambda)}{a_0}$$

and hence λ is an eigenvalue of C if and only if λ is a zero of $P(x)$.

Modified Bernoulli Method

There are a number of powerful methods available for solving matrix eigenvalue problems. Several of these are treated in Chapter 14. One could solve the polynomial equation (1.1) simply by using a general method for finding the eigenvalues of a matrix. However, this might be somewhat inefficient since most of the elements of the companion matrix are zero. Two matrix eigenvalue methods appear

---

* See 14-(2.16).

particularly well adapted to take advantage of the special form of the matrix. One such method is the *power method*.* A basic step in the power method is the following. Given a vector

7.6
$$x = \begin{pmatrix} x_1 \\ x_2 \\ x_3 \\ x_4 \end{pmatrix}$$

and a number β, determine x' such that

7.7
$$x' = (C - \beta I)x,$$

i.e., such that

7.8
$$\begin{pmatrix} x_1' \\ x_2' \\ x_3' \\ x_4' \end{pmatrix} = \begin{pmatrix} -\beta & 1 & 0 & 0 \\ 0 & -\beta & 1 & 0 \\ 0 & 0 & -\beta & 1 \\ -b_4 & -b_3 & -b_2 & -b_1 - \beta \end{pmatrix} \begin{pmatrix} x_1 \\ x_2 \\ x_3 \\ x_4 \end{pmatrix}$$

or

7.9
$$\begin{cases} x_1' = -\beta x_1 + x_2 \\ x_2' = -\beta x_2 + x_3 \\ x_3' = -\beta x_3 + x_4 \\ x_4' = -a_0^{-1}\{a_4 x_1 + a_3 x_2 + a_2 x_3 + (a_1 + \beta a_0)x_4\}. \end{cases}$$

We thus define the modified Bernoulli method by

7.10
$$x^{(k+1)} = (C - \beta I)x^{(k)}$$

where $x^{(0)}$ is arbitrary. After each iteration we estimate the root by the Rayleigh quotient

7.11
$$R_{k+1} = R(x^{(k+1)}) = \frac{(Cx^{(k+1)}, x^{(k+1)})}{(x^{(k+1)}, x^{(k+1)})}$$

where in general, for any two vectors x and y the inner product† is

7.12
$$(x, y) = \sum_{i=1}^{n} x_i \bar{y}_i.$$

---

* See Section 14.3.
† See 11-(14.4).

If desired, one can multiply $x^{(k)}$ at any stage by a factor to keep the numbers within range. This will not affect the convergence of the method. It can be shown that unless $x^{(0)}$ is chosen very badly, the method will converge to the root of (1.1) which is farthest away from β provided no other root is equally far away.* In the case $\beta = 0$, we have

7.13
$$\begin{cases} x_1' = x_2 \\ x_2' = x_3 \\ x_3' = x_4 \\ x_4' = -\dfrac{1}{a_0}(a_4 x_1 + a_3 x_2 + a_2 x_3 + a_1 x_4) \end{cases}$$

which is equivalent to the Bernoulli method. Actually, in the Bernoulli method

7.14
$$x_{s+1} = \frac{-1}{a_0}(a_4 x_{s-3} + a_3 x_{s-2} + a_2 x_{s-1} + a_1 x_s).$$

We continue until the values of x_{s+1}/x_s converge to a limit, say r. The limiting value, r, will be the root of largest modulus.

We now show that, given a vector x, the value of the Rayleigh quotient

7.15
$$R(x) = \frac{(Cx, x)}{(x, x)}$$

is a weighted average with positive weights of the ratios (x_2/x_1), (x_3/x_2), (x_4/x_3), (x_5/x_4) where x_5 is the value produced by the unmodified Bernoulli method (7.14). Thus

7.16
$$R(x) = \frac{\bar{x}_1 x_2 + \bar{x}_2 x_3 + \bar{x}_3 x_4 + \bar{x}_4 q}{|x_1|^2 + |x_2|^2 + |x_3|^2 + |x_4|^2}$$
$$= \frac{|x_1|^2(x_2/x_1) + |x_2|^2(x_3/x_2) + |x_3|^2(x_4/x_3) + |x_4|^2(q/x_4)}{|x_1|^2 + |x_2|^2 + |x_3|^2 + |x_4|^2}$$

where

7.17
$$q = -\frac{1}{a_0}(a_4 x_1 + a_3 x_2 + a_2 x_3 + a_1 x_4)$$
$$= x_5.$$

It is desirable to choose β so that β is farther away from one root, say λ_1, than from all others. On the other hand, we also desire to choose β so that

---

* See Section 14.3.

$\mu = |(\lambda_1 - \beta)/(\lambda_2 - \beta)|$ is as large as possible, where λ_2 is the next closest root to β. Thus, for example, suppose $\lambda_1 = 1$, $\lambda_2 = \frac{1}{2}$, $\lambda_3 = 0$. The best choice of β would be $\frac{1}{4}$, and the corresponding value of μ would be 3.

Choice of Starting Vector

We have already stated that the modified Bernoulli method will converge for suitable β provided the initial vector is not badly chosen. To make this statement more precise, let us consider the eigenvectors and principal vectors of C. If λ is an eigenvalue of C, then by (7.3)

7.18
$$C\begin{pmatrix} 1 \\ \lambda \\ \lambda^2 \\ \lambda^3 \end{pmatrix} = \begin{pmatrix} \lambda \\ \lambda^2 \\ \lambda^3 \\ -b_4 - b_3\lambda - b_2\lambda^2 - b_1\lambda^3 \end{pmatrix}$$
$$= \begin{pmatrix} \lambda \\ \lambda^2 \\ \lambda^3 \\ -\dfrac{1}{a_0}(P(\lambda) - a_0\lambda^4) \end{pmatrix} = \begin{pmatrix} \lambda \\ \lambda^2 \\ \lambda^3 \\ \lambda^4 \end{pmatrix} = \lambda \begin{pmatrix} 1 \\ \lambda \\ \lambda^2 \\ \lambda^3 \end{pmatrix}.$$

Hence $(1, \lambda, \lambda^2, \lambda^3)^T$ is an eigenvector of C. On the other hand, if $(x_1, x_2, x_3, x_4)^T$ is an eigenvector of C associated with the eigenvalue λ, then

7.19
$$\begin{cases} x_2 = \lambda x_1 \\ x_3 = \lambda x_2 = \lambda^2 x_1 \\ x_4 = \lambda x_3 = \lambda^3 x_1 \end{cases}$$

and

7.20 $-b_4 x_1 - b_3 x_2 - b_2 x_3 - b_1 x_4 = -\dfrac{x_1}{a_0}(a_4 + a_3\lambda + a_2\lambda^2 + a_1\lambda^3)$

$$= -\dfrac{x_1}{a_0}[P(\lambda) - a_0\lambda^4]$$

$$= \lambda^4 x_1.$$

Hence, x is a multiple of $(1, \lambda, \lambda^2, \lambda^3)^T$. Thus if λ is a repeated eigenvalue, there can be only one eigenvector to within a constant factor. Moreover, if λ is an m-fold eigenvalue, then there exist $(m - 1)$ principal vectors* $v^{(2)}, v^{(3)}, \ldots, v^{(m)}$ in addition to the eigenvector

7.21
$$v^{(1)} = \begin{pmatrix} 1 \\ \lambda \\ \lambda^2 \\ \vdots \\ \lambda^{n-1} \end{pmatrix}$$

such that

7.22
$$Cv^{(k)} = \lambda v^{(k)} + v^{(k-1)}, \qquad k = 2, 3, \ldots, m.$$

The principal vectors are

7.23
$$v^{(2)} = \begin{pmatrix} 0 \\ 1 \\ 2\lambda \\ 3\lambda^2 \\ \vdots \\ (n-1)\lambda^{n-2} \end{pmatrix}, \quad v^{(3)} = \begin{pmatrix} 0 \\ 0 \\ 1 \\ 3\lambda \\ \vdots \\ \dfrac{(n-1)(n-2)}{2}\lambda^{n-3} \end{pmatrix}, \text{etc.}$$

In any case, our initial vector $x^{(0)}$ can be expanded† uniquely in terms of the eigenvectors and principal vectors of C. Thus we have

7.24
$$x^{(0)} = c_1 v^{(1)} + c_2 v^{(2)} + \cdots + c_n v^{(n)}$$

where the $v^{(i)}$ now represent eigenvectors or principal vectors of C. Suppose $v^{(1)}$ is the eigenvector associated with the eigenvalue λ_1 which is farther from β than from any other eigenvalue. Then the modified Bernoulli method will converge if $c_1 \neq 0$. As a matter of fact, it will converge if the coefficient corresponding to any principal vector associated with the eigenvalue λ_1 does not vanish.

We now show how one can choose $x^{(0)}$ so that $c_1 \neq 0$. Let $\lambda_1, \lambda_2, \ldots, \lambda_n$ denote the roots of (1.1). We do not assume the roots to be distinct. Let us consider the sums $s_0, s_1, \ldots$ defined by

7.25
$$s_i = \sum_{j=1}^{n} \lambda_j^i.$$

---

* See 11-(21.62).
† See 11-(21.63).

We seek to show that

7.26
$$\begin{cases} s_0 = n \\ a_0 s_1 + a_1 = 0 \\ a_0 s_2 + a_1 s_1 + 2a_2 = 0 \\ a_0 s_3 + a_1 s_2 + a_2 s_1 + 3a_3 = 0 \\ \qquad \vdots \\ a_0 s_n + a_1 s_{n-1} + \cdots + a_{n-1} s_1 + na_n = 0 \end{cases}$$

and for $k > n$ we have

7.27
$$a_0 s_k + a_1 s_{k-1} + \cdots + a_{n-1} s_{k-n+1} + a_n s_{k-n} = 0.$$

Thus for the case $n = 4$ we have

7.28
$$\begin{cases} s_0 = 4 \\ a_0 s_1 + a_1 = 0 \\ a_0 s_2 + a_1 s_1 + 2a_2 = 0 \\ a_0 s_3 + a_1 s_2 + a_2 s_1 + 3a_3 = 0 \\ a_0 s_4 + a_1 s_3 + a_2 s_2 + a_3 s_1 + 4a_4 = 0 \end{cases}$$

and for $k \geq 5$

7.29
$$a_0 s_k + a_1 s_{k-1} + a_2 s_{k-2} + a_3 s_{k-3} + a_4 s_{k-4} = 0.$$

We give a derivation for the case $n = 4$. Evidently the roots of the equation

7.30
$$Q(y) = a_4 y^4 + a_3 y^3 + a_2 y^2 + a_1 y + a_0 = 0$$

are $\lambda_1^{-1}, \lambda_2^{-1}, \lambda_3^{-1}, \lambda_4^{-1}$. Moreover,

7.31
$$Q(y) = a_4 (y - \lambda_1^{-1})(y - \lambda_2^{-1})(y - \lambda_3^{-1})(y - \lambda_4^{-1})$$

7.32
$$\log Q(y) = \log a_4 + \sum_{i=1}^{4} \log(y - \lambda_i^{-1})$$

7.33
$$\frac{Q'(y)}{Q(y)} = \sum_{i=1}^{4} \frac{1}{y - \lambda_i^{-1}}$$
$$= \sum_{i=1}^{4} \frac{\lambda_i}{\lambda_i y - 1}$$

$$= -\sum_{i=1}^{4} \lambda_i(1 + \lambda_i y + \lambda_i^2 y^2 + \cdots)$$

$$= -\sum_{j=0}^{\infty} y^j \left(\sum_{i=1}^{4} \lambda_i^{j+1} \right)$$

$$= -\sum_{j=0}^{\infty} s_{j+1} y^j.$$

The above series converges for y sufficiently small. Thus we have

7.34 $Q'(y) = 4a_4 y^3 + 3a_3 y^2 + 2a_2 y + a_1$
$$= -(s_1 + s_2 y + s_3 y^2 + \cdots)(a_4 y^4 + a_3 y^3 + a_2 y^2 + a_1 y + a_0)$$
$$= (-a_0 s_1) + (-a_0 s_2 - a_1 s_1)y + (-a_0 s_3 - a_1 s_2 - a_2 s_1)y^2$$
$$+ (-a_0 s_4 - a_1 s_3 - a_2 s_2 - a_3 s_1)y^3$$
$$+ (-a_0 s_5 - a_1 s_4 - a_2 s_3 - a_3 s_2 - a_4 s_1)y^4 + \cdots$$

The results (7.28) and (7.29) follow by equating coefficients of like powers of y.
We now let $x^{(0)}$ be given by

7.35
$$x^{(0)} = \begin{pmatrix} s_0 \\ s_1 \\ s_2 \\ s_3 \end{pmatrix} = \begin{pmatrix} 1 + 1 + 1 + 1 \\ \lambda_1 + \lambda_2 + \lambda_3 + \lambda_4 \\ \lambda_1^2 + \lambda_2^2 + \lambda_3^2 + \lambda_4^2 \\ \lambda_1^3 + \lambda_2^3 + \lambda_3^3 + \lambda_4^3 \end{pmatrix}$$

$$= \begin{pmatrix} 1 \\ \lambda_1 \\ \lambda_1^2 \\ \lambda_1^3 \end{pmatrix} + \begin{pmatrix} 1 \\ \lambda_2 \\ \lambda_2^2 \\ \lambda_2^3 \end{pmatrix} + \begin{pmatrix} 1 \\ \lambda_3 \\ \lambda_3^2 \\ \lambda_3^3 \end{pmatrix} + \begin{pmatrix} 1 \\ \lambda_4 \\ \lambda_4^2 \\ \lambda_4^3 \end{pmatrix}.$$

In the expansion (7.24), if $v^{(i)}$ is an eigenvector, then the coefficient c_i is m_i where m_i is the multiplicity of the associated eigenvalue. If $v^{(i)}$ is a principal vector, then $c_i = 0$. In any case, we have $c_1 \neq 0$, and the modified Bernoulli method converges provided λ_1 is further from β than from any other eigenvalue.

EXERCISES 5.7

1. For the polynomial $P(x) = x^4 - 3x^3 + x^2 + 3x - 2$ construct the companion matrix and find its eigenvalues and eigenvectors. Determine the sums s_i by (7.26) and verify (7.25). With $x^{(0)}$ as determined by (7.35) carry out 5 iterations of the Bernoulli method and compute the Rayleigh quotient.

2. Perform one iteration of the modified Bernoulli method for the polynomial $P(x) = x^4 - 5x^3 + 10x^2 - 10x + 4$ with $\alpha = \frac{1}{2}$ and the starting vector

$$x^{(0)} = \begin{pmatrix} 1 \\ \frac{1}{2} \\ (\frac{1}{2})^2 \\ (\frac{1}{2})^3 \end{pmatrix}.$$

Compute the Rayleigh quotient (7.11).

3. For the polynomial $P(x)$ in Exercise 2, determine a starting vector from (7.28) and perform one iteration of the modified Bernoulli method with $\alpha = \frac{1}{2}$. Verify (7.35) knowing that the zeros of $P(x)$ are $1, 2, 1 + i, 1 - i$.

5.8 MATRIX RELATED METHODS: THE *IP* METHOD

In many cases the *inverse* power method for finding eigenvalues of matrices is much more powerful than the *direct* power method especially when one has a good estimate of an eigenvalue. For a general matrix, however, a great deal more computational effort is required per iteration with the inverse power method than with the direct power method. Hence, the usual practice when seeking matrix eigenvalues is to perform a number of iterations with the direct power method, or to use some other procedure to get a good estimate of an eigenvalue, and then to use the inverse power method to obtain an accurate value. However, if the matrix is the companion matrix of some polynomial, the inverse power method can be carried out in essentially the same number of steps as the direct power method; and so it is feasible to use it from the outset.

A basic step* in the *IP* method is the following: given a vector x and a number α, determine x' such that

8.1 $(C - \alpha I)x' = x,$

where C is given by (7.1). Evidently $x' = (C - \alpha I)^{-1}x$, and hence the procedure is the same as the direct power method for the matrix $(C - \alpha I)^{-1}$.

Let us solve (8.1) for the case $n = 4$. We have, by (7.3),

8.2
$$\begin{pmatrix} -\alpha & 1 & 0 & \\ 0 & -\alpha & 1 & 0 \\ 0 & 0 & -\alpha & 1 \\ -b_4 & -b_3 & -b_2 & (-b_1 - \alpha) \end{pmatrix} \begin{pmatrix} x'_1 \\ x'_2 \\ x'_3 \\ x'_4 \end{pmatrix} = \begin{pmatrix} x_1 \\ x_2 \\ x_3 \\ x_4 \end{pmatrix}.$$

---

* See 14-(3.19) and 14-(3.23).

We consider the augmented matrix

8.3
$$M = \begin{pmatrix} -\alpha & 1 & 0 & 0 & x_1 \\ 0 & -\alpha & 1 & 0 & x_2 \\ 0 & 0 & -\alpha & 1 & x_3 \\ -b_4 & -b_3 & -b_2 & (-b_1 - \alpha) & x_4 \end{pmatrix}.$$

Adding α times the first row to the second, and then α times the new second row to the third, we obtain

8.4
$$M' = \begin{pmatrix} -\alpha & 1 & 0 & 0 & x_1 \\ -\alpha^2 & 0 & 1 & 0 & x_2 + \alpha x_1 \\ -\alpha^3 & 0 & 0 & 1 & x_3 + \alpha(x_2 + \alpha x_1) \\ -b_4 & -b_3 & -b_2 & (-b_1 - \alpha) & x_4 \end{pmatrix}.$$

Next, we add b_3 times the first row, plus b_2 times the second row, plus $(b_1 + \alpha)$ times the third row to the fourth row obtaining

8.5
$$M'' = \begin{pmatrix} -\alpha & 1 & 0 & 0 & x_1 \\ -\alpha^2 & 0 & 1 & 0 & x_2 + \alpha x_1 \\ -\alpha^3 & 0 & 0 & 1 & x_3 + \alpha(x_2 + \alpha x_1) \\ -\dfrac{P(\alpha)}{a_0} & 0 & 0 & 0 & q \end{pmatrix}$$

where

8.6
$$q = x_4 + b_3 x_1 + b_2(x_2 + \alpha x_1) + (b_1 + \alpha)(x_3 + \alpha(x_2 + \alpha x_1))$$

$$= \frac{1}{a_0}\{a_0 x_4 + (a_1 + \alpha a_0)x_3 + (a_2 + \alpha(a_1 + \alpha a_0))x_2$$

$$+ (a_3 + \alpha(a_2 + \alpha(a_1 + \alpha a_0)))x_1\}.$$

From the augmented matrices M'' and M we have

8.7
$$\begin{cases} x_1' = -\dfrac{a_0 q}{P(\alpha)} \\ x_i' = \alpha x_{i-1}' + x_{i-1}, \quad i = 2, 3, 4. \end{cases}$$

The above procedure can clearly be carried out as long as $P(\alpha) \neq 0$. (If $P(\alpha) = 0$, then α is a root of (1.1).) To actually carry out the calculations we let

8.8 $qa_0 = x_4S_0(\alpha) + x_3S_1(\alpha) + x_2S_2(\alpha) + x_1S_3(\alpha)$

where the $S_i(\alpha)$ are determined by the following procedure, which is similar to synthetic division,

8.9
$$\begin{cases} S_0(\alpha) = a_0 \\ S_i(\alpha) = \alpha S_{i-1}(\alpha) + a_i, \qquad i = 1, 2, 3. \end{cases}$$

Therefore, we have

8.10 $x_1' = -\dfrac{1}{P(\alpha)}[x_4S_0(\alpha) + x_3S_1(\alpha) + x_2S_2(\alpha) + x_1S_3(\alpha)].$

Having obtained x_1', we then use a procedure, similar to synthetic division, based on (8.7), to get x_2', x_3', and x_4'.

If x, α and the coefficients of the polynomial are real, then the above process requires $3n - 2$ multiplications, $3n - 3$ additions, and one division. For, to get the $S_i(\alpha)$ requires $n - 1$ multiplications and $n - 1$ additions. Then, n multiplications, $n - 1$ additions, and one division are needed to compute x_1'. Finally, $n - 1$ multiplications and $n - 1$ additions are needed to get $x_2', x_3', \ldots, x_n'$.

We thus define the *IP* method by

8.11 $(C - \alpha I)x^{(k+1)} = x^{(k)}$

where $x^{(0)}$ is arbitrary. Here, it is desirable to choose α as close as possible to the desired root λ_1. As a matter of fact, it can be shown* that the method will converge, unless $x^{(0)}$ is chosen very badly, if α is closer to λ_1 than to any other root.

After each iteration one can estimate the root λ_1 by computing the Rayleigh quotient R_{k+1} using (7.11). The convergence can often be accelerated by replacing α by α_k in (8.11) where

8.12 $\alpha_k = \dfrac{(Cx^{(k)}, x^{(k)})}{(x^{(k)}, x^{(k)})}.$

This formula can even be used for α_0 if desired. (This might be appropriate if $x^{(0)}$ is a good estimate of the eigenvector $v^{(1)}$ corresponding to λ_1.)

The convergence of the scheme where α_k is determined by (8.12) for $k = 0, 1, 2, \ldots,$ can be proved, if $x^{(0)}$ is sufficiently close to $v^{(1)}$ and if λ_1 is a simple eigenvalue. It does not appear to be known whether convergence holds if λ_1 is a repeated eigenvalue. However, numerical experiments indicate that the accelerated method does converge in many cases, especially when the starting vector is chosen as described below (see, for instance, B. Wilkinson [1969]).

---

* See, for instance, Young, *et al.* [1971].

With fixed α the method is guaranteed to converge if $x^{(0)}$ is chosen in accordance with the identities (7.26) unless there are two or more eigenvalues which are closest to α. This cannot happen unless α lies on the perpendicular bisector of a line segment joining a pair of eigenvalues closest to α.

Starting Values and Vectors

We now consider the choice of the starting value of α and the starting vector. It seems desirable to introduce some randomness into the choice of α so that we shall be very unlikely to have α lie on a perpendicular bisector of a line segment joining the closest pair of roots. In particular, if $P(x)$ has real coefficients we want to avoid choosing α real since α would be equally close to each complex conjugate pair of roots.

The overall procedure is to determine one root and then calculate the reduced polynomial. Then we determine a root of the next reduced polynomial, etc. It can be shown (see, for instance, J. Wilkinson [1963, p. 64]) that there is less accumulation of rounding errors in this process if the smaller roots are determined first. Thus we seek an α which is closest to one of the smallest, and preferably the smallest, root.

Our procedure is to first determine, by one of the schemes described in Section 5.5, a circle $C_\rho : |x| \leqq \rho$ so that there is no zero of $P(x)$ in C_ρ. We then generate two random numbers s and t in the interval $[0, 1]$ and let

8.13 $$\alpha = (s\rho)e^{2\pi i t}.$$

With this choice of α we are reasonably sure that α will be closest to one of the small roots. Certainly, for some α in C_ρ, α will be closest to the smallest root.

As an example let us consider the equation

8.14 $$P(x) = x^4 - 3x^3 + 2x^2 + 2x - 4 = 0$$

whose roots are $-1, 2, 1 + i, 1 - i$. By (5.84) we know that no root of (8.14) lies inside the circle $|x| \leqq \rho$ where

8.15 $$\rho = \left[\max\left\{ 1, \sum_{j=0}^{3} \left| \frac{a_j}{a_4} \right| \right\} \right]^{-1} = \frac{1}{2}.$$

If we select the "random" numbers $s = 0.3$ and $t = 0.7$ we obtain, by (8.13) the starting value

8.16 $$\alpha = (0.3)(\tfrac{1}{2})e^{2\pi i(0.7)} = (0.15)(\cos 252° + i \sin 252°)$$

$$\doteq 0.15(-0.309 - 0.951i) \doteq -0.0464 - 0.1426i.$$

We now consider the choice of starting vector. As in the case of the modified Bernoulli method we could use the identities (7.26). However, we now use the backward polynomial and turn the vector v upside down. Thus for the case $n = 4$ we determine t_0, t_1, t_2, t_3 so that

8.17
$$\begin{cases} t_0 = 4 \\ a_4 t_1 + a_3 = 0 \\ a_4 t_2 + a_3 t_1 + 2a_2 = 0 \\ a_4 t_3 + a_3 t_2 + a_2 t_1 + 3a_1 = 0. \end{cases}$$

Evidently, we have

8.18
$$t_i = \sum_{j=1}^{4} \lambda_j^{-i}, \qquad i = 0, 1, 2, 3.$$

Moreover,

8.19
$$w = \begin{pmatrix} t_3 \\ t_2 \\ t_1 \\ t_0 \end{pmatrix} = \begin{pmatrix} \lambda_1^{-3} \\ \lambda_1^{-2} \\ \lambda_1^{-1} \\ 1 \end{pmatrix} + \begin{pmatrix} \lambda_2^{-3} \\ \lambda_2^{-2} \\ \lambda_2^{-1} \\ 1 \end{pmatrix} + \begin{pmatrix} \lambda_3^{-3} \\ \lambda_3^{-2} \\ \lambda_3^{-1} \\ 1 \end{pmatrix} + \begin{pmatrix} \lambda_4^{-3} \\ \lambda_4^{-2} \\ \lambda_4^{-1} \\ 1 \end{pmatrix}$$

$$= \lambda_1^{-3} \begin{pmatrix} 1 \\ \lambda_1 \\ \lambda_1^2 \\ \lambda_1^3 \end{pmatrix} + \lambda_2^{-3} \begin{pmatrix} 1 \\ \lambda_2 \\ \lambda_2^2 \\ \lambda_2^3 \end{pmatrix} + \lambda_3^{-3} \begin{pmatrix} 1 \\ \lambda_3 \\ \lambda_3^2 \\ \lambda_3^3 \end{pmatrix} + \lambda_4^{-3} \begin{pmatrix} 1 \\ \lambda_4 \\ \lambda_4^2 \\ \lambda_4^3 \end{pmatrix}.$$

Thus we have an expansion of w in terms of the eigenvectors of C. Now, however, if we normalize the eigenvectors so that the largest coefficient in modulus has a modulus of unity, the coefficient of the eigenvector associated with λ has modulus $|\lambda|^{-n}$ if $|\lambda| \leq 1$ and has modulus unity if $|\lambda| \geq 1$. Consequently, we are much more likely to converge to the eigenvector associated with smallest modulus than if we had used the sums s_i of (7.25).

If for the smallest eigenvalue in modulus, say λ_1, we have $|\lambda_1| \geq 1$, then the coefficients of all of the normalized eigenvectors will be unity. In order to accelerate the convergence to the smallest eigenvalue we scale the polynomial. We let

8.20
$$y = 2^{-p} x$$

and obtain

8.21 $P(x) = 2^{4p}[a_0 y^4 + (2^{-p}a_1)y^3 + (2^{-2p}a_2)y^2 + (2^{-3p}a_3)y + (2^{-4p}a_4)].$

This scaling of the coefficients by an integral power of two should not introduce very much rounding error. We choose p as the smallest integer so that $2^{-p}|\lambda_1| \leq \frac{1}{2}$.

By this procedure, we know that the coefficient c_1 relative to the normalized eigenvector $v^{(1)}$ corresponding to λ_1 will in modulus be either at least 2^n times $|c_2|$, if $2^{-p}|\lambda_2| \geq 1$ or else $|\lambda_2/\lambda_1|^n$ times as large as $|c_2|$ if $2^{-p}|\lambda_2| < 1$. Here c_2 is the coefficient corresponding to $v^{(2)}$, the normalized eigenvector associated with the eigenvalue λ_2 of next largest modulus. In either case, we would expect the convergence to be quite rapid unless $|\lambda_2|$ is nearly as small as $|\lambda_1|$.

An upper bound for $|\lambda_1|$ can be found by using any of the bounds (5.34). For the polynomial (8.14) we obtain from (5.34) with $k = 4$ and $\alpha = 0$ the bound

8.22 $|\lambda_1| \leq \sqrt[4]{4} = \sqrt{2} \doteq 1.414.$

Hence if we let $p = 2$ we have $2^{-p}|\lambda_1| < \frac{1}{2}$. We then consider the scaled polynomial equation

8.23 $y^4 - \frac{3}{4}y^3 + \frac{1}{8}y^2 + \frac{1}{32}y - \frac{1}{64} = 0$

obtained from (8.14) by letting $y = x/4$. We then determine the reciprocal sums

8.24 $t_i' = \sum_{j=1}^{4} \left[\frac{1}{\lambda_j/4} \right]^i, \quad i = 0, 1, 2, 3$

by the equations

8.25
$$\begin{cases} t_0' = 4 \\ -\frac{1}{64}t_1' + \frac{1}{32} = 0; \quad t_1' = 2 \\ -\frac{1}{64}t_2' + \frac{1}{32}(2) + 2(\frac{1}{8}) = 0; \quad t_2' = 20 \\ -\frac{1}{64}t_3' + \frac{1}{32}(20) + \frac{1}{8}(2) + 3(-\frac{3}{4}) = 0; \quad t_3' = -88. \end{cases}$$

Thus we use the normalized starting vector

8.26 $w = \frac{1}{\sqrt{2041}} \begin{pmatrix} -44 \\ 10 \\ 1 \\ 2 \end{pmatrix}.$

Before applying the IP method we scale α as given by (8.16) by 2^{-2} obtaining

8.27 $\alpha' \doteq -0.0116 - 0.0356i.$

While the convergence of the IP method for variable α has only been proved if the desired eigenvalue is simple and if the vector $x^{(0)}$ is very close to the eigenvector $v^{(1)}$, nevertheless, experience has shown that the method nearly always does converge when our choice of initial vector $x^{(0)}$ is based on the "backward" identities (8.17). This is probably due to the fact that the coefficients of the principal vectors in the expansion of $x^{(0)}$ in terms of the eigenvectors and principal vectors vanish. Thus, while these coefficients might become different from zero and grow during the calculation because of rounding errors, it appears that convergence usually occurs before this happens.

If a reasonable estimate α of a root is available, we can determine a good starting vector as follows. First we shift the origin, by determining $\hat{a}_i, i = 0, 1, 2, 3, 4$ so that

8.28 $P(x) = \hat{a}_0(x - \alpha)^4 + \hat{a}_1(x - \alpha)^3 + \hat{a}_2(x - \alpha)^2 + \hat{a}_3(x - \alpha) + \hat{a}_4.$

We then use the backward identities (8.17) with the $\hat{a}_i$ to determine $\hat{t}_0, \hat{t}_1, \hat{t}_2, \hat{t}_3$ and we let

8.29
$$\hat{t} = \begin{pmatrix} \hat{t}_3 \\ \hat{t}_2 \\ \hat{t}_1 \\ \hat{t}_0 \end{pmatrix}.$$

Next we determine the vector

8.30 $t = Q^{-1}\hat{t}$

where

8.31 $Q = \begin{pmatrix} 1 & 0 & 0 & 0 \\ -\alpha & 1 & 0 & 0 \\ \alpha^2 & -2\alpha & 1 & 0 \\ -\alpha^3 & 3\alpha^2 & -3\alpha & 1 \end{pmatrix}, \qquad Q^{-1} = \begin{pmatrix} 1 & 0 & 0 & 0 \\ \alpha & 1 & 0 & 0 \\ \alpha^2 & 2\alpha & 1 & 0 \\ \alpha^3 & 3\alpha^2 & 3\alpha & 1 \end{pmatrix}.$

It can be shown that the companion matrix $\hat{C}$ of the shifted polynomial becomes

8.32 $\hat{C} = \begin{pmatrix} 0 & 1 & 0 & 0 \\ 0 & 0 & 1 & 0 \\ 0 & 0 & 0 & 1 \\ -\hat{b}_4 & -\hat{b}_3 & -\hat{b}_2 & -\hat{b}_1 \end{pmatrix}$

where

8.33 $\hat{b}_i = \dfrac{\hat{a}_i}{\hat{a}_0}, \qquad i = 1, 2, 3, 4$

and the companion matrix C of the original polynomial, given by (7.3), are related by

8.34 $$\hat{C} = Q(C - \alpha I)Q^{-1}.$$

Hence, if $\hat{t}$ is an eigenvector of $\hat{C}$ associated with the eigenvalue $\hat{\lambda}$, then $t = Q^{-1}\hat{t}$ is an eigenvector of C associated with the eigenvalue $\hat{\lambda} + \alpha$.

Our procedure is to iterate with the IP method based on the original polynomial with a starting value of α and a starting vector of t. As an alternative, we can iterate with the IP method based on the shifted polynomial with a starting value of zero and a starting vector of $\hat{t}$. When convergence has been achieved say to a vector $\hat{t}*$ and to a value $\hat{\lambda}*$, then $\lambda* = \hat{\lambda}* + \alpha$ is an approximate zero of $P(x)$. If greater accuracy is desired one can iterate with the IP method based on the original polynomial with a starting value of $\hat{\lambda}*+ \alpha$ and with either the starting vector $t = Q^{-1}\hat{t}*$ or with the starting vector

8.35 $$t = (\hat{\lambda}* + \alpha)^{-3} = \begin{pmatrix} 1 \\ \hat{\lambda}* + \alpha \\ (\hat{\lambda}* + \alpha)^2 \\ (\hat{\lambda}* + \alpha)^3 \end{pmatrix}.$$

EXERCISES 5.8

1. Carry out one iteration of the IP method for the polynomial $P(x) = x^4 - 5x^3 + 10x^2 - 10x + 4$ with $\alpha = \tfrac{1}{2}$ and the starting vector

$$x^{(0)} = \begin{pmatrix} 1 \\ \tfrac{1}{2} \\ (\tfrac{1}{2})^2 \\ (\tfrac{1}{2})^3 \end{pmatrix}.$$

   Then compute the Rayleigh quotient (8.12).

2. For the polynomial of the preceding exercise find ρ such that there are no roots inside the circle $|z| \leq \rho$. With the "random" numbers $s = 0.4$, $t = 0.6$, determine a starting value of α.

3. For the polynomial of Exercise 1 scale the polynomial approximately so that the root of smallest modulus has modulus not greater than $\tfrac{1}{2}$. Also determine a suitable starting vector for use with the IP method. How would the starting value of α, determined by Exercise 2, be modified?

4. For the polynomial $P(x)$ of Exercise 1 construct a starting vector using (8.17). Verify (8.19) using the fact that the zeros of $P(x)$ are 1, 2, 1 + i, 1 − i.

5. Verify (8.34) and show that if $C\hat{v} = \hat{\lambda}\hat{v}$ and if $v = Q^{-1}\hat{v}$ where Q is given by (8.31) then $Cv = (\hat{\lambda} + \alpha)v$.

6. Obtain a good starting vector for finding a zero of $P(x) = x^4 - 5x^3 + 10x^2 - 10x + 4$ using the estimate $\frac{1}{2}$ for a root as follows. Determine a starting vector $\hat{t}$ using (8.17) based on the shifted polynomial (8.28), with $\alpha = \frac{1}{2}$. Then determine $t = Q^{-1}\hat{t}$ where Q^{-1} is given by (8.31). Iterate with the IP method starting with α and t.

   Also, iterate with the IP method with the shifted polynomial using an initial value of $\alpha = 0$ and an initial vector of $\hat{t}$. When convergence has been achieved, perform additional iterations of the IP method with the original polynomial with an initial value of $\lambda^* = \hat{\lambda}^* + \alpha$, where $\hat{\lambda}^*$ is the value obtained with the shifted polynomial, and with the two alternative choices of starting vectors given in the text.

7. For $P(x)$ as in Exercise 1 determine a starting value of α as in (8.16) except use (5.85) instead of (5.84). Use the same "random" numbers. Do the same for the estimate (5.86).

8. Consider the polynomial equation $P(x) = x^3 - 2x + 4 = 0$. Apply the bound (5.86) to find ρ such that no root lies in the circle $|z| \leq \rho$. Also, using (5.34)–(5.35) find σ such that there is at least one root in the circle $|z| \leq \sigma$. Scale the polynomial by a suitable power of two so that there will be a root of the scaled polynomial in the circle $|z| \leq \frac{1}{2}$. Then construct an initial vector based on the identities (8.17). Determine the coefficients c_1, c_2, and c_3 of this initial vector relative to the eigenvectors $w^{(1)}$, $w^{(2)}$, $w^{(3)}$ of the companion matrix for the scaled polynomial.

   Let α for the scaled polynomial correspond to $0.5 + 0.3i$ for the original polynomial. Apply the IP method with α as the scaled value of $0.5 + 0.3i$ for the initial iteration and the Rayleigh quotient for subsequent iterations. After each iteration compute the modulus of the scaled polynomial and the coefficients of the (normalized) vector relative to $w^{(1)}$, $w^{(2)}$, $w^{(3)}$. The process should be terminated when the modulus of the scaled polynomial is less than 10^{-30}. (The scaled polynomial should be $y^3 - \frac{1}{8}y + \frac{1}{16} = 0$.)

   Repeat the computation procedure keeping α fixed and equal to the scaled value of $0.5 + 0.3i$ and compare the number of iterations required.

5.9 POLYALGORITHMS

We now describe some of the features of a good polyalgorithm for solving the general polynomial equation. One such polyalgorithm, known as POLYHOOK was developed by Champagne and Kripke (see Champagne [1964]) and is based on the use of the Newton method and Lehmer's method. Another polyalgorithm, based on the use of the IP method, is known as POLLIF and is described by Young, et al. [1971]. The initial phase of POLLIF, known as POLLIN, was developed by B. Wilkinson [1969].

We consider a polyalgorithm as made up of two phases, an initial phase and an assessment and refinement phase. In the initial phase one seeks to determine the roots, in ascending order of magnitude where possible. Each time a root is accepted it is divided out and a reduced polynomial is obtained. Where possible, when a root is found an attempt is made to find a multiple root nearby, as described in Section 5.6. If the polynomial has real coefficients and if a complex root is found, one modifies it if necessary so that it is an acceptable zero of the current polynomial and so that its conjugate is an acceptable zero of the associated reduced polynomial (see Section 5.6).

In the assessment and refinement phase we seek to obtain estimates of the accuracy of the roots found in the initial phase, based on the original polynomial, and to modify the roots where possible and appropriate to improve their accuracy.

The Initial Phase

The basic step in the initial phase is to locate a single root, preferably one of the roots of smallest magnitude, of a given polynomial. A suggested procedure is as follows:

a) Find the radius of a circle about the origin containing no zero of $P(x)$ (see (5.84)–(5.86)).

b) Choose an initial value α as a random point in the circle (see (8.13)).

c) Choose an initial vector $x^{(0)}$ using the identities (8.17). Scale the coefficients and the initial α if necessary as described in Section 5.8 so that the smallest root will not exceed $\frac{1}{2}$ in magnitude. To estimate the magnitude of the smallest root use one of the bounds (5.34) with $\alpha = 0$.

d) Iterate with the *IP* method, first with α fixed and later with α varying in accordance with the Rayleigh quotient (8.12).

e) After each iteration, test whether the Rayleigh quotient satisfies the convergence test (6.2). If the Rayleigh quotient, R, is very small, see if there is a root close enough to R based on the bounds (5.34) or similar bounds.

f) If an acceptable root is found, seek a nearby multiple root. If $P(x)$ has real coefficients and a complex root β is found, seek a nearby value, say β', if necessary, so that β' is an acceptable zero of $P(x)$ and $\bar{\beta}'$ is an acceptable zero of the reduced polynomial.

g) If the convergence is too slow, repeat first with the same initial α and with α fixed. If this fails, choose α in a larger circle. To preclude the unlikely possibility that a root would never be found, a limit should be placed on the number of attempts to find a root.

The Assessment and Refinement Phase

Having obtained approximate values of the roots as described above, we now wish to assess the accuracy and to refine the roots. Ideally, we would like to find n small non-overlapping circles each containing one approximate root and each guaranteed to contain one zero of the original polynomial. In such a situation one can perform additional iterations based on the original polynomial to improve the accuracy if desired, as long as one remains inside the circle. Moreover, knowledge of the location of the other roots can be used to obtain improved bounds for the location of the root in question (see Section 5.5). If two of the circles overlap, however, then we are not sure whether there are two zeros of the original polynomial in the union of the two circles or just one. It may be possible to determine a number of small circles each containing a specified number of roots (i.e., a "cluster") with the total adding up to n. In particularly bad cases, it may not be possible to account for all roots in terms of the original polynomial except by taking a circle containing all roots. In such cases it is not possible to get all of the roots correct to within a guaranteed, small error. One could still accept the values produced by the initial phase, and these would in all probability be good enough for most purposes.

We now describe a procedure for isolating clusters of roots. This procedure which we describe for the case $n = 4$ was suggested by Champagne and Kripke (see Champagne [1964]). Let α be close to a root r, and let $Q(x)$ be the true reduced polynomial based on the root r, i.e., let $Q(x)$ be defined by

9.1 $$P(x) = (x - r)Q(x).$$

Let

9.2 $$P(x) = \hat{a}_0(x - \alpha)^4 + \hat{a}_1(x - \alpha)^3 + \hat{a}_2(x - \alpha)^2 + \hat{a}_3(x - \alpha) + \hat{a}_4$$

where

9.3 $$\hat{a}_i = \frac{P^{(n-i)}(\alpha)}{(n-i)!}, \quad i = 0, 1, 2, 3, 4.$$

(The $\hat{a}_i$ can be determined by repeated synthetic division.) Moreover, let

9.4 $$Q(x) = \hat{b}_0(x - \alpha)^3 + \hat{b}_1(x - \alpha)^2 + \hat{b}_2(x - \alpha) + \hat{b}_3$$

where

9.5 $$\hat{b}_i = \frac{Q^{(n-i)}(\alpha)}{(n-i)!}, \quad i = 0, 1, 2, 3.$$

Since $P(x) = [(x - \alpha) - (r - \alpha)]Q(x)$ we have

9.6 $\hat{a}_0(x - \alpha)^4 + \hat{a}_1(x - \alpha)^3 + \hat{a}_2(x - \alpha)^2 + \hat{a}_3(x - \alpha) + \hat{a}_4$

$$= [(x - \alpha) - (r - \alpha)][\hat{b}_0(x - \alpha)^3 + \hat{b}_1(x - \alpha)^2 + \hat{b}_2(x - \alpha) + \hat{b}_3].$$

Equating coefficients of like powers of $x - \alpha$ we obtain

9.7
$$\begin{cases} \hat{b}_0 = \hat{a}_0 \\ \hat{b}_1 = \hat{a}_1 + \hat{b}_0(r - \alpha) \\ \hat{b}_2 = \hat{a}_2 + \hat{b}_1(r - \alpha) \\ \hat{b}_3 = \hat{a}_3 + \hat{b}_2(r - \alpha). \end{cases}$$

Thus we have

9.8
$$\begin{cases} |\hat{b}_0| = |\hat{a}_0| \\ |\hat{b}_1| \leqq |\hat{a}_1| + |\hat{b}_0| \, |r - \alpha| \\ |\hat{b}_2| \leqq |\hat{a}_2| + |\hat{b}_1| \, |r - \alpha| \\ |\hat{b}_3| \leqq |\hat{a}_3| + |\hat{b}_2| \, |r - \alpha|. \end{cases}$$

Using any of the error bounds of Section 5.5, for example (5.34)–(5.35), we can give a bound on $|r - \alpha|$. If the $\hat{a}_i$ are computed from the formulas for the derivatives of $P(x)$, we can compute upper bounds for the errors using (6.9). Alternatively, we can perform a rounding error analysis for the shift of origin process based on synthetic division. In any case, upper and lower bounds for the $|\hat{a}_i|$ can be obtained. We can thus calculate an upper bound for $|\hat{b}_3|$, and lower bounds for $|\hat{b}_2|$, $|\hat{b}_1|$, and $|\hat{b}_0|$. For example,

9.9
$$\begin{cases} |\hat{b}_2| \geqq |\hat{a}_2| - |\hat{b}_1| \, |r - \alpha| \\ |\hat{b}_1| \geqq |\hat{a}_1| - |\hat{b}_0| \, |r - \alpha|. \end{cases}$$

Thus we can again use (5.34) with $P(\alpha)$ replaced by $Q(\alpha)$ to obtain a bound on the zero of $Q(x)$ closest to α. Thus, we have a circle guaranteed to contain at least two zeros of $P(x)$. Naturally, this process would normally be used if $|\hat{a}_3|$ is small, so that a double root or a cluster is suspected. The process can be continued to determine circles with α as a center containing $3, 4, \ldots$ roots. Details of this procedure can be found in Champagne [1964] and Young, et al. [1971]. Champagne and Kripke used (5.34) with $k = n$, whereas smaller values of k seem superior in many cases. We remark that if other roots are known, one can use improved bounds such as (5.45) and (5.46) for the cases $k = 1$ and $k = n$, respectively.

Once the roots have been isolated, one can refine them using the original polynomial. One can iterate with the IP method, always requiring that α stay inside of the circle involved. If a circle contains more than one root, one can use a set of reduced polynomials, but only for the roots in that circle.

As a means of isolating the roots one can use the error bounds given in Section 5.5, using bounds on other roots to obtain improved bounds. Indeed, the process of determining the error bounds is an iterative process. Hopefully, by obtaining more accurate error bounds, one can achieve a high degree of separation of the roots.

The process just described is admittedly complicated, and the reader is referred to Young, et al. [1971] for details.

EXAMPLE

Let us again consider the polynomial (5.49). Suppose $\alpha = 1$ is an approximation to one of the roots. We first seek two circles around α containing at least one and at least two circles, respectively.

We first shift the origin to $\alpha = 1$, obtaining

9.10 $P(x) = (x - 1)^3 + 2(x - 1)^2 - 0.01(x - 1) - 0.02.$

(The reader should verify this both by the repeated use of synthetic division and also by the use of the formulas for the derivatives of $P(x)$.) From (5.52) we have

9.11 $|r - \alpha| \leqq 0.173.$

Next, let us get bounds on the $|\hat{b}_i|$. Evidently,

9.12 $\hat{b}_0 - \hat{a}_0, \qquad \hat{b}_1 = \hat{a}_1 + (r - \alpha)\hat{b}_0, \qquad \hat{b}_2 = \hat{a}_2 + (r - \alpha)\hat{b}_1$

so that

9.13 $$\begin{cases} |\hat{b}_0| = 1 \\ 1.827 = 2 - 1(0.173) \leq |\hat{b}_1| \leq 2 + 0.173 = 2.173 \\ |\hat{b}_2| \leq 0.01 + (2.173)(0.173) \doteq 0.386. \end{cases}$$

Therefore, there is a zero r' of the reduced polynomial such that

9.14 $$|r' - \alpha| \leq \begin{cases} \sigma_1 = 2 \left| \dfrac{0.386}{1.827} \right| \doteq 0.423 \\ \\ \sigma_2 = \sqrt{\dfrac{0.386}{1}} \doteq 0.621. \end{cases}$$

Hence, there are at least two zeros of $P(x)$ in the circle

9.15 $|x - 1| \leqq 0.423.$

As an estimate of the third root we let $\alpha = -0.9$. From (5.34) we find that

9.16 $|r_3 + 0.9| \leqq \min\{0.335, 0.540, 0.711\} = 0.335.$

Hence, we have completely isolated the root at -1. (See Fig. 9.20.) We could then iterate using the original polynomial to get this root more accurately.

Even without doing so, however, we can use a knowledge of the location of the third root, namely -1, to improve our bounds on the location of the other two. Thus, going back to our derivation of (5.35) we have

9.17 $$\frac{\hat{a}_1}{\hat{a}_3} = \frac{1}{(r_1 - \alpha)(r_2 - \alpha)} + \frac{1}{(r_1 - \alpha)(r_3 - \alpha)} + \frac{1}{(r_2 - \alpha)(r_3 - \alpha)}.$$

Let $\rho = \min\{|r_1 - \alpha|, |r_2 - \alpha|\}$. Then, since $|r_3 - \alpha| \geq 1.9 - 0.335 = 1.565$, we have

9.18
$$\left|\frac{\hat{a}_1}{\hat{a}_3}\right| \leqq \frac{1}{\rho^2} + \frac{2}{1.565\rho}$$

or

9.19
$$\rho \doteq 0.108,$$

which is very close to the best possible bound, namely, 0.1. Using this bound for the first root, we can get improved bounds for the second root, and thus a smaller circle containing both of the roots near 1. These improved bounds can then be used to get a more accurate bound for the third root. This process could be repeated several times.

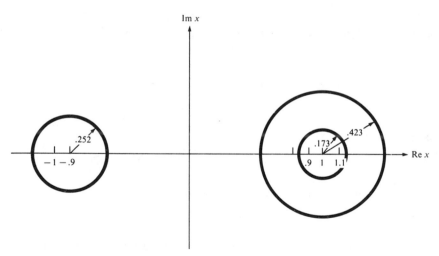

9.20 Fig. $P(x) = x^3 - x^2 - 1.01x + 0.99 = (x - 1.1)(x - 0.9)(x + 1)$.

Alternatively, one could iterate in the circle with radius 0.108 around 1 to get an accurate value of the root at 1.1 (or 0.9). One could "divide out" this root and use the resulting reduced polynomial to get the other root near $x = 1$. Eventually, one could obtain error bounds sufficiently good that all three roots would be separated.

We remark that the scheme suggested by McDonald for getting bounds on the roots in terms of derivatives of $P'(x)/P(x)$ is better adapted to using knowledge of some of the roots than are the bounds (5.34). Hence, as described by Young, et al. [1971], McDonald's bounds are actually used in POLLIF.

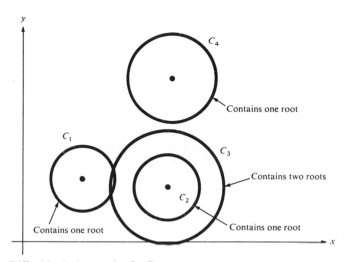

9.21 Fig. Difficulties in Accounting for Roots.

As already mentioned, in pathological cases it may not be possible to achieve any appreciable separation of the roots. It may happen that by enlarging the circles so as to guarantee that they contain more roots, we may cause them to cover other roots or to overlap with circles known to contain other roots. Thus in Fig. 9.21 if the circles C_1, C_2, and C_4 are known to contain one root, and the circle C_3 is known to contain two roots, it is not certain that the union of C_1 and C_3 contain three roots. To obtain a circle containing three roots might require a much larger circle which could overlap C_4. In some cases, by repeated application of the bounds on the roots the circles can be reduced so that overlapping will not occur. When overlapping is avoided, considerably smaller circles can often be obtained.

EXERCISES 5.9

1. For the polynomial $P(x) = 4x^5 - 16x^4 + 17x^3 + 6x^2 - 21x + 10$, find the radius of a circle about $x = \frac{1}{2}$ which contains at least 2 roots. Do the same thing if it is known that there is a root within 0.01 of $x = 2$.

5.10 OTHER METHODS

In the space available it has only been possible to treat a few of the many good methods which can be used to solve polynomial equations. In this section we mention some of the methods not treated earlier. Further information can be obtained from the references listed in the bibliography. See especially the recent book of Householder [1970] and the book edited by Dejon and Henrici [1970] which contains the proceedings of a conference on the solution of polynomial equations held in Zürich-Rüschlikon, Switzerland, in June 1967.

Laguerre's Method

An excellent method for determining zeros of a polynomial having only real zeros is Laguerre's method defined by

10.1
$$x_{k+1} = x_k - \frac{nP(x_k)}{P'(x_k) \pm \sqrt{(n-1)\{(n-1)[P'(x_k)]^2 - nP(x_k)P''(x_k)\}}} \; .$$

(See, for instance, Ralston [1965], pp. 368–370.) The sign of the square root is chosen to agree with the sign of $P'(x_k)$. If all of the zeros are real, the method converges for any real starting value x_0. The convergence is very rapid (third order) near simple zeros and is first order near multiple zeros. There is no guarantee of convergence when the polynomial has complex zeros but the method has been reported to work well in many such cases. Parlett [1964] has shown that when the method converges to a simple complex zero the convergence is third order.

Root Squaring and the Resultant Procedure

Bareiss [1960], [1967] developed a program based on the use of the Graeffe root-squaring process combined with a resultant procedure. With the root-squaring method one constructs a sequence of polynomials whose zeros are powers of the zeros of the original polynomial. By repeated application of the root-squaring algorithm one can determine the moduli of the roots of the original polynomial. On the other hand, it is difficult to separate complex roots with equal moduli. This difficulty is overcome by the use of the resultant procedure which enables one to find all zeros with a given modulus.

The Quotient Difference Algorithm

The quotient-difference algorithm of Rutishauser [1957] can be used to solve equations involving polynomials and other functions. A description of the algorithm is given by Henrici [1967] who also describes a computer program where the algorithm is used to obtain initial approximations to the roots, and then Newton's method and the Lin-Bairstow method are used to obtain more accurate values. Henrici does not recommend that the quotient-difference algorithm be used for the final accurate determination of the roots since it is slow (converging only linearly), but he does recommend it for getting good first approximations to the roots for use with faster methods. One of the advantages of the quotient-difference method is that no starting values or other information are needed except for the coefficients of the polynomial.

A Three-stage Variable Shift Procedure

Jenkins and Traub have developed a number of programs for solving polynomial equations (see Traub [1966a], [1966b], [1967], Jenkins [1969], Jenkins and Traub [1967], [1968], [1970]). The methods used are similar to the IP method except that the transpose of the companion matrix considered in Section 5.8 is used. A three-stage procedure is used, each stage being characterized by the type of shifting involved. Thus no shift, a fixed shift, and a variable shift are used in the first, second, and third stages, respectively. These correspond to letting the α used with the IP method in Section 5.2 be zero, fixed or variable, respectively.

A Family of Globally Convergent, Multiplicity Independent Methods

We now discuss briefly a family of globally convergent, multiplicity independent methods due to Schröder [1870]. These methods can be defined for the general equation

10.2 $$f(z) = 0.$$

Given integers p and λ such that $p \geq 1$, $\lambda \geq 0$, let us define the functions

10.3 $$\phi_{p,\lambda}(z) = z + (p-1)\frac{\left[z^{\lambda}\frac{f'(z)}{f(z)}\right]^{[p-2]}}{\left[z^{\lambda}\frac{f'(z)}{f(z)}\right]^{[p-1]}}.$$

One set of methods can be obtained by fixing p and λ and determining the sequence $z_0, z_1, \ldots$ by

10.4 $$z_{n+1} = \phi_{p,\lambda}(z_n).$$

One can also fix λ and define the sequence $z_0, z_1, z_2, \ldots$ by

10.5 $$z_n = \phi_{n,\lambda}(z_0).$$

We remark that if $p = 2$ and $\lambda = 0$, the method (10.4) reduces to the Newton method applied to the equation

10.6 $$Q(z) = f(z)/f'(z) = 0.$$

This follows since

10.7 $$z - \frac{Q(z)}{Q'(z)} = z - \frac{f'(z)f(z)}{(f'(z))^2 - f(z)f''(z)} = \phi_{2,0}(z).$$

The use of $f(z)/f'(z)$ rather than $f(z)$ has the advantage that there are no multiple roots. Thus the methods are, in a sense, multiplicity independent.

Let us now consider the case where $f(z)$ is a polynomial of degree n. It can be shown (see, for instance, McDonald [1970]) that if $\lambda \leqq p - 1$, then

10.8
$$\phi_{p,\lambda}(z) = \frac{\displaystyle\sum_{j=1}^{n} \frac{\alpha_j^{\lambda+1}}{(z - \alpha_j)^p}}{\displaystyle\sum_{j=1}^{n} \frac{\alpha_j^{\lambda}}{(z - \alpha_j)^p}} .$$

From this it can be shown that if z_0 is closer to a zero, say α_j, of $f(z)$ than to any other zero, then

10.9
$$\lim_{p \to \infty} \phi_{p,\lambda}(z) = \alpha_j.$$

Thus the iterative method defined by (10.5) converges.

It can also be shown (see, for instance, McDonald [1970]) that the method (10.4) converges to α_j, provided p is sufficiently large and provided z_0 is closer to α_j than to any other zero of $f(z)$.

Numerical studies in the application of these methods to the solution of polynomial equations are described by McDonald [1970].

EXERCISES 5.10

1. Carry out two iterations of Laguerre's method for the polynomial equation $P(x) = x^3 - 6x^2 + 11x - 6 = 0$. Let $x_0 = 0$.

2. Let α and β be the roots of the quadratic equation $x^2 + bx + c = 0$. Show that $|\alpha|/|\beta|$ is small if and only if $|b^2/c|$ is large and that if $|b^2/c|$ is large then

$$\alpha \sim -\frac{c}{b}, \qquad \beta \sim -b.$$

Apply this result to the equation $x^2 - 100x + 1 = 0$. Compute the actual roots.

3. Construct the cubic equation with leading coefficient unity whose roots are the squares of the roots of $P(x) = x^3 + bx^2 + cx + d = 0$. (Consider $Q(x) = -P(x)P(-x)$ and let $y = x^2$.) Verify the result for the case $P(x) = x^3 - 6x^2 + 11x - 6$. In the general case, given that one of the three zeros of $Q(x)$ is much larger in magnitude than the other two, give an approximate value for it.

4. Prove the statement preceding (10.6) for the case $\lambda = 0$.

5. For the polynomial equation $P(z) = z^3 - 6z^2 + 11z - 6$ carry out 3 iterations of the iteration method (10.5) with $z_0 = 0$. Also carry out 3 iterations of (10.4) with $p = 2$ and $z_0 = 0$. In each case let $\lambda = 0$.

SUPPLEMENTARY DISCUSSION 5.2

Our development of the theory of polynomial equations has been done in such a way as to prove as much as possible without the use of the fundamental theorem of algebra (see Theorem A-8.6).

SUPPLEMENTARY DISCUSSION 5.5

For a proof of Descartes' rule of signs, see Householder [1970, p. 82].
For an interesting application of Sturm sequences, see Givens [1953].
Delves and Lyness [1967] have developed a procedure for determining how many zeros of a polynomial lie inside a given circle. The procedure also can be used for analytic functions. The bounds (5.34)–(5.35) were derived by Schmidt and Dressel [1967]. A similar family of bounds was given by Marden [1966].
The Cauchy bound, determined by (5.70), for the moduli of the roots of a polynomial equation was used by Jenkins and Traub [1967]. Similar bounds on the moduli of the roots were obtained by Simeunović [1967].

SUPPLEMENTARY DISCUSSION 5.6

An alternative scheme for finding multiple roots is given by McDonald [1970].

SUPPLEMENTARY DISCUSSION 5.7

The derivation of the formula (7.26) for the sums s_i given by (7.25) is based on the analysis given by Householder [1953] and by Ralston [1965].

SUPPLEMENTARY DISCUSSION 5.8

For a discussion of the convergence of the inverse power method for finding the eigenvalues of a matrix, see the series of six papers by Ostrowski [1958–1959].

SUPPLEMENTARY DISCUSSION 5.10

For an analysis of Schröder's method see the translation of the paper of Schröder [1870] by Stewart [1968], and McDonald [1970]. Stewart [1969] studied the relation between the methods of Schröder and various other methods for solving polynomial equations, including many matrix power methods.

INTERPOLATION AND APPROXIMATION

6.1 INTRODUCTION

In this chapter we shall be concerned with methods for the approximate evaluation of a given function $f(x)$. The function $f(x)$ may be defined in a number of ways:

a) By one or more explicit formulas, for example,

   i) $f(x) = \sin x + 2$

   ii) $f(x) = \begin{cases} 1 + x, & x \geq 0 \\ 2, & -1 \leqq x < 0 \\ 3x + \sin x, & x < -1. \end{cases}$

b) By an infinite series, for example,

$$f(x) = 1 + x + \frac{x^2}{2!} + \frac{x^3}{3!} + \cdots .$$

c) As the solution of a differential equation.

d) By a numerical algorithm, for example,

   i) a finite algorithm

$$z_0 = a_0$$
$$z_k = xz_{k-1} + a_k, \qquad k = 1, 2, \ldots, n$$

   where $a_0, a_1, \ldots, a_n$ are given;

   ii) an infinite algorithm

$$f(x) = \lim_{n \to \infty} z_n$$

   where

$$z_0 = (1 + x)/2, \qquad x \geq 0$$
$$z_{n+1} = \frac{1}{2}\left(z_n + \frac{x}{z_n}\right), \qquad n = 0, 1, 2, \ldots .$$

e) By a table giving values of $f(x)$ for several values of x. The tabulated values may be highly accurate, as is usually the case with a mathematical table, or the values may be the result of experimental observations, and hence in general inaccurate.

In the evaluation of $f(x)$ we seek to determine another function $F(x)$ such that for any desired value x the value of $F(x)$ will be close enough to $f(x)$ for the desired purpose. Frequently $F(x)$ will be a relatively simple function such as a polynomial or a rational function. In some cases different formulas are used to evaluate $F(x)$ in different ranges of the argument.

In this chapter we shall be primarily concerned with *interpolation* of a function of one variable. Given a function $f(x)$, one chooses a function $F(x)$ from among a certain class of functions (frequently, but not always, the class of polynomials of degree n or less, for some n) such that $F(x)$ agrees with $f(x)$ at certain values of x. These values of x are often referred to as "interpolation points." One may also specify that certain derivatives of $F(x)$ agree with the corresponding derivatives of $f(x)$ at some of the interpolation points. It should be noted that if $F(x)$ is the sum of $n + 1$ terms of the Taylor's series for $f(x)$ at the point $x = a$, then $F(x)$ can be considered as an interpolating polynomial for $f(x)$ of degree n or less, since

1.1 $$F^{(k)}(a) = f^{(k)}(a), \qquad k = 0, 1, \ldots, n.$$

In Sections 6.2 through 6.9 we are concerned with polynomial interpolation for a function of one variable. The familiar process of linear interpolation is treated in Sections 6.2 and 6.3. Here the function $F(x)$ is a linear function and agrees with $f(x)$ at two points of interpolation. Lagrangian interpolation, where $F(x)$ may be a higher-degree polynomial which agrees with $f(x)$ at several points of interpolation, is treated in Sections 6.4 and 6.5. In Section 6.6 we consider the Gregory-Newton interpolation formula and other formulas which can be used when the interpolation points are evenly spaced. Section 6.7 is concerned with Hermite interpolation where $F(x)$ agrees with $f(x)$ at the interpolation points and $F'(x)$ also agrees with $f'(x)$ at some or all of the interpolation points. In Section 6.8 we indicate some of the limitations in polynomial interpolation and show how smooth interpolation can often be effectively used. Section 6.9 is concerned with inverse interpolation.

Instead of requiring that $F(x)$ agree with $f(x)$ at certain points, as in the case of interpolation, one may require that $F(x)$ be "close" enough to $f(x)$, in some sense, so that for certain purposes $F(x)$ may be used instead of $f(x)$ for a certain range of the independent variable. The determination of a suitable function $F(x)$ is a problem in *approximation*. In Sections 6.10 and 6.11 we consider approximation by polynomials. In Section 6.10 we seek to minimize the maximum deviation $|f(x) - F(x)|$ over a certain range of x. In Section 6.11 we seek to minimize the integral

1.2
$$\int_a^b (f(x) - F(x))^2 \, dx$$

where $a \leqq x \leqq b$ is the interval of interest. We also consider the case where we minimize

1.3
$$\sum_{i=1}^{M} (f(x_i) - F(x_i))^2$$

i.e., when the sum of the squares of the deviations over a finite point set is minimized. This latter criterion is useful in fitting experimental data.

In Sections 6.12 and 6.13 we consider the use of functions other than polynomials. Section 6.12 is concerned with approximation by rational functions. Section 6.13 is concerned with interpolation and approximation by trigonometric polynomials.

Chapter 6 is concluded by an elementary discussion of interpolation of a function of two variables, which is given in Section 6.14.

6.2 LINEAR INTERPOLATION

Before considering the general problem of polynomial interpolation let us discuss the familiar process of linear interpolation. The reader has undoubtedly used linear interpolation with tables of logarithms and trigonometric functions. We shall first illustrate the mechanics of the linear interpolation process for a specific function and then formalize the process.

Suppose we wish to determine $e^{0.26}$ given a table of the function $f(x) = e^x$ for $x = 0(0.1)2.0$. A portion of the table is given by

x	$f(x)$
0	1.00000
0.1	1.10517
0.2	1.22140
0.3	1.34986
0.4	1.49182
0.5	1.64872

According to the well-known linear interpolation algorithm we first subtract $f(0.2)$ from $f(0.3)$ obtaining

2.1 $f(0.3) - f(0.2) \doteq 1.34986 - 1.22140 = 0.12846$

Since 0.26 is $\frac{6}{10}$ of the distance between 0.2 and 0.3 we add $\frac{6}{10}$ of the above difference to $f(0.2)$ obtaining

2.2 $f(0.2) + \frac{6}{10}(0.12846) \doteq 1.22140 + \frac{6}{10}(0.12846) \doteq 1.22140 + 0.07708$

$$= 1.29848$$

Since 1.29693 is the value of $e^{0.26}$, correct to 5 decimal places, the error in linear interpolation is

2.3 $f(0.26) - 1.29848 \doteq 1.29693 - 1.29848 = -0.00155.$

In the next section we shall obtain a theoretical bound for the error in linear interpolation.

We now seek to formalize the linear interpolation process. Given a function $f(x)$ defined in an interval $a \leqq x \leqq b$ we seek to determine a linear function $F(x)$ such that

2.4 $f(a) = F(a),   f(b) = F(b).$

Since $F(x)$ has the form

2.5 $F(x) = Ax + B$

for some constants A and B we have

2.6 $F(a) = Aa + B,   F(b) = Ab + B.$

Solving for A and B we get

2.7 $A = \dfrac{f(b) - f(a)}{b - a},   B = \dfrac{bf(a) - af(b)}{b - a}$

and, by (2.5),

2.8 $F(x) = \left(\dfrac{f(b) - f(a)}{b - a}\right)x + \dfrac{bf(a) - af(b)}{b - a}$

$$= f(a) + \frac{x - a}{b - a}(f(b) - f(a)).$$

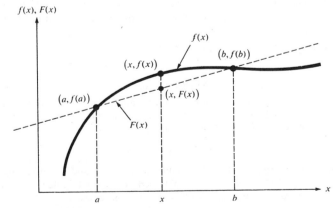

2.9 Fig. Linear Interpolation.

One can verify directly that for each x the point $(x, F(x))$ lies on the line joining $(a, f(a))$ and $(b, f(b))$ (see Fig. 2.9).

If we apply (2.8) to the case where $f(x) = e^x$, $a = 0.2$, $b = 0.3$, $x = 0.26$, $f(a) = 1.22140$, $f(b) = 1.34986$ we get

2.10 $$F(0.26) = 1.22140 + \frac{(0.26 - 0.2)}{(0.3 - 0.2)}(1.34986 - 1.22140)$$

$$= 1.22140 + \tfrac{6}{10}(0.12846) \doteq 1.29848$$

as before.

We remark that we can write (2.8) in the form

2.11 $$F(x) = w_0(x)f(a) + w_1(x)f(b)$$

where the "weights" $w_0(x)$ and $w_1(x)$ are given by

2.12 $$w_0(x) = \frac{b - x}{b - a}, \qquad w_1(x) = \frac{x - a}{b - a}.$$

If x lies in the interval $a \leqq x \leqq b$, then the weights are nonnegative and

2.13 $$w_0(x) + w_1(x) = 1.$$

The fact that the weights are nonnegative, in this case, implies that, except for possible rounding errors in carrying out (2.11), if the tabulated values of $f(a)$ and $f(b)$ are correct to within an accuracy of ε, in absolute value, then the computed value of $F(x)$ and the value of $F(x)$ corresponding to the exact values of $f(a)$ and $f(b)$ will differ by at most ε in absolute value.

Evidently, if $f(x)$ is a linear function, then the linear interpolation process is exact. In this case $f''(x) = 0$. In the general case it is clear that the size of the difference $f(x) - F(x)$ in the interval $[a, b]$ depends on the curvature of the graph of $f(x)$, and hence it depends on $f''(x)$. In the next section we shall derive a bound on $|F(x) - f(x)|$ in terms of $b - a$ and a bound on $|f''(x)|$.

EXERCISES 6.2

1. Using the values of e^x given in Section 6.2 find an approximate value of $e^{0.28}$ by linear interpolation. Also find the linear function $F(x)$ such that $F(0.2) = e^{0.2}$ and $F(0.3) = e^{0.3}$ and evaluate $F(0.28)$. Find the weights w_0 and w_1 such that

$$F(0.28) = w_0 e^{0.2} + w_1 e^{0.3}.$$

2. In the preceding exercise show that if the values used for $e^{0.2}$ and $e^{0.3}$ are changed by ε_1 and ε_2 respectively, then the value of $e^{0.28}$ obtained by linear interpolation is changed by an amount ε where $|\varepsilon| \leqq \max(|\varepsilon_1|, |\varepsilon_2|)$. Work out the cases: $\varepsilon_1 = 10^{-4}$, $\varepsilon_2 = 10^{-4}$; $\varepsilon_1 = 10^{-4}$, $\varepsilon_2 = -10^{-4}$; $\varepsilon_1 = -10^{-4}$, $\varepsilon_2 = 10^{-4}$; and $\varepsilon_1 = -10^{-4}$, $\varepsilon_2 = -10^{-4}$.

3. Construct a table of $f(x) = 2x + 1$ for $x = 0(0.1)0.5$. Determine $f(0.26)$ by linear interpolation and verify that the result is exact. What if $f(x) = 10x^2 + 1$? Draw a graph in each case.

4. Using the values of e^x given in Section 6.2 find an approximate value of $e^{0.56}$ by *linear extrapolation* (where one replaces e^x by a linear function $F(x)$ which agrees with e^x at $x = 0.4$ and $x = 0.5$).

6.3 CONVERGENCE AND ACCURACY OF LINEAR INTERPOLATION

Suppose that $f \in C[a, b]$. For a given integer M we construct the function $F(M; x)$ which is piecewise linear and which agrees with $f(x)$ at the $M + 1$ interpolation points

3.1
$$x_k = a + \frac{k(b - a)}{M}, \qquad k = 0, 1, \ldots, M.$$

In each subinterval $[x_k, x_{k+1}]$ we determine $F(M; x)$ by linear interpolation. Thus we have, for $x \in [x_k, x_{k+1}]$

3.2
$$F(M; x) = f(x_k) + \frac{x - x_k}{x_{k+1} - x_k} (f(x_{k+1}) - f(x_k)).$$

We now seek to show that

3.3
$$\lim_{M \to \infty} F(M; x) = f(x)$$

uniformly for $x \in [a, b]$. But by (3.2) we have

3.4
$$F(M; x) = w_0(x)f(x_k) + w_1(x)f(x_{k+1})$$

where

3.5
$$w_0(x) = \frac{x_{k+1} - x}{x_{k+1} - x_k}, \qquad w_1(x) = \frac{x - x_k}{x_{k+1} - x_k}$$

and

3.6
$$w_0(x) + w_1(x) = 1.$$

By (3.4) and (3.6) we have

3.7 $f(x) - F(M; x) = w_0(x)(f(x) - f(x_k)) + w_1(x)(f(x) - f(x_{k+1})).$

Since $f(x) \in C[a, b]$, it is uniformly continuous by Theorem A-3.21, and it follows that $|f(x) - f(x_k)|$ and $|f(x) - f(x_{k+1})|$ can be made arbitrarily small by choosing M large enough. Therefore,

3.8 $|f(x) - F(M;x)| \leq |w_0(x)| \, |f(x) - f(x_k)| + |w_1(x)| \, |f(x) - f(x_{k+1})|$

$$\leq \max(|f(x) - f(x_k)|, |f(x) - f(x_{k+1})|)$$

since $w_0(x) \geq 0$, $w_1(x) \geq 0$ and since (3.6) holds. Hence (3.3) holds. We remark that for convergence it is not necessary that the interpolation points be evenly spaced as long as the maximum distance between any two consecutive points converges to zero as the number of interpolation points increases without limit.

We have just seen that given a continuous function defined on a closed interval we can attain any desired accuracy using linear interpolation provided that we use enough points of interpolation. We now investigate the question as to how accurate the linear interpolation process is for a given function and for a given spacing of the interpolation points.

It is sufficient to consider the accuracy of the linear function $F(x)$ treated in Section 6.2 which agrees with the given function $f(x)$ at two points, say a and b. We prove

3.9 Theorem. Let $f(x) \in C^{(1)}[a, b]$ and $f \in D^{(2)}(a, b)$. If $F(x)$ is given by (2.8) then for some $c \in (a, b)$ we have

3.10 $$f(x) - F(x) = \frac{(x-a)(x-b)}{2} f''(c).$$

Moreover, if for all $x \in (a, b)$ we have

3.11 $$|f''(x)| \leq M_2$$

for some constant M_2 then

3.12 $$|f(x) - F(x)| \leq \frac{(b-a)^2}{8} M_2$$

for all x in $[a, b]$.

Proof. Let us define the function $R(x)$ by

3 13 $$R(x) = f(x) - F(x)$$

and let

3.14 $$P(x) = \frac{R(x)}{(x-a)(x-b)}, \qquad a < x < b.$$

We remark that while $P(x)$ is not defined for $x = a$ or $x = b$, nevertheless, by L'Hospital's rule (see Theorem A-3.52) we have

3.15
$$\begin{cases} \lim_{x \to a+} P(x) = \dfrac{R'(a)}{a - b} \\[3mm] \lim_{x \to b-} P(x) = \dfrac{R'(b)}{b - a} . \end{cases}$$

Hence if we extend the definition of $P(x)$ by letting

3.16
$$P(a) = \frac{R'(a)}{a - b}, \qquad P(b) = \frac{R'(b)}{b - a}$$

we obtain a function which is continuous in $[a, b]$.

Given $x \in (a, b)$, let us consider the function

3.17
$$\phi(z) = \phi(z; x) = f(z) - F(z) - (z - a)(z - b)P(x).$$

Evidently, for fixed x, $\phi(z; x)$ is a continuous function of z in I and has a continuous first derivative in I. Moreover, $\phi''(z; x)$ exists for $z \in (a, b)$. Since

3.18
$$\phi(a; x) = \phi(b; x) = \phi(x; x) = 0$$

it follows from Rolle's theorem (Theorem A-3.47) that there exist c_1 and c_2 such that $c_1 \in (a, x)$, $c_2 \in (x, b)$ and such that

3.19
$$\phi'(c_1; x) = \phi'(c_2; x) = 0.$$

Again applying Rolle's theorem to $\phi'(z; x)$, we find that there exists c in the interval (c_1, c_2) such that $\phi''(c; x) = 0$, i.e., such that

3.20
$$\begin{aligned} \phi''(c; x) &= f''(c) - F''(c) - 2P(x) \\ &= f''(c) - 2P(x) \\ &= 0 \end{aligned}$$

since $F''(c) = 0$. Therefore we have

3.21
$$P(x) = \frac{f''(c)}{2}.$$

From (3.13), (3.14), and (3.16) we have, for $x \in [a, b]$

3.22
$$f(x) - F(x) = (x - a)(x - b)P(x),$$

and hence (3.10) follows.

To get (3.12), we note that in I the function $(x - a)(x - b)$ has a maximum absolute value at $(a + b)/2$ and

3.23 $$\max_{a \leq x \leq b} |(x - a)(x - b)| = |(x - a)(x - b)|_{x = (a+b)/2} = \frac{(b - a)^2}{4}.$$

The result (3.12) follows from (3.10) and (3.23), and the proof of Theorem 3.9 is complete.

Let us now apply Theorem 3.9 to the interpolation problem considered in Section 6.2 involving the approximate determination of $e^{0.26}$ by interpolation in the points $a = 0.20$ and $b = 0.30$. Evidently $f''(x) = e^x$ so that

3.24 $$|f''(x)| \leq e^{0.30} \doteq 1.34986, \qquad 0.20 \leq x \leq 0.30.$$

Hence by (3.10) we have

3.25 $$|e^{0.26} - F(0.26)| \leq \frac{(0.06)(0.04)}{2}(1.34986) \doteq 0.00162$$

which agrees closely with the actual value of $|e^{0.26} - F(0.26)|$ found in Section 6.2. If we had used (3.12) instead of (3.10) we would have obtained the bound

3.26 $$|e^{0.26} - F(0.26)| \leq \frac{(0.1)^2}{8}(1.34986) \doteq 0.00169.$$

If we wished to give a bound on the error of linear interpolation for e^x for $x = 0(0.1)2.0$ we would use (3.12) with

3.27 $$M_2 = e^{2.0} \doteq 7.38906$$

and obtain

3.28 $$|f(x) - F(x)| \leq 0.00924.$$

Let us now study the proof of Theorem 3.9 in the light of the above example with the calculations being carried out to greater accuracy. First let us evaluate $P(0.26)$ accurately. From a table of e^x given to 10 decimals we have

3.29 $$e^{0.20} \doteq 1.2214027582, \qquad e^{0.30} \doteq 1.3498588076$$

so that by linear interpolation

3.30 $$F(0.26) \doteq 1.2984763878.$$

Therefore since $e^{0.26} \doteq 1.2969300867$ we have

3.31 $$R(0.26) \doteq -0.0015463011$$

and

3.32 $$P(0.26) = \frac{R(0.26)}{(0.26 - 0.20)(0.26 - 0.30)} \doteq 0.644292.$$

Next let us tabulate $\phi(z, 0.26)$, $\phi'(z; 0.26)$ and $\phi''(z; 0.26)$ where

3.33 $$\begin{cases} \phi(z; 0.26) = R(z) - (z - 0.20)(z - 0.30)P(0.26) \\ \phi'(z; 0.26) = R'(z) - (2z - 0.50)P(0.26) \\ \quad = e^z - (e^{0.3} - e^{0.2})(0.1)^{-1} - (2z - 0.50)P(0.26) \\ \phi''(z; 0.26) = R''(z) - 2P(0.26) \\ \quad = e^z - 2P(0.26) \end{cases}$$

3.34 Table

z	e^z	$F(z)$	$R(z)$	$\phi(z; x)$	$\phi'(z; x)$	$\phi''(z; x)$
0.20	1.2214027582	1.2214027582	0.0000000000	0.0000000000	0.00127148	−0.06718149
0.21	1.2336780600	1.2342483631	−0.0005703031	0.0000095598	0.00066094	−0.05490619
0.22	1.2460767306	1.2470939681	−0.0010172375	0.0000136299	0.00017376	−0.04250752
0.23	1.2586000099	1.2599395730	−0.0013395631	0.0000134504	−0.00018880	−0.02998424
0.24	1.2712491503	1.2727851780	−0.0015360277	0.0000102734	−0.00042550	−0.01733510
0.25	1.2840254167	1.2856307829	−0.0016053662	0.0000053641	−0.00053508	−0.00455883
0.26	1.2969300867	1.2984763878	−0.0015463011	0.0000000000	−0.00051625	0.00834584
0.27	1.3099644507	1.3113219928	−0.0013575421	−0.0000045286	−0.00036773	0.02138020
0.28	1.3231298123	1.3241675977	−0.0010377854	−0.0000069180	−0.00008821	0.03454556
0.29	1.3364274880	1.3370132027	−0.0005857147	−0.0000058518	0.00032362	0.04784324
0.30	1.3498588076	1.3498588076	0.0000000000	0.0000000000	0.00086910	0.06127456

(Values of $\phi(z; x)$, $\phi'(z; x)$, $\phi''(z; x)$ for $z = 0.2(0.01)0.3$).

The functions $\phi(z; 0.26)$, $\phi'(z; 0.26)$, and $\phi''(z; 0.26)$ are plotted in Fig. 3.39. Evidently $\phi'(z; 0.26)$ vanishes for z approximately 0.224 and 0.282. Also $\phi''(z; 0.26)$ vanishes for z approximately 0.254.

To get c_1 and c_2 more accurately we apply the Newton method to $\phi'(z; 0.26)$ with $z_0 = 0.22$ and also with $z_0 = 0.28$. We obtain

3.35 $$c_1 = 0.224368, \quad c_2 = 0.282439.$$

To get c we set $\phi''(z; 0.26)$ equal to zero and obtain

3.36 $$c = \log(2P(0.26)) = \log 1.288584 = 0.253544.$$

(We compute $\log 1.288584$ by using linear interpolation in a table of $\log x$ for $x = 0(0.0001)5$.)

Let us now verify (3.10). As we have already seen,

3.37 $R(0.26) = -0.0015463011.$

On the other hand for $x = 0.26$ and $c = 0.253544$ we have

3.38 $\dfrac{(x - a)(x - b)}{2} f''(c) = \dfrac{(0.06)(-0.04)}{2} e^{0.253544}$

$$= -(0.0012)(1.288584) = -0.0015463008$$

which agrees with $R(0.26)$ to within 0.0000000003. (Here we use linear interpolation in a table of e^x for $x = 0(0.0001)2.5$.)

3.39 Fig. Graphs of $\phi(z; x)$, $\phi'(z; x)$, $\phi''(z; x)$ for $f(x) = e^x$, $a = 0.20$, $b = 0.30$, $x = 0.26$.

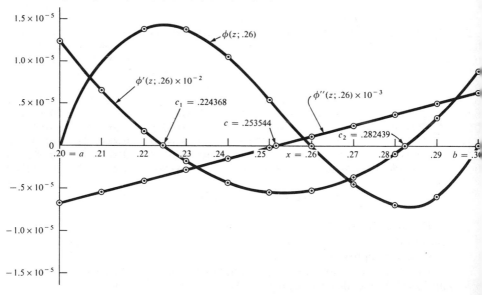

We now give an alternative proof of (3.10) under the added assumption that $f''(x)$ is continuous on (a, b). For fixed x we have by Taylor's theorem (Theorem A-3.64)

3.40 $f(t) = f(x) + (t - x)f'(x) + \dfrac{(t - x)^2}{2} f''(c(t))$

where $c(t)$ lies between x and t. Thus

3.41
$$\begin{cases} f(a) = f(x) + (a - x)f'(x) + \dfrac{(a - x)^2}{2} f''(c_1) \\[2mm] f(b) = f(x) + (b - x)f'(x) + \dfrac{(b - x)^2}{2} f''(c_2) \end{cases}$$

where $a < c_1 < x < c_2 < b$. Therefore, substituting (3.41) into (2.11), we obtain

3.42
$$F(x) = f(x) + \alpha_1 f''(c_1) + \alpha_2 f''(c_2)$$

where

3.43
$$\alpha_1 = \frac{(a - x)^2(b - x)}{2(b - a)}, \qquad \alpha_2 = \frac{(b - x)^2(x - a)}{2(b - a)}.$$

Consider the function

3.44
$$\psi(x) = \alpha_1 f''(c_1) + \alpha_2 f''(c_2) - (\alpha_1 + \alpha_2)f''(x).$$

Since

3.45
$$\psi(c_1)\psi(c_2) = -\alpha_1\alpha_2(f''(c_1) - f''(c_2))^2$$

and since $\alpha_1 \geq 0, \alpha_2 \geq 0$ it follows from Theorem A-3.24 that $\psi(c) = 0$ for some c such that $c \in [c_1, c_2]$. Thus we have

3.46
$$f(x) = F(x) + \frac{(x - a)(x - b)}{2} f''(c)$$

and (3.10) holds.

EXERCISES 6.3

1. Use (3.10) for the approximate determination of the error involved in using the linear interpolation process of Exercise 1, Section 6.2. Give upper and lower bounds for the error. Compare with the actual error.

2. What accuracy can be obtained using linear interpolation in tables for which functions $f(x)$ are given with the following characteristics?

$f(x)$	Arguments	No. of decimals given for $f(x)$
$\log x$	1(0.001)5	8
$\log_{10} x$	1(0.001)10	5
$\sin x$	1(0.0001)2	12
$\arctan x$	0(0.0001)2	12
$\sin(\pi x/180)$	0(1)90	6
$\dfrac{2}{\pi} \displaystyle\int_0^x e^{-x^2}\, dx$	0(0.001)2.5	5

3. What interval size should be used in the tabulation of $f(x) = (1 + x^2)^{-1}$ in the interval $[0, 2]$ so that linear interpolation will be accurate to four decimal places?

4. What interval size should be used to tabulate $\exp(\sin x)$ in the interval $[0, 2]$ so that linear interpolation will be accurate to four decimal places?

5. Let $F(x)$ be the linear interpolating polynomial for $f(x)$ corresponding to a and b. Find c such that $a < c < b$ and

$$f(x) - F(x) = \frac{(x - a)(x - b)}{2} f''(c)$$

for each of the following cases:

	$f(x)$	a	b	x
(i)	x^4	0	1	$\frac{1}{2}$
(ii)	$\sin\left(\dfrac{\pi}{180}x\right)$	30	40	33
(iii)	e^x	0.20	0.30	0.24
(iv)	$\sin x$	0.40	0.50	0.44

6.4 LAGRANGIAN INTERPOLATION

Let us now assume that we are given $f(x)$ at $n + 1$ distinct values of x, say x_0, $x_1, \ldots, x_n$. We seek to determine a polynomial $F(x)$ of degree n or less such that

4.1 $$F(x_i) = f(x_i), \qquad i = 0, 1, 2, \ldots, n.$$

We shall show that the polynomial $F(x)$ exists and is unique. Two alternative proofs will be given.

4.2 Theorem. If $x_0, x_1, \ldots, x_n$ are distinct, then for any $y_0, y_1, \ldots, y_n$ there exists a unique polynomial $F(x)$, of degree n or less, such that

4.3 $$F(x_i) = y_i, \qquad i = 0, 1, 2, \ldots, n.$$

Proof. Let us define $F(x)$ in the following way:

4.4 $$F(x) = F_0(x) + F_1(x) + \cdots + F_n(x)$$

where

4.5 $$F_i(x) = \left\{ \prod_{\substack{j=0 \\ j \neq i}}^{n} \frac{x - x_j}{x_i - x_j} \right\} y_i, \qquad i = 0, 1, 2, \ldots, n.$$

Evidently $F(x)$ is a polynomial of degree n or less which satisfies (4.3). To prove uniqueness, consider any polynomial $G(x)$ of degree n or less which satisfies (4.3). Then $H(x) = F(x) - G(x)$ is a polynomial of degree n or less which vanishes at

$n + 1$ distinct values of x. Hence, by Corollary 5-2.26, it follows that $H(x) \equiv 0$ and hence $G(x) \equiv F(x)$. This completes the proof of Theorem 4.2.

As an example, let us determine the polynomial $F(x)$ of degree 2 or less such that $F(0) = 1$, $F(1) = 2$, $F(2) = 4$. By (4.4), we have

4.6 $$F(x) = F_0(x) + F_1(x) + F_2(x),$$

and it follows from (4.5) that

4.7 $$F(x) = \frac{(x - x_1)(x - x_2)}{(x_0 - x_1)(x_0 - x_2)} y_0 + \frac{(x - x_0)(x - x_2)}{(x_1 - x_0)(x_1 - x_2)} y_1 + \frac{(x - x_0)(x - x_1)}{(x_2 - x_0)(x_2 - x_1)} y_2.$$

We note that $F(x)$ is of degree 2 or less and that $F(x_0) = y_0$, $F(x_1) = y_1$, $F(x_2) = y_2$. Substituting the values for the x_i and y_i, we have

4.8 $$F(x) = \frac{(x - 1)(x - 2)}{(0 - 1)(0 - 2)} \cdot 1 + \frac{(x - 0)(x - 2)}{(1 - 0)(1 - 2)} \cdot 2 + \frac{(x - 0)(x - 1)}{(2 - 0)(2 - 1)} \cdot 4,$$
$$= \tfrac{1}{2}(x - 1)(x - 2) - 2x(x - 2) + 2x(x - 1),$$
$$= \tfrac{1}{2}x^2 + \tfrac{1}{2}x + 1.$$

One can easily verify that $F(0) = 1$, $F(1) = 2$, $F(2) = 4$.

The Method of Undetermined Coefficients

We now give an alternative method for determining the interpolating polynomial based on the use of undetermined coefficients. Let us choose any convenient value of α and let

4.9 $$F(x) = a_0(x - \alpha)^n + a_1(x - \alpha)^{n-1} + \cdots + a_n.$$

We seek to determine the coefficients $a_0, a_1, \ldots, a_n$ such that (4.3) holds. Evidently the a_k must satisfy the following system of linear algebraic equations

4.10 $$\begin{cases} y_0 = a_0(x_0 - \alpha)^n + a_1(x_0 - \alpha)^{n-1} + \cdots + a_n \\ y_1 = a_0(x_1 - \alpha)^n + a_1(x_1 - \alpha)^{n-1} + \cdots + a_n \\ \quad \cdot \qquad\qquad \cdot \qquad\qquad\quad \cdot \\ y_n = a_0(x_n - \alpha)^n + a_1(x_n - \alpha)^{n-1} + \cdots + a_n. \end{cases}$$

The above system of $n + 1$ equations with $n + 1$ unknowns has a unique solution if the determinant of the system does not vanish (see Theorem A-7.2). The determinant is

4.11
$$\Delta = \begin{vmatrix} (x_0 - \alpha)^n & (x_0 - \alpha)^{n-1} \cdots (x_0 - \alpha) & 1 \\ (x_1 - \alpha)^n & (x_1 - \alpha)^{n-1} \cdots (x_1 - \alpha) & 1 \\ \cdot & \cdot & \cdots & \cdot & \cdot \\ (x_n - \alpha)^n & (x_n - \alpha)^{n-1} \cdots (x_n - \alpha) & 1 \end{vmatrix}.$$

The above determinant is a *Vandermonde determinant* whose value is given by

4.12
$$\Delta = \prod_{\substack{i,j=0 \\ i<j}}^{n} (x_i - x_j).$$

We shall verify this for the special case $\alpha = 0$, $n = 3$. We have

4.13
$$\Delta = \begin{vmatrix} x_0^3 & x_0^2 & x_0 & 1 \\ x_1^3 & x_1^2 & x_1 & 1 \\ x_2^3 & x_2^2 & x_2 & 1 \\ x_3^3 & x_3^2 & x_3 & 1 \end{vmatrix}.$$

Subtracting the first row from each of the other rows we obtain

4.14
$$\Delta = \begin{vmatrix} x_0^3 & x_0^2 & x_0 & 1 \\ x_1^3 - x_0^3 & x_1^2 - x_0^2 & x_1 - x_0 & 0 \\ x_2^3 - x_0^3 & x_2^2 - x_0^2 & x_2 - x_0 & 0 \\ x_3^3 - x_0^3 & x_3^2 - x_0^2 & x_3 - x_0 & 0 \end{vmatrix}.$$

Next, we multiply the second column by x_0 and subtract from the first, obtaining

4.15
$$\Delta = \begin{vmatrix} 0 & x_0^2 & x_0 & 1 \\ x_1^2(x_1 - x_0) & x_1^2 - x_0^2 & x_1 - x_0 & 0 \\ x_2^2(x_2 - x_0) & x_2^2 - x_0^2 & x_2 - x_0 & 0 \\ x_3^2(x_3 - x_0) & x_3^2 - x_0^2 & x_3 - x_0 & 0 \end{vmatrix}.$$

Now subtract x_0 times the third column from the second column and then x_0

times the fourth column from the third column. We have

4.16
$$\Delta = \begin{vmatrix} 0 & 0 & 0 & 1 \\ x_1^2(x_1 - x_0) & x_1(x_1 - x_0) & x_1 - x_0 & 0 \\ x_2^2(x_2 - x_0) & x_2(x_2 - x_0) & x_2 - x_0 & 0 \\ x_3^2(x_3 - x_0) & x_3(x_3 - x_0) & x_3 - x_0 & 0 \end{vmatrix}$$

$$= -(x_1 - x_0)(x_2 - x_0)(x_3 - x_0) \begin{vmatrix} x_1^2 & x_1 & 1 \\ x_2^2 & x_2 & 1 \\ x_3^2 & x_3 & 1 \end{vmatrix}.$$

The above procedure can be repeated on the smaller determinant, yielding

4.17
$$\Delta = -(x_1 - x_0)(x_2 - x_0)(x_3 - x_0)(x_2 - x_1)(x_3 - x_1) \begin{vmatrix} x_2 & 1 \\ x_3 & 1 \end{vmatrix}$$

$$= (x_1 - x_0)(x_2 - x_0)(x_3 - x_0)(x_2 - x_1)(x_3 - x_1)(x_3 - x_2)$$

$$= (x_0 - x_1)(x_0 - x_2)(x_0 - x_3)(x_1 - x_2)(x_1 - x_3)(x_2 - x_3).$$

Thus (4.12) is verified for this case. Since the x_i are distinct, it follows that $\Delta \neq 0$ and the a_k can be uniquely determined.

Let us apply the method of undetermined coefficients to the example given above. If we let $\alpha = 0$ we have

4.18
$$F(x) = a_0 x^2 + a_1 x + a_2$$

and

4.19
$$\begin{cases} 1 = a_2 \\ 2 = a_0 + a_1 + a_2 \\ 4 = 4a_0 + 2a_1 + a_2. \end{cases}$$

Solving, we get

4.20
$$a_0 = \tfrac{1}{2}, \qquad a_1 = \tfrac{1}{2}, \qquad a_2 = 1$$

and

4.21
$$F(x) = \tfrac{1}{2}x^2 + \tfrac{1}{2}x + 1$$

which agrees with the previous result.

Method of Undetermined Weights

Formula (4.4) can be written in the form

4.22
$$F(x) = \sum_{i=0}^{n} w_i(x)y_i = \sum_{i=0}^{n} w_i(x)F(x_i),$$

where the polynomials $w_i(x)$ have degree n or less. In fact, by (4.5) we have

4.23
$$w_i(x) = \prod_{\substack{j=0 \\ j \neq i}}^{n} \left(\frac{x - x_j}{x_i - x_j}\right), i = 0, 1, 2, \ldots, n.$$

For a given value of x, the numbers $w_0(x), w_1(x), \ldots, w_n(x)$ can be considered as "weighting coefficients." Thus $F(x)$ is a linear combination of values of $F(x_i)$ with weights $w_0(x), w_1(x), \ldots, w_n(x)$. Clearly, the weights can be determined by (4.23), and $F(x)$ can be evaluated without actually determining the coefficients a_0, $a_1, \ldots, a_n$ of $F(x)$, considered as a polynomial.

In the case of simple Lagrangian interpolation, the weights are given explicitly by (4.23). However, in more complicated cases to be considered later, such an explicit expression may not be available. Nevertheless, for a particular value of x, one can determine the weights by solving a system of linear algebraic equations similar to that encountered in the method of undetermined coefficients given above. In the method of undetermined weights we assume that $F(x)$ has the form

4.24
$$F(x) = \sum_{i=0}^{n} w_i(x)F(x_i)$$

and we choose $w_i(x)$ so that the above equation holds for the special cases: $F(x) = 1$, $F(x) = (x - \alpha)$, $F(x) = (x - \alpha)^2, \ldots, F(x) = (x - \alpha)^n$. If (4.24) is exact for such functions, then for any coefficients $a_0, a_1, \ldots, a_n$ we have

4.25 $a_0(x - \alpha)^n + a_1(x - \alpha)^{n-1} + \cdots + a_n$

$$= \sum_{i=0}^{n} w_i(x)\{a_0(x - \alpha)^n + a_1(x - \alpha)^{n-1} + \cdots + a_n\}.$$

Hence (4.24) will be exact for any polynomial in $(x - \alpha)$ of degree n or less. As a matter of fact, since any polynomial of degree n or less in $x - \alpha$ is also a polynomial of degree n or less in $x(=(x - \alpha) + \alpha)$ it follows that (4.24) will be exact for any polynomial of degree n or less in x.

The requirement that (4.24) hold for the special cases $F(x) = 1$, $F(x) = x - \alpha$, etc., implies

4.26
$$
\begin{cases}
1 = w_0(x) + w_1(x) + \cdots + w_n(x) \\
x - \alpha = w_0(x)(x_0 - \alpha) + w_1(x)(x_1 - \alpha) + \cdots + w_n(x)(x_n - \alpha) \\
\quad\vdots \\
(x - \alpha)^n = w_0(x)(x_0 - \alpha)^n + w_1(x)(x_1 - \alpha)^n + \cdots + w_n(x)(x_n - \alpha)^n.
\end{cases}
$$

For fixed x, we can determine the $w_i(x)$ uniquely provided the determinant Δ of the system does not vanish. But this determinant is to within sign, the same as the determinant of the system (4.10). Moreover, by Cramer's rule,

4.27
$$
w_k(x) = \frac{\Delta_k}{\Delta}, k = 0, 1, 2, \ldots, n
$$

where

4.28
$$
\begin{cases}
\Delta = (-1)^k \prod_{\substack{i,j=0 \\ i>j \\ i \neq k, j \neq k}}^{n} (x_i - x_j) \prod_{\substack{j=0 \\ j \neq k}}^{n} (x_k - x_j) \\[2em]
\Delta_k = (-1)^k \prod_{\substack{i,j=0 \\ i>j \\ i \neq k, j \neq k}}^{n} (x_i - x_j) \prod_{\substack{j=0 \\ j \neq k}}^{n} (x - x_j).
\end{cases}
$$

Hence, we have shown that Δ_k/Δ is equal to the right member of (4.23).

Let us now apply the above procedure to the example where $n = 2$ and where $F(0) = 1$, $F(1) = 2$, $F(2) = 4$, where $x = \frac{1}{2}$. Let $w_i(\frac{1}{2}) = w_i$, $i = 0, 1, 2$. Then we have, letting $\alpha = 0$,

4.29
$$
\begin{cases}
1 = w_0 + w_1 + w_2 \\
\frac{1}{2} = (0)w_0 + (1)w_1 + (2)w_2 \\
(\frac{1}{2})^2 = (0)w_0 + (1)w_1 + (4)w_2.
\end{cases}
$$

Solving for w_0, w_1, w_2, we get

4.30
$$
w_0 = \tfrac{3}{8}, \qquad w_1 = \tfrac{3}{4}, \qquad w_2 = -\tfrac{1}{8}.
$$

Hence

4.31
$$
\begin{aligned}
F(\tfrac{1}{2}) &= (\tfrac{3}{8})F(0) + (\tfrac{3}{4})F(1) - (\tfrac{1}{8})F(2) \\
&= (\tfrac{3}{8})(1) + (\tfrac{3}{4})(2) - (\tfrac{1}{8})(4) = \tfrac{11}{8}.
\end{aligned}
$$

This agrees with the result obtained by letting $x = \frac{1}{2}$ in (4.8).

Aitken's Method

The Lagrange interpolating polynomial can be evaluated using a series of linear interpolations by the method of Aitken [1932]. We shall illustrate by considering the case of 5 points of interpolation. We develop the following table:

$$
\begin{array}{ccccc}
f(x_0) \\
 & I_{0,1} \\
f(x_1) & & I_{0,1,2} \\
 & I_{0,2} & & I_{0,1,2,3} \\
f(x_2) & & I_{0,1,3} & & I_{0,1,2,3,4} \\
 & I_{0,3} & & I_{0,1,2,4} \\
f(x_3) & & I_{0,1,4} \\
 & I_{0,4} \\
f(x_4)
\end{array}
$$

Here we let

$$
\textbf{4.32} \quad
\begin{cases}
I_{0,j} = I_{0,j}(x) = \dfrac{x_j - x}{x_j - x_0} f(x_0) + \dfrac{x - x_0}{x_j - x_0} f(x_j), \quad j = 1, 2, 3, 4 \\[2ex]
I_{0,1,j} = I_{0,1,j}(x) = \dfrac{x_j - x}{x_j - x_1} I_{0,1} + \dfrac{x - x_1}{x_j - x_1} I_{0,j}, \quad j = 2, 3, 4 \\[2ex]
I_{0,1,2,j} = I_{0,1,2,j}(x) = \dfrac{x_j - x}{x_j - x_2} I_{0,1,2} + \dfrac{x - x_2}{x_j - x_2} I_{0,1,j}, \quad j = 3, 4 \\[2ex]
I_{0,1,2,3,4} = I_{0,1,2,3,4}(x) = \dfrac{x_4 - x}{x_4 - x_3} I_{0,1,2,3} + \dfrac{x - x_3}{x_4 - x_3} I_{0,1,2,4}.
\end{cases}
$$

It is easy to show that $I_{0,j}(x_k) = f(x_k)$, $k = 0, j$. Next we seek to show that $I_{0,1,j}(x_k) = f(x_k)$, $k = 0, 1, j$. But we have

$$
\textbf{4.33} \quad
\begin{cases}
I_{0,1,j}(x_0) = \dfrac{x_j - x_0}{x_j - x_1} I_{0,1}(x_0) + \dfrac{x_0 - x_1}{x_j - x_1} I_{0,j}(x_0) = f(x_0) \\[2ex]
I_{0,1,j}(x_1) = I_{0,1}(x_1) = f(x_1) \\[2ex]
I_{0,1,j}(x_j) = I_{0,j}(x_j) = f(x_j), \quad j = 2, 3, 4.
\end{cases}
$$

In the same way we can show that

$$
\textbf{4.34} \quad
\begin{cases}
I_{0,1,2,j}(x_k) = f(x_k), & k = 0, 1, 2, j \\
I_{0,1,2,3,4}(x_k) = f(x_k), & k = 0, 1, 2, 3, 4.
\end{cases}
$$

Since $I_{0,1,2,3,4}(x)$ is clearly a polynomial of degree 4 or less, it then follows that it is the same as the Lagrange interpolating polynomial.

Let us now consider the example

4.35 $f(0) = 1,$ $f(1) = 2,$ $f(2) = 4.$

We have

4.36
$$\left\{ \begin{aligned} I_{0,1} &= \frac{1-x}{1-0}(1) + \frac{x-0}{1-0}(2) = 1 + x \\ I_{0,2} &= \frac{2-x}{2-0}(1) + \frac{x-0}{2-0}(4) = 1 + \tfrac{3}{2}x \\ I_{0,1,2} &= \frac{2-x}{2-1}(1+x) + \frac{x-1}{2-1}(1+\tfrac{3}{2}x) = 1 + \tfrac{1}{2}x + \tfrac{1}{2}x^2 \end{aligned} \right.$$

which is the same as was obtained by Lagrangian interpolation. In the actual application of Aitken's method one would presumably have a specific value of x and would construct a specific value of $I_{0,1}, I_{0,2}, \ldots, I_{0,1,2}, \ldots$, etc.

This procedure of using repeated linear interpolation is somewhat analogous to the Romberg method for numerical integration which will be described in Chapter 7.

EXERCISES 6.4

1. Find the unique polynomial $F(x)$ of degree 2 or less such that

$$F(1) = 1,\quad F(2) = 4,\quad F(3) = 10$$

   using each of the following methods: the Lagrangian interpolation formula; the method of undetermined coefficients; the method of undetermined weights. Evaluate $F(1.6)$.

2. Find the unique polynomial $F(x)$ of degree 3 or less such that

$$F(0.5) = 2,\quad F(0.6) = 8,\quad F(0.7) = -2,\quad F(0.8) = 5$$

   and evaluate $F(0.56)$ by each of the schemes used in the previous exercise.

3. Verify the formula (4.4) for the case $x_0 = 1$, $x_1 = 3$, $x_2 = -1$, $x_3 = 2$, and $n = 3$.

4. Given the following table, find log 3.53 by linear interpolation, by three-point Lagrangian interpolation, and also by four-point Lagrangian interpolation.

x	$\log x$
3.50	1.252763
3.60	1.280934
3.70	1.308333
3.80	1.335001

5. Carry out each of the interpolations of the preceding exercise using Aitkin's method.

6. Write a computer program for carrying out the following process. Given $x_0, x_1, \ldots, x_N$ and $y_0, y_1, \ldots, y_N$ and given x and $n \leq N$,

a) find the $n + 1$ points x_i nearest to x;

b) evaluate at x the Lagrangian interpolating polynomial of degree n or less such that $F(x_i) = y_i, i = 0, 1, \ldots, n$.

Then generate a table of $f(x) = \sin(\pi x/180)$ for $x = 0(10)90$ and apply the program to the cases $x = 24, 40, 45, 88$ and $n = 0(1)5$. In each case print the exact value of $f(x)$, the approximate value, and the error.

6.5 CONVERGENCE AND ACCURACY OF LAGRANGIAN INTERPOLATION

Suppose we are given a function $f(x)$ which is continuous in an interval $I = [a, b]$. We shall generate a sequence of functions based on the use of Lagrangian interpolation and study the convergence of this sequence. First we subdivide I into M subintervals of length h by considering the interpolation points

5.1 $$x_k = a + kh, \qquad k = 0, 1, \ldots, M$$

where

5.2 $$h = \frac{b - a}{M}.$$

Next we select an integer n and for each $M \geq n$ we use Lagrangian $(n + 1)$-point interpolation in the following way. In any subinterval

5.3 $$I_k = [x_{k-1}, x_k], \qquad k = 1, 2, \ldots, M$$

we use Lagrangian $(n + 1)$-point interpolation based on x_{k-1}, x_k and $n - 1$ additional points as close as possible to x. Thus if $n = 3$ we would use $x_{k-2}, x_{k-1}, x_k, x_{k+1}$ if $x \in I_k$ except if $x \in I_1$ we would use x_0, x_1, x_2, x_3. Also, if $x \in I_M$ we would use $x_{M-3}, x_{M-2}, x_{M-1}, x_M$.

Given x, let us label the $n + 1$ points of interpolation which are used as $t_0, t_1, \ldots, t_n$ where $t_0 < t_1 < \cdots < t_n$. Evidently we can construct a sequence composed of the functions

5.4 $$F_M(x) = \sum_{j=0}^{n} w_j(x) f(t_j)$$

by the Lagrangian interpolation formula, where

5.5 $$w_j(x) = \prod_{\substack{s=0 \\ s \neq j}}^{n} \frac{x - t_s}{t_j - t_s}.$$

Since

5.6 $$\sum_{j=0}^{n} w_j(x) = 1$$

we have

5.7
$$f(x) - F_M(x) = \sum_{j=0}^{n} w_j(x)[f(x) - f(t_j)]$$

and

5.8
$$|f(x) - F_M(x)| \leq K(n) \max_{0 \leq j \leq n} |f(x) - f(t_j)|$$

where the $K(n)$ are constants depending on n which are bounds for $\sum_{j=0}^{n} |w_j(x)|$. We assume for the moment that $K(n)$ exists. By the uniform continuity of $f(x)$ it follows that

5.9
$$\max_{0 \leq j \leq n} |f(x) - f(t_j)| \to 0$$

as $M \to \infty$, uniformly on $[a, b]$, and hence

5.10
$$\lim_{M \to \infty} |f(x) - F_M(x)| = 0$$

uniformly on $[a, b]$.

We now show that the constants $K(n)$ exist. Each factor of the denominator of (5.5) is at least h; hence the denominator is at least h^n. Each factor of the numerator is at most nh. Hence the numerator is at most $(nh)^n$ so that we have $|w_j(x)| \leq n^n$ for each j. Therefore,

5.11
$$K(n) = \sum_{j=0}^{n} |w_j(x)| \leq (n + 1)n^n.$$

Let us now consider the accuracy of Lagrangian interpolation. We now state the following result which is a direct extension of Theorem 3.9. The proof is similar to that of Theorem 3.9 and will be left to the reader.

5.12 Theorem. Let $f(x) \in C^{(n)}[x_0, x_n]$ and $f(x) \in D^{(n+1)}(x_0, x_n)$. Let $x_0, x_1, \ldots, x_n$ be distinct numbers such that $x_0 < x_1 < \cdots < x_n$ and let $F(x)$ be the unique polynomial of degree n or less such that

5.13
$$F(x_i) = f(x_i), \qquad i = 0, 1, 2, \ldots, n.$$

For any $x \in [x_0, x_n]$ we have

5.14
$$f(x) - F(x) = \frac{(x - x_0)(x - x_1)\cdots(x - x_n)}{(n + 1)!} f^{(n+1)}(c)$$

for some $c \in (x_0, x_n)$.

For each $k = 0, 1, 2, \ldots$ let us define

5.15
$$M_k = \max_{x \in I} |f^{(k)}(x)|$$

where $I = [x_0, x_n]$. We have

5.16 Corollary. Under the hypotheses of Theorem 5.12, if $f^{(n+1)}(x)$ is continuous in I and if $x_1 - x_0 = x_2 - x_1 = \cdots = x_n - x_{n-1} = h$, then

5.17
$$|f(x) - F(x)| \leqq \frac{h^{n+1}}{4(n+1)} M_{n+1}.$$

Proof. It is easy to show that the largest value of $|(x - x_0)(x - x_1)\ldots(x - x_n)|$ is assumed in one of the intervals $x_0 \leqq x \leqq x_1$ and $x_{n-1} \leqq x \leqq x_n$. Moreover, this maximum value does not exceed

5.18
$$\max_{x_0 \leqq x \leqq x_1} |(x - x_0)(x - x_1)| \max_{x_0 \leqq x \leqq x_1} |(x - x_2)(x - x_3)\ldots(x - x_n)| \leqq \frac{n!}{4} h^{n+1}.$$

The result (5.17) now follows from (5.14).

More accurate bounds can be obtained in special cases. Thus we have

5.19 Corollary. Under the hypotheses of Corollary 5.16 we have

5.20
$$|f(x) - F(x)| \leqq \frac{\sqrt{3}}{27} h^3 M_3$$

if $n = 2$ and

5.21
$$|f(x) - F(x)| \leqq \frac{h^4}{24} M_4$$

if $n = 3$.

Proof. For the case $n = 2$, if we let $y = x - x_1$ we have

5.22
$$Q_2(x) = (x - x_0)(x - x_1)(x - x_2) = y(y^2 - h^2).$$

Evidently $Q_2'(x)$ vanishes for

5.23
$$y = \pm \frac{h}{\sqrt{3}},$$

i.e., for

5.24
$$x = x_1 \pm \frac{h}{\sqrt{3}}$$

and we have

5.25
$$|Q_2|_{\max} = \left| Q_2\left(x_1 \pm \frac{h}{\sqrt{3}}\right) \right| = \frac{2\sqrt{3}}{9} h^3$$

from which (5.20) follows.

For the case $n = 3$, we let $y = x - (x_1 + x_2)/2$ and we have

5.26 $Q_3(x) = (x - x_0)(x - x_1)(x - x_2)(x - x_3) = y^4 - \frac{5}{2}h^2 y^2 + \frac{9}{16}h^4.$

Evidently $Q_3'(x)$ vanishes for

5.27 $y = 0, \pm\sqrt{\frac{5}{4}}h$

or

5.28 $x = \dfrac{x_1 + x_2}{2} \qquad \dfrac{x_1 + x_2}{2} \pm \sqrt{\frac{5}{4}}h$

and we have

5.29 $\begin{cases} \left| Q_3\left(\dfrac{x_1 + x_2}{2}\right) \right| = \frac{9}{16}h^4 \\[2mm] \left| Q_3\left(\dfrac{x_1 + x_2}{2} \pm \sqrt{\frac{5}{4}}h\right) \right| = h^4 \end{cases},$

The result (5.21) follows.

Evidently, with four-point interpolation the largest values of

$$|(x - x_0)(x - x_1)(x - x_2)(x - x_3)|$$

occur in the outer intervals $x_0 \leqq x \leqq x_1$ and $x_2 \leqq x \leqq x_3$. This suggests that whenever possible one should choose, for given x, two interpolation points greater than x and two interpolation points less than x. Thus, for instance, if $f(x)$ is tabulated for $x = 0(0.1)1$ and one wishes to compute $f(0.37)$ by four-point interpolation, one should use the interpolation points $\{0.2, 0.3, 0.4, 0.5\}$ rather than $\{0.3, 0.4, 0.5, 0.6\}$.

We remark that a bound on M_{n+1} can often be estimated in terms of the $(n + 1)$-st difference as described in the next section.

EXERCISES 6.5

1. Find $\log 0.54$ using three-point Lagrangian interpolation given the following table.

x	$\log x$
0.4	-0.91629
0.5	-0.69315
0.6	-0.51083
0.7	-0.35667
0.8	-0.22314

Find the exact error given that $\log 0.54 \doteq -0.61619$. Also, compute a bound on the error. Do the same for four-point Lagrangian interpolation.

2. For each case considered in Exercise 4, Section 6.4, estimate the error and compare with the exact error. (Note: $\log 3.53 \doteq 1.261298$.)

3. Obtain more accurate values of $K(n)$ for $n = 1, 2, 3, 4$ than those given by (5.11).

4. Carry out the details of the proof of Theorem 5.12 for the cases $n = 2$ and $n = 3$.

5. What accuracy can be obtained using three-point Lagrangian interpolation with a table of e^x given to eight decimal places for $x = 0(0.1)2$? What if four-point Lagrangian interpolation were used? Do the same for the function $\log x$ tabulated for $x = 1(0.001)5$.

6. What accuracy can be obtained using three-point Lagrangian interpolation with the tables specified in Exercise 2, Section 6.3? What if four-point Lagrangian interpolation were used?

7. Find $\max |(x - x_0)(x - x_1)(x - x_2)(x - x_3)(x - x_4)|$ for $x_0 \leqq x \leqq x_4$ where $x_1 - x_0 = x_2 - x_1 = x_3 - x_2 = x_4 - x_3 = h$. Thus obtain a formula analogous to (5.21) for the case $n = 4$. Give an improved bound if we require that $x_1 \leqq x \leqq x_3$.

8. What spacing of the values of x in a table of e^x, given to 15 decimals, would be necessary so that three-point Lagrangian interpolation would be accurate to 8 decimals in the interval $[0, 2]$?

9. It is desired to tabulate the function

$$\int_0^x e^{-t^2/2} \, dt$$

in the interval $[0, 2]$ so that three-point Lagrangian interpolation will be correct to 6 decimal places. How small must the tabulation interval be and how many significant figures should be given in the table?

6.6 INTERPOLATION WITH EQUAL INTERVALS

We now assume that the points of interpolation $x_0, x_1, \ldots, x_n$ are evenly spaced and we let

6.1 $$h = x_1 - x_0 = x_2 - x_1 = \cdots = x_n - x_{n-1}.$$

The Lagrange interpolating polynomial can be written in the form

6.2 $$F(x) = \sum_{k=0}^{n} w_k(x) f(x_k)$$

where

6.3 $$w_k(x) = \prod_{\substack{j=0 \\ j \neq k}}^{n} \frac{u - j}{k - j}, \qquad k = 0, 1, 2, \ldots, n,$$

and where

6.4
$$u = \frac{x - x_0}{h}$$

Tables of the Lagrangian interpolation coefficients $w_k(x)$ have been constructed for values of n up to 10 so that for any given value of u one can look up the values of the $w_k(x)$. (See WPA [1944].)

Difference Tables

If the interpolation points are equally spaced, then for manual computation the interpolation process is usually more effectively carried out using formulas involving finite differences.

Let us define forward differences as follows:

$$\Delta f(x) = f(x + h) - f(x)$$
$$\Delta^2 f(x) = \Delta(\Delta f(x)) = \Delta(f(x + h) - f(x))$$
$$= f(x + 2h) - 2f(x + h) + f(x).$$

$$\vdots$$

$$\Delta^n f(x) = \Delta(\Delta^{n-1} f(x)).$$

With the use of the above definitions, one can construct the following table:

x	$f(x)$	$\Delta f(x)$	$\Delta^2 f(x)$	$\Delta^3 f(x)$	$\Delta^4 f(x)$	$\Delta^5 f(x)$
x_0	$f(x_0)$					
		$\Delta f(x_0)$				
x_1	$f(x_1)$		$\Delta^2 f(x_0)$			
		$\Delta f(x_1)$		$\Delta^3 f(x_0)$		
x_2	$f(x_2)$		$\Delta^2 f(x_1)$		$\Delta^4 f(x_0)$	
		$\Delta f(x_2)$		$\Delta^3 f(x_1)$		$\Delta^5 f(x_0)$
x_3	$f(x_3)$		$\Delta^2 f(x_2)$		$\Delta^4 f(x_1)$	
		$\Delta f(x_3)$		$\Delta^3 f(x_2)$		
x_4	$f(x_4)$		$\Delta^2 f(x_3)$			
		$\Delta f(x_4)$				
x_5	$f(x_5)$					

In each case, the element $\Delta^k f(x_p)$ is obtained by subtracting $\Delta^{k-1} f(x_p)$ from $\Delta^{k-1} f(x_{p+1})$ where $\Delta^k f(x_p) = f(x_p)$ for $k = 0$.

As an example, let us construct a difference table for the function $f(x) = x^3$ given for $x = 0(1)5$.

x	$f(x)$	$\Delta f(x)$	$\Delta^2 f(x)$	$\Delta^3 f(x)$	$\Delta^4 f(x)$
0	0				
		1			
1	1		6		
		7		6	
2	8		12		0
		19		6	
3	27		18		0
		37		6	
4	64		24		
		61			
5	125				

The above example illustrates the following theorem:

6.5 Theorem. If $f(x)$ is a polynomial of degree n or less, then $\Delta^{n+1} f(x) \equiv 0$.

Proof. Evidently the operator $\Delta f(x)$ is linear in the sense that

6.6
$$\begin{cases} \Delta(f(x) + g(x)) = \Delta f(x) + \Delta g(x) \\ \Delta(cf(x)) = c\Delta f(x). \end{cases}$$

Therefore, if

$$f(x) = a_0 x^n + a_1 x^{n-1} + \cdots + a_n$$

we have

$$\Delta f(x) = a_0 \Delta(x^n) + a_1 \Delta(x^{n-1}) + \cdots + \Delta(a_n).$$

But

6.7
$$\Delta(x^k) = (x + h)^k - x^k = khx^{k-1} + \cdots + h^k$$

which is a polynomial of degree $k - 1$. Therefore $\Delta f(x)$ is a polynomial of degree $n - 1$. Similarly, $\Delta^2 f(x)$ is a polynomial of degree $n - 2$, etc. Finally, we have $\Delta^n f(x)$ is a polynomial of degree zero (i.e., a constant) and $\Delta^{n+1} f(x) \equiv 0$. This completes the proof of Theorem 6.5.

We now show that

6.8 $\quad \Delta^k f(x) = f(x + kh) - \binom{k}{1} f(x + (k - 1)h) + \binom{k}{2} f(x + (k - 2)h)$

$$+ \cdots + (-1)^s \binom{k}{s} f(x + (k - s)h) + \cdots + (-1)^k f(x)$$

where the binomial coefficients $\binom{k}{s}$ are given by

6.9 $$\binom{k}{s} = \frac{k(k - 1) \ldots (k - s + 1)}{s!} = \frac{k!}{s!(k - s)!}$$

for nonnegative integers k and s. Evidently, the above result is true for $k = 1$ since $\Delta f(x) = f(x + h) - f(x)$. Assume it is true for k and then consider

6.10 $\quad \Delta^{k+1} f(x) = \Delta(\Delta^k f(x)) = \Delta f(x + kh) - \binom{k}{1} \Delta f(x + (k - 1)h)$

$$+ \binom{k}{2} \Delta f(x + (k - 2)h) + \cdots + (-1)^s \binom{k}{s} \Delta f(x + (k - s)h)$$

$$+ \cdots + (-1)^k \Delta f(x)$$

$$= f(x + (k + 1)h) - f(x + kh) - \binom{k}{1} [f(x + kh) - f(x + (k - 1)h)]$$

$$+ \binom{k}{2} [f(x + (k - 1)h) - f(x + (k - 2)h)]$$

$$+ \cdots + (-1)^s \binom{k}{s} [f(x + (k - s + 1)h) - f(x + (k - s)h)]$$

$$+ \cdots + (-1)^k f(x + h) + (-1)^{k+1} f(x).$$

The result (6.8) now follows since

6.11 $$\binom{k}{s} + \binom{k}{s - 1} = \binom{k + 1}{s}.$$

We can use (6.8) to help locate errors in a table. Thus, if an error of ε is made at any tabulated value, this error will be propagated through the table as indicated in Tables 6.12 and 6.13.

6.12 Table

x	f	Δf	$\Delta^2 f$	$\Delta^3 f$	$\Delta^4 f$	$\Delta^5 f$	$\Delta^6 f$	$\Delta^7 f$	$\Delta^8 f$
x_0	f_0								
		Δf_0							
x_1	f_1		$\Delta^2 f_0$						
		Δf_1		$\Delta^3 f_0$					
x_2	f_2		$\Delta^2 f_1$		$\Delta^4 f_0 + \varepsilon$				
		Δf_2		$\Delta^3 f_1 + \varepsilon$		$\Delta^5 f_0 - 5\varepsilon$			
x_3	f_3		$\Delta^2 f_2 + \varepsilon$		$\Delta^4 f_1 - 4\varepsilon$		$\Delta^6 f_0 + 15\varepsilon$		
		$\Delta f_3 + \varepsilon$		$\Delta^3 f_2 - 3\varepsilon$		$\Delta^5 f_1 + 10\varepsilon$		$\Delta^7 f_0 - 35\varepsilon$	
x_4	$f_4 + \varepsilon$		$\Delta^2 f_3 - 2\varepsilon$		$\Delta^4 f_2 + 6\varepsilon$		$\Delta^6 f_1 - 20\varepsilon$		$\Delta^8 f_0 + 70\varepsilon$
		$\Delta f_4 - \varepsilon$		$\Delta^3 f_3 + 3\varepsilon$		$\Delta^5 f_2 - 10\varepsilon$		$\Delta^7 f_1 + 35\varepsilon$	
x_5	f_5		$\Delta^2 f_4 + \varepsilon$		$\Delta^4 f_3 - 4\varepsilon$		$\Delta^6 f_2 + 15\varepsilon$		
		Δf_5		$\Delta^3 f_4 - \varepsilon$		$\Delta^5 f_3 + 5\varepsilon$			
x_6	f_6		$\Delta^2 f_5$		$\Delta^4 f_4 + \varepsilon$				
		Δf_6		$\Delta^3 f_5$					
x_7	f_7		$\Delta^2 f_6$						
		Δf_7							
x_8	f_8								

We note from the preceding table that the coefficients of the error terms in each column are the coefficients of the binomial expansion. We have

6.13 Table

f	Δf	$\Delta^2 f$	$\Delta^3 f$	$\Delta^4 f$	$\Delta^5 f$	$\Delta^6 f$	$\Delta^7 f$	$\Delta^8 f$
				$+\varepsilon$				
			$+\varepsilon$		-5ε			
		$+\varepsilon$		-4ε		$+15\varepsilon$		
	$+\varepsilon$		-3ε		$+10\varepsilon$		-35ε	
$+\varepsilon$		-2ε		$+6\varepsilon$		-20ε		$+70\varepsilon$
	$-\varepsilon$		$+3\varepsilon$		-10ε		$+35\varepsilon$	
		$+\varepsilon$		-4ε		$+15\varepsilon$		
			$-\varepsilon$		$+5\varepsilon$			
				$+\varepsilon$				

Evidently, the maximum error occurs directly opposite the entry whose functional value is in error.

The following example is presented to illustrate the detection of an error in a table.

6.14   Example: $f(x) = x^3$

By Theorem 6.5, the values in the $\Delta^4 f(x)$ column should be close to zero. Furthermore, -30 is the greatest deviation from zero; if there were only one error in the

6.15 Table

x	$f(x)$	$\Delta f(x)$	$\Delta f^2(x)$	$\Delta^3 f(x)$	$\Delta^4 f(x)$
1	1				
		7			
2	8		12		
		19		6	
3	27		18		−5
		37		1	
4	64		19		20
		56		21	
5	120		40		−30
		96		−9	
6	216		31		20
		127		11	
7	343		42		
		169			
8	512				

table, the error would be in the functional value opposite the entry -30, namely, 120. From Table 6.13, we have that the coefficient of the error term is $+6$. Therefore, $-30 - +6\varepsilon$ and $\varepsilon = -5$. Thus, if a is the correct value for $x = 5$, then $a + \varepsilon = 120$, and $a = 125$. Changing the 120 to a 125, we obtain

x	f	Δf	$\Delta^2 f$	$\Delta^3 f$	$\Delta^4 f$
1	1				
		7			
2	8		12		
		19		6	
3	27		18		0
		37		6	
4	64		24		0
		61		6	
5	125		30		0
		91		6	
6	216		36		0
		127		6	
7	343		42		
		169			
8	512				

Suppose now, however, that the table only extends through $x = 6$. The portion below the dotted line in Table 6.15 would be missing. The largest entry under $\Delta^4 f(x)$ would be 20, but the error would not be in the entry opposite the 20, namely, that corresponding to $x = 4$. One would have to recall that the error pattern for fourth differences would be

$$
\begin{array}{c}
\varepsilon \\
-4\varepsilon \\
6\varepsilon \\
-4\varepsilon \\
\varepsilon.
\end{array}
$$

One recognizes that the actual pattern

$$
\begin{array}{c}
-5 \\
20
\end{array}
$$

fits a portion of the theoretical pattern

$$
\begin{array}{c}
\varepsilon \\
-4\varepsilon
\end{array}
$$

provided $\varepsilon = -5$. Moreover, the error clearly occurs in the functional value opposite where the 6ε would appear.

The preceding analysis was based on the case where there is only one error in the tabulated values. A similar analysis, although more complicated, can be made when there is more than one error. The reader is referred to pages 42–43 of Kunz [1957] for an example illustrating the effect and detection of two errors in a table.

The Gregory-Newton Interpolation Formula

Let us introduce the operator

6.16 $Ef(x) = f(x + h)$.

Evidently,

6.17 $\Delta f(x) = Ef(x) - f(x) = (E - I)f(x)$

where I is the identity operator, i.e.,

6.18 $If(x) = f(x)$.

Evidently

$$E^2 f(x) = E(Ef(x)) = f(x + 2h), \text{ etc.}$$

Thus we can formally derive (6.8) by using the fact that

6.19 $\Delta = E - I$

and writing

6.20 $$\Delta^k f(x) = (E - I)^k f(x).$$

Expanding (6.19) by the binomial theorem, we have

6.21 $$(E - I)^k = E^k - \binom{k}{1} E^{k-1} + \binom{k}{2} E^{k-2} + \cdots + (-1)^k I$$

and this leads to (6.8). We again note that this is a *formal* derivation of (6.8). A rigorous derivation has already been given.

Since $E = I + \Delta$, we have

6.22 $$E^k = I + \binom{k}{1} \Delta + \binom{k}{2} \Delta^2 + \cdots + \binom{k}{k} \Delta^k$$

and we are led to the formula

6.23 $$f(x + kh) = f(x) + \binom{k}{1} \Delta f(x) + \binom{k}{2} \Delta^2 f(x) + \cdots + \binom{k}{k} \Delta^k f(x).$$

A rigorous proof can be given by induction. Clearly, the result is true if $k = 1$ since $f(x + h) = f(x) + \Delta f(x)$. Assuming it is true for k, then

6.24 $$f(x + (k + 1)h) = f(x + kh) + \Delta f(x + kh)$$

$$= \left[f(x) + \binom{k}{1} \Delta f(x) + \binom{k}{2} \Delta^2 f(x) + \cdots + \binom{k}{k} \Delta^k f(x) \right]$$

$$+ \left[\Delta f(x) + \binom{k}{1} \Delta^2 f(x) + \cdots + \binom{k}{k-1} \Delta^k(x) \right]$$

$$+ \binom{k}{k} \Delta^{k+1} f(x).$$

The result now follows by (6.11).

Consider now the problem of interpolating $f(x)$ based on values at $x_0, x_1, \ldots, x_n$. Evidently the values of $f(x_0), f(x_1), \ldots, f(x_n)$ can be used to determine $\Delta f(x_0)$, $\Delta^2 f(x_0), \ldots, \Delta^n f(x_0)$. Consider now the function

6.25 $$F(x) = f(x_0) + \binom{u}{1} \Delta f(x_0) + \binom{u}{2} \Delta^2 f(x_0) + \cdots + \binom{u}{n} \Delta^n f(x_0),$$

where u is given by (6.4) and for each positive integer s

6.26 $$\binom{u}{s} = \frac{u(u - 1) \ldots (u - s + 1)}{s!}.$$

Evidently, by (6.23) we have

6.27 $F(x_k) = f(x_k), \qquad k = 0, 1, 2, \ldots, n.$

Moreover, $F(x)$ is a polynomial in x of degree n or less. Hence $F(x)$ is the same as the Lagrange interpolation polynomial.

As an example, let us again consider the case $f(0) = 1$, $f(1) = 2$, $f(2) = 4$. We construct the difference table

n	x	$f(x)$	$\Delta f(x)$	$\Delta^2 f(x)$
0	0	1		
			1	
1	1	2		1
			2	
2	2	4		

Since $h = 1$ and $x_0 = 0$, we have $u = (x - x_0)/h = x$, and

6.28 $F(x) = f(x_0) + x\Delta f(x_0) + \dfrac{x(x-1)}{2}\Delta^2 f(x_0) = 1 + x(1) + \dfrac{x(x-1)}{2}(1)$

$= 1 + \tfrac{1}{2}x + \tfrac{1}{2}x^2$

which agrees with the result previously found by Lagrangian interpolation.

In practice, one would not express $F(x)$ in powers of x. For example, if $F(\tfrac{1}{2})$ were desired, one would compute

6.29 $F(\tfrac{1}{2}) = f(x_0) + \tfrac{1}{2}\Delta(x_0) + \dfrac{\frac{1}{2}(\frac{1}{2} - 1)}{2}\Delta^2 f(x_0)$

$= 1 + \tfrac{1}{2}(1) + (-\tfrac{1}{8})(1) = \tfrac{11}{8} = 1.375.$

We now give an alternate derivation of the Gregory-Newton formula. Let us introduce the notation

6.30 $u^{[k]} = \begin{pmatrix} u \\ k \end{pmatrix} k! = \begin{cases} u(u-1)\ldots(u-k+1), & k \geq 1 \\ 1, & k = 0. \end{cases}$

Since $u^{[k]} = u^k + P_{k-1}(u)$ where $P_{k-1}(u)$ is a polynomial in u of degree $k - 1$, it follows that for each integer k we can express u^k as a linear combination of

$u^{[0]}, u^{[1]}, \ldots, u^{[k]}$. Thus, for instance, we have

6.31 $\begin{cases} u^{[0]} = 1, & 1 = u^{[0]} \\ u^{[1]} = u, & u = u^{[1]} \\ u^{[2]} = u(u-1) = u^2 - u, & u^2 = u^{[2]} + u^{[1]} \\ u^{[3]} = u(u-1)(u-2) = u^3 - 3u^2 + 2u, & u^3 = u^{[3]} + 3u^{[2]} + u^{[1]} \\ \text{etc.} \end{cases}$

Thus, our polynomial $F(x)$ can be written in the form

6.32 $$F(x) = a_0 u^{[n]} + a_1 u^{[n-1]} + \cdots + a_n$$

where u is given by (6.4). It is easily verified that $u(x + h) = u(x) + 1$ and that

6.33 $$\Delta u^{[k]} = (u + 1)^{[k]} - u^{[k]} = k u^{[k-1]}.$$

Letting $u = 0$ in (6.32) we have

6.34 $$a_n = F(x_0) = f(x_0).$$

Next, applying Δ to both sides of (6.32) and letting $u = 0$ we get

6.35 $$a_{n-1} = \Delta F(x_0) = \Delta f(x_0).$$

Continuing this process we have

6.36 $$a_k = \frac{\Delta^{n-k} F(x_0)}{(n-k)!} = \frac{\Delta^{n-k} f(x_0)}{(n-k)!}, \qquad k = 0, 1, \ldots, n.$$

Substitution in (6.32) and using (6.30) we get (6.25).

For the actual evaluation of (6.32) we can use an analog of the nesting procedure for evaluating polynomials. Thus, to evaluate (6.32) we have

6.37 $\quad F(x) = (\ \ (\ \cdots (\ (\ (a_0)(u-n+1) + a_1)(u-n+2) + a_2) \cdots + a_{n-1})u + a_n)$
$\qquad\qquad {\scriptstyle n\ n-1} \qquad {\scriptstyle 2\ 1\ 0\ \ 0} \qquad\qquad {\scriptstyle 1} \qquad\qquad {\scriptstyle 2} \qquad\qquad\quad {\scriptstyle n-1} \qquad {\scriptstyle n}$

and we can use the following formulas

6.38 $\begin{cases} z_0 = a_0 \\ z_k = (u - n + k)z_{k-1} + a_k, \qquad k = 1, 2, \ldots, n \end{cases}$

where

6.39 $$F(x) = z_n.$$

Normally, if one is given a table of values of a function $f(x)$, the function is given to a fixed number of decimal places. Moreover, the number of significant digits is usually less than that which can be represented on the machine (except for the loss of accuracy in converting from decimal to binary). Thus, there will in general not be a significant loss of accuracy in the computation of the differences $\Delta f(x_0)$, $\Delta^2 f(x_0)$, ... in spite of the fact that one is subtracting nearly equal quantities. Thus the result of using (6.32) can, with the nesting procedure, be expected to be quite accurate in most cases.

In some cases it is useful to express $F(x)$ as a polynomial in $(x - x_0)$, i.e., to determine the coefficients $\hat{a}_0, \hat{a}_1, \ldots, \hat{a}_n$ such that

6.40 $$F(x) = \hat{a}_0(x - x_0)^n + \hat{a}_1(x - x_0)^{n-1} + \cdots + \hat{a}_n.$$

One procedure for finding the $\hat{a}_0$ would be to first compute the a_k in (6.32) using (6.36), after having computed the differences $\Delta f(x_0)$, $\Delta^2 f(x_0)$, ..., $\Delta^n f(x_0)$. We can express each $u^{[k]}$ in terms of powers of u and then group powers of u. Thus, in the case $n = 3$, we have by (6.31)

6.41 $$F(x) = a_0 u^{[3]} + a_1 u^{[2]} + a_2 u^{[1]} + a_3 u^{[0]}$$

$$= a_0(u^3 - 3u^2 + 2u) + a_1(u^2 - u) + a_2(u) + a_3$$

$$= a_0 u^3 + (a_1 - 3a_0)u^2 + (a_2 - a_1 + 2a_0)u + a_3$$

so that, by (6.26),

6.42
$$
\begin{cases}
h^3 \hat{a}_0 = a_0 = \dfrac{\Delta^3 f(x_0)}{6} = \dfrac{f(x_3) - 3f(x_2) + 3f(x_1) - f(x_0)}{6} \\[3mm]
h^2 \hat{a}_1 = a_1 - 3a_0 = \dfrac{\Delta^2 f(x_0)}{2} - 3\dfrac{\Delta^3 f(x_0)}{6} = \dfrac{-f(x_3) + 4f(x_2) - 5f(x_1) + 2f(x_0)}{2} \\[3mm]
h \hat{a}_2 = a_2 - a_1 + 2a_0 = \Delta f(x_0) - \dfrac{\Delta^2 f(x_0)}{2} + \dfrac{\Delta^3 f(x_0)}{3} \\[3mm]
\qquad = \dfrac{2f(x_3) - 9f(x_2) + 18f(x_1) - 11f(x_0)}{6} \\[3mm]
\hat{a}_3 = f(x_0).
\end{cases}
$$

It may be more convenient to compute the $\hat{a}_k$ using (6.42) rather than computing the a_k first.

Central Difference Interpolation Formulas

If one were to do extensive computations by hand, one would probably wish to use interpolation formulas based on central differences rather than the Gregory-

Newton formula. This alone would not justify mention of central difference formulas in this book. However, central differences are used frequently in books and papers on numerical analysis and some familiarity with them seems desirable.

As before, assume that $f(x)$ is given for equally spaced values of x, namely, $x_0, x_1, \ldots, x_n$ where (6.1) holds. We define the operator δ by

6.43
$$\delta f(x) = f(x + h/2) - f(x - h/2).$$

Evidently $\delta f(x)$ is only defined for $x_{k+1/2}, k = 0, 1, \ldots, n - 1$. Similarly, we let

6.44
$$\delta^k f(x) = \delta(\delta^{k-1} f(x)).$$

The construction of a central difference table is identical with that of a forward difference table in that one obtains the same numbers. However, the *labeling* of the numbers in the difference table is different. Thus, if $n = 5$ we have the table

k	x	$f(x)$	$\delta f(x)$	$\delta^2 f(x)$	$\delta^3 f(x)$	$\delta^4 f(x)$	$\delta^5 f(x)$
0	x_0	$f(x_0)$					
			$\delta f(x_{1/2})$				
1	x_1	$f(x_1)$		$\delta^2 f(x_1)$			
			$\delta f(x_{3/2})$		$\delta^3 f(x_{3/2})$		
2	x_2	$f(x_2)$		$\delta^2 f(x_2)$		$\delta^4 f(x_2)$	
			$\delta f(x_{5/2})$		$\delta^3 f(x_{5/2})$		$\delta^5 f(x_{5/2})$
3	x_3	$f(x_3)$		$\delta^2 f(x_3)$		$\delta^4 f(x_3)$	
			$\delta f(x_{7/2})$		$\delta^3 f(x_{7/2})$		
4	x_4	$f(x_4)$		$\delta^2 f(x_4)$			
			$\delta f(x_{9/2})$				
5	x_5	$f(x_5)$					

We note that elements in the table on the same horizontal line have the same subscript.

We now consider two central difference formulas, Stirling's formula and Bessel's formula. To decide which to use for given x, first find the argument $\bar{x}$ in the table which is closest to x, and compute

6.45
$$u = \frac{x - \bar{x}}{h}.$$

If $|u| \leq \frac{1}{4}$, use Stirling's formula and label $\bar{x} = x_0$. Then consider the interpolation points $x_{-p}, x_{-(p-1)}, \ldots, x_{-1}, x_0, x_1, \ldots, x_{p-1}, x_p$ for some $p \geq 0$. Stirling's formula for interpolation is

6.46 $$F(x) = f(x_0) + uN_1f(x_0) + \frac{u^2}{2!}\delta^2f(x_0) + \frac{u(u^2-1)}{3!}N_3f(x_0) + \cdots$$

$$+ \frac{u(u^2-1)(u^2-2^2)\cdots(u^2-(p-1)^2)}{(2p-1)!}N_{2p-1}f(x_0)$$

$$+ \frac{u^2(u^2-1)(u^2-2^2)\cdots(u^2-(p-1)^2)}{(2p)!}\delta^{2p}f(x_0)$$

where

6.47 $$N_kf(x_0) = \tfrac{1}{2}[\delta^kf(x_{1/2}) + \delta^kf(x_{-1/2})], \qquad k = 1, 3, 5, \ldots$$

We illustrate by giving the difference table for the case $p = 2$.

k	x	$f(x)$	$\delta f(x)$	$\delta^2 f(x)$	$\delta^3 f(x)$	$\delta^4 f(x)$
-2	x_{-2}	$f(x_{-2})$				
			$\delta f(x_{-3/2})$			
-1	x_{-1}	$f(x_{-1})$		$\delta^2 f(x_{-1})$		
			$\boxed{\delta f(x_{-1/2})}$		$\boxed{\delta^3 f(x_{-1/2})}$	
0	x_0	$\boxed{f(x_0)}$		$\boxed{\delta^2 f(x_0)}$		$\boxed{\delta^4 f(x_0)}$
			$\boxed{\delta f(x_{1/2})}$		$\boxed{\delta^2 f(x_{1/2})}$	
1	x_1	$f(x_1)$		$\delta^2 f(x_1)$		
			$\delta f(x_{3/2})$			
2	x_2	$f(x_2)$				

The values which are circled are those used in the interpolation formula.

If $\frac{1}{4} < |u| \leq \frac{1}{2}$, we use Bessel's interpolation formula. We label the interpolation points nearest to x as x_0 and x_1 with $x_0 < x_1$. Evidently x is in the interval $x_0 \leq x \leq x_1$. We consider the interpolation points $x_{-p}, x_{-(p-1)}, \ldots, x_{-1}, x_0, x_1, \ldots, x_{p+1}$. Bessel's interpolation formula is given by

6.48 $$F(x) = N_0f(x_{1/2}) + v\delta f(x_{1/2}) + \frac{(v^2-\frac{1}{4})}{2!}N_2f(x_{1/2}) + \frac{v(v^2-\frac{1}{4})}{3!}\delta^3f(x_{1/2})$$

$$+ \cdots + \frac{(v^2-\frac{1}{4})(v^2-\frac{9}{4})\cdots(v^2-(2p-1)^2/4)}{(2p)!}N_{2p}f(x_{1/2})$$

$$+ \frac{v(v^2-\frac{1}{4})(v^2-\frac{9}{4})\cdots(v^2-(2p-1)^2/4)}{(2p+1)!}\delta^{2p+1}f(x_{1/2})$$

where

6.49 $$v = u - \tfrac{1}{2}$$

6.50 $$\dot{N}_k f(x_{1/2}) = \tfrac{1}{2}[\delta^k f(x_0) + \delta^k f(x_1)], \qquad k = 0, 2, 4, \dots .$$

We illustrate by giving the difference table for the case $p = 2$.

k	x	$f(x)$	$\delta f(x)$	$\delta^2 f(x)$	$\delta^3 f(x)$	$\delta^4(x)$	$\delta^5 f(x)$
-2	x_{-2}	$f(x_{-2})$					
			$\delta f(x_{-3/2})$				
-1	x_{-1}	$f(x_{-1})$		$\delta^2 f(x_{-1})$			
			$\delta f(x_{-1/2})$		$\delta^3 f(x_{-1/2})$		
0	x_0	$\boxed{f(x_0)}$		$\boxed{\delta^2 f(x_0)}$		$\boxed{\delta^4 f(x_0)}$	
			$\boxed{\delta f(x_{1/2})}$		$\boxed{\delta^3 f(x_{1/2})}$		$\boxed{\delta^5 f(x_{1/2})}$
1	x_1	$\boxed{f(x_1)}$		$\boxed{\delta^2 f(x_1)}$		$\boxed{\delta^4 f(x_1)}$	
			$\delta f(x_{3/2})$		$\delta^3 f(x_{3/2})$		
2	x_2	$f(x_2)$		$\delta^2 f(x_2)$			
			$\delta f(x_{5/2})$				
3	x_3	$f(x_3)$					

The values which are circled are those used in the interpolation formulas.
Suppose we are given the following table of values

x	0	5	10	15	20
$f(x)$	0	0.08716	0.17365	0.25882	0.34202

If we desire $f(12)$, then $\bar{x} = 10$ and $u = \tfrac{2}{5} = 0.40$. This means that Bessel's formula would be used with $x_0 = 10$, $x_1 = 15$. One could let $p = 0$ and have just two interpolation points or one could let $p = 1$ and let $x_{-1} = 5$, $x_0 = 10$, $x_1 = 15$, $x_2 = 20$. The number v would equal -0.1.

If we desired $f(9)$, then $\bar{x} = 10$ and $u = -0.2$. Hence we would use Stirling's formula with one of the following sets of points

$p = 0:\ \ x_0 = 10$

$p = 1: x_{-1} = \ \ 5, \qquad x_0 = 10, \qquad x_1 = 15$

$p = 2: x_{-2} = \ \ 0, \qquad x_{-1} = \ \ 5, \qquad x_0 = 10, \qquad x_1 = 15, \qquad x_2 = 20.$

Central difference interpolation formulas have several advantages as compared with the Gregory-Newton formulas. It is true that interpolation through a fixed set of points should yield the same result no matter which interpolation formula is used. However, in the case of the forward or backward Gregory-Newton

formula, the value of u would have to be large, in general, if the points of interpolation are to be approximately symmetrically spaced about the value x for
which $f(x)$ is desired. This means that the terms would tend to zero much slower
than in the case of central difference formulas where u, or v, is small when the
interpolation points are approximately symmetrically spaced about x. If, on the
other hand, one chooses the interpolation points so that the u used in the Gregory-
Newton formula will be small, then the value of x will be near the end of the range
covered by the points of interpolation. This will result in a larger error since the
quantity

6.51 $$Q_n(x) = (x - x_0)(x - x_1) \cdots (x - x_n)$$

which appears in the error formula (5.14) has larger extrema in intervals
$x_i \leq x \leq x_{i+1}$ near the end of the range than in the middle intervals. Thus the
overall error given by

6.52 $$\frac{(x - x_0)(x - x_1) \cdots (x - x_n)}{(n + 1)!} f^{(n+1)}(c)$$

would normally* be larger for the Gregory-Newton formula where u is small than
for the central difference formulas which use a different set of x_i, where u or v is
small.

Estimation of the Accuracy of Interpolation

Let us now consider the accuracy of the interpolating polynomial $F(x)$ as a
representation of $f(x)$ when $F(x)$ is a polynomial of degree n or less which agrees
with $f(x)$ at $n + 1$ evenly spaced points $x_k = x_0 + kh, k = 0, 1, \ldots, n$. As we have
seen in Section 6.5 for various values of n one can obtain good bounds on $|Q_n(x)|$
in the interval $x_0 \leq x \leq x_n$, where

6.53 $$Q_n(x) = \prod_{k=0}^{n} (x - x_k).$$

The most difficult problem is that of estimating a bound on the $(n + 1)$-st derivative of $f(x)$ which appears in (5.14). If $f(x)$ is a polynomial of degree $n + 1$ or less,

6.54 $$f(x) = a_0 x^{n+1} + a_1 x^n + \cdots + a_{n+1}$$

then, as in the proof of Theorem 6.5, we have

6.55 $$\Delta^{n+1} f(x) = h^{n+1}(n + 1)! a_0 = h^{n+1} f^{(n+1)}(x)$$

---

* Assuming that $f^{(n+1)}(x)$ does not vary appreciably.

and hence

6.56
$$f^{(n+1)}(x) \equiv \frac{\Delta^{n+1} f(x)}{h^{n+1}}.$$

Thus in this case we can determine $f^{(n+1)}(x)$ exactly by dividing the $(n+1)$-st difference of $f(x)$ by h^{n+1}.

In the more general case, if $f(x) \in C^{(n+1)}$ for $x_0 \leq x \leq x_0 + h$ for some $h > 0$ then we can show that

6.57
$$\lim_{h \to 0} \frac{\Delta^{n+1} f(x_0)}{h^{n+1}} = f^{(n+1)}(x_0).$$

This provides some justification for replacing M_{n+1} by $\Delta^{n+1} f(x) h^{-(n+1)}$ in Corollary 5.16.

We give a proof of (6.57) for the case $n = 2$. By Taylor's theorem we have for $k = 0, 1, 2, 3$

6.58
$$f(x_0 + kh) = f(x_0) + khf'(x_0) + \frac{k^2 h^2}{2} f^{(2)}(x_0) + \frac{k^3 h^3}{6} f^{(3)}(c_k)$$

where $x_0 \leq c_k \leq x_0 + kh$. Therefore, by (6.8),

6.59
$$\Delta^3 f(x_0) = \frac{h^3}{6} [27 f^{(3)}(c_3) - 24 f^{(3)}(c_2) + 3 f^{(3)}(c_1)],$$

and by the continuity of $f^{(3)}(x)$ we have

6.60
$$\lim_{h \to 0} \frac{\Delta^3 f(x_0)}{h^3} = f^{(3)}(x_0).$$

Thus (6.57) holds for the case $n = 2$.

Some caution must be exercised however. Merely because the values of $\Delta^{n+1} f(x)$ are small it does not necessarily follow that M_{n+1} is small. For instance, if the function

6.61
$$f(x) = 1 + \sin 10\pi x$$

is tabulated for $x = 0(0.1)2$, we would find that $f(x) = 1$ for all tabulated values of x. Hence if we were to use linear interpolation, we might erroneously conclude that the result was exact since $\Delta^2 f(x) = 0$. Of course, it is unlikely that one would encounter a table in practice where there is so much variation of the function between interpolation points. Nevertheless, the reader should be aware that the possibility exists.

EXERCISES 6.6

1. Given the table

$$x:\quad 0\quad 5\quad\quad 10\quad\quad 15\quad\quad 20$$
$$f(x):\quad 0\quad 0.08716\quad 0.17365\quad 0.25882\quad 0.34202$$

find $f(11)$ by the Gregory-Newton interpolation formula and also by the most appropriate central difference interpolation formula.

2. Given the table

$$x:\quad 50\quad\quad 60\quad\quad 70\quad\quad 80$$
$$f(x):\quad 3.684\quad 3.915\quad 4.121\quad 4.309$$

find $f(63)$ by the Gregory-Newton interpolation formula and by linear interpolation.

3. Let $f(x)$ be a function such that

$$f(0) = 1.25,\quad f(\tfrac{1}{2}) = 1.75,\quad f(1) = 2.10.$$

Let $F(x)$ be a polynomial of degree 2 or less such that $F(0) = f(0)$, $F(\tfrac{1}{2}) = f(\tfrac{1}{2})$, $F(1) = f(1)$. Find $F(x)$ and $F(\tfrac{1}{4})$

a) by the Lagrangian interpolation formula,
b) by the Gregory-Newton interpolation formula.

4. Using the values of $\log x$ given in Exercise 4, Section 6.4, find $\log 3.53$ by the Gregory-Newton interpolation formula using all four values given. Give an estimate for the accuracy of three-point Lagrangian interpolation based on the use of a suitable difference quotient for the third derivative. Compare the error estimate with the exact error.

5. Tabulate $f(x) = 3x^3 + 2x^2 + 5$ for $x = 0(1)6$. Construct a difference table and verify that all differences of order 4 and higher are zero.

6. Find any error that may be present in the following table

x	$f(x)$
0.0	50.326
0.1	50.113
0.2	49.757
0.3	49.263
0.4	48.837
0.5	47.885
0.6	47.019
0.7	46.033
0.8	44.951
0.9	43.780
1.0	42.532

7. Given the following part of a difference table, fill in the blank entries.

x	$f(x)$	Δ	Δ^2	Δ^3	Δ^4
0	16.121	____			
0.1	____		-0.021		
0.2	____	____		-0.001	
0.3	____	0.032	____	0.005	
0.4	____	____			

8. Prove (6.11).

9. Estimate the error in linear interpolation in computing $f(0.26)$ using a table of $f(x) = e^x$ with $x = 0(0.1)1.0$. Estimate M_2 of (3.12) using the appropriate second-difference quotient.

10. Verify that if

$$f(x) = 3x^3 - 2x + 5$$

then

$$f^{(3)}(x) \equiv \frac{\Delta^3 f(x)}{h^3}$$

by constructing a table of $f(x)$ for $x = 0(0.1)0.5$ and computing $\Delta^3 f(x)$. Let $F(x)$ be a polynomial of degree two or less which agrees with $F(x)$ for $x = 0, 0.1, 0.2$. Compute $F(0.28)$ and compare $|f(0.28) - F(0.28)|$ with the value based on the use of Corollary 5.19 and (6.56).

11. Verify Stirling's formula (6.46) for the case $p = 2$ and verify Bessel's formula (6.48) for the case $p = 2$.

12. Show that if $F(x) = a_0 u^{[n]} + a_1 u^{[n-1]} + \cdots + a_n$ where $u = (x - x_0)/h$ then $\Delta F(x) = w_n(x)$ where

$$w_0 = 0$$
$$w_k(x) = (u - n + k + 1)w_{k-1}(x) + z_{k-1}(x), \qquad k = 1, 2, \ldots, n$$

and where

$$z_0(x) = a_0$$
$$z_k(x) = (u - n + k)z_{k-1}(x) + a_k, \qquad k = 1, 2, \ldots, n.$$

13. Evaluate the function

$$2u^{[4]} - 3u^{[2]} + 5u^{[1]} + 6$$

for $u = \frac{1}{2}$ by the nesting procedure.

14. Express u^4 in terms of $u^{[0]}, u^{[1]}, \ldots, u^{[4]}$.

15. Using the Gregory-Newton interpolation formula, express the unique polynomial of degree 3 or less such that

$$F(0.5) = 2, \quad F(0.6) = 8, \quad F(0.7) = -2, \quad F(0.8) = 5$$

in the form

$$F(x) = \hat{a}_0(x - 0.5)^3 + \hat{a}_1(x - 0.5)^2 + \hat{a}_2(x - 0.5) + \hat{a}_3.$$

Use (6.42) and also the procedure based on determining the a_k of (6.32) by (6.36) and representing the $u^{[k]}$ in powers of u.

16. Derive (6.42) by the use of the Lagrangian interpolation formula.

6.7 HERMITE INTERPOLATION

With Hermite interpolation one requires not only that $F(x)$ agree with $f(x)$ at the interpolation points $x_0, x_1, \ldots, x_n$ but also requires that certain derivatives of $F(x)$ agree with those of $f(x)$ at some of the points. More precisely, one chooses nonnegative integers $\gamma_0, \gamma_1, \ldots, \gamma_n$ and requires that for each $k = 0, 1, \ldots, n$, $F(x)$ and all derivatives up to and including the γ_k-th agree with $f(x)$ and the corresponding derivatives, respectively. Thus we have

7.1 $F(x_k) = f(x_k), \qquad k = 0, 1, 2, \ldots, n$

and for each k such that $\gamma_k \geqq 1$, we also have

7.2 $F^{(j)}(x_k) = f^{(j)}(x_k), \qquad j = 1, 2, \ldots, \gamma_k.$

We now show* that there exists a unique polynomial $F(x)$ of degree m or less where

7.3 $$m = n + \sum_{j=0}^{n} \gamma_j$$

and which satisfies (7.1) and (7.2). Consider the problem of finding the $m + 1$ coefficients $a_0, a_1, \ldots, a_m$ such that the $m + 1$ conditions (7.1) and (7.2) hold for

7.4 $$F(x) = \sum_{j=0}^{m} a_j(x - \alpha)^{m-j}$$

where α is any constant. We have a system of $m + 1$ linear equations with $m + 1$ unknowns. Consider the homogeneous system formed by replacing all right-hand sides of (7.1) and (7.2) by zeros. For any solution $a_0, a_1, \ldots, a_m$ of the homogeneous system let $F(x)$ be given by (7.4). Evidently $F(x)$ has a zero of multiplicity $\gamma_k + 1$ at x_k for $k = 0, 1, \ldots, n$. Hence $F(x)$, which is a polynomial of degree m or less, has $m + 1$ zeros and $F(x) \equiv 0$. Therefore $a_0 = a_1 = \cdots = a_m = 0$. By Theorem A-8.7 the fact that the homogeneous system has only the trivial solution $a_0 = a_1 = a \ldots a_m = 0$ implies that a unique solution of the nonhomogeneous system exists. (See Theorem A-7.9.)

_____

* See Davis [1963, p. 28] where various alternative sets of conditions are considered.

We now give an explicit expression for the case where each γ_k is either zero or one. In this case the solution is

7.5 $\quad F(x) = \sum_{k=0}^{n} \frac{Q_k(x)}{Q_k(x_k)} \left\{ \left[1 - \gamma_k(x - x_k) \frac{Q_k'(x_k)}{Q_k(x_k)} \right] f(x_k) + \gamma_k(x - x_k) f'(x_k) \right\}$

where

7.6 $\qquad\qquad\qquad Q_k(x) = \prod_{\substack{j=0 \\ j \neq k}}^{n} (x - x_j)^{\gamma_j + 1}$

Evidently $Q_k(x)$ is a polynomial of degree

7.7 $\qquad\qquad \sum_{j=0}^{n} \gamma_j + n + 1 - (\gamma_k + 1) = n + \sum_{j=0}^{n} \gamma_j - \gamma_k.$

Hence $F(x)$ is a polynomial of degree m or less where

7.8 $\qquad\qquad\qquad\qquad m = n + \sum_{j=0}^{n} \gamma_j.$

Moreover,

7.9 $\qquad\qquad\qquad F(x_s) = f(x_s), \qquad s = 0, 1, 2, \ldots, n$

and

7.10 $\quad F'(x) = \sum_{k=0}^{n} \frac{Q_k'(x)}{Q_k(x_k)} \left\{ \left[1 - \gamma_k(x - x_k) \frac{Q_k'(x_k)}{Q_k(x_k)} \right] f(x_k) + \gamma_k(x - x_k) f'(x_k) \right\}$

$\qquad\qquad + \sum_{k=0}^{n} \frac{Q_k(x)}{Q_k(x_k)} \left\{ - \gamma_k \frac{Q_k'(x_k)}{Q_k(x_k)} f(x_k) + \gamma_k f'(x_k) \right\}.$

Therefore, if $\gamma_s = 1$ we have $Q_k'(x_s) = 0$ if $s \neq k$, and

7.11 $\qquad\qquad\qquad F'(x_s) = f'(x_s), \qquad s = 0, 1, 2, \ldots, n.$

The polynomial $F(x)$ could be determined by the method of undetermined coefficients where we actually solve (7.1) and (7.2) for the coefficients $a_0, a_1, \ldots, a_m$ of (7.4).

Still another method is to use the method of undetermined weights. By (7.5) it follows that there exist weights $w_k(x)$ and $w_k^*(x)$, $k = 0, 1, \ldots, n$, such that

7.12 $\qquad\qquad\qquad F(x) = \sum_{k=0}^{n} (w_k(x) f(x_k) + w_k^*(x) f'(x_k)).$

In the case of the more general problem defined by (7.1)–(7.2) there exist weights $w_{k,j}(x)$, $k = 0, 1, \ldots, n, j = 0, 1, \ldots, \gamma_k$ such that

7.13

$$F(x) = \sum_{j=0}^{\gamma_0} w_{0,j}(x) f^{(j)}(x_0) + \sum_{j=0}^{\gamma_1} w_{1,j}(x) f^{(j)}(x_1)$$

$$+ \cdots + \sum_{j=0}^{\gamma_n} w_{n,j}(x) f^{(j)}(x_n)$$

$$= \sum_{k=0}^{n} \sum_{j=0}^{\gamma_k} w_{k,j}(x) f^{(j)}(x_k).$$

We determine the weights so that the formula is exact for $f(x) \equiv 1$, $(x - \alpha)$, $(x - \alpha)^2, \ldots, (x - \alpha)^m$. This leads to a system of $m + 1$ linear algebraic equations in $m + 1$ unknowns. The determinant of the system is the same as that of the system obtained using the method of undetermined coefficients and does not vanish. Hence the weights can be uniquely determined.

As an example, let us consider the case where $n = 1$, $\gamma_0 = \gamma_1 = 1$. Let $a = x_0$, $b = x_1$. By (7.6) we have $Q_1(x) = (x - b)^2$, $Q_2(x) = (x - a)^2$, and by (7.5)

7.14

$$F(x) = \frac{(x - b)^2}{(a - b)^2} \left\{ \left[1 - (x - a) \frac{2(a - b)}{(a - b)^2} \right] f(a) + (x - a) f'(a) \right\}$$

$$+ \frac{(x - a)^2}{(b - a)^2} \left\{ \left[1 - (x - b) \frac{2(b - a)}{(b - a)^2} \right] f(b) + (x - b) f'(b) \right\}$$

$$= \frac{(x - b)^2}{(b - a)^2} \left\{ \left[1 + \frac{2(x - a)}{b - a} \right] f(a) + (x - a) f'(a) \right\}$$

$$+ \frac{(x - a)^2}{(b - a)^2} \left\{ \left[1 - \frac{2(x - b)}{b - a} \right] f(b) + (x - b) f'(b) \right\}.$$

Evidently $F(x)$ is a polynomial of degree 3 or less and $F(a) = f(a)$, $F(b) = f(b)$, $F'(a) = f'(a)$, and $F'(b) = f'(b)$.

Let us now apply the method of undetermined coefficients. We have

7.15

$$F(x) = a_0(x - \alpha)^3 + a_1(x - \alpha)^2 + a_2(x - \alpha) + a_3$$

and

7.16

$$F'(x) = 3a_0(x - \alpha)^2 + 2a_1(x - \alpha) + a_2.$$

Letting $\alpha = a$ and $h = b - a$, we have

7.17
$$\begin{cases} f(a) = a_3 \\ f(b) = a_0 h^3 + a_1 h^2 + a_2 h + a_3 \\ f'(a) = a_2 \\ f'(b) = 3a_0 h^2 + 2a_1 h + a_2. \end{cases}$$

Solving, we get

7.18
$$\begin{cases} a_0 = \dfrac{2}{h^3}[f(a) - f(b)] + \dfrac{1}{h^2}[f'(a) + f'(b)] \\[2mm] a_1 = \dfrac{3}{h^2}[f(b) - f(a)] - \dfrac{1}{h}[2f'(a) + f'(b)] \\[2mm] a_2 = f'(a) \\[2mm] a_3 = f(a). \end{cases}$$

The reader should show that if one substitutes the coefficients thus obtained into (7.15) with $\alpha = a$ one obtains (7.14).

To apply the method of undetermined weights we seek to determine the weights $w_0(x)$, $w_1(x)$, $w_0^*(x)$, and $w_1^*(x)$ such that the relation

7.19 $f(x) = w_0(x)f(a) + w_1(x)f(b) + w_0^*(x)f'(a) + w_1^*(x)f'(b)$

is exact for the functions 1, $x - \alpha$, $(x - \alpha)^2$, and $(x - \alpha)^3$. Here, as in the case of the method of undetermined coefficients, it is convenient to let $\alpha = a$ rather than $(a + b)/2$. Evidently the weights must satisfy

7.20
$$\begin{cases} 1 = w_0(x) + w_1(x) \\ x - a = \quad w_1(x)h + w_0^*(x) + w_1^*(x) \\ (x - a)^2 = \quad w_1(x)h^2 + \quad\quad 2hw_1^*(x) \\ (x - a)^3 = \quad w_1(x)h^3 + \quad\quad 3h^2 w_1^*(x). \end{cases}$$

Solving, we obtain

7.21
$$\begin{cases} w_0(x) = h^{-3}(x - b)^2(b - 3a + 2x) \\ w_1(x) = h^{-3}(x - a)^2(3b - a - 2x) \\ w_0^*(x) = h^{-2}(x - a)(x - b)^2 \\ w_1^*(x) = h^{-2}(x - a)^2(x - b) \end{cases}$$

which agrees with (7.14).

An expression for the error for the Hermite interpolation formulas can be obtained by methods similar to those used for the case of Lagrangian interpolation. It can be shown that for some continuous function $P(x)$ we have

7.22
$$f(x) - F(x) = \prod_{i=0}^{n} (x - x_i)^{\gamma_i + 1} P(x)$$

and that

7.23
$$P(x) = \frac{1}{(m + 1)!} f^{(m+1)}(\xi)$$

where ξ lies in the interval spanned by the interpolation points. Thus we have

7.24
$$f(x) - F(x) = \frac{1}{(m + 1)!} \prod_{i=0}^{n} (x - x_i)^{\gamma_i + 1} f^{(m+1)}(\xi).$$

We give a proof of (7.22), (7.23), and (7.24) for the case of three interpolation points x_0, x_1, x_2 where

7.25
$$\gamma_0 = \gamma_2 = 0, \qquad \gamma_1 = 1.$$

Evidently by (7.5) we have

7.26
$$F(x) = \frac{(x - x_1)^2(x - x_2)}{(x_0 - x_1)^2(x_0 - x_2)} f(x_0) + \frac{(x - x_0)(x - x_2)}{(x_1 - x_0)(x_1 - x_2)}$$
$$\cdot \left\{ \left[1 - (x - x_1)\left(\frac{1}{x_1 - x_0} + \frac{1}{x_1 - x_2} \right) \right] f(x_1) + (x - x_1)f'(x_1) \right\}$$
$$+ \frac{(x - x_0)(x - x_1)^2}{(x_2 - x_0)(x_2 - x_1)^2} f(x_2).$$

Let us now define the function $P(x)$ by

7.27
$$P(x) = \begin{cases} \dfrac{f(x) - F(x)}{(x - x_0)(x - x_1)^2(x - x_2)}, & x \neq x_0, x \neq x_1, x \neq x_2 \\[2ex] \dfrac{f'(x_0) - F'(x_0)}{(x_0 - x_1)^2(x_0 - x_2)}, & x = x_0 \\[2ex] \dfrac{f''(x_1) - F''(x_1)}{2(x_1 - x_0)(x_1 - x_2)}, & x = x_1 \\[2ex] \dfrac{f'(x_2) - F'(x_2)}{(x_2 - x_0)(x_2 - x_1)^2}, & x = x_2. \end{cases}$$

By L'Hospital's rule (see Theorem A-3.52) it follows that $P(x)$ is a continuous function in the interval $I = [x_0, x_2]$.

We now seek to show that

7.28
$$P(x) = \frac{f^{(4)}(\xi)}{24}$$

for some ξ in I. For each x we consider the function

7.29
$$\phi(z; x) = f(z) - F(z) - (z - x_0)(z - x_1)^2(z - x_2)P(x).$$

If $x \neq x_0, x_1, x_2$, then $\phi(z; x)$ vanishes at the distinct abscissas x_0, x_1, x_2, and x. By Rolle's theorem (Theorem A-3.47), $\phi'(z; x)$ vanishes at the distinct abscissas ξ_1, ξ_2, and ξ_3 each of which is different from x_1. Since $\phi'(x_1; x)$ also vanishes, it follows by repeated application of Rolle's theorem that for some ξ in I we have

7.30
$$\phi^{(4)}(\xi; x) = 0$$

and hence (7.28) holds.

We therefore have

7.31
$$f(x) - F(x) = (x - x_0)(x - x_1)^2(x - x_2)P(x)$$

for some continuous function $P(x)$ such that (7.28) holds for some ξ in I. Moreover,

7.32
$$f(x) - F(x) = \tfrac{1}{24}(x - x_0)(x - x_1)^2(x - x_2)f^{(4)}(\xi).$$

The reader should carry out the above analysis for the case $n = 1$, $\gamma_0 = \gamma_1 = 1$ and show that if $F(x)$ is given by (7.5) then

7.33
$$f(x) - F(x) = \frac{(x - x_0)^2(x - x_1)^2}{24} f^{(4)}(\xi)$$

for some ξ in $[x_0, x_1]$.

EXERCISES 6.7

1. Find the polynomial $F(x)$ of degree 2 or less such that $F(1) = 1$, $F'(1) = 1$. $F(2) = 1$ by: the Hermite interpolation formula (7.5); the method of undetermined coefficients; the method of undetermined weights. Evaluate $F(\tfrac{3}{2})$.

2. Show that (7.5) can be written in the form

$$F(x) = \sum_{k=0}^{n} \frac{Q_k(x)}{Q_k(x_k)} \left\{ \left[1 + \gamma_k \sum_{\substack{j=0 \\ j \neq k}}^{n} \frac{(\gamma_j + 1)(x - x_k)}{x_j - x_k} \right] f(x_k) + \gamma_k(x - x_k)f'(x_k) \right\}.$$

3. Find the polynomial $F(x)$ of degree 3 or less such that $F(1) = 1$, $F'(1) = 2$, $F''(1) = 4$, and $F(2) = 5$. Find $F(1.6)$.

4. Verify directly that $F(x)$ given by (7.14) satisfies the conditions $F(a) = f(a)$, $F'(a) = f'(a)$, $F(b) = f(b)$, $F'(b) = f'(b)$.

5. Verify that substitution of the coefficients a_i given by (7.18) into (7.15) leads to (7.14).

6. Derive the weights given by (7.21) using $\alpha = (a + b)/2$.

7. Show that the unique solvability of the problem defined by (7.1) and (7.2) implies that there exist weights $w_{k,j}(x)$, $k = 0, 1, \ldots, n$; $j = 0, 1, \ldots, \gamma_k$, such that (7.13) holds.

8. Derive (7.5) for the case $\gamma_0 - \gamma_1 = \cdots = \gamma_n = 1$ as follows. Determine polynomials $w_k(x)$ and $w_k^*(x)$, $k = 0, 1, \ldots, n$, of degree $2n + 1$ or less such that

$$w_k(x_j) = \begin{cases} 0 & j \neq k \\ 1 & j = k \end{cases}$$

$$w_k'(x_j) = 0, \qquad j = 0, 1, \ldots, n$$

$$w_k^*(x_j) = 0, \qquad j = 0, 1, \ldots, n$$

$$(w_k^*)'(x_j) = \begin{cases} 0, & j \neq k \\ 1, & j = k. \end{cases}$$

9. Derive (7.14) by writing the Lagrangian interpolation formula for the four distinct points a, b, c, and d and then letting $c \to a$ and $d \to b$.

10. Prove (7.33).

11. Verify (7.24) in the general case.

12. Derive a formula similar to (7.5) for the unique polynomial $F(x)$ of degree $3n + 2$ or less corresponding to a function $f(x)$, where

$$F(x_k) = f(x_k)$$

$$F'(x_k) = f'(x_k)$$

$$F''(x_k) = f''(x_k)$$

for $k = 0, 1, \ldots, n$, and where the points $x_0, x_1, \ldots, x_n$ are distinct. Apply this formula to the case $n = 1$ and $x_0 = 0$, $x_1 = 1$, $f(x_0) = 1$, $f'(x_0) = 2$, $f''(x_0) = 1$, $f(x_1) = 3$, $f'(x_1) = 0$, $f''(x_1) = -2$. Also, for the special case find $F(\frac{1}{2})$ both directly and by the method of undetermined weights.

13. Let a and b be distinct real numbers. Show that for any α and β there is a unique polynomial $F(x)$ of degree one or less such that $F(a) = \alpha$, $F'(b) = \beta$.

14. Find the unique polynomial of degree 7 or less such that $F^{(k)}(x_i) = f^{(k)}(x_i)$, $k = 0, 1, 2, 3$ and $i = 0, 1$ where $x_0 \neq x_1$.

15. Show that there does not exist a polynomial $F(x)$ of degree two or less such that

$$F(0) = 0, \quad F(1) = 1, \quad F'(\tfrac{1}{2}) = 2.$$

More generally, show that we *can* find a polynomial $F(x)$ of degree two or less such that $F(a)$, $F(b)$, and $F'(c)$ are prescribed where a, b, and c are distinct unless

$$c = \frac{a + b}{2}.$$

and

$$F'(c) \neq \frac{F(b) - F(a)}{2}.$$

16. Given the set of numbers x_0, x_1, x_2, ρ_0, σ_1, ρ_2, where x_0, x_1, and x_2 are distinct, under what conditions does there exist a polynomial $F(x)$ of degree 2 or less such that $F(x_0) = \rho_0$, $F'(x_1) = \sigma_1$, $F(x_2) = \rho_2$?

17. Show that given x_0, x_1, and x_2 such that $x_0 < x_1 < x_2$ and given ρ_0, ρ_1, σ_0, σ_1, and σ_2, there exists a unique polynomial of degree four or less such that $F(x_0) = \rho_0$, $F(x_1) = \rho_1$, $F'(x_0) = \sigma_0$, $F'(x_1) = \sigma_1$, $F'(x_2) = \sigma_2$. Find $F(x_2)$ in the case $x_2 - x_1 = x_1 - x_0 = h$.

18. Find the unique polynomial of degree four or less such that $F(0) = 1$, $F(1) = -2$, $F'(0) = 2$, $F'(1) = 3$, and $F'(2) = 2$.

19. Show that if $x_1 < x_2 < \cdots < x_n < x_{n+1}$ then there exists a unique polynomial $F(x)$ of degree $2n$ or less such that

$$F(x_i) = f(x_i), \qquad F'(x_i) = f'(x_i), \qquad i = 1, 2, \ldots, n$$

$$F'(x_{n+1}) = f'(x_{n+1}).$$

6.8 LIMITATIONS ON POLYNOMIAL INTERPOLATION: SMOOTH INTERPOLATION

Let us suppose that one is interested in determining a function $F(x)$ which approximates a given function $f(x)$ in an interval $I = [a, b]$. One method would be to subdivide the interval into N subintervals $I_1 = [x_0, x_1]$; $I_2 = [x_1, x_2]$; $\ldots$; $I_N = [x_{N-1}, x_N]$ where $a = x_0$, $b = x_N$, and $x_0 < x_1 < \cdots < x_N$. One could then determine by Lagrangian interpolation, or if the intervals are of equal length, by Gregory-Newton interpolation, a polynomial $F(x)$ of degree N or less such that $F(x_i) = f(x_i)$, $i = 0, 1, \ldots, N$. However, for certain functions the approximate representation of $f(x)$ by a single polynomial throughout the interval is not satisfactory.

For example, consider the function

8.1
$$f(x) = \frac{1}{1 + x^2}$$

in the interval $[-5, 5]$. Assume that for $n = 2, 3, 4, \ldots$ one determines the polynomial $P_{2n}(x)$ of degree $2n$ or less such that

8.2
$$P_{2n}\left(\frac{5j}{n}\right) = f\left(\frac{5j}{n}\right), \qquad j = 0, \pm 1, \ldots, \pm n.$$

As n increases, the maximum difference between $P_{2n}(x)$ and $f(x)$ increases in spite of the fact that the distance between consecutive interpolation points decreases. (See Exercise 4.) As a matter of fact, as $n \to \infty$, the maximum difference between

$P_{2n}(x)$ and $f(x)$ becomes infinite. This phenomenon was noted by Runge [1901] and is sometimes known as the *Runge phenomenon*. The Runge phenomenon is also discussed by Montel [1910] and Steffensen [1927]. It can be shown that the interpolating process converges if one remains in the interval $[-\alpha, \alpha]$ where $\alpha \leq \alpha_0 \doteq 3.63$, but diverges if $\alpha > \alpha_0$.

An alternative approach is to use a different interpolating polynomial of lower degree in each subinterval. For example, we might use linear interpolation in each subinterval. This would clearly lead to a continuous function in the entire interval, though the first derivative would in general not be continuous. One could also use a higher-degree polynomial in each subinterval, determining the polynomial on the basis of values of $f(x)$ and possibly of its derivatives at other points of interpolation. However, even then in most cases one is led to a function which has discontinuities in the first derivative at the points $x_0, x_1, \ldots, x_N$.

One could obtain a function with a continuous first derivative throughout the entire interval using Hermite interpolation as follows. In each interval $I_k = [x_{k-1}, x_k]$ let $F(x)$ be a polynomial of degree three or less such that

$$F(x_i) = f(x_i), F'(x_i) = f'(x_i), \qquad i = k - 1, k.$$

Clearly, this will lead to a function of class $C^{(1)}$ over the entire interval.

If one is willing to use a cubic polynomial in each subinterval, it is possible to obtain a function $S(x)$ which interpolates to $f(x)$ at the $\{x_i\}$ and is of Class $C^{(2)}$ in the entire interval. Such a function is known as a *cubic spline function*. We shall describe here only the simplest case. For more information on splines the reader is referred to the literature, for example, to Ahlberg, Nilson, and Walsh [1967].

In using cubic spline interpolation we shall not insist that $F'(x)$ and $f'(x)$ agree at the points of interpolation. We seek to determine a function $F_k(x)$ in the interval I_k such that

8.3 $F_k(x_i) = f(x_i), \qquad i = k - 1, k.$

We also require that for $k = 1, 2, \ldots, N - 1$

8.4 $\begin{cases} F_k'(x_k -) = F_{k+1}'(x_k +) \\ F_k''(x_k -) = F_{k+1}''(x_k +) \end{cases}$

The procedure involves determining M_k, where

8.5 $M_k = F_k''(x_k -) = F_{k+1}''(x_k +).$

Since $F_k(x)$ is a cubic polynomial, $F_k''(x)$ is a linear function of x in I_k, i.e.,

8.6 $F_k''(x) = M_{k-1} \dfrac{x_k - x}{x_k - x_{k-1}} + M_k \dfrac{x - x_{k-1}}{x_k - x_{k-1}}.$

Integrating, we have

8.7
$$F_k'(x) = -M_{k-1}\frac{(x_k - x)^2}{2h_k} + M_k\frac{(x - x_{k-1})^2}{2h_k} + C_1$$

where

8.8
$$h_k = x_k - x_{k-1}$$

and where C_1 is a constant of integration to be determined. Integrating again, we have

8.9
$$F_k(x) = M_{k-1}\frac{(x_k - x)^3}{6h_k} + M_k\frac{(x - x_{k-1})^3}{6h_k} + C_1x + C_2.$$

Letting $y_k = f(x_k)$, we have

8.10
$$\begin{cases} y_{k-1} = M_{k-1}\frac{h_k^2}{6} + C_1x_{k-1} + C_2 \\ \\ y_k = M_k\frac{h_k^2}{6} + C_1x_k + C_2; \end{cases}$$

hence,

8.11
$$\begin{cases} C_1 = \dfrac{(y_k - y_{k-1}) - (M_k - M_{k-1})(h_k^2/6)}{h_k} \\ \\ C_2 = \dfrac{(x_k y_{k-1} - x_{k-1}y_k) - (x_k M_{k-1} - x_{k-1}M_k)(h_k^2/6)}{h_k}. \end{cases}$$

Substituting in (8.9) we get

8.12
$$F_k(x) = M_{k-1}\left(\frac{(x_k - x)((x_k - x)^2 - h_k^2)}{6h_k}\right) + M_k\left(\frac{(x - x_{k-1})((x - x_{k-1})^2 - h_k^2)}{6h_k}\right)$$
$$+ \frac{1}{h_k}y_{k-1}(x_k - x) + \frac{1}{h_k}y_k(x - x_{k-1}).$$

Differentiating, we have

8.13
$$F_k'(x) = M_{k-1}\left(\frac{h_k^2 - 3(x_k - x)^2}{6h_k}\right) + M_k\left(\frac{3(x - x_{k-1})^2 - h_k^2}{6h_k}\right) + \frac{1}{h_k}(y_k - y_{k-1}).$$

If we require that $F'_k(x_k-) = F'_{k+1}(x_k+)$ we have

8.14 $\quad \dfrac{h_k}{6}M_{k-1} + \dfrac{h_k}{3}M_k + \dfrac{1}{h_k}(y_k - y_{k-1}) = -\dfrac{h_{k+1}}{3}M_k - \dfrac{h_{k+1}}{6}M_{k+1} + \dfrac{1}{h_{k+1}}(y_{k+1} - y_k)$

or

8.15 $\quad \dfrac{h_k}{6}M_{k-1} + \dfrac{h_k + h_{k+1}}{3}M_k + \dfrac{h_{k+1}}{6}M_{k+1} = \left\{ \dfrac{1}{h_{k+1}}(y_{k+1} - y_k) - \dfrac{1}{h_k}(y_k - y_{k-1}) \right\}$

$$(k = 1, 2, \ldots, N-1).$$

This is a system of $N-1$ linear algebraic equations with unknowns M_0, $M_1, \ldots, M_N$. If we arbitrarily assign values to M_0 and M_N, say zero, we can uniquely determine $M_1, M_2, \ldots, M_{N-1}$. This follows since the determinant of the matrix A of the system,

8.16 $\quad A = \begin{pmatrix} \dfrac{h_1 + h_2}{3} & \dfrac{h_2}{6} & 0 \cdots & & & \cdot \\[2ex] \dfrac{h_2}{6} & \dfrac{h_2 + h_3}{3} & \dfrac{h_3}{6}\cdots & \cdot & & \cdot \\[2ex] \cdot & \cdot & \cdot \cdots & \cdot & & \vdots \\[2ex] \cdot & \cdot & \cdots & \dfrac{h_{N-2} + h_{N-1}}{3} & & \dfrac{h_{N-1}}{6} \\[2ex] \cdot & \cdot & \cdots & \dfrac{h_{N-1}}{6} & & \dfrac{h_{N-1} + h_N}{3} \end{pmatrix}$,

does not vanish. As will be shown in Chapter 16, this follows since the matrix is "strictly diagonally dominant," i.e., in each row the absolute value of the diagonal element is greater than the sum of the absolute values of all other elements. Such a matrix is "nonsingular" and has a nonvanishing determinant.

We remark that the solution of a system of linear algebraic equations with a tri-diagonal matrix can easily be carried out, as shown in Chapter 10.

Numerical Example

As an example, let us consider the case

8.17 $$f(x) = \dfrac{1}{1 + x^2}$$

in the interval $0 \leq x \leq 1$. We consider the six points of interpolation

8.18 $$x_k = \frac{k}{5}, \qquad k = 0, 1, 2, \ldots, 5.$$

By direct computation we have

8.19
$$\begin{cases} y_0 = 1.00000000 \\ y_1 = 0.96153846 \\ y_2 = 0.86206896 \\ y_3 = 0.73529412 \\ y_4 = 0.60975610 \\ y_5 = 0.50000000. \end{cases}$$

Since $h_1 = h_2 = \cdots = h_5 = h = \frac{1}{5}$ we have by (8.15), after letting $M_0 = M_5 = 0$ and multiplying both sides by $6/h$,

8.20
$$\begin{pmatrix} 4 & 1 & 0 & 0 \\ 1 & 4 & 1 & 0 \\ 0 & 1 & 4 & 1 \\ 0 & 0 & 1 & 4 \end{pmatrix} \begin{pmatrix} M_1 \\ M_2 \\ M_3 \\ M_4 \end{pmatrix} = \frac{6}{h^2} \begin{pmatrix} y_2 - 2y_1 + y_0 \\ y_3 - 2y_2 + y_1 \\ y_4 - 2y_3 + y_2 \\ y_5 - 2y_4 + y_3 \end{pmatrix} = \begin{pmatrix} -9.151194 \\ -4.095803 \\ 0.185524 \\ 2.367288 \end{pmatrix}.$$

Solving we get

8.21 $\qquad M_1 = -2.165813, \; M_2 = -0.487940, \; M_3 = 0.021771, \; M_4 = 0.586379.$

From (8.12) we compute the $F_k(x)$ for $x = 0(0.02)1.0$ obtaining the following results:

x	k	$F_k(x)$	$f(x)$	$F_k(x) - f(x)$
0.00	1	1.000000	1.000000	0.000000
0.02	1	0.997583	0.999600	−0.002017
0.04	1	0.995080	0.998403	−0.003323
0.06	1	0.992403	0.996413	−0.004010
0.08	1	0.989467	0.993641	−0.004174
0.10	1	0.986184	0.990099	−0.003915
0.12	1	0.982468	0.985804	−0.003337
0.14	1	0.978232	0.980777	−0.002545
0.16	1	0.973389	0.975039	−0.001650
0.18	1	0.967854	0.968617	−0.000763
0.20	1, 2	0.961538	0.961538	0.000000
0.22	2	0.954383	0.953834	0.000548
0.24	2	0.946427	0.945537	0.000890

0.26	2	0.937740	0.936680	0.001060
0.28	2	0.928388	0.927300	0.001088
0.30	2	0.918438	0.917431	0.001007
0.32	2	0.907957	0.907112	0.000846
0.34	2	0.897013	0.896379	0.000634
0.36	2	0.885672	0.885269	0.000403
0.38	2	0.874002	0.873820	0.000181
0.40	2, 3	0.862069	0.862069	0.000000
0.42	3	0.849933	0.850051	-0.000118
0.44	3	0.837623	0.837802	-0.000179
0.46	3	0.825158	0.825355	-0.000197
0.48	3	0.812559	0.812744	-0.000184
0.50	3	0.799847	0.800000	-0.000153
0.52	3	0.787041	0.787154	-0.000112
0.54	3	0.774163	0.774234	-0.000071
0.56	3	0.761232	0.761267	-0.000035
0.58	3	0.748269	0.748279	-0.000010
0.60	3, 4	0.735294	0.735294	0.000000
0.62	4	0.722328	0.722335	-0.000006
0.64	4	0.709394	0.709421	-0.000027
0.66	4	0.696514	0.696573	-0.000059
0.68	4	0.683710	0.683807	-0.000098
0.70	4	0.671005	0.671141	-0.000136
0.72	4	0.658421	0.658588	-0.000167
0.74	4	0.645982	0.646162	-0.000180
0.76	4	0.633710	0.633874	-0.000164
0.78	4	0.621627	0.621736	-0.000109
0.80	4, 5	0.609756	0.609756	0.000000
0.82	5	0.598112	0.597943	0.000169
0.84	5	0.586679	0.586304	0.000375
0.86	5	0.575434	0.574845	0.000589
0.88	5	0.564353	0.563571	0.000782
0.90	5	0.553412	0.552486	0.000926
0.92	5	0.542589	0.541594	0.000994
0.94	5	0.531860	0.530898	0.000961
0.96	5	0.521201	0.520400	0.000801
0.98	5	0.510589	0.510100	0.000489
1.00	5	0.500000	0.500000	0.000000

Since the errors are so small, a graph of $F_k(x)$ would be nearly indistinguishable from that of $f(x)$. Clearly, the interpolation is smooth. A graph of the error function $F_k(x) - (1 + x^2)^{-1}$ is given in Fig. 8.22.

EXERCISES 6.8

1. In the example given in the text, compute $F(x), f(x)$, and $F(x) - f(x)$ for $x = 0.26$ and $x = 0.27$. Find the values of $F'(x)$ at the points $x = k/5, k = 1, 2, 3, 4$.

2. For the function $f(x) = (1 + x^2)^{-1}$ in the interval $[0, 1]$, let $P_5(x)$ be a polynomial of degree 5 or less such that $P_5(x) = f(x)$ for $x = k/5, k = 0, 1, \ldots, 5$. Compute $P_5(x), f(x)$, and $P_5(x) - f(x)$ for $x = 0(0.02)1.0$. How does this interpolation process compare with the cubic spline interpolation process?

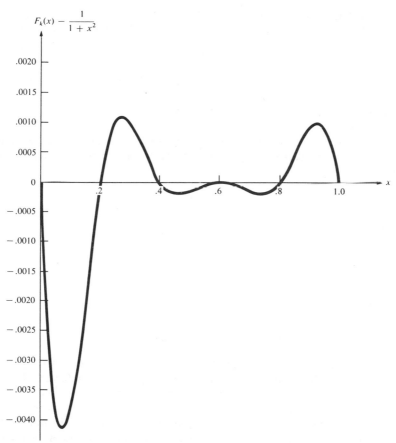

8.22 Fig. The Error Function $F_k - 1/(1 + x^2)$.

3. Carry out the cubic spline interpolation process for the function $f(x) = \sin(\pi/2)x$, $x = 0(0.2)1.0$. Then compute $F(x), f(x)$, and $F(x) - f(x)$ for $x = 0(0.02)1.0$.

4. Consider the approximation of the function $f(x) = (1 + x^2)^{-1}$ in the interval $-5 \leqq x \leqq 5$. For $h = 5/M$, $M = 2, 4, 6, 8, 10$, determine the polynomial of interpolation in the points jh where $j = 0, \pm 1, \pm 2, \ldots, \pm M$. Also, determine the cubic spline approximation in each case. Plot the results along with $f(x)$.

6.9 INVERSE INTERPOLATION

We now consider the following problem. Given a table of a function $f(x)$ for various values of x and given $\bar{y}$, find an approximate value of x such that

9.1 $$f(x) = \bar{y}.$$

We will be primarily concerned here with a method based on the construction of
an interpolating polynomial $F(x)$ and the solution of the polynomial equation

9.2 $F(x) = \bar{y}.$

In principle, once $F(x)$ has been constructed, any of the methods in Chapter 5 could
be used to solve the polynomial equation (9.2). However, because we are seeking
a real root for which a good initial approximation is known, it is usually best to
use special methods.

Inverse linear interpolation is frequently used. Here one scans the table to
determine two consecutive interpolation points which we label as x_0 and x_1 such
that

9.3 $(f(x_0) - \bar{y})(f(x_1) - \bar{y}) \leqq 0.$

One then constructs the linear function

9.4 $F(x) = \dfrac{x_1 - x}{x_1 - x_0} f(x_0) + \dfrac{x - x_0}{x_1 - x_0} f(x_1).$

Solving (9.2) we obtain

9.5 $x = \left(\dfrac{f(x_1) - \bar{y}}{f(x_1) - f(x_0)}\right)x_0 + \left(\dfrac{\bar{y} - f(x_0)}{f(x_1) - f(x_0)}\right)x_1.$

If one desires to use a higher order interpolation procedure involving $n + 1$
points and if one has found two consecutive interpolation points such that (9.3)
holds, then one can find a solution of (9.2) in the interval $x_0 \leqq x \leqq x_1$ as follows.
Apply Muller's method, the secant method, the method of false position, and the
method of bisection, in that order, passing from one method to the next when a
method fails to converge, is too slow, or yields an iterant outside of the interval.
Alternatively, one can construct the coefficients $\hat{a}_0, \hat{a}_1, \ldots, \hat{a}_n$ such that

9.6 $F(x) = \hat{a}_0(x - x_0)^n + \hat{a}_1(x - x_0)^{n-1} + \cdots + \hat{a}_n,$

as described in Section 6.6, and then use any of the methods of Chapter 5, such as
the IP method, to solve (9.2).

For manual computation the following procedure is suggested. We illustrate
it for the case of four-point interpolation. Let us relabel the points if necessary so
that we have $x_3 - x_2 = x_2 - x_1 = x_1 - x_0 = h$ and the root of (9.2) lies some-
where in the interval $[x_0, x_3]$. Let $u = (x - x_0)/h$. By the Gregory-Newton interpolation formula we have

9.7 $F(x) = f(x_0) + u\Delta f(x_0) + \dfrac{u(u-1)}{2}\Delta^2 f(x_0) + \dfrac{u(u-1)(u-2)}{3!}\Delta^3 f(x_0).$

We first compute

9.8
$$u_0 = \frac{\bar{y} - f(x_0)}{\Delta f(x_0)}.$$

We then iterate using the formula

9.9 $u_{n+1} = \dfrac{\bar{y} - f(x_0)}{\Delta f(x_0)} - \dfrac{1}{\Delta f(x_0)} \left\{ \dfrac{u_n(u_n - 1)}{2} \Delta^2 f(x_0) + \dfrac{u_n(u_n - 1)(u_n - 2)}{6} \Delta^3 f(x_0) \right\}.$

Convergence will normally occur after one or two iterations.

In the event that (9.3) does not hold for any pair of consecutive interpolation points, then (9.2) cannot be satisfied if we are using linear interpolation. However, if we are using a higher order interpolation procedure it is still possible for (9.2) to be satisfied. Thus, suppose $\bar{y}$ is greater than all tabulated values of $f(x)$. We seek the interpolation point where $f(x)$ is largest and select n additional nearby interpolation points. We then determine the coefficients $\hat{a}_k$ in (9.6) and apply any polynomial equation solving scheme to determine whether (9.2) has a solution within the interval spanned by the $n + 1$ points.

The analysis of Section 4.2 is applicable to the estimation of the accuracy of inverse interpolation. The accuracy of $F(x)$ as a representation of $f(x)$ now depends on the number of significant figures given for $f(x)$ in the table as well as on the accuracy of the interpolation process. Thus, suppose one uses a small enough value of h and a sufficiently accurate interpolation formula so that the error in the interpolation is less than $\frac{1}{2}10^{-t}$ where the values of $f(x)$ are given to t decimal places. The inverse interpolation process will by no means be exact. For, as we saw in Section 4.2, the accuracy attainable would be approximately

9.10
$$\delta = \frac{\frac{1}{2}10^{-t}}{|f'(\bar{x})|}$$

where $\bar{x}$ is the desired zero of $f(x) - \bar{y}$.

As an example, suppose we are given the following (inaccurate) table of the function e^x

x	$f(x)$
0.5	1.65
0.6	1.82
0.7	2.01
0.8	2.23
0.9	2.46
1.0	2.72

and suppose we wish to find x such that $e^x = 2.00$. By Theorem 3.9, we verify at once that the accuracy of linear interpolation near the desired value of x is approximately

9.11 $$\frac{h^2}{8}(2.00) = 0.0025$$

so that linear interpolation is entirely adequate considering the accuracy of $f(x)$. With the interpolation points $x_0 = 0.6$ and $x_1 = 0.7$ we obtain by (9.5)

9.12 $$x = \left(\frac{2.01 - 2.00}{2.01 - 1.82}\right)(0.6) + \left(\frac{2.00 - 1.82}{2.01 - 1.82}\right)(0.7)$$

$$= \frac{1}{19}(0.6) + \frac{18}{19}(0.7) = \frac{13.2}{19} \doteq 0.6947.$$

Since the solution, $\bar{x}$, is given to four decimal places by

9.13 $$\bar{x} = 0.6932$$

we see that the error is

9.14 $$\bar{x} - x = -0.0015.$$

This agrees reasonably well with the expected error given by (9.10) which is

9.15 $$\frac{0.005}{2.00} = 0.0025$$

since $f'(\bar{x}) = 2.00$.

Given a table of values of $f(x)$ for various values of x we could construct a table of values of $x = g(y)$ for various values of y, where $y = f(x)$. The function $g(y)$ is defined by

9.16 $$x = g(y) = g(f(x)).$$

(For example, if $f(x) = x^2$, then $g(y) = \sqrt{y}$.) Of course, the values of the independent variable y for $g(y)$ will not be equally spaced. Nevertheless we could solve the inverse interpolation problem by using the Lagrange interpolation formula to construct a polynomial $G(y)$ of a certain degree which agrees with $g(y)$ at several values of y. We could then evaluate $\bar{x} = G(\bar{y})$ directly.

If we assume that $G(y)$ is a linear function of y we have

9.17 $$G(y) = \frac{y_1 - y}{y_1 - y_0} g(y_0) + \frac{y - y_0}{y_1 - y_0} g(y_1)$$

and

9.18
$$\bar{x} = G(\bar{y}) = \left(\frac{y_1 - \bar{y}}{y_1 - y_0}\right)x_0 + \left(\frac{\bar{y} - y_0}{y_1 - y_0}\right)x_1$$

which agrees with (9.5).

As an example let us consider the function $f(x) = x^2$ for $x = 0(0.1)0.2$. Evidently

9.19
$$f(0) = 0, \quad f(0.1) = 0.01, \quad f(0.2) = 0.04.$$

We seek $\bar{x}$ such that

9.20
$$f(\bar{x}) = 0.02.$$

Using linear interpolation in the points $x_1 = 0.1, x_2 = 0.2$, we get

9.21
$$F(x) = \frac{0.2 - x}{0.1}(0.01) + \frac{x - 0.1}{0.1}(0.04)$$
$$= -0.02 + 0.3x.$$

Solving $F(x) = 0.02$ we get

9.22
$$\bar{x} \doteq 0.1333.$$

On the other hand, we have for the inverse function $g(0) = 0$, $g(0.01) = 0.1$, $g(0.04) = 0.2$). Using linear interpolation in the points $y_1 = 0.01$, $y_2 = 0.04$, we have

9.23
$$\bar{x} = \frac{0.04 - \bar{y}}{0.04 - 0.01}(0.1) + \frac{\bar{y} - 0.01}{0.04 - 0.01}(0.2)$$
$$= \frac{0.02}{0.03}(0.1) + \frac{0.01}{0.03}(0.2) = \frac{4}{3}(0.1) \doteq 0.1333$$

as before.

It is perhaps logical to ask why one does not always use direct interpolation in the inverse function rather than inverse interpolation. The answer lies in the fact that even though $f(x)$ may be well represented by a polynomial, and hence determined to good accuracy by direct interpolation, the inverse function may not be well represented by a polynomial. For example, if $f(x) = x^2$, the inverse function

$x = \sqrt{y}$ is not well represented by a polynomial near $x = y = 0$. Thus in the above example, involving the function $f(x) = x^2$, we have

x	$f(x)$	$\Delta f(x)$	$\Delta^2 f(x)$
0	0		
		0.01	
0.1	0.01		0.02
		0.03	
0.2	0.04		

and by the Gregory-Newton interpolation formula with $u = x/0.1 = 10x$,

9.24
$$F(x) = f(0) + u\Delta f(x_0) + \frac{u(u-1)}{2}\Delta^2 f(x_0)$$

$$= 0 + (10x)(0.01) + \frac{(10x)(10x-1)}{2}(0.02)$$

$$= 0.1x + x^2 - 0.1x = x^2.$$

Hence

9.25
$$\bar{x} = \sqrt{0.02} \doteq 0.1414.$$

This value is exact. On the other hand, for the inverse function we have by Lagrange's formula

9.26
$$x = \frac{(\bar{y} - 0.01)(\bar{y} - 0.04)}{(0 - 0.01)(0 - 0.04)}(0) + \frac{(\bar{y} - 0)(\bar{y} - 0.04)}{(0.01 - 0)(0.01 - 0.04)}(0.1)$$

$$+ \frac{(\bar{y} - 0)(\bar{y} - 0.01)}{(0.04 - 0)(0.04 - 0.01)}(0.2),$$

which gives $\bar{x} \doteq 0.1666$ for $\bar{y} = 0.02$. Thus the value of $\bar{x}$ is *less* accurate than that obtained using inverse linear interpolation.

EXERCISES 6.9

1. Given the following table of the function e^x

x	$f(x)$
0.4	1.4918
0.5	1.6487
0.6	1.8221
0.7	2.0138
0.8	2.2255
0.9	2.4596
1.0	2.7183

find $\bar{x}$ such that $f(\bar{x}) = 2$ using inverse linear interpolation and inverse four-point interpolation. For inverse four-point interpolation, use the method based on (9.8) and (9.9) and also the method based on the determination of the interpolation polynomial $F(x)$ in the form (9.6). The polynomial equation can be solved by the Newton method. How accurate are the results in each case?

2. Given the table

x:	2.0	2.4	2.8	3.2	3.6
$f(x)$:	-3.00	-1.24	0.84	3.24	5.96

find $\bar{x}$ such that $f(\bar{x}) = 0$ to as great an accuracy as possible. Estimate the error.

3. Given a table of $\log x$ for $1(0.1)2$, how accurately can one determine $\bar{x}$ so that $\log \bar{x} = 0.5$ using $(n + 1)$-point inverse interpolation for $n = 1, 2, 3$, if $\log x$ is given to four decimal places? Actually carry out the calculation in each case and compare the results with the exact value $\bar{x} = e^{0.5} \doteq 1.6487$.

4. Solve the problem of the preceding exercise if the table is given for $x = 0.1(0.1)1.0$ and if $\log \bar{x} = -1.5$.

5. Given the table of Exercise 1, Section 6.6, find $\bar{x}$ such that $f(\bar{x}) = 0.20000$ using: inverse linear interpolation; inverse quadratic interpolation; inverse four-point interpolation. Estimate the accuracy of the results in each case.

6. Using the following table, find $\bar{x}$ such that $f(\bar{x}) = 0.98$ using two-point and also three-point inverse interpolation.

x:	1.74	1.76	1.78	1.80
$f(x)$:	0.98571918	0.98215432	0.97819661	0.97384763

Estimate the accuracy of the result in each case.

7. Write a computer program for carrying out inverse $(n + 1)$-point interpolation, with $n \leq 4$, based on a table of $f(x)$ given for $x = a(h)b$, where $h = (b - a)M$ for some M, $(n < M)$. Use the following scheme:

a) If (9.3) holds for two consecutive interpolation points use the method of bisection.

b) If (9.3) does not hold, let $\hat{x}$ be the interpolation point such that $f(\hat{x})$ is closest to $\bar{y}$ and select n additional interpolation points near $\hat{x}$. Then construct the coefficients $\hat{a}_0, \hat{a}_1, \ldots, \hat{a}_n$ of (9.6), and solve (9.2), thus determining whether or not the inverse problem can be solved.

Apply the program to the example of Exercise 1 (with $n = 1$ and $n = 3$) and also to the case where

$$f(x) = xe^{-ax},$$

with $a = \frac{4}{3}$, is tabulated for $x = 0(0.1)1$. Let $n = 3$ and $\bar{y} = 0.138$. Also let $\bar{y} = 0.140$. Also apply the program to Exercises 2, 3, 4, 5, 6, and to the case $f(x) = \sin(\pi x/180) + \cos(\pi x/180)$, $x = 0(10)90$ with $\bar{y} = 1.2001$ and also with $\bar{y} = 1.4140$.

8. Find a condition such that if $F(x)$ is the polynomial of degree two or less such that $F(x_i) = f(x_i)$, $i = 0, 1, 2$, where $x_2 - x_1 = x_1 - x_0 = h$, then $F(x) = \bar{y}$ has a solution in the interval $x_0 \leq x \leq x_2$. Apply the condition to the cases considered in Exercise 7 involving $f(x) = xe^{-ax}$.

9. Apply the condition found in Exercise 8 to the cases:
   a) $f(x_0) = 1$, $f(x_1) = 1$, $f(x_2) = 0.5$, $\bar{y} = 1.1$
   b) $f(x_0 = 1$, $f(x_1) = 1$, $f(x_2) = -1$, $\bar{y} = 1.1$
   Find the solution of $F(x) = \bar{y}$ in each case if possible.

10. Find $\bar{x}$ such that $\bar{x}^3 = 0.002$ using inverse interpolation in the three points $x_0 = 0$, $x_1 = 0.1$, $x_2 = 0.2$. Also use Lagrangian interpolation assuming x is a polynomial in y.

6.10 APPROXIMATION BY POLYNOMIALS

So far we have primarily been concerned with representing a given function $f(x)$ by an interpolating polynomial $F(x)$. We now consider the representation of a function $f(x)$ which is known to be continuous in an interval by a polynomial $P_n(x)$ of degree n or less. There are various measures of the closeness of the approximation of $f(x)$ by $P_n(x)$. In this section we shall be primarily interested in finding a polynomial $P_n(x)$ which minimizes the quantity

10.1
$$\max_{a \le x \le b} |P_n(x) - f(x)|$$

which is sometimes called the "uniform norm" of $P_n - f$ and is frequently denoted by $\|P_n - f\|$ or $\|P_n - f\|_\infty$. Another measure which is often used is the L_2-norm given by

10.2
$$\left\{ \int_a^b (P_n(x) - f(x))^2 \, dx \right\}^{1/2}$$

which is frequently denoted by $\|P_n - f\|_2$. We shall consider this measure of approximation in Section 6.11.

We now state without proof two mathematical theorems which relate to the approximation of functions in the uniform norm.

10.3 Theorem. If $f(x) \in C[a, b]$, then given any $\varepsilon > 0$ there exists a polynomial $P_n(x)$ such that for $x \in [a, b]$ we have

10.4
$$|P_n(x) - f(x)| \le \varepsilon.$$

This is the Weierstrass approximation theorem. Several proofs are given in Todd [1963].

10.5 Theorem. If $f(x) \in C[a, b]$, for a given integer n there is a unique polynomial $\pi_n(x)$ of degree n or less such that

10.6
$$\delta_n = \max_{a \le x \le b} |\pi_n(x) - f(x)| \le \max_{a \le x \le b} |P_n(x) - f(x)|$$

for any polynomial $P_n(x)$ of degree n or less. Moreover, there exists a set of $n + 2$ points $x_0, x_1, \ldots, x_{n+1}$ in $[a, b]$ where $x_0 < x_1 < \cdots < x_{n+1}$, such that either

10.7 $$\pi_n(x_i) - f(x_i) = (-1)^i \delta_n, \qquad i = 0, 1, \ldots, n$$

or

10.8 $$\pi_n(x_i) - f(x_i) = (-1)^{i+1} \delta_n, \qquad i = 0, 1, \ldots, n$$

For a proof of Theorem 10.5 see Todd [1963]. We remark that by Theorem 10.3 it follows that $\delta_n \to 0$ as $n \to \infty$.

No general finite algorithm is known for finding the polynomial $\pi_n(x)$ of best approximation (Todd [1963]). However, if we replace the interval $[a, b]$ by a discrete set of points, J, in the interval, then in a finite number of steps we can find a polynomial $\pi_n(x)$ of degree n or less such that

10.9 $$\max_{x \in J}|\pi_n(x) - f(x)| \leq \max_{x \in J}|P_n(x) - f(x)|$$

for any polynomial $P_n(x)$ of degree n or less. We now describe an algorithm for finding $\pi_n(x)$ presented by Stiefel [1959] and based on earlier work of Remes [1934, 1934a].

Much of our analysis depends on the following theorem (see Stiefel [1959]).

10.10 Lemma. Given $n + 2$ distinct points $x_0, x_1, \ldots, x_{n+1}$ such that $x_0 < x_1 < \cdots < x_{n+1}$ there exist non-vanishing constants $\lambda_0, \lambda_1, \ldots, \lambda_{n+1}$ of alternating sign in the sense that $\lambda_0\lambda_1, \lambda_1\lambda_2, \ldots, \lambda_n\lambda_{n+1}$ are negative and such that for any polynomial $P_n(x)$ of degree n or less we have

10.11 $$\sum_{i=0}^{n+1} \lambda_i P_n(x_i) = 0.$$

Moreover,

10.12 $$\lambda_i = \prod_{\substack{j=0 \\ j \neq i}}^{n+1} \frac{1}{x_i - x_j}.$$

Proof. By the Lagrange interpolation formula we have

10.13 $$P_n(x) = \sum_{i=0}^{n} \left\{ \prod_{\substack{j=0 \\ j \neq i}}^{n} \frac{x - x_j}{x_i - x_j} \right\} P_n(x_i).$$

Letting $x = x_{n+1}$ and dividing by $\prod_{j=0}^{n}(x_{n+1} - x_j)$ we get (10.11). Clearly the λ_i do not vanish and have alternating signs.

Following Stiefel [1959] we introduce the following terminology.*

1. A *reference set* is a set of $n + 2$ distinct numbers $x_0, x_1, \ldots, x_{n+1}$ in J.

2. A *reference-polynomial* with respect to the function $f(x)$ defined on J and with respect to the reference set $x_0, x_1, \ldots, x_{n+1}$ where $x_0 < x_1 < \cdots < x_{n+1}$ is a polynomial $P_n(x)$ of degree n or less such that the h_i have alternating signs, where

10.14 $$h_i = P_n(x_i) - f(x_i), \qquad i = 0, 1, \ldots, n + 1.$$

3. A *levelled reference-polynomial* is a reference-polynomial such that all h_i have the same magnitude.

   We show below that for a given reference set and for a given function $f(x)$ there is a unique levelled reference-polynomial.

4. The *reference deviation* of a reference set is the value $H = |h_0| = |h_1| = \cdots = |h_{n+1}|$ for the levelled reference-polynomial.

We seek to show the existence and uniqueness of a levelled reference-polynomial corresponding to a given reference set $x_0, x_1, \ldots, x_{n+1}$. We prove

10.15 Lemma. Given a function $f(x)$ and a reference set $R: x_0, x_1, \ldots, x_{n+1}$ where $x_0 < x_1 < \cdots < x_{n+1}$ there exists a unique levelled reference-polynomial $P_n(x)$ of degree n or less. Moreover, we have

10.16 $$P_n(x_i) = f(x_i) + (-1)^i \delta, \qquad i = 0, 1, \ldots, n + 1$$

where

10.17 $$\delta = -\operatorname{sgn} \lambda_0 \frac{\displaystyle\sum_{i=0}^{n+1} \lambda_i f(x_i)}{\displaystyle\sum_{i=0}^{n+1} |\lambda_i|}$$

and where the λ_i are given by (10.12). The reference deviation is given by

10.18 $$H = |\delta| = \frac{\left| \displaystyle\sum_{i=0}^{n+1} \lambda_i f(x_i) \right|}{\displaystyle\sum_{i=0}^{n+1} |\lambda_i|}.$$

If $Q_n(x)$ is any polynomial of degree n or less such that

10.19 $$\max_{x \in R} |Q_n(x) - f(x)| = H,$$

* Stiefel considered a more general class of approximating functions than polynomials. We have specialized the definitions to the case of polynomials.

then

10.20 $$Q_n(x) \equiv P_n(x).$$

Proof. If $P_n(x)$ is a levelled reference-polynomial for R then (10.16) holds for some δ. From (10.11) it follows that

10.21 $$\sum_{i=0}^{n+1} \lambda_i P_n(x_i) = \sum_{i=0}^{n+1} \lambda_i (f(x_i) + (-1)^i \delta) = 0$$

and hence δ satisfies (10.17).

By using the Lagrangian interpolation formula one can show that there exists a unique polynomial, say $P_n(x)$, of degree n or less such that (10.16) holds for $i = 0, 1, 2, \ldots, n$ though not necessarily for $i = n + 1$. By (10.11) and (10.17) we have

10.22 $$0 = \sum_{i=0}^{n+1} \lambda_i P_n(x_i) = \sum_{i=0}^{n} \lambda_i (f(x_i) + (-1)^i \delta) + \lambda_{n+1} P_n(x_{n+1})$$

$$= \sum_{i=0}^{n+1} \lambda_i (f(x_i) + (-1)^i \delta) + \lambda_{n+1}[P_n(x_{n+1}) - (f(x_{n+1}) + (-1)^{n+1}\delta)]$$

$$= \lambda_{n+1}[P_n(x_{n+1}) - (f(x_{n+1}) + (-1)^{n+1}\delta)].$$

Since $\lambda_{n+1} \neq 0$ it follows that

10.23 $$P_n(x_{n+1}) = f(x_{n+1}) + (-1)^{n+1}\delta.$$

Therefore $P_n(x)$ is a levelled reference-polynomial. Clearly, it is the only possible levelled reference-polynomial.

Suppose $Q_n(x)$ is a polynomial of degree n or less such that (10.19) holds and let

10.24 $$R_n(x) = Q_n(x) - P_n(x).$$

Evidently $R_n(x_i)R_n(x_{i+1}) \leq 0$ for $i = 0, 1, \ldots, n$ so that by Lemma 10.10 we have

10.25 $$\sum_{i=0}^{n+1} \lambda_i R_n(x_i) = \pm \sum_{i=0}^{n+1} |\lambda_i| |R_n(x_i)| = 0.$$

Since none of the λ_i vanishes we must have $R_n(x_i) = 0$, $i = 0, 1, \ldots, n + 1$, or

10.26 $$Q_n(x_i) = P_n(x_i), \qquad i = 0, 1, \ldots, n + 1.$$

Since the polynomials $Q_n(x)$ and $P_n(x)$ agree at $n + 2$ points, it follows that they are identically equal. This completes the proof of Lemma 10.15.

Suppose, now, that $x_0, x_1, \ldots, x_{n+1}$, where $x_0 < x_1 < \cdots < x_{n+1}$ is a reference set and that $P_n(x)$ is a reference-polynomial, not necessarily a levelled reference-polynomial with respect to $f(x)$. Let

10.27
$$h_i = P_n(x_i) - f(x_i), \qquad i = 0, 1, \ldots, n + 1.$$

We seek to show that the reference deviation, H, can be written in the form

10.28
$$H = \sum_{i=0}^{n+1} w_i |h_i|$$

where the weights w_i are positive and are given by

10.29
$$w_i = \frac{|\lambda_i|}{\sum\limits_{j=0}^{n+1} |\lambda_j|}$$

and where the λ_i are given by (10.12). Since no λ_i vanishes the w_i are positive. Thus H is a weighted average of the $|h_i|$ with positive weights. To prove (10.28) we have by (10.11) and (10.27)

10.30
$$\sum_{i=0}^{n+1} \lambda_i(f(x_i) + h_i) = 0.$$

By (10.18) and the fact that both the h_i and the λ_i have alternating signs we have

10.31
$$H = \frac{\sum\limits_{i=0}^{n+1} |\lambda_i| \, |h_i|}{\sum\limits_{i=0}^{n+1} |\lambda_i|}$$

and hence (10.28) and (10.29) follow.

We are now able to define the following computational procedures.

1. Choose any $n + 2$ points $x_0, x_1, \ldots, x_{n+1}$ in J such that $x_0 < x_1 < \cdots < x_{n+1}$, i.e., choose any reference set, R.

2. Determine the levelled reference-polynomial $P_n(x)$ and the reference deviation H for R.

3. Find $\hat{x}$ in J such that

10.32
$$|P_n(\hat{x}) - f(\hat{x})| = \max_{x \in J} |P_n(x) - f(x)| = \Delta.$$

If $\Delta = H$, terminate the process.

4. Find a new reference set which contains $\hat{x}$ and is such that $P_n(x)$ is a reference-polynomial with respect to the new reference set. This can be done as follows.

If $\hat{x}$ lies in an interval $x_i \leqq x \leqq x_{i+1}$, $i = 0, 1, \ldots, n$, replace x_i by $\hat{x}$ if $P_n(\hat{x}) - f(\hat{x})$ has the same sign as $P_n(x_i) - f(x_i)$. Otherwise replace x_{i+1} by $\hat{x}$. If $\hat{x} < x_0$ then replace x_0 by $\hat{x}$ if $P_n(x_0) - f(x_0)$ has the same sign as $P_n(\hat{x}) - f(\hat{x})$. Otherwise replace x_{n+1} by $\hat{x}$. A similar procedure is used if $\hat{x} > x_{n+1}$.

5.   With the new reference set, return to step (2).

The reference deviations are strictly increasing unless, at a given stage, the process terminates. For the reference deviation H' of the new reference set R', found from R by replacing one value of x by another for which the absolute deviation is strictly greater, is greater than H because of (10.28). Since there are only a finite number of reference sets and since no reference set can occur more than once, the process must eventually terminate.

When the process terminates we have a reference set R^* with a reference deviation H^* and a polynomial $\pi_n(x)$ which is a levelled reference-polynomial for R^*. If $\hat{\pi}_n(x)$ is any polynomial of degree n or less, then by Lemma 10.15 either $\hat{\pi}_n(x) \equiv \pi_n(x)$ or else $\hat{\pi}_n(x)$ has a maximum absolute deviation greater than H^* on R^*, and hence on J. Thus $\pi_n(x)$ is the polynomial of best approximation.

Numerical Example

As an example let us consider the problem of finding the polynomial of degree one or less which gives the best approximation to the function

10.33 $f(x) = e^x$

in the interval $[0, 1]$. As we shall see below, this problem can be solved analytically, and to the accuracy indicated the polynomial of best approximation is

10.34 $\pi_1(x) = 1.718282x + 0.894067$

and

10.35 $\max_{0 \leqq x \leqq 1} |\pi_1(x) - f(x)| = 0.105933.$

Let J be the set

10.36 $t_i = \dfrac{i}{100}, \qquad i = 0, 1, 2, \ldots, 100$

and let us start with the reference set

10.37 $x_0 = 0, \quad x_1 = 0.5, \quad x_2 = 1.0.$

From a table of e^x we have

10.38 $f(x_0) = 1.0000, \quad f(x_1) = 1.6487, \quad f(x_2) = 2.7183.$

Let us determine the levelled reference-polynomial $P(x)$ and the reference deviation H. From (10.12) we have

10.39
$$\begin{cases} \lambda_0 = \dfrac{1}{(x_0 - x_1)(x_0 - x_2)} = 2 \\[2mm] \lambda_1 = \dfrac{1}{(x_1 - x_0)(x_1 - x_2)} = -4 \\[2mm] \lambda_2 = \dfrac{1}{(x_2 - x_0)(x_2 - x_1)} = 2. \end{cases}$$

We then obtain δ from (10.17) getting

10.40 $\delta = -\dfrac{2(1.0000) - 4(1.6487) + 2(2.7183)}{2 + 4 + 2} = -0.1052.$

Thus the reference deviation is

10.41 $H = 0.1052.$

Let us suppose that instead of starting with the reference set (10.37) we had started with some other reference set and that for this reference set we had the levelled reference-polynomial

10.42 $Q(x) = 0.9 + 1.6x.$

Since

10.43
$$\begin{cases} h_0 = Q(x_0) - f(x_0) = 0.9 - 1.0000 = -0.1000 \\ h_1 = Q(x_1) - f(x_1) = 1.7 - 1.6487 = 0.0513 \\ h_2 = Q(x_2) - f(x_2) = 2.5 - 2.7183 = -0.2183 \end{cases}$$

it follows that $Q(x)$ is a reference-polynomial for the reference-set (10.37). We now verify (10.28). Evidently, by (10.29) we have

10.44 $w_1 = \tfrac{1}{4}, \quad w_2 = \tfrac{1}{2}, \quad w_3 = \tfrac{1}{4},$

and hence by (10.43),

10.45 $H = 0.1052 = \tfrac{1}{4}|h_0| + \tfrac{1}{2}|h_1| + \tfrac{1}{4}|h_2|$

$= \tfrac{1}{4}(0.1000) + \tfrac{1}{2}(0.0513) + \tfrac{1}{4}(0.2183).$

We now determine $P(x)$ by linear interpolation in the points x_0 and x_1 of (10.37) with

10.46 $P(x_0) = f(x_0) + \delta = 0.8948, \quad P(x_1) = f(x_1) - \delta = 1.7539.$

Evidently we have

10.47 $P(x) = P(x_0) + \dfrac{x - x_0}{x_1 - x_0}(P(x_1) - P(x_0)) \doteq 0.8948 + 1.7182x.$

and

10.48 $P(x_2) = 2.6130 = f(x_2) - 0.1053 \doteq f(x_2) + \delta.$

We verify that

10.49 $\lambda_0 P(x_0) + \lambda_1 P(x_1) + \lambda_2 P(x_2) \doteq 2(0.8948) - 4(1.7539) + 2(2.6130) = 0$

so that (10.11) holds.

Since the $P(x_i) - f(x_i)$, $i = 0, 1, 2$ have alternating signs and equal magnitudes, it follows that $P(x)$ is the levelled reference-polynomial for the reference set (10.37).

By direct computation it is found that

10.50 $\max\limits_{x \in J} |P(x) - f(x)| = |P(\hat{x}) - f(\hat{x})| = 0.106656$

where $\hat{x} = 0.54$. Since $P(\hat{x}) - f(\hat{x}) = 0.106656$ we replace the reference set $\{0, 0.5, 1\}$ by $\{0, 0.54, 1\}$. We then determine the levelled reference-polynomial

10.51 $\pi_1(x) = 0.894067 + 1.718282x$

for this new reference set. Since the reference deviation is $H = 0.105933$ and since

10.52 $\max\limits_{J} |\pi_1(x) - f(x)| = |\pi_1(1) - f(1)| = 0.105933$

the process is terminated and $\pi_1(x)$ is our desired polynomial of best approximation.

From a computer program based on the above procedure, the following results were obtained

n	$\pi_n(x)$	x	$\max\limits_{J}|\pi_n(x) - f(x)|$
0	1.859141	0	0.859141
1	$1.718282x + 0.894067$	1	0.105933
2	$0.846024x^2 + 0.854752x + 1.008753$	1	0.008753
3	$0.279979x^3 + 0.421699x^2 + 1.016603x + 0.999455$	1	0.000545

Evidently the accuracy of the approximation increases rapidly as the degree of the polynomial increases.

For the case $n = 1$ the polynomial $\pi_1(x)$ agrees exactly with the polynomial of best approximation with respect to the continuous interval $[0, 1]$. The latter can be found analytically as follows. We seek constants α and β such that

10.53
$$\max_{0 \leq x \leq 1} |\alpha x + \beta - e^x|$$

is minimized. Evidently the extreme values of $\alpha x + \beta - e^x$ occur at $x = 0$, $x = 1$, and at $x = \log \alpha$. From Theorem 10.3 we seek to choose α and β so that

10.54
$$\begin{cases} [\alpha x + \beta - e^x]_{x=0} = \beta - 1 = \delta \\ [\alpha x + \beta - e^x]_{x=\log \alpha} = \alpha \log \alpha + \beta - \alpha = -\delta \\ [\alpha x + \beta - e^x]_{x=1} = \alpha + \beta - e = \delta. \end{cases}$$

This leads to the conditions

10.55
$$\begin{cases} \alpha = e - 1 \doteq 1.718282 \\ \beta = \tfrac{1}{2}[1 + \alpha - \alpha \log \alpha] \doteq 0.894067. \end{cases}$$

Thus the polynomial $\pi_1(x)$ of best approximation is given by (10.34) and the maximum absolute deviation on the interval $[a, b]$ is given by (10.35) and is the same as that found by the Remes algorithm.

Let us now discuss the choice of the initial reference set. One procedure is simply to let $x_0 = a$, $x_{n+1} = b$ and to choose the remaining n points uniformly, or nearly uniformly, spaced in the interval $[a, b]$. Another procedure, suggested by Stiefel [1959] is to let*

10.56
$$x_i = \frac{b + a}{2} + \frac{b - a}{2} \cos \frac{\pi i}{n + 1}, \qquad i = 0, 1, \ldots, n + 1.$$

In the case $a = -1$, $b = 1$, this choice corresponds to the extreme values of the Chebyshev polynomial of degree $n + 1$, which is given by

10.57
$$T_{n+1}(x) = \cos((n + 1)\cos^{-1} x).$$

Another choice, also suggested by Stiefel [1959], is to determine the polynomial $Q_n(x)$ of degree n or less which minimizes

10.58
$$\sum_{x \in J} (P_n(x) - f(x))^2$$

with respect to all polynomials $P_n(x)$ of degree n or less. Methods for finding

---

* Actually, we use points of J as close as possible to those given by (10.56).

$Q_n(x)$ are given in Section 6.11. Stiefel [1959a] has shown that there is a reference set R such that $Q_n(x)$ is a reference-polynomial for R. Thus, assuming that $Q_n(x)$ is reasonably close to minimizing (10.58), we would have a good start for the application of the Remes algorithm.

The convergence of the algorithm can often be accelerated by the following modification. The resulting method is known as the *second algorithm of Remes* (see Remes [1934a], Hart, *et al.* [1968, p. 45], Rice [1964, p. 176], and Fanett [1963]). Given a reference set $x_0, x_1, \ldots, x_{n+1}$ we proceed to construct the levelled reference-polynomial $P_n(x)$ as before. Since the function

10.59
$$h(x) = P_n(x) - f(x)$$

changes sign in each of the intervals $[x_0, x_1], [x_1, x_2], \ldots, [x_n, x_{n+1}]$ it has at least one zero in each interval. Let z_i be any zero of $h(x)$ in the interval $[x_i, x_{i+1}]$, $i = 0, 1, 2, \ldots, n$. In each of the $n + 2$ intervals

$$I_0 = [a, z_0]$$
$$I_1 = [z_0, z_1]$$
$$\vdots$$
$$I_{n+1} = [z_n, b]$$

find a value of x, say v_i, such that $h(v_i)$ is an extremum in the interval and $h(v_i)$ has the same sign as $h(x_i)$. Clearly $P_n(x)$ is a reference-polynomial for the reference set $v_0, v_1, \ldots, v_{n+1}$. If for some v_i we have

10.60
$$|P_n(v_i) - f(v_i)| = \max_{x \in J} |P_n(x) - f(x)|.$$

then we accept the reference set $v_0, v_1, \ldots, v_{n+1}$ and proceed as before. Otherwise we find $\hat{x}$ such that

10.61
$$|P_n(\hat{x}) - f(\hat{x})| = \max_{x \in J} |P_n(x) - f(x)|$$

and "exchange" $\hat{x}$ with one of the v_i so that $P_n(x)$ is a reference-polynomial for the resulting reference set. The exchange is carried out as in Step (4) of the algorithm described above.

We remark that it is not necessary to find the zero z_i exactly. If one can find two consecutive values z_i', z_i'' of J such that $h(z_i')h(z_i'') < 0$ then the interval I_i can have z_i' as a right end point and I_{i+1} can have z_i'' as a left end point.

EXERCISES 6.10

1. Give a direct proof that the choice of α and β given by (10.55) actually yields the polynomial of best approximation of degree one or less.

2. Consider the approximation of the function $f(x) = \sqrt{x}$ in the interval $0.5 \leq x \leq 1.5$. For the case $n = 1$ and the reference set $R: x_0 = 0.5$, $x_1 = 0.75$, $x_2 = 1.50$ determine the levelled reference-polynomial $P(x)$ and the reference deviation. Using the second Remes algorithm (assume J is the set $x = 0.5(0.01)1.5$) find a new reference set and compute its reference deviation. Verify (10.11) and (10.28) for $P(x)$ and for the new reference set.

3. In the previous exercise determine the optimum polynomial of degree one or less relative to the continuous interval $\frac{1}{2} \leq x \leq \frac{3}{2}$.

4. If $f(x)$ is continuous in $a \leq x \leq b$, show that for any reference set the reference deviation does not exceed $\max_{a \leq x \leq b}|f(x)|$.

5. Write a computer program to determine the polynomial $\pi_n(x)$ of degree n or less which best approximates $f(x)$, in the uniform norm, over the set J:

$$x_i = a + \frac{i}{M}(b - a), \qquad i = 0, 1, \ldots, M.$$

Consider the following special cases:

$f(x)$	a	b	M	n
e^x	0	1	50,100	$1, 2, 3, 4, 5$
$\sin\frac{\pi}{2}x$	0	1	50,100	$1, 2, 3, 4, 5$
$\sqrt{x}$	$\frac{1}{2}$	$\frac{3}{2}$	50,100	$1, 2, 3, 4, 5$

Find

$$\max_{x \in J}|\pi_n(x) - f(x)|.$$

Choose the initial reference set by letting $x_0 = a$, $x_{n+1} = b$, and by letting the remaining points be as nearly as possible uniformly spaced in $a \leq x \leq b$.

6. Modify the program in the previous exercise to determine the initial reference set using points as near as possible to those given in (10.56) and to use the second algorithm of Remes.

7. Find the polynomial $\pi_5(x)$ of degree 5 or less which is the best approximation to e^x in the uniform norm over the set $t_i = i/100$, $i = 0, 1, 2, \ldots, 100$.

8. How many terms of the Taylor series expansion of e^x at $x = 0$ would be needed to give the same accuracy in the interval $0 \leq x \leq 1$ as can be obtained by using the polynomial of degree one of best approximation with respect to the uniform norm?

6.11 LEAST SQUARES APPROXIMATION BY POLYNOMIALS

We now consider the problem of approximating a given function $f(x)$ continuous on the interval $[a, b]$ by a polynomial $\tau_n(x)$ of degree n or less in such a way that the L_2-norm of $P_n(x) - f(x)$, namely,

11.1 $$\|P_n - f\|_2 = \left[\int_a^b (P_n(x) - f(x))^2\, dx\right]^{1/2},$$

is minimized over all polynomials $P_n(x)$ of degree n or less. Instead of the continuous interval $a \leq x \leq b$, we may wish to consider a finite set of points J: $x_1, x_2, \ldots, x_M$ and seek to minimize

11.2 $$\|P_n - f\|_{2,J} = \left[\sum_{i=1}^{M} (P_n(x_i) - f(x_i))^2 \right]^{1/2}.$$

We omit the subscript J when no confusion will arise. For this modified problem the values of $f(x_i)$ may correspond to observed "data" rather than being the values of a continuous function at the x_i.

Let us now introduce the *inner product* (f, g) of two functions $f(x)$ and $g(x)$ by

11.3 $$(f, g) = \begin{cases} \displaystyle\int_a^b f(x)g(x)dx, & \text{for the continuous case} \\ \displaystyle\sum_{i=1}^{M} f(x_i)g(x_i), & \text{for the discrete case.} \end{cases}$$

We seek to construct a family of polynomials $\phi_0(x), \phi_1(x), \ldots$ of exact degree zero, one, two, etc. which are pairwise orthogonal in the sense that

11.4 $$(\phi_i, \phi_j) = 0 \quad \text{if} \quad i \neq j.$$

We note that if $M > n$, then

11.5 $$(\phi_i, \phi_i) = \|\phi_i\|_2^2 > 0, \qquad i = 0, 1, 2, \ldots, n.$$

For otherwise $\phi_i(x) = 0$ on $[a, b]$ (or on J). But if a polynomial of degree n or less vanishes at more than n points, it must vanish identically.

We note that in the continuous case

11.6 $$(\phi, \phi) = \|\phi\|_2^2 > 0$$

unless $\phi(x) \equiv 0$ in $[a, b]$. Similarly, (11.6) holds in the discrete case unless $\phi(x) \equiv 0$ on J. Moreover, if $M > n$ and if $\phi(x)$ is a polynomial of degree n or less, then either (11.6) holds or else

11.7 $$\phi(x) \equiv 0,$$

not only on J but for all x. This follows since if a polynomial of degree n or less vanishes at more than n points it vanishes identically.

The construction of the polynomials $\phi_0(x), \phi_1(x), \ldots$ is carried out by the following recurrence relation

11.8
$$\begin{cases} \phi_0(x) = 1 \\[2mm] \phi_1(x) = x - \dfrac{(x\phi_0, \phi_0)}{(\phi_0, \phi_0)}\phi_0(x) = x - \dfrac{(x, 1)}{(1, 1)} \\[3mm] \phi_{k+1}(x) = x\phi_k(x) - \alpha_k\phi_k(x) - \beta_k\phi_{k-1}(x), \qquad k = 1, 2, \ldots \end{cases}$$

where

11.9
$$\alpha_k = \frac{(x\phi_k, \phi_k)}{(\phi_k, \phi_k)}$$

11.10
$$\beta_k = \frac{(\phi_k, \phi_k)}{(\phi_{k-1}, \phi_{k-1})}.$$

We readily verify that $(\phi_0, \phi_1) = 0$. Assume that (11.4) holds for all $i, j \leq k$. It is easily seen that $(\phi_k, \phi_{k+1}) = 0$. Moreover,

11.11
$$(\phi_{k-1}, \phi_{k+1}) = (\phi_{k-1}, x\phi_k) - \alpha_k(\phi_{k-1}, \phi_k) - \beta_k(\phi_{k-1}, \phi_{k-1})$$
$$= (x\phi_{k-1}, \phi_k) - \beta_k(\phi_{k-1}, \phi_{k-1})$$

since $(\phi_{k-1}, \phi_k) = 0$ by assumption. But by (11.8) we have if $k \geq 2$,

11.12
$$x\phi_{k-1}(x) = \phi_k(x) + \alpha_{k-1}\phi_{k-1}(x) + \beta_{k-1}\phi_{k-2}(x)$$

so that

11.13
$$(x\phi_{k-1}, \phi_k) = (\phi_k, \phi_k).$$

By (11.10) it follows that $(\phi_{k-1}, \phi_{k+1}) = 0$ for $k \geq 2$. If $k = 1$, we have, by (11.8)

11.14
$$x\phi_{k-1}(x) = x\phi_0(x) = \phi_1(x) + \frac{(x, 1)}{(1, 1)}\phi_0(x)$$

and

11.15
$$(x\phi_0, \phi_1) = \left(\phi_1 + \frac{(x, 1)}{(1, 1)}, \phi_1\right)$$
$$= (\phi_1, \phi_1) + \frac{(x, 1)}{(1, 1)}(1, \phi_1) = (\phi_1, \phi_1)$$

since $(1, \phi_1) = 0$. Thus, by (11.11), $(\phi_0, \phi_2) = 0$, and therefore $(\phi_{k-1}, \phi_{k+1}) = 0$ for $k \geq 1$.

Next we have

11.16 $\qquad (\phi_{k-2}, \phi_{k+1}) = (\phi_{k-2}, x\phi_k) - \alpha_k(\phi_{k-2}, \phi_k) - \beta_k(\phi_{k-2}, \phi_{k-1})$

$\qquad\qquad\qquad\qquad = (x\phi_{k-2}, \phi_k).$

If $k \geqq 3$ we have

11.17 $\qquad x\phi_{k-2}(x) = \phi_{k-1}(x) + \alpha_{k-2}\phi_{k-2}(x) + \beta_{k-2}\phi_{k-3}(x)$

so that $(x\phi_{k-2}, \phi_k) = 0$ and, by (11.16),

11.18 $\qquad\qquad\qquad\qquad (\phi_{k-2}, \phi_{k+1}) = 0.$

If $k = 2$, then, as before

11.19 $\qquad\qquad\qquad x\phi_0(x) = \phi_1(x) + \dfrac{(x, 1)}{(1, 1)}\phi_0(x)$

so that $(\phi_{k-2}, \phi_{k+1}) = 0$. In a similar way we can show that

11.20 $\qquad (\phi_{k-3}, \phi_{k+1}) = (\phi_{k-4}, \phi_{k+1}) = \cdots = (\phi_0, \phi_{k+1}) = 0;$

hence (11.4) follows by induction.

As an example, consider the continuous case with $a = -1, b = 1$. We get the polynomials

11.21 $\qquad \begin{cases} \phi_0(x) = 1 \\ \phi_1(x) = x \\ \phi_2(x) = x^2 - \frac{1}{3} \\ \phi_3(x) = x^3 - \frac{3}{5}x \end{cases}$

etc., which are proportional to the Legendre polynomials.* One can easily verify that the polynomials are pairwise orthogonal.

Evidently each $\phi_k(x)$, $k = 0, 1, \ldots, n$ is a polynomial of degree k with the coefficient of x^k equal to unity. Hence x^k can be expressed as a linear combination of $\phi_0(x), \phi_1(x), \ldots, \phi_k(x)$. Therefore, for any polynomial $P_n(x)$ of degree n or less we have

11.22 $\qquad\qquad\qquad\qquad P_n(x) = \sum_{k=0}^{n} c_k\phi_k(x).$

We now seek to choose the c_k to minimize

11.23 $\qquad\qquad\qquad\qquad I = \|(P_n(x) - f(x)\|_2^2.$

---

* The Legendre polynomials are discussed in Section 7.8.

By (11.5), (11.22), and (11.4) we have

11.24 $$I = \sum_{k=0}^{n} c_k^2(\phi_k, \phi_k) - 2 \sum_{k=0}^{n} c_k(\phi_k, f) + (f, f).$$

Therefore,

11.25 $$\frac{\partial I}{\partial c_k} = 2(c_k(\phi_k, \phi_k) - (\phi_k, f))$$

and

11.26 $$\frac{\partial^2 I}{\partial c_k \partial c_s} = \begin{cases} 2(\phi_k, \phi_k) & \text{if } k = s \\ 0 & \text{if } k \neq s. \end{cases}$$

Consequently we have an absolute minimum of I provided

11.27 $$c_k = \frac{(\phi_k, f)}{(\phi_k, \phi_k)}, \qquad k = 0, 1, \ldots, n.$$

With this choice of the c_k we have

11.28 $$I = (f, f) - \sum_{k=0}^{n} \frac{(\phi_k, f)^2}{(\phi_k, \phi_k)}.$$

As an example, consider the approximation of the function $f(x) = 1 + x^3$ by a polynomial of degree 2 or less on the interval $-1 \leq x \leq 1$. By (11.21) and (11.27) we have

11.29 $$\tau_2(x) = c_0 + c_1 x + c_2(x^2 - \tfrac{1}{3})$$

where

11.30 $$\begin{cases} c_0 = \dfrac{(1, 1 + x^3)}{(1, 1)} = 1 \\[2mm] c_1 = \dfrac{(x, 1 + x^3)}{(x, x)} = \dfrac{3}{5} \\[2mm] c_2 = \dfrac{(x^2 - \frac{1}{3}, 1 + x^3)}{(x^2 - \frac{1}{3}, x^2 - \frac{1}{3})} = 0. \end{cases}$$

Thus the polynomial of best approximation is

11.31 $$\tau_2(x) = 1 + \tfrac{3}{5}x$$

and by (11.28)

11.32 $\|\tau_2(x) - f(x)\|_2^2 = \int_{-1}^{1} [\tau_2(x) - (1 + x^3)]^2 \, dx$

$$= (f, f) - \sum_{k=0}^{1} \frac{(\phi_k, f)^2}{(\phi_k, \phi_k)} = \frac{16}{7} - 2 - \frac{6}{25} = \frac{8}{175} \doteq 0.0457.$$

We remark that in practice it is generally better to compute the $\phi_k(x)$ using the recurrence relation (11.8) as needed rather than writing $\tau_n(x)$ as a polynomial in x (Forsythe [1957]).

For the case of the discrete set J one could, alternatively, proceed as follows. Let

11.33 $P_n(x) = \sum_{i=0}^{n} d_i x^i$

and choose the coefficients d_i so that

11.34 $\|P_n - f\|_2^2$

is minimized. This leads to the "normal" equations

11.35 $\sum_{j=0}^{n} \alpha_{i,j} d_j = \beta_i, \qquad i = 0, 1, \ldots, n$

where

11.36 $\alpha_{i,j} = (x^i, x^j), \qquad \beta_i = (x^i, f), \qquad i, j = 0, 1, 2, \ldots, n.$

Thus, for example, if $n = 2$ we have

11.37 $\begin{cases} (1, 1)d_0 + (1, x)d_1 + (1, x^2)d_2 = (1, f) \\ (x, 1)d_0 + (x, x)d_1 + (x, x^2)d_2 = (x, f) \\ (x^2, 1)d_0 + (x^2, x)d_1 + (x^2, x^2)d_2 = (x^2, f). \end{cases}$

If $\|\tau_n\|_2^2 = 0$, then $\tau_n(x)$ vanishes for all x in J. If J contains at least $n + 1$ points, this implies that $\tau_n(x) \equiv 0$ and hence all of the coefficients d_i vanish. Hence $\|\tau_n\|_2^2 > 0$ unless all $d_i = 0$. Therefore, as shown in Chapter 11, the expression $\|\tau_n\|_2^2$, which can be considered as a *quadratic form* in the d_i, is positive definite. Hence the matrix

11.38 $\begin{pmatrix} \alpha_{0,0} & \alpha_{0,1} \cdots \alpha_{0,n} \\ \alpha_{1,0} & \alpha_{1,1} \cdots \alpha_{1,n} \\ . & . \quad \cdots \quad . \\ \alpha_{n,0} & \alpha_{n,1} \cdots \alpha_{n,n} \end{pmatrix}$

is positive definite, and is therefore *nonsingular*. That implies that there exists a unique solution of the system (11.37). On the other hand, the matrix is frequently very "ill-conditioned," in the sense that when one tries to solve the system (11.37), a severe loss of accuracy may result. (See Forsythe [1957].) This may be explained as follows. If the x_i are uniformly distributed in the interval $0 \leqq x \leqq 1$, then the matrix (11.38) is very close to a multiple of the Hilbert matrix of order $n + 1$, which is given by

11.39
$$H = \begin{pmatrix} 1 & \dfrac{1}{2} & \dfrac{1}{3} & \cdots & \dfrac{1}{n+1} \\ \dfrac{1}{2} & \dfrac{1}{3} & \dfrac{1}{4} & \cdots & \dfrac{1}{n+2} \\ \cdot & \cdot & \cdot & \cdots & \cdot \\ \dfrac{1}{n} & \dfrac{1}{n+1} & \dfrac{1}{n+2} & \cdots & \dfrac{1}{2n} \\ \dfrac{1}{n+1} & \dfrac{1}{n+2} & \dfrac{1}{n+3} & \cdots & \dfrac{1}{2n+1} \end{pmatrix}.$$

This follows since

11.40 $$\alpha_{i,j} = (x^i, x^j) = \sum_{k=0}^{M} x^{i+j} \sim M \int_0^1 x^{i+j}\,dx = M(i + j + 1)^{-1}.$$

The Hilbert matrix is known to be very ill-conditioned. With single precision on most machines, it is impractical to solve accurately a system involving a Hilbert matrix of order more than 8. (See discussion in Chapter 12.)

On the other hand, the procedure described above, based on the use of orthogonal polynomials in effect involves only a diagonal matrix because of the orthogonality of the $\phi_k(x)$. Hence there is no concern about the condition of the matrix. In practice, even if the $\phi_k(x)$ are computed based on the recurrence relation (11.8) the $\phi_k(x)$ may not actually be exactly orthogonal, i.e., the computed values of the (ϕ_i, ϕ_j) may not vanish exactly for $i \neq j$. In this case the c_k can be determined by the system

11.41 $$\sum_{s=0}^{n} \beta_{k,s} c_s = (\phi_k, f), \qquad k = 0, 1, \ldots, n$$

where

11.42 $$\beta_{k,s} = (\phi_k, \phi_s).$$

Even though the matrix

11.43
$$\begin{pmatrix} (\phi_0, \phi_0) & (\phi_0, \phi_1) \cdots (\phi_0, \phi_n) \\ (\phi_1, \phi_0) & (\phi_1, \phi_1) \cdots (\phi_1, \phi_n) \\ \cdot & \cdot \quad \cdots \quad \cdot \\ (\phi_n, \phi_0) & (\phi_n, \phi_1) \cdots (\phi_n, \phi_n) \end{pmatrix}$$

is no longer a diagonal matrix, it is nearly so and one would not expect any serious difficulty in finding the c_k.

EXERCISES 6.11

1. Compute $\phi_n(x)$ for $n = 0(1)5$ from (11.8) for the continuous case with $a = -1, b = 1$.

2. The *Legendre polynomials* $P_n(x)$ are defined by

$$P_n(x) = \frac{1}{2^n n!} \frac{d^n}{dx^n} (x^2 - 1)^n.$$

   Verify that for $n = 1(1)5$ we have

$$P_n(x) = a_n \phi_n(x)$$

   where $a_0 = a_1 = 1$ and $a_{n+1} = (2n + 1)a_n/(n + 1)$, $n = 1, 2, \ldots$. Also verify that for $n = 1(1)5$ the Legendre polynomials satisfy the differential equation

$$(1 - x^2)y'' - 2xy' + n(n + 1)y = 0$$

   and the recurrence relation $P_0(x) = 1$, $P_1(x) = x$, $(n + 1)P_{n+1}(x) = (2n + 1)xP_n(x) - nP_{n-1}(x)$, $n = 1, 2, \ldots$. Finally, show that the polynomials $\phi_0(x), \phi_1(x), \ldots, \phi_5(x)$ satisfy the recurrence relation

$$\phi_0(x) = 1, \phi_1(x) = x$$

$$\phi_{n+1}(x) = x\phi_n(x) - \frac{n^2}{4n^2 - 1}\phi_{n-1}(x).$$

3. Using the last result of Exercise 2, find (ϕ_n, ϕ_n) for $n = 1(1)5$ where the $\phi_n(x)$ are given by Exercise 1. Also verify the result by direct integration for $n = 1, 2, 3$.

4. For $n = 0(1)5$ express x^n in terms of Legendre polynomials of degree n or less.

5. Find the polynomial of degree two or less which gives the best approximation to e^x in the L_2-norm over the interval $0 \le x \le 1$.

6. Given the following experimental data, determine the polynomials $\tau_1(x)$ and $\tau_2(x)$ of degree one or less and two or less, respectively, which best represent the data in the L_2-norm.

x:	0	0.4	0.8	1.2	1.6	2.0
$f(x)$:	0.21	1.25	2.31	2.70	2.65	3.20

6.12 RATIONAL APPROXIMATION

Other functions besides polynomials can be used for interpolation and for approximation of a given function.   In the next two sections we consider two classes of functions, namely rational functions and trigonometric polynomials.   Our discussion in each case will be rather brief.   The reader should consult the references given for further information.

Rational Functions

In spite of the fact that any function $f(x)$ which is continuous on a closed interval can be approximated to an arbitrary degree of accuracy on the interval by some polynomial (see Theorem 10.3), nevertheless, for many functions the use of polynomial approximation may not be satisfactory.   For such functions the degree of the polynomial required to attain a given accuracy of approximation may be excessively high.   In some cases, the function may have a pole near the interval under consideration, as, for example, in the case of the function $\tan x$ in the interval $0 \leqq x \leqq \pi/4$.   One might expect that in such a case a rational function would give a more accurate approximation than a polynomial of corresponding degree.   As a matter of fact, it has been found that in many cases considerably greater accuracy can be obtained for a given amount of computational effort by the use of rational functions than by the use of polynomials.   This is borne out by an inspection of the tables of polynomial and rational function approximations given by Hart, et $al.$ [1968] for a variety of mathematical functions.

Let us, then, consider the approximate representation of a given function $f(x)$ by a rational function of the form

12.1
$$F(x) = F_{m,k}(x) = \frac{P_m(x)}{Q_k(x)}$$

where $P_m(x)$ and $Q_k(x)$ are polynomials of degree m and k respectively, which have no common zero.   With $Padé$ $rational$ $approximation$, Padé [1892], given m, k, and α, one chooses $P_m(x)$ and $Q_k(x)$ such that $F(\alpha) = f(\alpha)$ and such that as many derivatives as possible of $F(x)$ at $x = \alpha$ are equal to the corresponding derivatives of $f(\alpha)$.   Thus Padé rational approximation is a kind of generalization of Taylor's series.   It is assumed that $f(x)$ has a Taylor's series expansion near $x = \alpha$ given by

12.2
$$f(x) = \sum_{j=0}^{\infty} \hat{c}_j(x - \alpha)^j$$

where

12.3
$$\hat{c}_j = \frac{f^{(j)}(\alpha)}{j!} \qquad j = 0, 1, \dots .$$

We seek polynomials $P_m(x)$ and $Q_k(x)$ of the form

12.4
$$\begin{cases} P_m(x) = \hat{a}_0(x - \alpha)^m + \hat{a}_1(x - \alpha)^{m-1} + \cdots + \hat{a}_m \\ Q_m(x) = \hat{b}_0(x - \alpha)^k + \hat{b}_1(x - \alpha)^{k-1} + \cdots + \hat{b}_k. \end{cases}$$

We clearly should require that $\hat{b}_k \neq 0$. Otherwise $F(x)$ would have a pole at $x = \alpha$, unless $\hat{a}_m = 0$. If $\hat{a}_m = \hat{b}_k = 0$, then the factor $x - \alpha$ would be common to $P_m(x)$ and $Q_k(x)$ and could be divided out. With $\hat{b}_k \neq 0$, there is no loss of generality in assuming that $\hat{b}_k = 1$. Thus there are $N + 1$ free coefficients available, where

12.5
$$N = m + k,$$

to be chosen so that $f(x) - F(x)$ and its first N derivatives vanish at $x = \alpha$. It is easy to show (see, for instance, Ralston [1965]) that the coefficients can be chosen so that in the expression

12.6
$$\sum_{j=0}^{\infty} \hat{c}_j(x - \alpha)^j Q_k(x) - P_m(x)$$

the constant term and the terms involving $(x - \alpha), (x - \alpha)^2, \ldots, (x - \alpha)^N$ all vanish. With this choice of the coefficients $f(x) - F(x)$ and its first N derivatives vanish at $x = \alpha$.

As an example consider the case $f(x) = e^x$ with $\alpha = 0$, $m = 1$, $k = 1$. Evidently,

12.7
$$e^x = 1 + x + \frac{x^2}{2} + \frac{x^3}{6} + \cdots.$$

We seek to choose a_0, a_1, and b_0 such that

12.8
$$e^x - \frac{a_0 x + a_1}{b_0 x + 1}$$

and its first 2 derivatives vanish at $x = 0$. We thus require that the constant term and the terms involving x and x^2 vanish in the expression

12.9
$$\left(1 + x + \frac{x^2}{2} + \frac{x^3}{6} + \cdots\right)(b_0 x + 1) - (a_0 x + a_1)$$

$$= (1 - a_1) + (1 + b_0 - a_0)x + (b_0 + \tfrac{1}{2})x^2 + \cdots$$

But this requirement implies that

12.10
$$b_0 = -\tfrac{1}{2}, \quad a_0 = \tfrac{1}{2}, \quad a_1 = 1.$$

Thus with $m = 1, k = 1$, the Padé approximation for e^x at $\alpha = 0$ is given by

12.11 $$F(x) = F_{1,1}(x) = \frac{1 + \frac{1}{2}x}{1 - \frac{1}{2}x}.$$

One can, of course, construct a Padé approximation $F_{m,k,\alpha}(x)$ for any pair of integers m and k and for any α. Empirical evidence indicates that for a given value of $N = m + k$ the most accurate approximations are usually obtained for $m = k$ or $k + 1$ (Ralston [1965]).

Like Taylor's series, the accuracy of Padé approximations usually decreases rapidly as $|x - \alpha|$ increases. Consequently, one is led to consider the possibility of choosing $P_m(x)$ and $Q_k(x)$ to minimize the maximum of the deviation $|P_m(x)/Q_k(x) - f(x)|$ over some interval. The existence of a unique rational approximation of best approximation was shown by Walsh [1931]. (For uniqueness one must require that $P_m(x)$ and $Q_k(x)$ have no common zeros, and one must impose a normalization condition, for example, by fixing $Q(\alpha)$ for a given value of α.) It was shown by Achiezer [1956] that for the rational function $F(x)$ of best approximation the error $F(x) - f(x)$ has a certain number of extrema where $F(x) - f(x)$ has equal magnitude and alternating signs. On the basis of this alternation property one can construct an algorithm, similar to the second algorithm of Remes, discussed in Section 6.10, to find $F(x)$ (see, for instance, Werner [1962], Ralston [1965], and Hart, *et al.* [1968]). Several difficulties are encountered which were not present in the polynomial case. First, the algorithm may not converge unless a sufficiently good initial estimate of $F(x)$ can be obtained. Second, at each step of the process one has to solve a set of nonlinear equations. Techniques for overcoming these difficulties are given by Hart, *et al.* [1968].

Many functions can be represented by an (infinite) *continued fraction*. A continued fraction is an expression of the form

12.12 $$F(x) = P_0(x) + \cfrac{P_1(x)}{Q_1(x) + \cfrac{P_2(x)}{Q_2(x) + \cfrac{P_3(x)}{Q_3(x) + \cdots}}}$$

which is often written in the compact form

12.13 $$F(x) = P_0(x) + \frac{P_1(x)}{\lfloor Q_1(x)} + \frac{P_2(x)}{\lfloor Q_2(x)} + \frac{P_3(x)}{\lfloor Q_3(x)} + \cdots$$

Normally the $P_i(x)$ and $Q_i(x)$ are simple functions such as linear functions or constants. In such a case it is clear that a finite, or truncated, continued fraction is equivalent to a rational function. (This is, of course, also true if each $P_i(x)$ and each $Q_i(x)$ is a polynomial in x.) Conversely, given a rational function one can construct an equivalent (truncated) continued fraction (see, for instance, Maehly

[1959], Ralston [1965], Fike [1968], and Hart, *et al.* [1968]). The use of the truncated continued fraction for evaluating a rational function may be more economical, from the standpoint of the number of arithmetic operations required, than the computation of the numerator and denominator polynomials separately and dividing. In the case where the $Q_i(x)$ have the form $x + \beta_i$ and where each $P_i(x)$ is a constant, say γ_i, then the evaluation of the continued fraction can be done by a sequence of additions and subtractions. The computational algorithm is analogous to the nesting procedure described in Chapter 2 for evaluating a polynomial where one alternately multiplies and adds. For a discussion of the numerical evaluation of continued fractions see Blanch [1964].

EXERCISES 6.12

1. For the function e^x construct a Padé rational approximation function $F(x)$ with $m = k = 2$ and $\alpha = 0$.

2. In the preceding exercise, compute the maximum deviation $|F(x) - e^x|$ in the interval $[0, 1]$ and compare with the maximum deviation of $|\hat{F}(x) - e^x|$ in $[0, 1]$ where $\hat{F}(x)$ is the polynomial of degree five or less which best approximates e^x in the minimax sense for $x \in J$. (Use the methods of Section 6.10 to find $\hat{F}(x)$ and let $J = 0(0.01)1$.)

3. For the function $\sin x$ construct a Padé rational approximation function $F(x)$ with $m = k = 2$ and $\alpha = 0$. Do the same with $m = 3, k = 2$.

4. Find a, b, and c so that

$$\max_{0 \le x \le 1} \left| e^x - \frac{ax + b}{cx + 1} \right| = \delta$$

is minimized. Compare the minimum value of δ with

$$\max_{0 \le x \le 1} \left| e^x - \frac{1 + \frac{1}{2}x}{1 - \frac{1}{2}x} \right|.$$

6.13 TRIGONOMETRIC INTERPOLATION AND APPROXIMATION

A *trigonometric polynomial* of degree M or less is an expression of the form

13.1 $$S_M(x) = \frac{a_0}{2} + \sum_{n=1}^{M} (a_n \cos nx + b_n \sin nx).$$

We consider the representation of a function $f(x)$ defined in the interval

13.2 $$0 \le x \le 2\pi$$

in terms of trigonometric polynomials. If $f(x)$ is defined in some other interval $[a, b]$, a simple linear change of variable can reduce the problem to the case of the interval (13.2).

A trigonometric polynomial can be obtained by truncating a *trigonometric series* of the form

13.3 $$S(x) = \frac{a_0}{2} + \sum_{n=1}^{\infty} (a_n \cos nx + b_n \sin nx).$$

Given a function $f(x)$ which is integrable over the interval $[0, 2\pi]$, we can define the *Fourier series* for $f(x)$ by

13.4 $$F(x) = \frac{\alpha_0}{2} + \sum_{n=1}^{\infty} (\alpha_n \cos nx + \beta_n \sin nx)$$

where

13.5 $$\begin{cases} \alpha_n = \frac{1}{\pi} \int_0^{2\pi} f(x) \cos nx \, dx, & n = 0, 1, 2, \ldots \\[2mm] \beta_n = \frac{1}{\pi} \int_0^{2\pi} f(x) \sin nx \, dx, & n = 1, 2, 3, \ldots. \end{cases}$$

We remark that the Fourier series for $f(x)$ need not converge (pointwise) for all values of x even if $f(x)$ is continuous. In fact, the series may not even converge for any x. Moreover, even if the series does converge for a particular value of x it need not converge to $f(x)$.

We now state without proof the following result (see, for instance, Jackson [1941]).

13.6 Theorem (Convergence Theorem). If $f(x)$ is periodic with period 2π and is continuous except possibly for a finite number of finite jumps in a period, then its Fourier series converges at every point where $f(x)$ has a right-hand and a left-hand derivative, whether these are the same or different. At a jump discontinuity, say x, the convergence is to

13.7 $$\frac{f(x+) + f(x-)}{2}.$$

Let $F_M(x)$ be defined by

13.8 $$F_M(x) = \frac{\alpha_0}{2} + \sum_{n=1}^{M} (\alpha_n \cos nx + \beta_n \sin nx)$$

where the α_i and β_i are given by (13.5). We show that $F_M(x)$ has the property that

13.9 $$\int_0^{2\pi} (f(x) - F_M(x))^2 \, dx \leqq \int_0^{2\pi} (f(x) - S_M(x))^2 \, dx$$

where $S_M(x)$ is any trigonometric polynomial of degree M or less. To see this we write

13.10
$$\int_0^{2\pi} (f(x) - S_M(x))^2\, dx = \int_0^{2\pi} [f(x)]^2\, dx$$
$$- 2\int_0^{2\pi} f(x)S_M(x)\, dx + \int_0^{2\pi} [S_M(x)]^2\, dx.$$

Using (13.1) and the fact that

13.11
$$\begin{cases} \int_0^{2\pi} \cos nx \cos mx\, dx = \begin{cases} 0, & m \neq n \\ \pi, & m = n,\, m \neq 0 \\ 2\pi, & m = n = 0 \end{cases} \\[2em] \int_0^{2\pi} \cos nx \sin mx\, dx = 0 \\[2em] \int_0^{2\pi} \sin nx \sin mx\, dx = \begin{cases} 0, & m \neq n \quad \text{or} \quad m = n = 0 \\ \pi, & m = n,\, m \neq 0 \end{cases} \end{cases}$$

we have

13.12
$$\int_0^{2\pi} (f(x) - S_M(x))^2\, dx = \int_0^{2\pi} [f(x)]^2\, dx - 2\pi\left(\frac{a_0\alpha_0}{2} + \sum_{n=1}^{M} (a_n\alpha_n + b_n\beta_n)\right)$$
$$+ \pi\left(\frac{a_0^2}{2} + \sum_{n=1}^{M} (a_n^2 + b_n^2)\right)$$
$$= \frac{\pi}{2}(a_0 - \alpha_0)^2 + \pi\left[\sum_{n=1}^{M} (a_n - \alpha_n)^2 + (b_n - \beta_n)^2\right]$$
$$+ \int_0^{2\pi} [f(x)]^2\, dx - \pi\left(\frac{\alpha_0^2}{2} + \sum_{n=1}^{M} (\alpha_n^2 + \beta_n^2)\right)$$

which is clearly minimized by letting

13.13
$$\begin{cases} a_n = \alpha_n & n = 0, 1, \ldots, M \\ b_n = \beta_n & n = 1, 2, \ldots, M. \end{cases}$$

Hence $S_M(x) \equiv F_M(x)$.

From the above result it follows that the truncated Fourier series $F_M(x)$ for $f(x)$ is the best trigonometric polynomial of degree M or less in the sense of minimizing the integral of the square of the deviation.

If the function $f(x)$ is periodic, then the evaluation of the Fourier coefficients α_n and β_n given by (13.5) can be effectively carried out using the formula

13.14
$$
\begin{cases}
\alpha_n \sim \dfrac{2}{N} \sum_{k=1}^{N} f(x_k) \cos nx_k = a_n \\[2mm]
\beta_n \sim \dfrac{2}{N} \sum_{k=1}^{N} f(x_k) \sin nx_k = b_n
\end{cases}
$$

where

13.15
$$
\begin{cases}
x_k = kh \\[2mm]
h = \dfrac{2\pi}{N}
\end{cases}
$$

(see Section 7.5). Here the integer N is chosen large enough to ensure that the a_n and b_n are sufficiently close to the α_n and β_n, respectively. From results given later in this section, it follows that if $n \leq N/2$ we have

13.16 $a_n - \alpha_n = \alpha_{N+n} + \alpha_{2N+n} + \alpha_{3N+n} + \cdots + \alpha_{N-n} + \alpha_{2N-n} + \alpha_{3N-n} + \cdots$

and

13.17 $b_n - \beta_n = \beta_{N+n} + \beta_{2N+n} + \beta_{3N+n} + \cdots - \beta_{N-n} - \beta_{2N-n} - \beta_{3N-n} - \cdots .$

The α_n and β_n are known to converge to zero as $n \to \infty$. If the convergence is sufficiently rapid and if N is large enough, the accuracy of the a_n and b_n as representations of the α_n and β_n, respectively, can be made as great as desired.

We remark that if the Fourier coefficients are represented by (13.14) and if $S_M^*(x)$ is given by

13.18 $S_M^*(x) = \tfrac{1}{2}a_0 + \displaystyle\sum_{n=1}^{M-1} (a_n \cos nx + b_n \sin nx) + \tfrac{1}{2}a_M \cos Mx$

where $M = N/2$ if N is even, then $S_M(x)$ agrees with $f(x)$ at the points $x_0, x_1, \ldots, x_N$. This will be shown below. We also show that if one truncates (13.18) to obtain a trigonometric polynomial of degree L, say $S_L(x)$, where $L < M$ then

13.19 $\displaystyle\sum_{k=1}^{N} [S_L(x_k) - f(x_k)]^2$

is minimized with respect to all trigonometric polynomials of degree L or less.

Trigonometric Interpolation

Given a function $f(x)$ defined in the interval $0 \leq x \leq 2\pi$ and given an integer N we seek a trigonometric polynomial $S(x)$ such that

13.20 $$S(x_k) = \hat{f}(x_k), \qquad k = 0, 1, \ldots, N$$

where x_k is given by (13.15). Here, for convenience, we define the function $\hat{f}(x)$ by

13.21 $$\hat{f}(x) = \begin{cases} f(x), & x \neq 0, 2\pi \\ \dfrac{f(0) + f(2\pi)}{2} & x = 0, 2\pi. \end{cases}$$

In the case N is even, we let $S(x)$ have the form $S_M^*(x)$ given by (13.18) where

13.22 $$M = \frac{N}{2}.$$

On the other hand, if N is odd, then we let $S(x) = S_M(x)$ where

13.23 $$S_M(x) = \tfrac{1}{2}a_0 + \sum_{n=1}^{M} (a_n \cos nx + b_n \sin nx)$$

where

13.24 $$M = \frac{N-1}{2}.$$

We first seek to show the existence of a unique solution of (13.20). Let us consider the case where N is even. Evidently by (13.18) we have N equations and N unknowns $a_0, a_1, \ldots, a_M$ and $b_1, b_2, \ldots, b_{M-1}$. We can show that this system has a unique solution if we can show that its determinant does not vanish. To show this, it is, by Theorem A-7.8, sufficient to show that the homogeneous system obtained from (13.20) by letting $\hat{f}(x) \equiv 0$ has only the trivial solution

13.25 $$a_0 = a_1 = \cdots = a_M = b_1 = b_2 = \cdots = b_{M-1} = 0.$$

We now prove

13.26 Lemma.

13.27 $$\sum_{k=1}^{N} \cos nx_k = \begin{cases} 0, & \text{if } n/N \text{ is not an integer,} \\ N, & \text{if } n/N \text{ is an integer,} \end{cases}$$

and

13.28 $$\sum_{k=1}^{N} \sin nx_k = 0.$$

Here the x_k are given by (13.15).

Proof. Evidently we have

13.29
$$\sum_{k=1}^{N} \cos nx_k + i \sum_{k=1}^{N} \sin nx_k = \sum_{k=1}^{N} e^{inx_k} = \sum_{k=1}^{N} e^{inkh}.$$

This is a geometric progression with ratio e^{inh}. Since $e^{inh} = 1$ if and only if $nh = 2\pi p$ for some integer p, i.e., if and only if n/N is an integer, it follows that

13.30
$$\sum_{k=1}^{N} e^{inkh} = \begin{cases} N, \text{if } n/N \text{ is an integer} \\ \dfrac{e^{inh} - e^{i(N+1)nh}}{1 - e^{inh}}, \quad \text{otherwise.} \end{cases}$$

But since $e^{iNnh} = 1$ the lemma follows.

We now prove

13.31 Lemma. If x_k is given for (13.15), then

13.32
$$\sum_{k=1}^{N} \cos nx_k \cos mx_k = \begin{cases} 0, & \text{if neither } \dfrac{n+m}{N} \text{ nor } \dfrac{n-m}{N} \text{ is an integer} \\ \dfrac{N}{2}, & \text{if either } \dfrac{n+m}{N} \text{ or } \dfrac{n-m}{N} \text{ but not both} \\ & \text{is an integer} \\ N, & \text{if both } \dfrac{n+m}{N} \text{ and } \dfrac{n-m}{N} \text{ are integers} \end{cases}$$

13.33
$$\sum_{k=1}^{N} \cos nx_k \sin mx_k = 0$$

13.34
$$\sum_{k=1}^{N} \sin nx_k \sin mx_k = \begin{cases} 0, & \text{if neither } \dfrac{n+m}{N} \text{ nor } \dfrac{n-m}{N} \text{ is an integer} \\ & \text{or if both } \dfrac{n+m}{N} \text{ and } \dfrac{n-m}{N} \text{ are integers} \\ -\dfrac{N}{2}, & \text{if } \dfrac{m+n}{N} \text{ is an integer and } \dfrac{m-n}{N} \text{ is not} \\ \dfrac{N}{2}, & \text{if } \dfrac{m-n}{N} \text{ is an integer and } \dfrac{m+n}{N} \text{ is not.} \end{cases}$$

Proof. The lemma follows from Lemma 13.26 and from the identities

13.35
$$\begin{cases} \cos nx_k \cos mx_k = \tfrac{1}{2} \cos(n+m)x_k + \tfrac{1}{2} \cos(n-m)x_k \\ \cos nx_k \sin mx_k = \tfrac{1}{2} \sin(m+n)x_k - \tfrac{1}{2} \sin(m-n)x_k \\ \sin nx_k \sin mx_k = \tfrac{1}{2} \cos(n-m)x_k - \tfrac{1}{2} \cos(n+m)x_k \end{cases}$$

Suppose now that (13.20) holds where $S(x)$ has the form $S_M^*(x)$ given by (13.18). Then by Lemma 13.31 we have for $n = 0, 1, \ldots, M$

13.36
$$\sum_{k=1}^{N} \hat{f}(x_k) \cos nx_k = \sum_{k=1}^{N} \left(\tfrac{1}{2}a_0 + \sum_{m=1}^{M-1} (a_m \cos mx_k + b_m \sin mx_k) \right.$$
$$\left. + \tfrac{1}{2}a_M \cos Mx_k \right) \cos nx_k$$
$$= \frac{N}{2} a_n.$$

Moreover, we also have for $n = 1, 2, \ldots, M - 1$

13.37
$$\sum_{k=1}^{N} \hat{f}(x_k) \sin nx_k = \frac{N}{2} b_n.$$

Thus, we have

13.38
$$\begin{cases} a_n = \dfrac{2}{N} \sum_{k=1}^{N} \hat{f}(x_k) \cos nx_k, & n = 0, 1, \ldots, M \\[2mm] b_n = \dfrac{2}{N} \sum_{k=1}^{N} \hat{f}(x_k) \sin nx_k, & n = 1, 2, \ldots, M - 1. \end{cases}$$

If $\hat{f}(x) \equiv 0$ then all the a_n and b_n vanish. Hence the determinant of the system (13.20) does not vanish and (13.20) has a unique solution. Since a solution, if it exists, must be given by (13.38) it follows that (13.38) is indeed the solution.

In the case N is odd, a similar analysis holds and we get

13.39
$$\begin{cases} a_n = \dfrac{2}{N} \sum_{k=1}^{N} \hat{f}(x_k) \cos nx_k, & n = 0, 1, 2, \ldots, M \\[2mm] b_n = \dfrac{2}{N} \sum_{k=1}^{N} \hat{f}(x_k) \sin nx_k, & n = 1, 2, \ldots, M. \end{cases}$$

13.40 Theorem. Let $f(x)$ be defined on the interval $0 \leq x \leq 2\pi$ and let $\hat{f}(x)$ be given by (13.21). If N is even and if $S(x) = S_M^*(x)$, where $S_M^*(x)$ is given by (13.18) and where the coefficients a_n and b_n are given by (13.38), then

13.41
$$S(x_k) = \hat{f}(x_k), \qquad k = 0, 1, \ldots, N$$

where x_k is given by (13.15). If N is odd and if $S(x) = S_M(x)$, where $S_M(x)$ is given by (13.23) and where the coefficients a_n and b_n are given by (13.39), then (13.41) holds. Conversely, if (13.41) holds for some trigonometric polynomial of the form (13.18), if N is even, and (13.23) if N is odd, then the coefficients a_n and b_n are given by (13.38) if N is even, and by (13.39) if N is odd.

As an example, consider the case $f(x) = x$, $N = 4$. Evidently the Fourier series for $f(x)$ is given by

13.42 $$f(x) = \pi - 2 \sin x - \sin 2x - \tfrac{2}{3} \sin 3x - \tfrac{1}{2} \sin 4x \ldots .$$

Moreover, since $1 - (\tfrac{1}{3}) + (\tfrac{1}{5}) - (\tfrac{1}{7}) + (\tfrac{1}{9}) + \cdots = \pi/4$ we have

13.43
$$\begin{cases} \hat{f}(x_0) = \hat{f}(x_4) = \tfrac{1}{2}(2\pi) = \pi \\[2mm] \hat{f}(x_1) = f(x_1) = \dfrac{\pi}{2} \\[2mm] \hat{f}(x_2) = f(x_2) = \pi \\[2mm] \hat{f}(x_3) = f(x_3) = \dfrac{3\pi}{2} . \end{cases}$$

By (13.22) $M = 2$ and

13.44 $$S_2^*(x) = \tfrac{1}{2}a_0 + a_1 \cos x + \tfrac{1}{2}a_2 \cos 2x + b_1 \sin x.$$

By (13.38) we have

13.45
$$\begin{cases} a_0 = \dfrac{2}{N} \sum_{k=1}^{4} \hat{f}(x_k) = \dfrac{2}{N}(4\pi) = 2\pi \\[3mm] a_1 = \dfrac{2}{N} \left\{ \dfrac{\pi}{2} \cos \dfrac{2\pi}{N} + \pi \cos \dfrac{4\pi}{N} + \dfrac{3\pi}{2} \cos \dfrac{6\pi}{N} + \pi \cos \dfrac{8\pi}{N} \right\} = 0 \\[3mm] a_2 = \dfrac{2}{N} \left\{ \dfrac{\pi}{2} \cos \dfrac{4\pi}{N} + \pi \cos \dfrac{8\pi}{N} + \dfrac{3\pi}{2} \cos \dfrac{12\pi}{N} + \pi \cos \dfrac{16\pi}{N} \right\} = 0 \\[3mm] b_1 = \dfrac{2}{N} \left\{ \dfrac{\pi}{2} \sin \dfrac{2\pi}{N} + \pi \sin \dfrac{4\pi}{N} + \dfrac{3\pi}{2} \sin \dfrac{6\pi}{N} + \pi \sin \dfrac{8\pi}{N} \right\} = -\dfrac{\pi}{2} . \end{cases}$$

Therefore the interpolating trigonometric polynomial is

13.46 $$S_4^*(x) = \pi - \dfrac{\pi}{2} \sin x.$$

We verify that

13.47
$$
\begin{cases}
S_4^*(0) &= \pi = \hat{f}(0) \\[2mm]
S_4^*\!\left(\dfrac{\pi}{2}\right) &= \dfrac{\pi}{2} = \hat{f}\!\left(\dfrac{\pi}{2}\right) \\[2mm]
S_4^*(\pi) &= \pi = \hat{f}(\pi) \\[2mm]
S_4^*\!\left(\dfrac{3\pi}{2}\right) &= \dfrac{3\pi}{2} = \hat{f}\!\left(\dfrac{3\pi}{2}\right) \\[2mm]
S_4^*(2\pi) &= \pi = \hat{f}(2\pi) \ .
\end{cases}
$$

Hence the required conditions are satisfied.

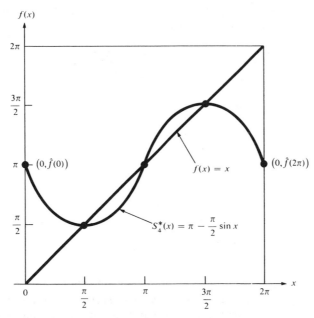

13.48 Fig. Trigonometric Interpolation for $f(x) = x$, $0 \leqq x \leqq 2\pi$.

It can be seen from Fig. 13.48 that the agreement between $S_4^*(x)$ and $f(x)$ is fairly good except near the ends of the interval $0 \leqq x \leqq 2\pi$. If the function were continuous and periodic with period 2π, this difficulty would not arise since we would have $f(0) = f(2\pi)$ and $\hat{f}(0) = f(0) = \hat{f}(2\pi) = f(2\pi)$.

We remark that if for all x in the interval $0 \leqq x \leqq 2\pi$ we have, for N even and $M = N/2$,

13.49
$$f(x) = \frac{\alpha_0}{2} + \sum_{n=1}^{\infty} (\alpha_n \cos nx + \beta_n \sin nx)$$

then

13.50
$$a_n = \alpha_n + \alpha_{N+n} + \alpha_{2N+n} + \cdots + \alpha_{N-n} + \alpha_{2N-n} + \cdots, \quad n = 0, 1, 2, \ldots, M$$

and

13.51 $b_n = \beta_n + \beta_{N+n} + \beta_{2N+n} + \cdots - \beta_{N-n} - \beta_{2N-n} - \cdots, \quad n = 1, 2, \ldots, M-1.$

Thus it can be seen that if N is large and if the coefficients α_k and β_k tend to zero reasonably rapidly, then the coefficients a_n and b_n are good approximations to the α_n and the β_n, respectively. (We remark that although it is possible for (13.49) to hold even if the α_n and β_n are not the Fourier coefficients for $f(x)$, nevertheless, if the convergence is uniform, then the α_n and β_n must be the Fourier coefficients.)
    In our example the Fourier coefficients are given by

13.52 $\qquad \alpha_0 = 2\pi, \qquad \alpha_1 = \alpha_2 = \cdots = 0, \qquad \beta_k = \dfrac{-2}{k} \qquad k = 1, 2, \ldots$

and

13.53
$$a_0 = \alpha_0 + \alpha_4 + \alpha_8 + \cdots$$
$$+ \alpha_4 + \alpha_8 + \cdots = 2\pi$$

13.54
$$a_1 = a_2 = \cdots = 0$$

13.55
$$b_1 = \beta_1 + \beta_5 + \beta_9 + \cdots$$
$$- \beta_3 - \beta_7 - \beta_{11} - \cdots$$
$$= (-2)(1 + \tfrac{1}{5} + \tfrac{1}{7} + \cdots - \tfrac{1}{3} - \tfrac{1}{7} - \tfrac{1}{11} - \cdots)$$
$$= (-2)\left(\frac{\pi}{4}\right) = -\frac{\pi}{2}$$

which is consistent with (13.52) and with (13.50)—(13.51). Thus (13.50) and (13.51) hold even though the series does not converge to $f(x)$ at $x = 0$ and 2π. (However, the series does converge to $\hat{f}(x)$ at $x = 0$ and 2π, though the convergence is not uniform and $\hat{f}(x)$ is not continuous.)

Least Squares Trigonometric Approximation Over a Finite Point Set

We now show that if we truncate the interpolating trigonometric polynomial (13.18) or (13.23) we get a polynomial

13.56
$$S_L(x) = \tfrac{1}{2}a_0 + \sum_{n=1}^{L} (a_n \cos nx + b_n \sin nx)$$

which has the property that

13.57 $$\sum_{k=1}^{N} (\hat{f}(x_k) - S_L(x_k))^2 \leq \sum_{k=1}^{N} (\hat{f}(x_k) - \hat{S}_L(x_k))^2$$

where $\hat{S}_L(x_k)$ is any other trigonometric polynomial of degree L or less of the form

13.58 $$\hat{S}_L(x_k) = \frac{\hat{a}_0}{2} + \sum_{n=1}^{L} (\hat{a}_n \cos nx + \hat{b}_n \sin nx).$$

We assume that $L < N/2$, and we let $\hat{f}(x)$ be given by (13.21). Evidently, we have

13.59 $$\sum_{k=1}^{N} (\hat{f}(x_k) - \hat{S}_L(x_k))^2 = \sum_{k=1}^{N} \hat{f}(x_k)^2 - \frac{N}{2} a_0 \hat{a}_0 - \sum_{n=1}^{L} N(a_n \hat{a}_n + b_n \hat{b}_n) + \frac{N}{4} \hat{a}_0^2$$

$$+ \sum_{n=1}^{L} \frac{N}{2} (\hat{a}_n^2 + \hat{b}_n^2)$$

$$= \sum_{k=1}^{N} \hat{f}(x_k)^2 + \frac{N}{4} (\hat{a}_0 - a_0)^2 + \frac{N}{2} \sum_{n=1}^{L} [(\hat{a}_n - a_n)^2$$

$$+ (\hat{b}_n - b_n)^2] - \frac{N}{4} a_0^2 - \frac{N}{2} \left(\sum_{n=1}^{L} (a_n^2 + b_n^2) \right)$$

which is clearly minimized when

13.60 $$\begin{cases} \hat{a}_n = a_n, & n = 0, 1, 2, \ldots, L \\ \hat{b}_n = b_n, & n = 1, 2, \ldots, L \end{cases}$$

Therefore $S_L(x)$ is the trigonometric polynomial of best approximation in the sense of (13.57).

EXERCISES 6.13

1. Verify directly that (13.20) holds if $S_N(x)$ is given by (13.18) where the a_n and b_n are given by (13.38).

2. Verify that the Fourier series for $f(x) = x$ in the interval $[0, 2\pi]$ is given by (13.42).

3. Consider the function $f(x) \equiv x, 0 \leq x \leq 2\pi$. Find the trigonometric polynomial of degree two or less which best approximates $\hat{f}(x)$ in the least squares sense over the point set

$$x_k = \frac{2\pi k}{N}, \qquad k = 0, 1, \ldots, N$$

where $N = 8$. Here $\hat{f}(x)$ is given by (13.21). What is the trigonometric polynomial of degree two or less which best approximates $\hat{f}(x)$ with respect to the continuous interval $0 \leq x \leq 2\pi$? Also find the trigonometric polynomial of the form (13.18) with $M = 4$ which interpolates to $\hat{f}(x)$ at the x_k.

4. Consider the function $f(x)$ defined by

$$f(x) = \begin{cases} x, & 0 \leq x \leq \pi \\ 2\pi - x, & \pi \leq x \leq 2\pi. \end{cases}$$

a) Find the Fourier series for $f(x)$.
b) Find the trigonometric polynomial of the form(13.18)with $M = 3$ which interpolates to $f(x)$ in the points $x_k = kh$, $k - 0, 1, 2, \ldots, 6$, where $h = \pi/3$.
c) Find the trigonometric polynomial of degree one which gives the best approximation to $f(x)$ in the least squares sense over the set $x_k = kh$, $k = 1, 2, \ldots, 6$. Compare the least squares approximation to that of the trigonometric polynomial of degree one obtained by truncating the Fourier series.
d) With $N = 12$ use (13.38) to get approximate values of the Fourier coefficients α_0, α_1, α_2, β_1, β_2.
e) Evaluate the Fourier series, the truncated Fourier series of degree two, the polynomial obtained by (d), and the polynomial obtained by (b), each for $x = \pi/2$, $3\pi/4$, π.

5. Let $f(x)$ be defined by $f(x) = x$, $0 \leq x \leq \pi$, and $f(x) = \pi - x$, $\pi < x < 2\pi$. Let $\hat{F}_2(x) = a_0/2 + \sum_{k=1}^{2} (a_k \cos kx + b_k \sin kx)$ where the a_k and b_k are given by (13.38) with $N = 10$. Determine $f(x) - \hat{F}_2(x)$ for $x = 0(\pi/10)2\pi$. Do the same for $F_2(x)$ which is the same as $\hat{F}_2(x)$ except that the a_k and b_k are the Fourier coefficients.

6.14 INTERPOLATION IN TWO VARIABLES

In this section we give some elementary procedures for interpolation for a function of two variables. The first procedure is applicable to cases where values of $u(x, y)$ are given for a number of not necessarily regularly spaced points (x_1, y_1), $(x_2, y_2), \ldots$. Within a triangle formed by three interpolation points, say (x_1, y_1), (x_2, y_2), and (x_3, y_3) we construct, as in Section 4.13, a function $U(x, y)$ of the form

14.1 $$U(x, y) = \alpha + \beta x + \gamma y.$$

If we require that

14.2 $$U(x_i, y_i) = u(x_i, y_i) = u_i, \qquad i = 1, 2, 3$$

we get

14.3 $$U(x, y) = \alpha_1 u_1 + \alpha_2 u_2 + \alpha_3 u_3$$

where

14.4 $$\begin{cases} \alpha_1 = \dfrac{(x_2 - x)(y_3 - y) - (y_2 - y)(x_3 - x)}{(x_2 - x_1)(y_3 - y_1) - (x_3 - x_1)(y_2 - y_1)} \\[2ex] \alpha_2 = \dfrac{(x_3 - x)(y_1 - y) - (x_1 - x)(y_3 - y)}{(x_2 - x_1)(y_3 - y_1) - (x_3 - x_1)(y_2 - y_1)} \\[2ex] \alpha_3 = \dfrac{(x_1 - x)(y_2 - y) - (x_2 - x)(y_1 - y)}{(x_2 - x_1)(y_3 - y_1) - (x_3 - x_1)(y_2 - y_1)}. \end{cases}$$

We note that the denominators in the above formulas do not vanish unless the three points lie in a straight line.

It is interesting to note that if we can choose nonoverlapping triangles which cover the region R of interest, then we obtain in this way a function which is continuous in R. For, if we have two nonoverlapping triangles $P_1P_2P_3$ and $P_1P_2P_4$ with a common side P_1P_2, then the interpolating functions for the two triangles have the same values on P_1P_2. This follows since on any straight line, in particular a side of a triangle of interpolation $u(x, y)$ is a linear function of the

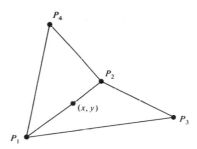

distance along the side. Thus, for instance, suppose (x, y) lies on the side P_1P_2. Then

14.5
$$\frac{y - y_1}{y_2 - y_1} = \frac{x - x_1}{x_2 - x_1} = s.$$

Since $U(x_1, y_1) = u_1$ we have

14.6
$$u_1 = \alpha + \beta x_1 + \gamma y_1.$$

Hence

14.7
$$U(x, y) = U(x_1 + s(x_2 - x_1), y_1 + s(y_2 - y_1))$$
$$= \alpha + \beta[x_1 + s(x_2 - x_1)] + \gamma[y_1 + s(y_2 - y_1)]$$

and

14.8
$$U(x, y) - u_1 = s[\beta(x_2 - x_1) + \gamma(y_2 - y_1)]$$

so that U is a linear function of s. Since the distance from P_1 to (x, y) is

14.9
$$d = \sqrt{(x - x_1)^2 + (y - y_1)^2} = s\sqrt{(x_2 - x_1)^2 + (y_2 - y)^2}$$

it follows that $U(x, y) - u_1$ is a linear function of d.

A second scheme which is particularly appropriate with the use of a rectangular grid is the use of *bilinear* interpolation. Thus, suppose that for some (x_0, y_0), we have values of $u(x, y)$ available for $x = x_0 + ih$, $y = y_0 + jk$, $i, j = 0, \pm 1, \pm 2, \ldots$, where $h > 0$, $k > 0$. Within a rectangle with vertices $(\bar{x}, \bar{y})$, $(\bar{x} + h, \bar{y})$, $(\bar{x}, \bar{y} + k)$, $(\bar{x} + h, \bar{y} + k)$ we construct a function $U(x, y)$ of the form

14.10 $$U(x, y) = \alpha + \beta(x - \bar{x}) + \gamma(y - \bar{y}) + \delta(x - \bar{x})(y - \bar{y}).$$

If we require that $U(x, y) = u(x, y)$ at the four vertices of the rectangle then we have

14.11 $$\alpha = u(\bar{x}, \bar{y}), \quad \beta = \frac{u(\bar{x} + h, \bar{y}) - u(\bar{x}, \bar{y})}{h}, \quad \gamma = \frac{u(\bar{x}, \bar{y} + k) - u(\bar{x}, \bar{y})}{k},$$

$$\delta = \frac{u(\bar{x} + h, \bar{y} + k) + u(\bar{x}, \bar{y}) - u(\bar{x} + h, y) - u(\bar{x}, \bar{y} + k)}{hk}.$$

We note that for fixed y the function $U(x, y)$ is a linear function of x while, for fixed x, $U(x, y)$ is a linear function of y. Hence we may perform bilinear interpolation in two stages.

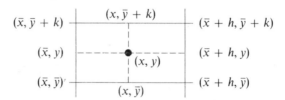

We may use linear interpolation in one variable to get $U(x, \bar{y})$ from $u(\bar{x}, \bar{y})$ and $u(\bar{x} + h, \bar{y})$ and also to get $U(x, \bar{y} + k)$ from $u(\bar{x}, \bar{y} + k)$ and $u(\bar{x} + h, \bar{y} + k)$. We then can use linear interpolation in one variable to get $U(x, y)$ in terms of $U(x, \bar{y})$ and $U(x, \bar{y} + k)$. This would yield the same result as though we had first obtained $U(\bar{x}, y)$ and $U(\bar{x} + h, y)$ and then $U(x, y)$.

Because the value of $U(x, y)$ on any side of the rectangle depends only on the values at the end points of the side it follows that given two nonoverlapping rectangles with a common side, the interpolating bilinear functions have the same values on the common side. Hence the use of bilinear interpolation yields a function which is continuous in the region covered by the rectangles of interpolation.

We remark that even if the region can be covered by rectangles the use of a form of triangular interpolation has certain advantages. This scheme was described in Section 4.13 and involves dividing each rectangle with four triangles formed by the two diagonals of the rectangle. The value at the intersection of the diagonals is taken as the average of the values at the four corners. Within each of the four triangles, triangular interpolation is used. This scheme has the advantage that the loci of constant values of the interpolating function are broken lines. With bilinear interpolation, on the other hand, there may be more than one branch of the locus in a given rectangle. This could complicate the automatic construction of a contour plot.

EXERCISES 6.14

1. Derive (14.3) and (14.4) and show that $\alpha_1 + \alpha_2 + \alpha_3 = 1$.

2. Suppose that $u(1, 1) = 2$, $u(2,1) = 3$, $u(1, 2) = 5$, $u(2, 2) = 7$. Find an approximate value of $u(\frac{7}{4}, \frac{9}{8})$ by each of the following schemes.

   a) Triangular interpolation in the points $(1, 1)$, (2.1), $(2, 2)$.
   b) Bilinear interpolation.
   c) Using triangular interpolation in one of the four triangles formed by the diagonals of the rectangle whose vertices are the given four points. The value at the center point of the rectangle is the average of the values at the corners.

   Also, find the locus of points such that $U(x, y) = 6$, where $U(x, y)$ is determined according to the scheme (c).

3. In Exercise 2, verify that if we use interpolation in the points $(1, 1)$, $(2, 1)$, and $(1, 2)$ we get the same values along the line segment joining $(1, 2)$ to $(2, 1)$ as when we use interpolation in the points $(1, 2)$, $(2, 2)$, $(2, 1)$.

SUPPLEMENTARY DISCUSSION

Section 6.9

We remark that the method of false position and the secant method for solving $f(x) = 0$ (see Section 4.6) are each equivalent to inverse linear interpolation. Similarly, Muller's method (Section 4.10) is equivalent to inverse quadratic interpolation, though the interpolation points are in general not equally spaced.

The example showing the danger of using Lagrangian interpolation in y for inverse interpolation is based on that given by Hartree [1952]. Hartree considered a five-point Lagrangian interpolation formula and showed by a graph the total dissimilarity between $g(y)$ and $G(y)$ in this case.

Section 6.10

For an extensive discussion of methods of approximation as well as a number of polynomial and rational approximations to a variety of functions, see Hart, *et al.* [1968].

Instead of using the Remes algorithm one can often obtain a near-optimum polynomial by expanding the function $f(x)$ as a series of Chebyshev polynomials. A number of such expansions are given by Clenshaw [1962]. For some functions such as e^x, $\cos x$, $\sinh x$, etc., the coefficients in the expansion can be determined in terms of tabulated functions such as Bessel's functions (see Hart, *et al.* [1968]). See also Snyder [1966] for a discussion of polynomial and rational approximation with Chebyshev polynomials.

A modification of the Remes method can be used to determine the polynomial having the smallest maximum relative error (see, for instance, Fike [1968]).

CHAPTER 7

NUMERICAL DIFFERENTIATION
AND QUADRATURE

7.1 INTRODUCTION

In this chapter we consider *numerical differentiation*, the approximate evaluation of certain derivatives of a function $f(x)$ in terms of the values of $f(x)$, and possibly the values of certain derivatives, at several selected values of x. We refer to the selected values of x as "interpolation points." We also consider *numerical quadrature*, the approximate evaluation of the integral

1.1
$$\int_a^b f(x)dx$$

in a similar manner.

Numerical differentiation and numerical quadrature formulas can be derived by replacing the given function $f(x)$ by another function $F(x)$ for which the desired derivative or integral can be found analytically. Frequently one selects a polynomial of a certain degree which agrees with $f(x)$ at the interpolation points and certain of whose derivatives agree with the corresponding derivatives of $f(x)$. In many cases $F(x)$ can be determined by the Lagrange interpolation formula, or if derivatives are prescribed, by the Hermite interpolation formula. The desired derivative or integral of $F(x)$ can then be expressed as a linear combination of the values of $f(x)$ and the derivatives of $f(x)$ at the interpolation points.

Another approach to the problem of determining numerical differentiation or numerical quadrature formulas is to determine "weights" such that, with those weights, a linear combination of values of $f(x)$ and derivatives of $f(x)$ will give a good approximation to the desired derivative or integral. Normally, one chooses the weights so that the resulting formula will be exact for any polynomial of degree p or less, where p is as large as possible. The *order* of a given formula is the largest p such that the formula is exact for any polynomial of degree p or less. We show that this method, known as the "method of undetermined weights" leads to the same formulas as would be obtained by differentiating or integrating the Lagrange, or Hermite, interpolating polynomial. In many cases, however, the method of undetermined weights gives the desired formula with much less effort.

Numerical differentiation and numerical quadrature formulas involving evenly spaced interpolation points are studied in Sections 7.3 and 7.4, respectively. One particular numerical quadrature formula, known as the Euler-Maclaurin formula, is treated in Section 7.5. By the use of this formula one can derive the Romberg integration method, as shown in Section 7.6. The Romberg method affords a means of obtaining high accuracy in many cases with relatively little effort.

In Section 7.7 we discuss methods for determining the error involved in numerical differentiation and numerical quadrature formulas. In nearly every case of interest, a result can be obtained which is analogous to those given in Chapter 6 for Lagrangian and for Hermite interpolation.

In Section 7.8 we discuss quadrature formulas based on the use of unevenly spaced interpolation points. With *Gaussian quadrature* one can obtain a high order of accuracy with relatively few interpolation points. With *Chebyshev quadrature* one can obtain a simpler formula using unevenly spaced points whose order of accuracy is at least as great as that of a less simple formula using evenly spaced points.

7.2 THE METHOD OF UNDETERMINED WEIGHTS

A basic tool for the development of formulas for numerical differentiation and numerical quadrature is the method of undetermined weights. We have already used this method for interpolation (see Sections 6.4 and 6.7). We let the quantity

2.1
$$L[f](x) \quad \text{or} \quad L[f],$$

as appropriate, represent the quantity which we are trying to determine. Thus, for instance,

2.2
$$L[f](x) = f(x),$$

in the case of interpolation,

2.3
$$L[f](x) = f^{(k)}(x),$$

in the case of numerical differentiation for the kth derivative, and

2.4
$$L[f] = \int_a^b f(x)dx,$$

in the case of numerical quadrature. We write $L[f]$ instead of $L[f](x)$ in (2.4) since the result is independent of x. We assume that $L[f](x)$ is *linear* in the sense that $L[c_1 f_1 + c_2 f_2](x) = c_1 L[f_1](x) + c_2 L[f_2](x)$.

Suppose we are given the $n + 1$ interpolation points $x_0, x_1, \ldots, x_n$ and nonnegative integers $\gamma_0, \gamma_1, \ldots, \gamma_n$. We desire to represent $L[f](x)$ in terms of the values of $f(x)$ at the interpolation points

2.5
$$f(x_k), \quad k = 0, 1, \ldots, n$$

and, for each point x_k such that $\gamma_k \geqq 1$, in terms of the values of the derivatives

2.6 $$f^{(j)}(x_k), \qquad j = 1, 2, \ldots, \gamma_k.$$

A very straightforward method for doing this is to let $F(x)$ be the unique polynomial of degree m or less, where

2.7 $$m = n + \sum_{k=0}^{n} \gamma_k,$$

such that

2.8 $$F^{(j)}(x_k) = f^{(j)}(x_k), \qquad k = 0, 1, \ldots, n; \qquad j = 0, 1, \ldots, \gamma_k.$$

From 6-(7.13) it follows that

2.9 $$F(x) = \sum_{k=0}^{n} \sum_{j=0}^{\gamma_k} w_{k,j}(x) f^{(j)}(x_k)$$

where the $w_{k,j}(x)$ are polynomials of degree m or less.

We now represent $L[f](x)$ by $\tilde{L}[f](x)$, where

2.10 $$\tilde{L}[f](x) = L[F](x)$$

$$= L\left[\sum_{k=0}^{n} \sum_{j=0}^{\gamma_k} w_{k,j}(x) f^{(j)}(x_k)\right](x)$$

$$= \sum_{k=0}^{n} \sum_{j=0}^{\gamma_k} f^{(j)}(x_k) L[w_{k,j}](x)$$

$$= \sum_{k=0}^{n} \sum_{j=0}^{\gamma_k} v_{k,j}(x) f^{(j)}(x_k)$$

and where, by the linearity of $L[f](x)$,

2.11 $$v_{k,j}(x) = L[w_{k,j}](x), \qquad k = 0, 1, \ldots, n; \qquad j = 0, 1, \ldots, \gamma_k.$$

Evidently with $\tilde{L}[f](x)$ as thus determined, the formula

2.12 $$L[f](x) \sim \tilde{L}[f](x)$$

is exact for any polynomial of degree m or less. Thus we say that the *order* of $\tilde{L}[f](x)$ as a representation of $L[f](x)$ is at least m. More generally, if we have any operator $\tilde{L}^*[f]$ of the form

2.13 $$\tilde{L}^*[f](x) = \sum_{k=0}^{n} \sum_{j=0}^{\gamma_k} r_{k,j}(x) f^{(j)}(x_k),$$

where the $r_{k,j}(x)$ are polynomials in x, we say that the *order* of $\tilde{L}^*[f](x)$ as a representation of $L[f](x)$ is p if p is the largest integer such that the representation is exact for any polynomial of degree p or less. This is evidently equivalent to the conditions

2.14
$$\begin{cases} \tilde{L}^*[x^s](x) = L[x^s](x), & s = 0, 1, \ldots, p \\ \tilde{L}^*[x^{p+1}](x) \neq L[x^{p+1}](x). \end{cases}$$

We note that the *order* of a given operator $\tilde{L}^*[f](x)$ may depend on x.

It is easy to verify that if, for all x, $\tilde{L}^*[f](x)$ represents $L[f](x)$ with order m or more, then

2.15 $$r_{k,j}(x) \equiv v_{k,j}(x); \qquad k = 0, 1, \ldots, n; \qquad j = 0, 1, \ldots, \gamma_k.$$

Hence

2.16 $$\tilde{L}^*[f](x) \equiv \tilde{L}[f](x).$$

Rather than finding the $v_{k,j}(x)$ by (2.11), it is often more convenient to find them from the first $m + 1$ equations of (2.14). With this procedure, known as the *method of undetermined weights*, we solve the system of $m + 1$ linear algebraic equations

2.17 $$\tilde{L}^*[(x - \alpha)^s](x) = L[(x - \alpha)^s](x), \qquad s = 0, 1, \ldots, m$$

where α is any convenient constant, for the m unknowns $r_{k,j}(x)$, $(k = 0, 1, \ldots, n;$ $j = 0, 1, \ldots, \gamma_k)$. (The result is clearly independent of α.)

As an example, let us consider the approximate determination of $f'(x) = L[f](x)$ in terms of $f(x_0), f(x_1), f'(x_0)$, and $f'(x_1)$. Using the first procedure based on the Hermite interpolation formula we have, by 6-(7.14)

2.18 $$F(x) = w_{0,0}(x)f(x_0) + w_{0,1}(x)f'(x_0) + w_{1,0}(x)f(x_1) + w_{1,1}(x)f'(x_1)$$

where

2.19
$$\begin{cases} w_{0,0}(x) = \dfrac{(x - x_1)^2}{h^2}\left[1 + \dfrac{2(x - x_0)}{h}\right], & w_{0,1}(x) = \dfrac{(x - x_1)^2}{h^2}(x - x_0) \\ \\ w_{1,0}(x) = \dfrac{(x - x_0)^2}{h^2}\left[1 - \dfrac{2(x - x_1)}{h}\right], & w_{1,1}(x) = \dfrac{(x - x_0)^2}{h^2}(x - x_1) \end{cases}$$

and $h = x_1 - x_0$. Therefore, by (2.11) we have

2.20
$$\begin{cases} v_{0,0}(x) = \dfrac{6(x - x_0)(x - x_1)}{h^3}, & v_{0,1}(x) = \dfrac{(x - x_1)}{h^2}[3x - 2x_0 - x_1] \\ \\ v_{1,0}(x) = -\dfrac{6(x - x_0)(x - x_1)}{h^3}, & v_{1,1}(x) = \dfrac{(x - x_0)}{h^2}[3x - x_0 - 2x_1]. \end{cases}$$

If $x = \alpha = (x_0 + x_1)/2$ we have

2.21
$$\begin{cases} v_{0,0}(\alpha) = -\dfrac{3}{2h}, & v_{0,1}(\alpha) = -\tfrac{1}{4} \\[4mm] v_{1,0}(\alpha) = \dfrac{3}{2h}, & v_{1,1}(\alpha) = -\tfrac{1}{4} \end{cases}$$

and hence

2.22 $\quad f'\left(\dfrac{x_0 + x_1}{2}\right) \sim \dfrac{3}{2h}[f(x_1) - f(x_0)] - \tfrac{1}{4}[f'(x_0) + f'(x_1)] = \tilde{L}[f](\alpha).$

It is easy to verify that the order of the method is four in this case while if $x \neq (x_0 + x_1)/2$, the order is three.

Let us now apply the method of undetermined weights for the case $x = (x_0 + x_1)/2 = \alpha$. Letting

2.23 $\qquad\qquad \tilde{L}^*[(x - \alpha)^s](\alpha) = L[(x - \alpha)^s](\alpha), \qquad s = 0, 1, 2, 3$

we have

2.24
$$\begin{cases} 0 = v_{0,0} + v_{1,0} \\ 1 = -\sigma v_{0,0} + v_{0,1} + \sigma v_{1,0} + v_{1,1} \\ 0 = \sigma^2 v_{0,0} - 2\sigma v_{0,1} + \sigma^2 v_{1,0} + 2\sigma v_{1,1} \\ 0 = -\sigma^3 v_{0,0} + 3\sigma^2 v_{0,1} + \sigma^3 v_{1,0} + 3\sigma^2 v_{1,1} \end{cases}$$

where we let

2.25 $\qquad\qquad\qquad\qquad\qquad \sigma = \dfrac{h}{2}.$

Solving, we get

2.26 $\qquad\qquad v_{0,0} = -\dfrac{3}{2h}, \quad v_{0,1} = -\tfrac{1}{4}, \quad v_{1,0} = \dfrac{3}{2h}, \quad v_{1,1} = -\tfrac{1}{4}$

which agrees with (2.21).

Let us define the remainder operator $R[f](x)$ by

2.27 $\qquad\qquad\qquad R[f](x) = L[f](x) - \tilde{L}[f](x).$

In Section 7.7 we shall show that, for a large class of methods, if $f(x) \in C^{(p+1)}$ then

2.28 $R[f](x) = R[x^{p+1}](x) \dfrac{f^{(p+1)}(\xi)}{(p+1)!} = \{L[x^{p+1}](x) - \tilde{L}[x^{p+1}](x)\} \dfrac{f^{(p+1)}(\xi)}{(p+1)!}$

for some ξ belonging to the smallest interval containing the interpolation points and all values of x appearing in $L[f](x)$. Here p is the order of the method. We shall refer to a method, or formula, for which (2.28) holds whenever $f(x) \in C^{(p+1)}$, as *definite*.

For example, in the case just considered we have $R[x^s](\alpha) = 0$ for $s = 0, 1, 2, 3, 4$ and

2.29 $$R[x^5](\alpha) = R[(x-\alpha)^5](\alpha) = \frac{h^4}{16}.$$

Hence the order of the method is 4 and from (2.28), (2.29), and (2.22) with $x = \alpha = (x_0 + x_1)/2$ we have

2.30 $f'\left(\dfrac{x_0 + x_1}{2}\right) = f'(\alpha) = \dfrac{3}{2h}[f(x_1) - f(x_0)] - \tfrac{1}{4}[f'(x_1) + f'(x_0)] + \dfrac{h^4}{1920} f^{(5)}(\xi).$

Actually, for this method and in practically every other case of interest the method is definite so that (2.28) holds. For simplicity in our subsequent discussion we shall assume without proof that all methods considered are definite unless the contrary is stated. We shall verify the definiteness in many cases in Section 7.7.

EXERCISES 7.2

1. Show that $L[f](x) = \tilde{L}[f](x)$ for any polynomial of degree m or less where $\tilde{L}[f](x)$ is given by (2.10).

2. Let $\tilde{L}^*[f](x)$ be given by (2.13) and let the order of $\tilde{L}^*[f](x)$ as a representation of $L[f](x)$ be at least m where m is given by (2.7). Show that $r_{k,j}(x) \equiv v_{k,j}(x)$, $k = 0, 1, \ldots, n$ and $j = 0, 1, \ldots, \gamma_k$, where the $v_{k,j}(x)$ are given by (2.11).

3. Show that the $r_{k,j}(x)$ as determined by the method of undetermined weights using the equations (2.17) are independent of the choice of α. Also show that the order of the method is independent of α.

4. Prove the following: given $x_0, x_1, \ldots, x_n$ such that $x_0 < x_1 < \cdots < x_n$ and given a there exist polynomials $v_0(x), v_1(x), \ldots, v_n(x)$ of degree $n + 1$ or less such that if $F(x)$ is any polynomial of degree $n + 1$ or less then

$$F(x) = F(a) + \sum_{i=0}^{n} v_i(x)F'(x_i).$$

Show that the $v_i(x)$ are unique.

5. Show that the order of accuracy of the numerical differentiation formula

$$f'(x) \sim v_{0,0}(x)f(x_0) + v_{1,0}(x)f'(x_0) + v_{1,0}(x)f(x_1) + v_{1,1}(x)f'(x_1)$$

where the $v_{i,j}(x)$ are given by (2.20) is three if $x \neq (x_0 + x_1)/2$.

6. Verify (2.29) and (2.30) assuming that (2.28) holds*

7. Express

$$\int_{x_0}^{x_1} f(x)dx$$

in terms of $f(x_0)$, $f'(x_0)$, $f(x_1)$, and $f'(x_1)$ by integrating (2.18) and also by the method of undetermined weights. Find an expression for the error assuming that (2.28) holds.

8. Show that (2.28) holds for the case of linear interpolation.

7.3 NUMERICAL DIFFERENTIATION

Let us first consider representations of $f'(x)$ involving $f(x_0)$ and $f(x_1)$. The unique polynomial $F(x)$ of degree one or less which agrees with $f(x)$ at the points x_0 and x_1 is

3.1
$$F(x) = f(x_0) + \frac{x - x_0}{x_1 - x_0}[f(x_1) - f(x_0)].$$

Therefore,

3.2
$$F'(x) = \frac{f(x_1) - f(x_0)}{x_1 - x_0} = \frac{\Delta f(x_0)}{h},$$

where $h = x_1 - x_0$ and

3.3
$$\Delta f(x_0) = f(x_1) - f(x_0).$$

Thus we have the *forward-difference numerical differentiation formula*

3.4
$$f'(x) \sim \frac{f(x_1) - f(x_0)}{x_1 - x_0} = \frac{\Delta f(x_0)}{h}.$$

In the notation of the previous section,

$$L[f](x) = f'(x),$$

$$\tilde{L}[f](x) = [f(x_1) - f(x_0)]/(x_1 - x_0).$$

It is easy to show that the order of (3.4) is one unless $x = (x_0 + x_1)/2$, in which case the order is two. From (2.28) we have

3.5
$$R[f](x) = \begin{cases} (2x - x_0 - x_1)\dfrac{f''(\xi)}{2}, & \text{if } x \neq \dfrac{x_0 + x_1}{2} \\[3mm] -\dfrac{1}{24}(x_1 - x_0)^2 f^{(3)}(\xi_1), & \text{if } x = \dfrac{x_0 + x_1}{2} \end{cases}$$

---

*Actually (2.28) does not hold in general. See the discussion on pages 396–398.

for some ξ in the smallest interval containing x_0, x_1, and x and for some $\xi_1 \in (x_0, x_1)$. Thus, in particular we have

3.6
$$f'(x_0) = \frac{f(x_1) - f(x_0)}{h} - \frac{h}{2} f''(\xi_2)$$

3.7
$$f'(x_1) = \frac{f(x_1) - f(x_0)}{h} + \frac{h}{2} f''(\xi_3)$$

3.8
$$f'\left(\frac{x_0 + x_1}{2}\right) = \frac{f(x_1) - f(x_0)}{h} - \frac{h^2}{24} f^{(3)}(\xi_1)$$

for some ξ_2 and ξ_3 in the interval (x_0, x_1).

As an example, let us consider the case

3.9
$$f(x) = e^x, \qquad x_0 = 1, x_1 = 1.2.$$

Since $f(1) \doteq 2.71828$, $f(1.2) \doteq 3.32012$, we have by (3.4)

3.10
$$f'(x) \sim \frac{3.32012 - 2.71828}{0.2} = 3.0092 = F'(x).$$

Since $f'(x) = e^x$, we have

3.11
$$f'(x_0) \doteq 2.71828$$

and

3.12
$$f'(x_0) - F'(x_0) \doteq 2.7183 - 3.0092 = -0.2909.$$

By (3.6), since $f''(x) = e^x$, the error lies between

3.13
$$-\frac{h}{2}e \quad \text{and} \quad -\frac{h}{2}e^{1.2},$$

i.e., between

3.14
$$-0.2718 \quad \text{and} \quad -0.3320.$$

For the case $x = (x_0 + x_1)/2 = 1.1$ we have

3.15
$$f'(1.1) = e^{1.1} \doteq 3.00417$$

and

3.16
$$f'(1.1) - F'(1.1) \doteq -0.0050.$$

By (3.8) the error is between

3.17 $$-\frac{h^2}{24}e \quad \text{and} \quad -\frac{h^2}{24}e^{1.2}$$

i.e., between

3.18 $-0.0045 \quad \text{and} \quad -0.0055.$

Evidently, then, in this case the accuracy of two-point numerical differentiation is much greater when x is midway between x_0 and x_1 than when $x = x_0$ or $x = x_1$. This is consistent with the theoretical results (3.6)–(3.8) and is also not unexpected on the basis of the diagram of Fig. 3.20. It certainly seems clear that the slope of the line AB, which is $F'(x)$, is a better representation of the slope of the tangent to the curve at the point C

3.19 $$\left(\frac{x_0 + x_1}{2}, \quad f\left(\frac{x_0 + x_1}{2} \right) \right)$$

than it is of the slope of the tangent at either $(x_0, f(x_0))$ or $(x_1, f(x_1))$.

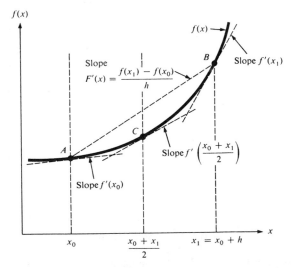

3.20 Fig. Two-point Numerical Differentiation.

For a given numerical differentiation formula the accuracy attainable is limited by the accuracy in the evaluation of $f(x)$. Thus, if $f(x)$ can be evaluated only to within $\pm\varepsilon$, then it does not pay to make h too small. For example, if $Q(h)$ is the computed value of the forward-difference quotient

3.21
$$\frac{f(a+h)-f(a)}{h}$$

then

3.22
$$Q(h) - \frac{f(a+h)-f(a)}{h} = \varepsilon_1(h)$$

may be as large as $2\varepsilon/h$ in absolute value. Therefore, by (3.6) we have

3.23
$$f'(a) - Q(h) \sim -\varepsilon_1(h) - \frac{h}{2}f''(\xi)$$

where

3.24
$$|\varepsilon_1(h)| \leq \frac{2\varepsilon}{h}$$

and*

3.25
$$\left|\frac{h}{2}f''(\xi)\right| \leq \frac{h}{2}M_2.$$

The function

3.26
$$\phi(h) = \frac{2\varepsilon}{h} + \frac{h}{2}M_2$$

decreases as h decreases for

3.27
$$h > \sqrt{\frac{4\varepsilon}{M_2}} = h_0$$

and then increases. Thus there is, in general, no purpose in choosing h less than h_0. Moreover, even if we use $h = h_0$, the error in the determination of $f'(a)$ may be as large as

3.28
$$\delta = 2\sqrt{\varepsilon M_2}.$$

As an example, let us consider the case

3.29
$$f(x) = x^2$$

* Here, as usual, M_2 is a bound for $|f''(x)|$ for $x \in [a, a+h]$.

where we assume that $f(x)$ can be computed correctly to two decimal places. Since $f''(x) = 2$ we have $M_2 = 2$. Also, $\varepsilon = 0.005$, and according to (3.27) there is no use using a value of h less than

3.30 $$h_0 = \sqrt{\frac{4\varepsilon}{M_2}} = 0.1.$$

Moreover, the error could be as large as

3.31 $$\delta = 2\sqrt{\varepsilon M_2} = 0.2.$$

In the case $x = 0.75$ we have

3.32 $$f'(0.75) = 1.50$$

while the computed value of $f(0.75)$ is 0.56. Also, we have

h	Computed $f(0.75 + h)$	Computed $\Delta f(0.75)$	Computed $Q(h) = \dfrac{\Delta f(0.75)}{h}$	$\lvert f'(0.75) - Q(h)\rvert$
0.5	1.56	1.00	2.00	0.50
0.4	1.32	0.76	1.90	0.40
0.3	1.10	0.54	1.80	0.30
0.2	0.90	0.34	1.70	0.20
0.1	0.72	0.16	1.60	0.10
0.05	0.64	0.08	1.60	0.10
0.01	0.58	0.02	2.00	0.50
0.005	0.57	0.01	2.00	0.50

Thus the minimum absolute value of the error occurs at approximately $h = 0.1$ and is 0.10. Hence, in the special case we are able to get somewhat better accuracy than that indicated by (3.31).

The above considerations suggest that, in many cases, in order to get high accuracy for $f'(x)$ one should seek a more accurate representation rather than use a smaller value of h. Thus, for instance, in the above example one can get arbitrarily high accuracy using the central difference formula

3.33 $$f'(a) \sim \frac{f(a + h/2) - f(a - h/2)}{h}$$

for h *large* enough. If $f(x) = x^3$, then there is a finite optimum value of h, but the minimum attainable error is much less than that corresponding to the forward difference quotient.

Let us now consider the use of three points x_0, x_1, x_2 where $x_2 - x_1 = x_1 - x_0 = h$. The polynomial $F(x)$ of degree two or less which agrees with $f(x)$ at the three points is

3.34
$$F(x) = \frac{(x - x_1)(x - x_2)}{(x_0 - x_1)(x_0 - x_2)} f(x_0) + \frac{(x - x_0)(x - x_2)}{(x_1 - x_0)(x_1 - x_2)} f(x_1)$$
$$+ \frac{(x - x_0)(x - x_1)}{(x_2 - x_0)(x_2 - x_1)} f(x_2)$$

and hence

3.35 $F'(x) = \dfrac{(2x - x_1 - x_2)}{2h^2} f(x_0) + \dfrac{(2x - x_0 - x_2)}{-h^2} f(x_1) + \dfrac{(2x - x_0 - x_1)}{2h^2} f(x_2)$

and

3.36
$$\begin{cases} F'(x_0) = -\dfrac{3}{2h} f(x_0) + \dfrac{2}{h} f(x_1) - \dfrac{1}{2h} f(x_2) \\[2mm] F'(x_1) = -\dfrac{1}{2h} f(x_0) + \dfrac{1}{2h} f(x_2) \\[2mm] F'(x_2) = \dfrac{1}{2h} f(x_0) - \dfrac{2}{h} f(x_1) + \dfrac{3}{2h} f(x_2). \end{cases}$$

It is easy to verify that the method defined by (3.35) is of order two for all values of x except at $x = x_1 \pm h/\sqrt{3}$ where $R[x^3](x) = 0$. Moreover, we have, by (2.28)

3.37
$$\begin{cases} f'(x_0) = F'(x_0) + \dfrac{h^2}{3} f^{(3)}(\xi_0) \\[2mm] f'(x_1) = F'(x_1) - \dfrac{h^2}{6} f^{(3)}(\xi_1) \\[2mm] f'(x_2) = F'(x_2) + \dfrac{h^2}{3} f^{(3)}(\xi_2) \end{cases}$$

for some ξ_0, ξ_1, and ξ_2 in the interval (x_0, x_2).

Let us again consider the case $f(x) = e^x$ where $x_0 = 1.0$, $x_1 = 1.1$, $x_2 = 1.2$. Evidently

3.38 $F'(x_0) \doteq \dfrac{1}{h}[-\tfrac{3}{2}(2.71828) + 2(3.00417) - \tfrac{1}{2}(3.32012)] = 2.7086.$

Therefore,

3.39 $f'(x_0) - F'(x_0) \doteq 2.7183 - 2.7086 = 0.0097.$

By (3.37), the error is between

3.40 $\dfrac{h^2}{3} e^{1.0}$ and $\dfrac{h^2}{3} e^{1.2},$

i.e., between

3.41 0.0091 and $0.0111.$

From the results for the two-point case we have $F'(x_1) \doteq 3.0092$ and

3.42 $f'(x_1) - F'(x_1) \doteq 3.0042 - 3.0092 = -0.0050$

as before.

An alternative procedure for deriving (3.35) is to use the Gregory-Newton interpolation formula. We have

3.43 $F(x) = f(x_0) + u \Delta f(x_0) + \dfrac{u(u-1)}{2} \Delta^2 f(x_0) = \psi(u)$

where

3.44 $u = \dfrac{x - x_0}{h}.$

Differentiating we have

3.45 $F'(x) = \psi'(u) \dfrac{du}{dx} = \dfrac{1}{h} \left[\Delta f(x_0) + \dfrac{2u-1}{2} \Delta^2 f(x_0) \right].$

Using (3.44), we get (3.35). In particular, if $u = 0$ or $u = 1$ we get, respectively,

3.46
$$\begin{cases} F'(x_0) = \dfrac{1}{h} [\Delta f(x_0) - \tfrac{1}{2}\Delta^2 f(x_0)] = \dfrac{1}{h} [-\tfrac{3}{2} f(x_0) + 2f(x_1) - \tfrac{1}{2} f(x_2)] \\[3mm] F'(x_1) = \dfrac{1}{h} [\Delta f(x_0) + \tfrac{1}{2}\Delta^2 f(x_0)] = \dfrac{1}{h} [\tfrac{1}{2} f(x_2) - \tfrac{1}{2} f(x_0)] \end{cases}$$

which agree with (3.36).

Still another procedure for deriving (3.35) is the method of undetermined weights where we assume that

3.47 $F'(x) = w_0(x)f(x_0) + w_1(x)f(x_1) + w_2(x)f(x_2)$

and choose $w_0(x)$, $w_1(x)$, and $w_2(x)$ so that

3.48 $$f'(x) = w_0(x)f(x_0) + w_1(x)f(x_1) + w_2(x)f(x_2)$$

for any polynomial of degree two or less. The reader should verify that this leads to

3.49 $$w_0(x) = \frac{2x - x_1 - x_2}{2h^2}, \quad w_1(x) = \frac{2x - x_0 - x_2}{-h^2}, \quad w_2(x) = \frac{2x - x_0 - x_1}{2h^2}$$

and hence (3.35) follows.

Table 3.54 gives the coefficients and the errors for numerical differentiation involving the equally spaced interpolation points $x_0, x_1, \ldots, x_n$ where

3.50 $$x_k = x_0 + kh, \quad k = 0, 1, \ldots, n.$$

In each case we have

3.51 $$f'(x_s) = \frac{1}{h} \sum_{k=0}^{n} \alpha_{s,k} f(x_k) + R[f](x).$$

Thus, for example, for the case $n = 3$, $s = 1$, we have

3.52 $$\alpha_{1,0} = -\tfrac{1}{3}, \quad \alpha_{1,1} = -\tfrac{1}{2}, \quad \alpha_{1,2} = 1, \quad \alpha_{1,3} = -\tfrac{1}{6}$$

and

3.53 $$f'(x_s) = \frac{1}{h}\left[-\tfrac{1}{3}f(x_0) - \tfrac{1}{2}f(x_1) + f(x_2) - \tfrac{1}{6}f(x_3)\right] + \frac{h^3}{12}f^{(4)}(\xi)$$

where ξ lies in the interval (x_0, x_3).

3.54 Table Coefficients for Numerical Differentiation Formulas for First Derivatives

	s	x	$\alpha_{s,0}$	$\alpha_{s,1}$	$\alpha_{s,2}$	$\alpha_{s,3}$	$\alpha_{s,4}$	$R[f](x_s)$
Two Points ($n = 1$)	0	x_0	-1	1	—	—	—	$-\tfrac{1}{2}hf^{(2)}$
	1	x_1	-1	1	—	—	—	$\tfrac{1}{2}hf^{(2)}$
Three Points ($n = 2$)	0	x_0	$-\tfrac{3}{2}$	2	$-\tfrac{1}{2}$	—	—	$\tfrac{1}{3}h^2 f^{(3)}$
	1	x_1	$-\tfrac{1}{2}$	0	$\tfrac{1}{2}$	—	—	$-\tfrac{1}{6}h^2 f^{(3)}$
	2	x_2	$\tfrac{1}{2}$	-2	$\tfrac{3}{2}$	—	—	$\tfrac{1}{3}h^2 f^{(3)}$
Four Points ($n = 3$)	0	x_0	$-\tfrac{11}{6}$	3	$-\tfrac{3}{2}$	$\tfrac{1}{3}$	—	$-\tfrac{1}{4}h^3 f^{(4)}$
	1	x_1	$-\tfrac{1}{3}$	$-\tfrac{1}{2}$	1	$-\tfrac{1}{6}$	—	$\tfrac{1}{12}h^3 f^{(4)}$
	2	x_2	$\tfrac{1}{6}$	-1	$\tfrac{1}{2}$	$\tfrac{1}{3}$	—	$-\tfrac{1}{12}h^3 f^{(4)}$
	3	x_3	$-\tfrac{1}{3}$	$\tfrac{3}{2}$	-3	$\tfrac{11}{6}$	—	$\tfrac{1}{4}h^3 f^{(4)}$
Five Points ($n = 4$)	0	x_0	$-\tfrac{25}{12}$	$\tfrac{48}{12}$	$-\tfrac{36}{12}$	$\tfrac{16}{12}$	$-\tfrac{3}{12}$	$\tfrac{1}{5}h^4 f^{(5)}$
	1	x_1	$-\tfrac{3}{12}$	$-\tfrac{10}{12}$	$\tfrac{18}{12}$	$-\tfrac{6}{12}$	$\tfrac{1}{12}$	$-\tfrac{1}{20}h^4 f^{(5)}$
	2	x_2	$\tfrac{1}{12}$	$-\tfrac{8}{12}$	0	$\tfrac{8}{12}$	$-\tfrac{1}{12}$	$\tfrac{1}{30}h^4 f^{(5)}$
	3	x_3	$-\tfrac{1}{12}$	$\tfrac{6}{12}$	$-\tfrac{18}{12}$	$\tfrac{10}{12}$	$\tfrac{3}{12}$	$-\tfrac{1}{20}h^4 f^{(5)}$
	4	x_4	$\tfrac{3}{12}$	$-\tfrac{16}{12}$	$\tfrac{36}{12}$	$-\tfrac{48}{12}$	$\tfrac{25}{12}$	$\tfrac{1}{5}h^4 f^{(5)}$

In Table 3.54 we note that the only coefficients which vanish are $\alpha_{1,1}$ (for $n = 2$) and $\alpha_{2,2}$ (for $n = 4$). We seek to show in general that

3.55 $$\alpha_{s,k} \neq 0, \quad \text{unless } n \text{ is even and } k = s = \frac{n}{2}.$$

By Lagrange's formula, see Section 6.4, we have

3.56 $$\frac{1}{h}\alpha_{s,k} = \frac{Q'_k(x_s)}{Q_k(x_k)}$$

where

3.57 $$Q_k(x) = \prod_{\substack{j=0 \\ j \neq k}}^{n} (x - x_j).$$

But since

3.58 $$\frac{Q'_k(x)}{Q_k(x)} = \sum_{\substack{j=0 \\ j \neq k}}^{n} \frac{1}{x - x_j}$$

it follows that

3.59 $$\frac{1}{h}\alpha_{s,k} = \begin{cases} \displaystyle\sum_{\substack{j=0 \\ j \neq k}}^{n} \frac{1}{x_k - x_j}, & \text{if } s = k \\[3ex] \displaystyle[Q_k(x_k)]^{-1} \prod_{\substack{j=0 \\ j \neq s,k}}^{n} (x_s - x_j), & \text{if } s \neq k. \end{cases}$$

Thus, if $\alpha_{s,k} = 0$ then we must have $s = k$ and

3.60 $$\sum_{\substack{j=0 \\ j \neq k}}^{n} \frac{1}{x_k - x_j} = 0.$$

Since the points are evenly spaced, this can hold only if n is even and

3.61 $$k = s = \frac{n}{2}.$$

It can be shown that the order of any numerical differentiation formula for $f'(x)$ based on Lagrangian interpolation with $n + 1$ evenly spaced interpolation points is exactly n if x is an interpolation point. Otherwise the order is at least n and may actually be greater than n in some cases (for instance, the case of two interpolation points x_0 and x_1 where $x = (x_0 + x_1)/2$).

It is, of course, possible to represent higher derivatives of a given function $f(x)$. For example, by (3.35) we can represent $f''(x)$ in terms of $f(x_0)$, $f(x_1)$, $f(x_2)$, where $x_2 - x_1 = x_1 - x_0 = h$, by

3.62
$$f''(x) \sim \frac{f(x_0) + f(x_2) - 2f(x_1)}{h^2}.$$

The order of the formula is two unless $x = x_1$, in which case it is three. Thus we have by (2.28)

3.63
$$f''(x) = \begin{cases} \dfrac{f(x_0) + f(x_2) - 2f(x_1)}{h^2} + (x - x_1)f^{(3)}(\xi), & x \neq x_1 \\[3mm] \dfrac{f(x_0) + f(x_2) - 2f(x_1)}{h^2} - \dfrac{h^2}{12} f^{(4)}(\xi_1), & x = x_1 \end{cases}$$

where ξ and ξ_1 lie in the interval (x_0, x_2).

Numerical differentiation can be used for the determination of extreme values of tabulated functions. In principle, one can determine the Lagrangian interpolating polynomial $F(x)$ corresponding to a specified set of interpolation points and then solve the equation

3.64
$$F'(x) = 0$$

using any of the methods of Chapters 4 and 5. However, one can also compute $F'(x)$ at each of the interpolation points using Table 3.54, or a larger table if necessary, and then carry out inverse interpolation using the methods of Section 6.9 as applied to the function $F'(x)$. (We note that the Lagrange interpolating polynomial for the values of $F'(x)$ at the interpolation points is just $F'(x)$ itself.)

One can, of course, use procedures other than those based on interpolation for numerical differentiation. For some purposes the use of spline functions is highly desirable for numerical differentiation. The use of splines usually results in a smoother value for the derivative than could be expected from the derivative of a high-order polynomial. The use of cubic splines as described in Section 6.8 would appear to be adequate for many purposes. As a matter of fact, the process of determining the second derivative at the interpolation points used for spline interpolation leads to a simple method for determining the first derivatives.

EXERCISES 7.3

1. Find $f'(0.5)$ in terms of $f(0), f(1)$, and $f(2)$ for the case $f(x) = x^3$. Represent the situation graphically.

2. Given the table of $f(x) = e^x$ for $x = 0(0.2)1.0$

x	0	0.2	0.4	0.6	0.8	1.0
$f(x)$	1.00000	1.22140	1.49182	1.82212	2.22554	2.71828

   find approximate values of $f'(0.4)$ using: the forward difference formula (3.4); the backward difference formula; each of the formulas (3.36). Compute the actual error in each case and compare with the theoretical bounds.

3. How accurate could one estimate $f'(0.75)$ if $f(x) = x^2$ using the numerical differentiation formula

$$f'(a) \sim \frac{f\left(a + \dfrac{h}{2}\right) - f\left(a - \dfrac{h}{2}\right)}{h}$$

   where, for any x, $f(x)$ can be evaluated correctly to two decimal places. What if $f(x) = x^3$?

4. Derive a formula for $f'(a + (h/2))$ in terms of $f(a), f'(a), f''(a), f(a + h)$ using the method of undetermined weights and give a formula for the error, assuming that (2.28) holds.

5. Find an approximate formula for $f'(a + (h/2))$ in terms of $f(a), f(a + h)$, and $f(a + 2h)$: by differentiating the polynomial $F(x)$ of degree two or less which agrees with $f(x)$ at $x = a, a + h, a + 2h$; by the method of undetermined weights. Find the order of accuracy and an expression for the error assuming that (2.28) holds.

6. Verify (3.35) and (3.36).

7. Verify that the order of the numerical differentiation formula (3.35) is two for all x except that it is three for $x = x_1 \pm h/\sqrt{3}$.

8. For the numerical differentiation formula (3.35) find $R[f](x)$, assuming (2.28) holds. Consider the special cases $x = x_0$ and $x = x_1$.

9. Verify (3.49).

10. Using the method of undetermined weights derive an approximate formula for $f'(a)$ in terms of $f(a + h), f(a), f(a - h)$, and $f(a - 2h)$. Find the order of accuracy and an expression for the error assuming that (2.28) holds.

11. Let $x_i = 2 + ih$, for $i = 0, 1, 2, 3$ and for $h = \frac{1}{2}$. Evaluate

$$\prod_{\substack{j=0 \\ j \neq 1}}^{3} (x_1 - x_j).$$

    Compare this with

$$R[(x - x_0)^4](x_1)$$

    where for any function $f(x)$ in $C^{(1)}[2, 3.5]$ we have

$$R[f(x)](x_1) = f'(x_1) - h^{-1}[-\tfrac{1}{3}f(x_0) - \tfrac{1}{2}f(x_1) + f(x_2) - \tfrac{1}{6}f(x_3)].$$

12. Derive the formulas in Table 3.54 from the Gregory-Newton interpolation formula. Also verify the formula for the error in each case assuming that (2.28) holds.

13. Show that the order of any numerical differentiation formula for $f'(x)$ based on Lagrangian interpolation with $n + 1$ evenly spaced interpolation points is n if x is an interpolation point. Is the result true if the points are not necessarily evenly spaced?

14. Compute approximate values of $f''(1.5)$ and $f''(2)$ in terms of $f(1.5)$, $f(2.0)$, and $f(2.5)$ for the function $f(x) = x^3 + x^4$. Verify (3.63).

15. Show that if $x_k = x_0 + kh$, $k = 0, 1, 2, 3, 4$ then

$$f''(x_2) = \frac{-f(x_0) + 16f(x_1) - 30f(x_2) + 16f(x_3) - f(x_4)}{12h^2} + \frac{h^4}{90} f^{(6)}(\xi)$$

where $x_0 < \xi < x_4$. Assume that (2.28) holds.

16. Let $F(x)$ be a polynomial of degree n or less such that $F(x) = f(x)$ at the $n + 1$ distinct points $x_0, x_1, \ldots, x_n$. Let $G(x)$ be the unique polynomial of degree n or less such that $G(x) = F'(x)$ at $x_0, x_1, \ldots, x_n$. Show that $G(x) \equiv F'(x)$.

17. Given $f(0) = 1$, $f(1) = 1.5$, $f(2) = 1.2$, and $f(3) = 2.0$, find an approximate value of $f'(2.5)$ by each of the following procedures:

   a) Compute $F'(2.5)$ where $F(x)$ is the polynomial of degree three or less which agrees with $f(x)$ at $x = 0, 1, 2, 3$.
   b) Compute $G(2.5)$ where $G(x)$ is the polynomial of degree three or less which agrees with $F'(x)$ at the points $x = 0, 1, 2, 3$.

18. Find an approximate value of x for which $f(x)$ is a maximum and find the corresponding approximate maximum value of $f(x)$ given the following:

x	0	0.5	1.0	1.5	2.0	2.5	3.0
$f(x)$	1.535	1.681	1.729	1.691	1.505	1.416	1.311

To do this find the value of x which maximizes $F(x)$ where $F(x)$ is defined by each of the following:

   a) $F(x)$ is a quadratic polynomial which best fits the data in the least squares sense.
   b) $F(x)$ is a quadratic polynomial which agrees with $f(x)$ at three consecutive interpolation points.
   c) $F(x)$ is a polynomial of degree six or less which agrees with $f(x)$ at all of the interpolation points.
   d) $F(x)$ is a cubic spline function such that $F(x) \in C^{(2)}[0, 3]$ which agrees with $f(x)$ at all of the interpolation points. Let $F''(0) = F''(3) = 0$.

19. Generate a table for the function $f(x) = xe^{-ax}$ for $x = 0(0.1)1$ where $a = \frac{4}{3}$. (See Exercise 7, Section 6.9.) Find an approximate value of the abscissa $\bar{x}$ which maximizes $f(x)$ by first determining an approximate value of $f'(x)$ at each interpolation point using an appropriate four-point formula and then using inverse four-point interpolation.

7.4 NUMERICAL QUADRATURE—EQUAL INTERVALS

Let us first consider representations of the integral

$$\textbf{4.1} \qquad\qquad\qquad L[f] = \int_a^b f(x)dx$$

in terms of a single value of $f(x)$, say $f(x_0)$. Since the polynomial $F(x)$ of degree zero which agrees with $f(x)$ at x_0 is

4.2
$$F(x) = f(x_0)$$

we have the *rectangle rule*

4.3
$$\int_a^b f(x)dx \sim \int_a^b F(x)dx = (b - a)f(x_0).$$

The order of the rectangle rule is zero if $x_0 \neq (b + a)/2$; otherwise it is one. By (2.28) we have

4.4
$$\int_a^b f(x)dx = \begin{cases} (b - a)f(x_0) + \dfrac{(b - a)(b + a - 2x_0)}{2} f'(\xi), & \text{if } x_0 \neq \dfrac{b + a}{2} \\[3mm] (b - a)f\left(\dfrac{b + a}{2}\right) + \dfrac{(b - a)^3}{24} f''(\xi_1), & \text{if } x_0 = \dfrac{b + a}{2} \end{cases}$$

for some ξ and ξ_1 in the smallest interval containing a, b, and x_0. In particular we have

4.5
$$\int_a^b f(x)dx = (b - a)f(a) + \frac{(b - a)^2}{2} f'(\xi_2)$$

for the (*forward*) *rectangle rule* and

4.6
$$\int_a^b f(x)dx = (b - a)f(b) - \frac{(b - a)^2}{2} f'(\xi_3)$$

for the *backward rectangle rule*. The formula

4.7
$$\int_a^b f(x)dx = (b - a)f\left(\frac{a + b}{2}\right) + \frac{(b - a)^3}{24} f''(\xi_1)$$

is known as the *midpoint rule*.

As an example, let us consider the evaluation of

4.8
$$L[f] = \int_{1.0}^{1.2} e^x \, dx.$$

With the forward rectangle rule we get*

4.9 $$\tilde{L}[f] = (0.2)e^{1.0} \doteq 0.54366.$$

Since* $L[f] = e^{1.2} - e^{1.0} \doteq 0.60184$ the error is

4.10 $$L[f] - \tilde{L}[f] \doteq 0.05818.$$

From (4.5) the error lies between

4.11 $$\frac{(0.2)^2}{2}e^{1.0} \quad \text{and} \quad \frac{(0.2)^2}{2}e^{1.2}$$

i.e., between

4.12 $$0.05437 \quad \text{and} \quad 0.06640.$$

With the midpoint rule* we get

4.13 $$\tilde{L}[f] \doteq 0.60083$$

so that the error is

4.14 $$L[f] - \tilde{L}[f] \doteq 0.00101.$$

The upper and lower bounds given by (4.7) are

4.15 $$0.00091 \quad \text{and} \quad 0.00111.$$

Let us now consider numerical quadrature formulas based on the use of two points x_0 and x_1. Evidently $F(x)$ is given by (3.1) and we have

4.16 $$\int_a^b f(x)dx \sim \int_a^b F(x)dx = \frac{b-a}{2(x_1-x_0)}\{(2x_1-(b+a))f(x_0)+(b+a-2x_0)f(x_1)\}.$$

We shall show in Section 7.8 that one can achieve an order of accuracy of *three* by suitable choice of x_1 and x_0. However, for the present we shall consider the choice

4.17 $$x_0 = a, \quad x_1 = b$$

which leads to the *trapezoidal rule*

4.18 $$\int_a^b f(x)dx \sim \frac{b-a}{2}(f(a) + f(b)).$$

---

* Here we use the values $e^{1.0} \doteq 2.71828$, $e^{1.1} \doteq 3.00417$, $e^{1.2} \doteq 3.32012$.

Evidently, the order of the trapezoidal rule is one and we have, by (2.28),

4.19 $$\int_a^b f(x)dx = \frac{b-a}{2}(f(a) + f(b)) - \frac{(b-a)^3}{12} f''(\xi)$$

where $\xi \in (a, b)$.

For example (4.8) the trapezoidal rule gives

4.20 $$\int_{1.0}^{1.2} f(x)dx \sim \frac{0.2}{2}[e^{1.0} + e^{1.2}] \doteq 0.60384$$

so that the error is

4.21 $$\int_{1.0}^{1.2} f(x)dx - 0.60384 \doteq 0.60184 - 0.60384 = -0.00200.$$

By (4.19) the bounds on the error are

$$-0.00181 \quad \text{and} \quad -0.00221.$$

Figure 4.22 gives a geometric interpretation of the elementary numerical quadrature formulas considered so far. Thus, the (forward) rectangle rule yields the area of the rectangle ABHI, while the backward rectangle rule gives the area of ADEI. The midpoint rule gives the area of the rectangle ACGI. (This is the same as that of the trapezoid AB'E'I where B'E' is tangent to the curve BFE at F.) The trapezoidal rule yields the area of the trapezoid ABEI. On the basis of Fig. 4.22 one might expect that the midpoint rule is the most accurate of the methods considered, the trapezoidal rule the next most accurate, and the rectangle rules the least accurate. This is borne out by the theoretical formulas given for the errors.

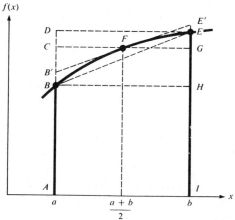

4.22 Fig. Geometric Interpretation of Elementary Numerical Quadrature Formulas.

Let us now consider numerical quadrature formulas based on the use of three points x_0, x_1, and x_2. The polynomial $F(x)$ of degree two or less which agrees with $f(x)$ at the interpolation points is given by (3.34). We could find

$$\textbf{4.23} \qquad \int_a^b F(x)dx$$

by integration. Moreover, the order of the accuracy of the formula

$$\textbf{4.24} \qquad \int_a^b f(x)dx \sim \int_a^b F(x)dx$$

would be at least *two*. On the other hand, as we show in Section 7.8, one can obtain an order of accuracy of *five* by a suitable choice of x_0, x_1, and x_2.

If the three points are evenly spaced, then it is probably simpler to evaluate (4.23) by using the Gregory-Newton interpolation formula

$$\textbf{4.25} \qquad F(x) = f(x_0) + u\Delta f(x_0) + \frac{u(u-1)}{2}\Delta^2 f(x_0)$$

where we let

$$\textbf{4.26} \qquad u = \frac{x - x_0}{h}, \qquad h = x_2 - x_1 = x_1 - x_0.$$

Evidently, we have

$$\textbf{4.27} \qquad \int_a^b F(x)dx = h\int_{u(a)}^{u(b)} \left(f(x_0) + u\Delta f(x_0) + \frac{u(u-1)}{2}\Delta^2 f(x_0)\right) du$$

$$= (b-a)\left\{ f(x_0) + \frac{b+a-2x_0}{2h}(\Delta f(x_0) - \tfrac{1}{2}\Delta^2 f(x_0))\right.$$

$$\left. + \frac{(b-x_0)^2 + (b-x_0)(a-x_0) + (a-x_0)^2}{6h^2}\Delta^2 f(x_0)\right\}.$$

If we let $x_0 = a$, $x_2 = b$ we get *Simpson's rule*

$$\textbf{4.28} \qquad \int_{x_0}^{x_2} f(x)dx \sim \frac{h}{3}[f(x_0) + 4f(x_1) + f(x_2)].$$

If we let $x_0 = a$, $x_1 = b$ we get the *five-eight rule*

$$\textbf{4.29} \qquad \int_{x_0}^{x_1} f(x)dx \sim \frac{h}{12}[5f(x_0) + 8f(x_1) - f(x_2)].$$

It is easy to verify that the order is three for Simpson's rule and two for the five-eight rule. Moreover, by (2.28) we have

4.30
$$\int_{x_0}^{x_2} f(x)dx = \frac{h}{3}[f(x_0) + 4f(x_1) + f(x_2)] - \frac{h^5}{90} f^{(4)}(\xi_1)$$

4.31
$$\int_{x_0}^{x_1} f(x)dx = \frac{h}{12}[5f(x_0) + 8f(x_1) - f(x_2)] + \frac{h^4}{24} f^{(3)}(\xi_2)$$

for some ξ_1 and ξ_2 in the interval (x_0, x_2).

As an example, let us consider the case (4.8) where $h = 0.1$. By Simpson's rule we get

4.32
$$\int_{1.0}^{1.2} e^x \, dx \sim \frac{0.1}{3} [e^{1.0} + 4e^{1.1} + e^{1.2}] \doteq 0.60184$$

which is correct to five decimals. The bounds for the theoretical error are

4.33 -0.00000030 and $-0.00000037.$

For the five-eight rule we get

4.34
$$\int_{1.0}^{1.1} e^x \, dx \sim \frac{0.1}{12} [5e^{1.0} + 8e^{1.1} - e^{1.2}] \doteq 0.28587$$

as compared with

4.35
$$\int_{1.0}^{1.1} e^x \, dx = e^{1.1} - e^{1.0} \doteq 0.28589.$$

Thus the error is 0.00002. Bounds for the theoretical error are

4.36   0.000011   and   0.000014.

(The fact that the actual error is larger than the upper bound is due to rounding.)

The method of undetermined weights is often very convenient for deriving numerical quadrature formulas. For example, to derive Simpson's rule we seek w_0, w_1, and w_2 such that

4.37
$$\int_{x_0}^{x_2} f(x)dx \sim w_0 f(x_0) + w_1 f(x_1) + w_2 f(x_2)$$

is exact for any polynomial of degree two or less. By successively letting $f(x) = 1$, $(x - x_1)$ and $(x - x_1)^2$ we get

4.38
$$\begin{cases} 2h = w_0 + w_1 + w_2 \\ 0 = -w_0 h + w_2 h \\ \dfrac{2h^3}{3} = w_0 h^2 + w_2 h^2 \end{cases}$$

Solving, we get

4.39
$$w_0 = w_2 = \frac{h}{3}, \quad w_1 = \frac{4h}{3}$$

which agree with (4.28).

As another example, let us determine

$$\int_{x_0}^{x_1} f(x)\,dx$$

in terms of $f(x_0)$, $f(x_1)$, $f'(x_0)$, $f'(x_1)$. We seek w_0, w_1, w_0^*, and w_1^* such that

4.40
$$\int_{x_0}^{x_1} f(x)\,dx \sim w_0 f(x_0) + w_1 f(x_1) + w_0^* f'(x_0) + w_1^* f'(x_1)$$

is exact for any polynomial of degree three or less. By successively letting $f(x)$ be $(x - \alpha)^0$, $x - \alpha$, $(x - \alpha)^2$, and $(x - \alpha)^3$ where

4.41
$$\alpha = \frac{x_0 + x_1}{2}$$

we get

4.42
$$\begin{cases} 2\sigma = w_0 + w_1 \\ 0 = -\sigma w_0 + \sigma w_1 + w_0^* + w_1^* \\ \dfrac{2\sigma^3}{3} = \sigma^2 w_0 + \sigma^2 w_1 - 2\sigma w_0^* + 2\sigma w_1^* \\ 0 = -\sigma^3 w_0 + \sigma^3 w_1 + 3\sigma^2 w_0^* + 3\sigma^2 w_1^* \end{cases}$$

where we let

4.43
$$\sigma = \frac{h}{2}.$$

Solving we have

4.44 $$w_0 = \frac{h}{2}, \quad w_1 = \frac{h}{2}, \quad w_0^* = \frac{h^2}{12}, \quad w_1^* = -\frac{h^2}{12}.$$

Evidently the order of the method is three and by (2.28) we have

4.45 $$\int_{x_0}^{x_1} f(x)dx = \frac{h}{2}[f(x_0) + f(x_1)] - \frac{h^2}{12}[f'(x_1) - f'(x_0)] + \frac{h^5}{720} f^{(4)}(\xi).$$

As an example, let us consider the evaluation of (4.8) using (4.45) with $h = 0.2$, we have

4.46 $$\int_{1.0}^{1.2} e^x\, dx \sim \frac{0.2}{2}[e^{1.0} + e^{1.2}] - \frac{(0.2)^2}{12}[e^{1.2} - e^{1.0}]$$

$$\doteq 0.60384 - 0.00201 = 0.60183.$$

Thus the error is

4.47 $$0.60184 - 0.60183 = 0.00001.$$

By (4.45) bounds for the error are

4.48 $$0.0000012 \quad \text{and} \quad 0.0000015.$$

Thus the error in the result is strictly due to rounding.

In Table 4.51 we give coefficients for a number of numerical quadrature formulas based on Lagrangian interpolation. We denote by Q.q.M.m the q-step formula

4.49 $$\int_{x_0}^{x_q} f(x)dx \sim h \sum_{i=m}^{M} \beta_i f(x_i)$$

involving values of $f(x)$ at the $M + 1 - m$ points $x_m, x_{m+1}, \ldots, x_M$. Thus, for example, we have the following designations:

(forward) rectangle rule: Q.1.0.0
backward rectangle rule: Q.1.1.1
midpoint rule: Q.2.1.1
Simpson's rule: Q.2.2.0
five-eight rule: Q.1.2.0.

In addition to the coefficients we give $R[f]$ in each case where $R[f]$ is given by (2.28). The argument for the indicated derivative is not given. It is some number in the interval (x_0, x_N) where

4.50 $$N = \max(q, M).$$

4.51 Table   Table of Coefficients for Numerical Quadrature Formulas with $m = 0$.

M	β_0	β_1	β_2	β_3	β_4	$R[f]$	Name	Symbol
Single-step $(q = 1)$								
$\int_{x_0}^{x_1} f(x)dx$ 0	1	—	—	—	—	$\frac{1}{2}h^2 f^{(1)}$	Rectangle Rule	Q.1.0.0
1	$\frac{1}{2}$	$\frac{1}{2}$	—	—	—	$-\frac{1}{12}h^3 f^{(2)}$	Trapezoidal Rule	Q.1.1.0
2	$\frac{5}{12}$	$\frac{8}{12}$	$-\frac{1}{12}$	—	—	$\frac{1}{24}h^4 f^{(3)}$	Five-eight Rule	Q.1.2.0
3	$\frac{9}{24}$	$\frac{19}{24}$	$-\frac{5}{24}$	$\frac{1}{24}$	—	$-\frac{19}{720}h^5 f^{(4)}$	—	Q.1.3.0
Double-step $(q = 2)$								
$\int_{x_0}^{x_2} f(x)dx$ 0	2	—	—	—	—	$2h^2 f^{(1)}$	—	Q.2.0.0
1	0	2	—	—	—	$\frac{1}{3}h^3 f^{(2)}$	Midpoint Rule	Q.2.1.0
2	$\frac{1}{3}$	$\frac{4}{3}$	$\frac{1}{3}$	—	—	$-\frac{1}{90}h^5 f^{(4)}$	Simpson's Rule	Q.2.2.0
3	$\frac{1}{3}$	$\frac{4}{3}$	$\frac{1}{3}$	0	—	$-\frac{1}{90}h^5 f^{(4)}$	Simpson's Rule	Q.2.3.0
Three-step $(q = 3)$								
$\int_{x_0}^{x_3} f(x)dx$ 1	$-\frac{3}{2}$	$\frac{9}{2}$	—	—	—	$\frac{9}{4}h^3 f^{(2)}$	—	Q.3.1.0
2	$\frac{3}{4}$	0	$\frac{9}{4}$	—	—	$\frac{3}{8}h^4 f^{(3)}$	—	Q.3.2.0
3	$\frac{3}{8}$	$\frac{9}{8}$	$\frac{9}{8}$	$\frac{3}{8}$	—	$-\frac{3}{80}h^5 f^{(4)}$	Simpson's Second Rule	Q.3.3.0
Four-step $(q = 4)$								
$\int_{x_0}^{x_4} f(x)dx$ 2	$\frac{8}{3}$	$-\frac{16}{3}$	$\frac{20}{3}$	—	—	$\frac{8}{3}h^4 f^{(3)}$	—	Q.4.2.0
3	0	$\frac{8}{3}$	$-\frac{4}{3}$	$\frac{8}{3}$	—	$\frac{14}{45}h^5 f^{(4)}$	—	Q.4.3.0
4	$\frac{14}{45}$	$\frac{64}{45}$	$\frac{24}{45}$	$\frac{64}{45}$	$\frac{14}{45}$	$-\frac{8}{945}h^7 f^{(6)}$	Newton-Cotes	Q.4.4.0

Formulas with $m = 1$

M	β_1	β_2	β_3	β_4	$R[f]$	Name	Symbol
Single-step $(q = 1)$							
$\int_{x_0}^{x_1} f(x)dx$ 1	1	—	—	—	$-\frac{1}{2}h^2 f^{(1)}$	Backward Rectangle Rule	Q.1.1.1
2	$\frac{3}{2}$	$-\frac{1}{2}$	—	—	$\frac{5}{12}h^3 f^{(2)}$	—	Q.1.2.1
3	$\frac{23}{12}$	$-\frac{16}{12}$	$\frac{5}{12}$	—	$-\frac{3}{8}h^4 f^{(3)}$	—	Q.1.3.1
4	$\frac{55}{24}$	$-\frac{59}{24}$	$\frac{37}{24}$	$-\frac{9}{24}$	$\frac{251}{720}h^5 f^{(4)}$	—	Q.1.4.1

(continued)

M	β_1	β_2	β_3	β_4	$R[f]$	Name	Symbol
Double-step $(q=2)$ $\int_{x_0}^{x_2} f(x)dx$							
1	2	—			$\frac{1}{3}h^3 f^{(2)}$	Midpoint Rule	Q.2.1.1
2	2	0	—	—	$\frac{1}{3}h^3 f^{(2)}$	Midpoint Rule	Q.2.2.1
3	$\frac{7}{3}$	$-\frac{2}{3}$	$\frac{1}{3}$	—	$-\frac{1}{3}h^4 f^{(3)}$	—	Q.2.3.1
4	$\frac{8}{3}$	$-\frac{5}{3}$	$\frac{4}{3}$	$-\frac{1}{3}$	$\frac{29}{90}h^5 f^{(4)}$	—	Q.2.4.1
Three-step $(q=3)$ $\int_{x_0}^{x_3} f(x)dx$							
2	$\frac{3}{2}$	$\frac{3}{2}$	—	—	$\frac{3}{4}h^3 f^{(2)}$	Newton-Cotes open	Q.3.2.1
3	$\frac{9}{4}$	0	$\frac{3}{4}$	—	$-\frac{3}{8}h^4 f^{(3)}$	—	Q.3.3.1
4	$\frac{21}{8}$	$-\frac{9}{8}$	$\frac{15}{8}$	$-\frac{3}{8}$	$\frac{27}{80}h^5 f^{(4)}$	—	Q.3.4.1
Four-step $(q=4)$ $\int_{x_0}^{x_4} f(x)dx$							
3	$\frac{8}{3}$	$-\frac{4}{3}$	$\frac{8}{3}$	—	$\frac{14}{45}h^5 f^{(4)}$	Newton-Cotes open	Q.4.3.1
4	$\frac{8}{3}$	$-\frac{4}{3}$	$\frac{8}{3}$	0	$\frac{14}{45}h^5 f^{(4)}$	—	Q.4.4.1

An important class of numerical quadrature formulas are the *Newton-Cotes formulas*. The *closed* formulas are

Q.1.1.0	the trapezoidal rule
Q.2.2.0	Simpson's rule
Q.3.3.0	Simpson's second rule
Q.4.4.0	—

etc. The *open* formulas are

Q.2.1.1	the midpoint rule
Q.3.2.1	—
Q.4.3.1	—

etc.

If we wish to integrate over an interval $a \leq x \leq b$ to a high degree of accuracy we can subdivide the interval into equal parts and then use a numerical quadrature formula. If we used a formula based on a single interpolating polynomial over the entire range we would obtain a closed Newton-Cotes formula. However, the use of such formulas corresponding to high-order polynomial interpolation is not usually satisfactory. It can, in fact, be shown that the sequence of results obtained by Newton-Cotes formulas need not converge even for a continuous function (see Kusman [1931]). One method of obtaining high accuracy without this difficulty is to use a *composite* numerical quadrature formula based on the use of different interpolating polynomials in each of several subintervals. Thus, for example, to evaluate (4.1) we could divide the interval $a \leqq x \leqq b$ into n subintervals of equal length,

4.52 $$x_i \leqq x \leqq x_{i+1} = x_i + h, \qquad i = 0, 1, \ldots, n - 1$$

where $x_0 = a, x_n = b$. One could then use the trapezoidal rule

4.53 $$\int_{x_i}^{x_{i+1}} f(x)dx \sim \frac{h}{2}[f(x_i) + f(x_{i+1})]$$

to evaluate the integral over each subinterval. Adding these approximate values we get

4.54 $$\int_a^b f(x)dx \sim h\left[\tfrac{1}{2}f(x_0) + \sum_{i=1}^{n-1} f(x_i) + \tfrac{1}{2}f(x_n)\right].$$

For any $\zeta_0, \zeta_1, \ldots, \zeta_{n-1}$ in the intervals $x_i \leqq x \leqq x_{i+1}$, $i = 0, 1, \ldots, n - 1$, respectively, we have

4.55 $$\sum_{i=0}^{n-1} f^{(2)}(\zeta_i) = nf^{(2)}(\zeta)$$

for some ζ with $a < \zeta < b$. This follows from the continuity of $f^{(2)}(x)$. Therefore by (4.19) we have, since $(b - a) = nh$,

4.56 $$\int_a^b f(x)dx = h\left[\tfrac{1}{2}f(x_0) + \sum_{i=1}^{n-1} f(x_i) + \tfrac{1}{2}f(x_n)\right] - \frac{(b-a)h^2}{12} f''(\zeta)$$

for some $\zeta \in (a, b)$.

As an example let us evaluate

4.57 $$\int_1^2 e^x \, dx$$

using a composite formula based on the trapezoidal rule with $n = 4$. We have $h = \tfrac{1}{4}$ and*

4.58 $$\int_1^2 e^x \, dx \sim \tfrac{1}{4}[\tfrac{1}{2}e^{1.0} + e^{1.25} + e^{1.50} + e^{1.75} + \tfrac{1}{2}e^{2.0}] \doteq 4.69508.$$

---

* Here we use the values $e^{1.0} \doteq 2.71828$, $e^{1.25} \doteq 3.49034$, $e^{1.50} \doteq 4.48169$, $e^{1.75} \doteq 5.75460$, and $e^{2.0} \doteq 7.38906$.

The value of the integral is $e^{2.0} - e^{1.0} \doteq 4.67078$ and hence the error is -0.02430. By (4.56) the error lies between

4.59 -0.01416 and -0.03848.

By the use of the midpoint rule one can achieve greater accuracy with fewer function evaluations. Thus, we again divide the interval $[a, b]$ into n subintervals of equal length, but we evaluate the interval over the subinterval $[x_i, x_{i+1}]$ by

4.60 $$\int_{x_i}^{x_{i+1}} f(x)dx \sim hf\left(\frac{x_i + x_{i+1}}{2}\right).$$

Thus we have the composite formula (see (4.7))

4.61 $$\int_a^b f(x)dx = h\sum_{i=0}^{n-1} f\left(\frac{x_i + x_{i+1}}{2}\right) + \frac{(b-a)h^2}{24} f''(\xi).$$

In the example (4.57) if we again let $n = 4$ we have

4.62 $$\int_1^2 e^x\, dx \sim (0.25)[e^{1.125} + e^{1.375} + e^{1.625} + e^{1.875}] \doteq 4.65863.$$

(We use the values $e^{1.125} \doteq 3.08022$, $e^{1.375} \doteq 3.95508$, $e^{1.625} \doteq 5.07842$, $e^{1.875} \doteq 6.52082$.) The error is

4.63 $4.67078 - 4.65863 = 0.01215.$

By (4.61) the error lies between

4.64 0.00708 and 0.01924.

Thus the accuracy is about twice as great as for the trapezoidal rule.

If we divide the interval $[a, b]$ into an even number of subintervals, then we can use a composite formula based on Simpson's rule. This composite formula has the form

4.65 $$\int_a^b f(x)dx = \frac{h}{3}[f(x_0) + 4f(x_1) + 2f(x_2) + 4f(x_3) + \cdots$$

$$+ 2f(x_{n-2}) + 4f(x_{n-1}) + f(x_n)] - \frac{(b-a)h^4}{180} f^{(4)}(\xi).$$

(The factor 180 appears rather than 90 since there are only $(b - a)/2h$ pairs of intervals.)

In the example (4.57) if we let $n = 4$ we have

4.66 $$\int_1^2 e^x \, dx \sim \frac{h}{3} [e^{1.0} + 4e^{1.25} + 2e^{1.5} + 4e^{1.75} + e^{2.0}]$$

$$\doteq \frac{0.25}{3} (56.05048) \doteq 4.67087.$$

Since the value of (4.57), correct to five places, is 4.67078, the error is

4.67 $-0.00009.$

By (4.65) the bounds on the error are

4.68 -0.00006 and $-0.00016.$

Sometimes one is given $f(x)$ in the form of a table and one wishes to integrate over a range spanned by an odd number of tabulated values. In this case one could use Simpson's second rule for three subintervals and then Simpson's rule over the remainder of the interval. This will give an overall error of order h^4.

EXERCISES 7.4

1. Evaluate the integral (4.8) using the backward rectangle rule with $h = 0.2$. Compare the actual error with the bounds given by (4.6).
2. Find x_1 such that if $x_0 = a$, the method (4.16) is of order two.
3. Show that for suitable choice of x_1 and x_2 we can achieve an order of accuracy of three with (4.16).
4. Apply the formula derived in the previous exercise to evaluate the integral (4.8).
5. Derive the formula for the five-eight rule by the method of undetermined weights. Find an expression for the error assuming that (2.28) holds.
6. Show that Simpson's rule is of order three and find an expression for the error assuming that (2.28) holds.
7. Derive Simpson's rule by integrating the Lagrangian interpolation formula (3.34) where $x_2 - x_1 = x_1 - x_0 = h$.
8. Derive the formula for method Q.1.2.(-1) by the method of undetermined weights. Find the error assuming (2.28). Apply the formula to compute

$$\int_0^1 (1 + x^2)^{-1} \, dx$$

with $h = \frac{1}{2}$. Find the actual error and theoretical bounds for the error.
9. Evaluate

$$\int_0^1 (1 + x^2)^{-1} \, dx$$

with the trapezoidal rule and by Simpson's rule with $h = \frac{1}{4}$. In each case compute the actual error and theoretical bounds in the error.

10. Calculate

$$\int_1^2 x^{-1}\,dx$$

by Simpson's rule with $h = \frac{1}{4}$. Compare the actual error with theoretical bounds for the error.

11. Verify the formula given in Table 4.51 for the error for Simpson's second rule.

12. Derive the formula for method Q.4.4.0 and the error.

13. Derive a formula for method Q.4.3.1 by the method of undetermined weights and find an expression for the error assuming that (2.28) holds. Apply the formula to evaluate

$$\int_0^{0.4} (5x^4 + 1)dx$$

with $h = 0.1$. Determine the actual error and the theoretical bounds for the error. Work to five decimal places.

14. Derive the formula for Q.1.1.(-2) by the method of undetermined weights and find an expression for the error assuming that (2.28) holds. Do the same for method Q.1.3.0.

15. Evaluate

$$\int_0^1 (1 + x^2)dx$$

with $h = 0.2$ using a composite formula based on Simpson's first and second rule. Find bounds in the error and compare with the actual error.

16. Show that the order of closed Newton-Cotes formula Q.n.n.0 is n if n is odd and $n + 1$ if n is even.

7.5 THE EULER-MACLAURIN FORMULA

Let us now develop a numerical quadrature formula in terms of $f(x_0), f'(x_0), \ldots,$ $f^{(2m-1)}(x_0), f(x_1), f'(x_1), \ldots, f^{(2m-1)}(x_1)$ for some positive integer m. We seek to show that

5.1 $$\int_{x_0}^{x_1} f(x)dx = \frac{h}{2}[f(x_0) + f(x_1)] - \sum_{k=1}^m \frac{B_{2k}}{(2k)!} h^{2k}[f^{(2k-1)}(x_1) - f^{(2k-1)}(x_0)]$$

$$- \frac{h^{2m+3}B_{2m+2}}{(2m+2)!} f^{(2m+2)}(\xi)$$

for some ξ in the interval (x_0, x_1) where

5.2 $$h = x_1 - x_0.$$

The numbers $B_2, B_4, B_6, \ldots$ are the *Bernoulli numbers* of even order which can be defined by

5.3
$$\coth x = \frac{1}{x}\left\{ B_0 + \sum_{k=1}^{\infty} B_{2k}\frac{(2x)^{2k}}{(2k)!} \right\}$$

$$= \frac{1}{x}\left\{ 1 + \tfrac{1}{3}x^2 - \tfrac{1}{45}x^4 + \tfrac{2}{945}x^6 + \cdots \right\}.$$

The first few Bernoulli numbers of even order are

5.4
$$B_0 = 1, \quad B_2 = \tfrac{1}{6}, \quad B_4 = -\tfrac{1}{30}, \quad B_6 = \tfrac{1}{42}, \quad B_8 = -\tfrac{1}{30}.$$

We have already proved (5.1) for the case $m - 1$ (see (4.45)). We now give a proof for the case $m = 2$. The technique used is applicable in the general case. We seek to prove that

5.5
$$\int_{x_0}^{x_1} f(x)dx = \frac{h}{2}[f(x_0) + f(x_1)] - \frac{h^2}{12}[f'(x_1) - f'(x_0)]$$

$$+ \frac{h^4}{720}[f^{(3)}(x_1) - f^{(3)}(x_0)] - \frac{h^7}{30{,}240}f^{(6)}(\xi).$$

If $f(x)$ is a polynomial of degree 5 or less, then for any α we have

5.6
$$\begin{cases}
f(x) = f(\alpha) + (x - \alpha)f'(\alpha) + \frac{(x - \alpha)^2}{2!}f''(\alpha) + \frac{(x - \alpha)^3}{3!}f^{(3)}(\alpha) \\[2mm]
\qquad + \frac{(x - \alpha)^4}{4!}f^{(4)}(\alpha) + \frac{(x - \alpha)^5}{5!}f^{(5)}(\alpha) \\[3mm]
f'(x) = f'(\alpha) + (x - \alpha)f''(\alpha) + \frac{(x - \alpha)^2}{2!}f^{(3)}(\alpha) + \frac{(x - \alpha)^3}{3!}f^{(4)}(\alpha) \\[2mm]
\qquad + \frac{(x - \alpha)^4}{4!}f^{(5)}(\alpha) \\[3mm]
\qquad \vdots \\[2mm]
f^{(4)}(x) = f^{(4)}(\alpha) + (x - \alpha)f^{(5)}(\alpha).
\end{cases}$$

If we let

5.7
$$\alpha = \frac{x_0 + x_1}{2} \quad \text{and} \quad \sigma = \frac{h}{2}$$

we have

$$
5.8 \quad
\begin{cases}
\frac{1}{2}(f(x_0) + f(x_1)) = f(\alpha) + \dfrac{\sigma^2}{2!} f''(\alpha) + \dfrac{\sigma^4}{4!} f^{(4)}(\alpha) \\[3mm]
\dfrac{1}{h} \displaystyle\int_{x_0}^{x_1} f(x)dx = f(\alpha) + \dfrac{\sigma^2}{3!} f''(\alpha) + \dfrac{\sigma^4}{5!} f^{(4)}(\alpha) \\[3mm]
\dfrac{1}{h}[f'(x_1) - f'(x_0)] = \qquad\qquad f''(\alpha) + \dfrac{\sigma^2}{3!} f^{(4)}(\alpha) \\[3mm]
\dfrac{1}{h}[f^{(3)}(x_1) - f^{(3)}(x_0)] = \qquad\qquad\qquad f^{(4)}(\alpha).
\end{cases}
$$

Therefore

$$
5.9 \quad \frac{1}{2}[f(x_0) + f(x_1)] - \frac{1}{h}\int_{x_0}^{x_1} f(x)dx - \frac{2}{3!}\frac{\sigma^2}{h}[f'(x_1) - f'(x_0)]
$$

$$
- \left(\frac{4}{5!} - \frac{2}{(3!)^2}\right)\frac{\sigma^4}{h}[f^{(3)}(x_1) - f^{(3)}(x_0)] = 0
$$

so that (5.5) holds.

Evidently the order of the method (5.5), without the remainder term, is five. By (2.28) the remainder term is

$$
5.10 \quad\quad\quad\quad R[f] = R[(x - \alpha)^6]\frac{f^{(6)}(\xi)}{6!}.
$$

But since

$$
5.11 \quad\quad\quad\quad R[(x - \alpha)^6] = -\frac{h^7}{42}
$$

we have

$$
5.12 \quad\quad\quad\quad R[f] = -\frac{h^7}{30,240} f^{(6)}(\xi).
$$

We note that the process of expressing $(\frac{1}{2})[f(x_0) + f(x_1)]$ in terms of

$$
\int_{x_0}^{x_1} f(x)dx, \quad f'(x_1) - f'(x_0), \quad f^{(3)}(x_1) - f^{(3)}(x_0),
$$

etc., is analogous to that of computing the power series expansion for $x \coth x$ by successive subtractions in the division of

5.13 $$\cosh x = 1 + \frac{x^2}{2!} + \frac{x^4}{4!} + \frac{x^6}{6!} + \cdots$$

by

5.14 $$\frac{1}{x} \sinh x = 1 + \frac{x^2}{3!} + \frac{x^4}{5!} + \frac{x^6}{7!} + \cdots.$$

Thus we have

$$
\begin{array}{l}
\left. 1 + \dfrac{x^2}{2!} + \dfrac{x^4}{4!} + \dfrac{x^6}{6!} + \cdots \quad\right| \quad 1 + \dfrac{x^2}{3!} + \dfrac{x^4}{5!} + \dfrac{x^6}{7!} + \cdots \\[2ex]
\underline{1 + \dfrac{x^2}{3!} + \dfrac{x^4}{5!} + \dfrac{x^6}{7!} + \cdots} \\[2ex]
\underline{\dfrac{2x^2}{3!} + \dfrac{4x^4}{5!} + \dfrac{6x^6}{7!} + \cdots} \\[2ex]
\underline{\dfrac{2x^2}{3!} + \dfrac{2}{(3!)^2}x^4 + \dfrac{2}{3!5!}x^6 + \cdots} \\[2ex]
\left(\dfrac{4}{5!} - \dfrac{2}{(3!)^2} \right)x^4 + \left(\dfrac{6}{7!} - \dfrac{2}{3!5!} \right)x^6 + \cdots.
\end{array}
$$

Hence the quotient is

5.15 $$1 + \frac{2}{3!}x^2 + \left(\frac{4}{5!} - \frac{2}{(3!)^2} \right)x^4 + \cdots = 1 + \tfrac{1}{3}x^2 - \tfrac{1}{45}x^4 + \cdots$$

and by (5.3) we have $B_2 = \frac{1}{6}$, $B_4 = -\frac{1}{30}$, etc.

Let us now consider the composite formula based on (5.1) for integrating between a and b. We let

5.16 $$x_k = x_0 + kh, \qquad k = 0, 1, \ldots, n$$

where

5.17 $$h = \frac{b - a}{n}$$

and we have, by (5.1), for any $m = 1, 2, \ldots$

5.18 $\displaystyle \int_a^b f(x)dx = h \sum_{j=0}^n {}'' f(x_j) - \sum_{k=1}^m \frac{B_{2k}}{(2k)!} h^{2k}[f^{(2k-1)}(b) - f^{(2k-1)}(a)]$

$$- \frac{(b-a)h^{2m+2}}{(2m+2)!} B_{2m+2} f^{(2m+2)}(\xi)$$

for some ξ in the interval (a, b). Here we let

5.19 $\displaystyle \sum_{j=0}^n {}'' f(x_j) = \tfrac{1}{2}f(x_0) + \sum_{j=1}^{n-1} f(x_j) + \tfrac{1}{2}f(x_n)$

As an example, let us consider the evaluation of (4.57) using the Euler-Maclaurin formula with $m = 1$ and $h = \tfrac{1}{4}$. We have*

5.20 $\displaystyle \int_1^2 e^x \, dx \sim \tfrac{1}{4}[\tfrac{1}{2}e^{1.0} + e^{1.25} + e^{1.50} + e^{1.75} + \tfrac{1}{2}e^{2.0}]$

$$- \frac{(\tfrac{1}{4})^2}{12}[e^{2.0} - e^{1.0}]$$

$$\doteq 4.695076 - \tfrac{1}{192}(4.670774) \doteq 4.670749.$$

Thus the error is $4.670774 - 4.670749 = 0.000025$. The error is between

5.21 $\displaystyle \frac{(\tfrac{1}{4})^4}{720}(2.718282)$ and $\displaystyle \frac{(\tfrac{1}{4})^4}{720}(7.389056)$

i.e., between

5.22 0.000015 and 0.000040.

Let us now introduce a word of caution concerning the Euler-Maclaurin formula. By (5.18) it is true that

5.23 $\displaystyle \int_a^b f(x)dx = h \sum_{j=0}^n {}'' f(x_j) - \sum_{k=1}^\infty \frac{B_{2k}}{(2k)!} h^{2k}[f^{(2k-1)}(b) - f^{(2k-1)}(a)]$

if $f(x)$ is a polynomial. One might be tempted to conclude that (5.23) is valid if $f(x)$ is an analytic function on $[a, b]$. However, even though the series on the right-hand side of (5.23) may converge, it need not converge to the left-hand side.

---

* Here we use the values of e^x to six decimals listed in the next section.

Thus, let us consider the case where $f(x)$ is a periodic function, with period 2π, which is analytic. Then, by (5.18) we have, for any m,

5.24 $$\int_0^{2\pi} f(x)dx = \frac{2\pi}{n}\sum_{j=1}^{n} f\left(\frac{2\pi j}{n}\right) - \frac{(2\pi)(2\pi/n)^{2m+2}}{(2m+2)!}B_{2m+2}f^{(2m+2)}(\xi)$$

for some ξ in the interval $0 < x < 2\pi$. If the remainder term were to converge to zero as $m \to \infty$, then we would have

5.25 $$\int_0^{2\pi} f(x)dx = \frac{2\pi}{n}\sum_{j=1}^{n} f\left(\frac{2\pi j}{n}\right).$$

That this is not the case can be seen by considering the function

5.26 $$f(x) = 1 + \cos 4x,$$

with n = 4. Evidently we have

5.27 $$\int_0^{2\pi}(1 + \cos 4x)dx = 2\pi \neq 4\pi = \frac{2\pi}{4}(2 + 2 + 2 + 2).$$

The reader should compute the remainder term for $m = 1, 2, 3, \ldots$ and verify that it does not converge to zero.

From (5.24) it follows that the error

5.28 $$E = \int_0^{2\pi} f(x)dx - h\sum_{j=1}^{n} f\left(\frac{2\pi j}{n}\right)$$

is $0(h^{2m+2})$ for any m provided $f(x) \in C^{(2m+2)}$. Thus, in a sense, one could say that the order of approximation is *infinite* (see Birkhoff, *et al.* [1951]). As a matter of fact, the formula is exact for any polynomial which is periodic. On the other hand, the only periodic polynomials are constant, so that the formula is not necessarily very accurate. However, we know from Section 6.13 that (5.25) is exact for any trigonometric polynomial of degree $n - 1$ or less.

The Euler-Maclaurin formula can be used to evaluate a sum of the form

5.29 $$S = \sum_{j=r}^{s} p(j)$$

where $p(j)$ is a polynomial in j of degree d. To do this we use (5.18) with $2m \geq d - 1$, $a = r, b = s$, and $h = 1$. We have

5.30 $$\int_r^s p(x)dx = S - \tfrac{1}{2}[p(r) + p(s)] - \sum_{k=1}^{m}\frac{B_{2k}}{(2k)!}[p^{(2k-1)}(s) - p^{(2k-1)}(r)].$$

Thus, for example, to show that

5.31
$$\sum_{j=1}^{N} j^3 = \left[\frac{N}{2}(N+1)\right]^2$$

we let $p(x) = x^3$, $r = 0$, $s = N$, $m = 1$, and we have

5.32
$$\int_0^N x^3\, dx = \sum_{j=1}^{N} j^3 - \tfrac{1}{2}N^3 - \tfrac{1}{12}[p'(N) - p'(0)].$$

Therefore,

5.33
$$\frac{N^4}{4} = \sum_{j=1}^{N} j^3 - \tfrac{1}{2}N^3 - \frac{N^2}{4}$$

and (5.31) follows.

EXERCISES 7.5

1. Obtain a recurrence relation for the Bernoulli numbers of even order by equating like powers of x in the identity

$$\left\{B_0 + \sum_{k=1}^{\infty} B_{2k} \frac{(2x)^{2k}}{(2k)!}\right\} \sinh x = x \cosh x$$

   Use this relation to get B_0, B_2, B_4, B_6, and B_8.

2. The Bernoulli numbers can be defined by

$$\frac{x}{e^x - 1} = \sum_{k=0}^{\infty} B_k \frac{x^k}{k!}.$$

   Show that all Bernoulli numbers of odd order, except B_1, vanish. Find the first eight Bernoulli numbers in this way. Reconcile this definition with that of (5.3).

3. Carry out the details of the derivation of (5.1) for the case $m = 3$.

4. Verify the formula

$$\int_1^2 f(x)dx = h\left[\tfrac{1}{2}f(1) + \sum_{j=1}^{N-1} f(1 + jh) + \tfrac{1}{2}f(2)\right]$$

$$- \frac{h^2}{12}[f'(2) - f'(1)] + \frac{h^4}{720}[f^{(3)}(2) - f^{(3)}(1)]$$

$$- \frac{h^6}{30,240} f^{(6)}(\xi)$$

   for the case $f(x) = e^x$, $n = 8$, $h = 1/N$. Find ξ.

5. Apply the Euler-Maclaurin formula (5.18) to find

$$\int_0^1 (1 + x^2)^{-1}\, dx$$

with $m = 1$ and $h = \frac{1}{4}$. Compare the exact error with the theoretical bounds on the error.

6. Compute the remainder term in the Euler-Maclaurin formula (5.18) for the case

$$\int_0^{2\pi} (1 + \cos 4x)\, dx$$

with $h = 2\pi/4$, for $m = 1, 2, 3$.

7. Using the Euler-Maclaurin formula, find

$$\sum_{n=2}^{25} (n^4 + 2n^2).$$

8. Derive an analog of the Euler-Maclaurin formula by expressing

$$\int_{x_0}^{x_1} f(x)dx - (x_1 - x_0)f\left(\frac{x_0 + x_1}{2}\right)$$

in terms of $f'(x_1) - f'(x_0)$, $f^{(3)}(x_1) - f^{(3)}(x_0)$, etc. Find the remainder term. Express

$$\int_{x_0}^{x_n} f(x)dx - h\left[f\left(\frac{x_0 + x_1}{2}\right) + f\left(\frac{x_1 + x_2}{2}\right) + \cdots + f\left(\frac{x_{n-1} + x_n}{2}\right) \right]$$

in terms of $f'(x_n) - f'(x_0)$, $f^{(3)}(x_n) - f^{(3)}(x_0), \ldots$, etc., where $x_1 - x_0 = x_2 - x_1 = \cdots = x_n - x_{n-1} = h$. Develop an analog for the Romberg integration scheme.

7.6 ROMBERG INTEGRATION

It follows from (5.18) that if $f(x) \in C^{2m+2}[a, b]$ then for some constants $c_1, c_2, \ldots, c_m$ we have

6.1 $T_0(h) = h \sum_{j=0}^{n}{}'' f(x_j) = \int_a^b f(x)dx + c_1 h^2 + c_2 h^4 + \cdots + c_m h^{2m} + 0(h^{2m+2}).$

Since

6.2 $T_0\left(\frac{h}{2}\right) = \int_a^b f(x)dx + \frac{1}{4}c_1 h^2 + \frac{1}{16}c_2 h^4 + \cdots + \frac{c_m}{2^{2m}} h^{2m} + 0(h^{2m+2})$

it follows that

6.3 $T_1(h) = \frac{4}{3}T_0\left(\frac{h}{2}\right) - \frac{1}{3}T_0(h) = \int_a^b f(x)dx + c_2' h^4 + c_3' h^6 + \cdots + c_m' h^{2m} + 0(h^{2m+2})$

where

6.4
$$c_2' = -\tfrac{1}{4}c_2, \quad c_3' = -\tfrac{5}{16}c_3,$$

etc. Since

6.5
$$T_1\!\left(\frac{h}{2}\right) = \int_a^b f(x)dx + c_2'\frac{h^4}{16} + c_3'\frac{h^6}{64} + \cdots + c_m'\frac{h^{2m}}{2^{2m}} + 0(h^{2m+2})$$

we have

6.6 $T_2(h) = \tfrac{16}{15}T_1\!\left(\dfrac{h}{2}\right) - \tfrac{1}{15}T_1(h) = \displaystyle\int_a^b f(x)dx + c_3^{(2)}h^6 + c_4^{(2)}h^8 + \cdots + c_m^{(2)}h^{2m} + 0(h^{2m+2})$

where

6.7
$$c_3^{(2)} = -\tfrac{1}{20}c_3', \quad c_4^{(2)} = -\tfrac{1}{16}c_4',$$

etc. Similarly, we can show that

6.8
$$T_3(h) = \tfrac{64}{63}T_2\!\left(\frac{h}{2}\right) - \tfrac{1}{63}T_2(h) = c_4^{(3)}h^8 + \cdots + c_m^{(3)}h^{2m} + 0(h^{2m+2}).$$

In general, if we let

6.9
$$T_k(h) = \frac{4^k}{4^k - 1}\, T_{k-1}\!\left(\frac{h}{2}\right) - \frac{1}{4^k - 1}\, T_{k-1}(h), \qquad k = 1, 2, \ldots,$$

then we have for $k = 0, 1, 2, \ldots,$

6.10
$$T_k(h) = \int_a^b f(x)dx + 0(h^{2(k+1)}).$$

In order to apply Romberg integration we choose h (which may equal $b - a$), and we compute

6.11
$$T_0(h), \; T_0\!\left(\frac{h}{2}\right), \ldots, T_0\!\left(\frac{h}{2^s}\right)$$

for some positive integer s. In the case $s = 3$ we construct the table

6.12

$$
\begin{array}{ccccccc}
T_0(h) & & & & & & \\
& T_1(h) & & & & & \\
T_0\left(\dfrac{h}{2}\right) & & T_2(h) & & & & \\
& T_1\left(\dfrac{h}{2}\right) & & T_3(h) & & & \\
T_0\left(\dfrac{h}{2^2}\right) & & T_2\left(\dfrac{h}{2}\right) & & & & \\
& T_1\left(\dfrac{h}{2^2}\right) & & & & & \\
T_0\left(\dfrac{h}{2^3}\right) & & & & & &
\end{array}
$$

using (6.9). The value $T_3(h)$ is accepted as the desired result.

It is easy to verify that $T_1(h)$ is the same as the composite integration procedure based on Simpson's rule with interval size $h/2$. Moreover, $T_2(h)$ is the same as the composite procedure based on the Newton-Cotes formula Q.4.4.0. However, $T_3(h)$ and all other $T_k(h)$ do not correspond to Newton-Cotes formulas.

It follows from a result of Bauer, *et al.* [1963] that if $f \in C^{(2m+2)}[a, b]$, then

6.13 $\quad T_m\left(\dfrac{b-a}{2^k}\right) = \displaystyle\int_a^b f(x)dx + \dfrac{(b-a)^{2m+3}}{2^{m(m+1)}} \dfrac{4^{-k(m+1)}}{(2m+2)!} |B_{2m+2}| f^{(2m+2)}(\xi)$

for some ξ in the interval (a, b). For the case $m = 1$, corresponding to Simpson's rule with interval size $(b-a)2^{-(k+1)}$, the error term is

6.14 $\quad \dfrac{(b-a)^5}{2^2} \dfrac{4^{-2k}}{4!} (\tfrac{1}{30}) f^{(4)}(\xi) = \dfrac{(b-a)\left(\dfrac{b-a}{2^{k+1}}\right)^4}{180} f^{(4)}(\xi)$

which agrees with the error for Simpson's rule. (See (4.65).)

As an example, let us consider the evaluation of the integral (4.57) with $h = 1$, and $s = 3$. By direct computation with a table of e^x correct to six decimal places we construct the table*

---

* Here we use the values

$e^{1.0} \doteq 2.718282$	$e^{1.375} \doteq 3.955077$	$e^{1.75} \doteq 5.754603$
$e^{1.125} \doteq 3.080217$	$e^{1.5} \doteq 4.481689$	$e^{1.875} \doteq 6.520819$
$e^{1.25} \doteq 3.490343$	$e^{1.625} \doteq 5.078419$	$e^{2.0} \doteq 7.389056$

$$T_0(1) \doteq 5.053669$$

$$T_1(1) \doteq 4.672349$$

$$T_0(\tfrac{1}{2}) \doteq 4.767679 \qquad\qquad T_2(1) \doteq 4.670777$$

6.15 $T_1(\tfrac{1}{2}) \doteq 4.670875 \qquad\qquad\qquad\qquad T_3(1) \doteq 4.670774.$

$$T_0(\tfrac{1}{4}) \doteq 4.695076 \qquad\qquad T_2(\tfrac{1}{2}) \doteq 4.670774$$

$$T_1(\tfrac{1}{4}) \doteq 4.670780$$

$$T_0(\tfrac{1}{8}) \doteq 4.676855$$

Thus $T_2(\tfrac{1}{2})$ and $T_3(1)$ are correct to six decimal places. The reader should verify that $T_1(1)$, $T_1(\tfrac{1}{2})$, and $T_1(\tfrac{1}{4})$ correspond to the values obtained by Simpson's rule for the interval sizes $\tfrac{1}{2}$, $\tfrac{1}{4}$, and $\tfrac{1}{8}$, respectively. He should also verify that $T_2(1)$ and $T_2(\tfrac{1}{2})$ correspond to the values obtained by Q.4.4.0 for the interval sizes $\tfrac{1}{4}$ and $\tfrac{1}{8}$, respectively. Finally, he should show that the error bound given by (6.13) is less than $\tfrac{1}{2} \times 10^{-6}$.

Many results concerning Romberg integration are given by Bauer, *et al.* [1963]. It is shown that if $f(x)$ is Riemann integrable, then not only does the sequence $T_0(h)$, $T_0(h/2)$, $T_0(h/2^2)$, ... converge but also the sequence $T_m(h)$, $T_m(h/2)$, $T_m(h/2^2)$, ... converges for each m, as does the diagonal sequence

6.16 $$T_0(h), \ T_1(h/2), \ T_2(h/2^2), \ \ldots \ .$$

Evidently, for each k and m, $T_m(h/2^k)$ is a linear combination of values of $T_0(h/2^j)$ for $j = k, k + 1, .., k + m$. The weights can be shown to have alternating signs but the sum of the absolute values is less than two. Hence, the procedure is stable. Moreover, $T_m(h/2^k)$ is also a linear combination of values of $f(x)$ at various values of x and the coefficients are nonnegative. Furthermore, no nonzero coefficient is more than three times as large as any other. This is in contrast to the situation for higher-order Newton-Cotes formulas (involving nine or more points) where some of the coefficients are negative.

While Romberg integration has many properties which make it attractive for numerical computation, a word of caution is appropriate. Consider the evaluation of

6.17 $$\int_0^{2\pi} f(x)dx$$

where

6.18 $$f(x) = 1 + \tfrac{1}{2}\cos x + \frac{1}{2^2}\cos 2x + \frac{1}{2^3}\cos 3x + \cdots.$$

Evidently

6.19 $$\int_0^{2\pi} f(x)dx = 2\pi$$

and by 6-(13.50)

6.20 $$T_0(h) = \frac{2\pi}{n} \sum_{j=1}^{n} f\left(\frac{2\pi j}{n}\right) = \pi\left(2 + \frac{2}{2^n} + \frac{2}{2^{2n}} + \cdots\right).$$

Hence for large n

6.21 $$\int_0^{2\pi} f(x)dx - T_0(h) \sim \frac{-2\pi}{2^n} = \frac{-2\pi}{2^{2\pi/h}}$$

The convergence of $T_0(h)$ to the desired integral is already more rapid than any power of h, and hence Romberg integration gives no improvement (Bauer, *et al.* [1963]).

EXERCISES 7.6

1. In the example in the text for the integral (4.57) show that $T_1(\frac{1}{4})$ corresponds to the composite method based on Simpson's rule with step size $\frac{1}{8}$ and that $T_2(\frac{1}{2})$ corresponds to the composite method based on method Q.4.4.0 with the interval size $\frac{1}{8}$.

2. Compute

$$\int_0^1 \frac{dx}{1 + x^2}$$

by Romberg integration based on $T_0(1)$, $T_0(\frac{1}{2})$, $T_0(\frac{1}{4})$, and $T_0(\frac{1}{8})$. Work to six decimal places. Verify that $T_1(\frac{1}{4})$ agrees with Simpson's rule with interval size $\frac{1}{8}$. Compute the exact error and compare with the theoretical error given by (6.13).

3. Show that $T_1(h)$ is the same as the composite method based on Simpson's rule with step size $h/2$ and that $T_2(h)$ is the same as the composite method based on Q.4.4.0 with step size $h/4$.

4. Verify (6.13) for the case $m = 2$. Note that the method is equivalent to the composite method based on the numerical quadrature formula Q.4.4.0 with $h = (b - a)/2^{k+2}$.

5. Show that $T_3(h)$ does not correspond to a Newton-Cotes formula.

6. Express $T_3(h)$ in terms of $T_0(h/2^j)$, $j = 0, 1, 2, \ldots$ and also in terms of $f(a + sh)$, $s = 0, 1, 2, \ldots$. Show that the coefficients in the first case have alternating signs but that the sum of their absolute values does not exceed two. Show that the coefficients in the second case are nonnegative.

7. Apply the Romberg integration scheme to the example (6.17) with $T_0(2\pi)$, $T_0(2\pi/2)$, $T_0(2\pi/2^2)$, $T_0(2\pi/2^3)$, and $T_0(2\pi/2^4)$ where

$$T_0\left(\frac{2\pi}{2^k}\right) = \frac{2\pi}{2^k} \sum_{j=1}^{2^k} f\left(\frac{2\pi j}{2^k}\right) \qquad k = 0, 1, \ldots .$$

7.7   ERROR DETERMINATION

In this section we show that for many of the numerical differentiation and numerical quadrature formulas considered above the error is given by (2.28) provided that

$f(x) \in C^{(p+1)}$ in a suitable interval where p is the order of the method. In some cases we can show that (2.28) holds under the slightly weaker assumption that $f(x) \in D^{(p+1)}$. We have referred to a method, or formula, for which (2.28) holds whenever $f \in C^{(p+1)}$ as *definite*.

We shall describe a procedure based on Peano's theorem which can be used to determine whether or not a given method is definite. This determination often involves tedious calculations, however, and it is often difficult to show that a general class of methods is definite. There are, however, a number of procedures which can be used in certain cases, including:

1. the use of Taylor's theorem with the Lagrangian form of the remainder;
2. differentiation of the error term in Lagrangian interpolation (for numerical differentiation);
3. integration of the error term in Lagrangian interpolation (for numerical quadrature).

We shall consider these special procedures before describing the general technique.

Error Formulas Based on Taylor's Theorem

In some simple cases we can obtain the error formulas directly from Taylor's theorem. Thus, for example, for the numerical differentiation formula

7.1
$$f'(a) \sim \frac{f(a + h) - f(a)}{h}$$

we can use the formula

7.2
$$f(a + h) = f(a) + hf'(a) + \frac{h^2}{2} f''(\xi)$$

where $\xi \in (a, a + h)$. Therefore we have

7.3
$$f'(a) - \frac{f(a + h) - f(a)}{h} = -\frac{h}{2} f''(\xi).$$

On the other hand, if we let

7.4
$$R[f](a) = f'(a) - \frac{f(a + h) - f(a)}{. h}$$

then

7.5
$$R[(x - a)^2](a) = -h$$

and hence (2.28) holds. Note that this result holds even if $f(x)$ is in $D^{(2)}[a, a + h]$ rather than in $C^{(2)}[a, a + h]$.

A slightly more sophisticated use of Taylor's theorem can be used for the formula

7.6
$$f'(a) \sim \frac{f(a + h) - f(a - h)}{2h}.$$

Here we have, by Taylor's theorem,

7.7
$$f(a + h) = f(a) + hf'(a) + \frac{h^2}{2} f''(a) + \frac{h^3}{6} f^{(3)}(\xi_1)$$

$$f(a - h) = f(a) - hf'(a) + \frac{h^2}{2} f''(a) - \frac{h^3}{6} f^{(3)}(\xi_2)$$

where ξ_1 and ξ_2 lie in the interval $(a - h, a + h)$. Therefore we have

7.8
$$f'(a) - \frac{f(a + h) - f(a - h)}{2h} = -\frac{h^2}{12} [f^{(3)}(\xi_1) + f^{(3)}(\xi_2)].$$

If we assume that $f(x) \in C^{(3)}[a - h, a + h]$, then for some ξ in the interval $(a - h, a + h)$ we have

7.9
$$f^{(3)}(\xi_1) + f^{(3)}(\xi_2) = 2f^{(3)}(\xi)$$

and

7.10
$$f'(a) - \frac{f(a + h) - f(a - h)}{2h} = -\frac{h^2}{6} f^{(3)}(\xi).$$

On the other hand,

7.11
$$R[(x - a)^3](a) = -h^2$$

so that (2.28) holds.

Error Formulas Based on Differentiating the Remainder in Lagrangian Interpolation

Let $x_0, x_1, \ldots, x_n$ be distinct interpolation points such that $x_0 < x_1 < \cdots < x_n$ and let $f(x) \in C^{(n+2)}$ in an interval I containing the interpolation points. Let $F(x)$ be the unique polynomial of degree n or less which agrees with $f(x)$ at the interpolation points. We showed in Chapter 6 that for some continuous function $P(x)$ we have, for x in I,

7.12
$$f(x) - F(x) = Q(x)P(x)$$

where

7.13
$$Q(x) = \prod_{j=0}^{n} (x - x_j).$$

Moreover,

7.14
$$P(x) = \frac{f^{(n+1)}(\xi)}{(n+1)!}$$

for some ξ in I.

It can be shown that $P'(x)$ exists and is continuous in I and moreover

7.15
$$P'(x) = \frac{f^{(n+2)}(\eta)}{(n+2)!}.$$

We prove the existence and continuity of $P'(x)$. Evidently, by (7.12) we have

7.16
$$P(x) = \frac{R(x)}{Q(x)}, \qquad x \neq x_0, x_1, \ldots, x_n$$

where

7.17
$$R(x) = f(x) - F(x).$$

Since $f'(x)$ is continuous in I we have by L'Hospital's rule

7.18
$$P(x_k) = \lim_{x \to x_k} \frac{R(x)}{Q(x)} = \frac{R'(x_k)}{Q'(x_k)}, \qquad k = 0, 1, \ldots, n.$$

Evidently $P'(x)$ exists and is continuous for $x_0 < x < x_1$, $x_1 < x < x_2, \ldots$, $x_{n-1} < x < x_n$. Let us show that for each k

7.19
$$\lim_{x \to x_k} P'(x)$$

exists. But

7.20
$$P'(x) = \frac{R'(x)Q(x) - R(x)Q'(x)}{Q(x)^2}$$

and, by L'Hospital's rule

7.21
$$\lim_{x \to x_k} P'(x) = \lim_{x \to x_k} \left\{ \frac{R''(x)Q(x) - R(x)Q''(x)}{2Q(x)Q'(x)} \right\} = \tfrac{1}{2} \lim_{x \to x_k} \left\{ \frac{R''(x) - P(x)Q''(x)}{Q'(x)} \right\}.$$

Thus the limit exists by (7.18) and because $f(x) \in C^{(2)}$ in I.

By the mean-value theorem we have

7.22
$$\frac{P(x_k + \delta) - P(x_k)}{\delta} = P'(\xi(\delta))$$

where $\xi(\delta)$ lies between x_k and $x_k + \delta$. Taking limits of both sides as $\delta \to 0$ we find that $P'(x_k)$ exists and that $P'(x)$ is continuous at x_k. This proves that $P(x) \in C^{(1)}$ in I. The proof of (7.15) will be left as an exercise. (See Ralston [1963, 1965] or Hayes and Rubin [1970].)

Let us now assume that x is an interpolation point, say x_s. Then by (7.12) we have

7.23 $f'(x_s) - F'(x_s) = Q'(x_s)P(x_s) + Q(x_s)P'(x_s) = Q'(x_s)P(x_s).$

Hence, by (7.14), we have

7.24
$$f'(x_s) - F'(x_s) = \left[\prod_{\substack{j=0 \\ j \neq s}}^{n} (x_s - x_j)\right] P(x_s)$$
$$= \left[\prod_{\substack{j=0 \\ j \neq s}}^{n} (x_s - x_j)\right] \frac{f^{(n+1)}(\xi)}{(n+1)!}$$

for some ξ in I. This result is the same as one would obtain by differentiating (7.12), with $P(x)$ treated as a constant.

We remark that the result (7.24) could be derived without first showing that $P'(x)$ exists. (See, for instance, Ford [1933] and Henrici [1962].) For, given $x = x_s$, one can define the function

7.25
$$\phi(z) = f(z) - F(z) - \lambda Q(z)$$

where λ is independent of z and is to be determined so that

7.26
$$\phi'(x_s) = 0.$$

Thus

7.27
$$\lambda = \frac{f'(x_s) - F'(x_s)}{Q'(x_s)} = \frac{f'(x_s) - F'(x_s)}{\prod_{\substack{j=0 \\ j \neq s}}^{n} (x_s - x_j)}.$$

By repeated use of Rolle's theorem one can show that for some ξ in I we have

7.28
$$\phi^{(n+1)}(\xi) = 0$$

and hence

7.29
$$\lambda = \frac{f^{(n+1)}(\xi)}{(n+1)!}.$$

Therefore (7.24) holds.
In order to show that the numerical differentiation formula

7.30
$$f'(x) \sim F'(x)$$

is definite when x is an interpolation point we need only show* that

7.31
$$R[x^{n+1}](x_s) = \prod_{\substack{j=0 \\ j \neq s}}^{n} (x_s - x_j).$$

But this follows by applying (7.24) to the case $f(x) = x^{n+1}$.

It is sometimes possible to obtain results similar to (7.24) for numerical differentiation for higher derivatives (see, for example, the formulas for $f''(x)$ involving three points treated above). However, in general, such results cannot be obtained. On the other hand, one can prove (see Isaacson and Keller [1966])

7.32 Theorem. Let $F(x)$ be the unique polynomial of degree n or less which agrees with $f(x)$ at the interpolation points $x_0, x_1, \ldots, x_n$ where $x_0 < x_1 < \cdots < x_n$. Let $f(x) \in C^{(n+1)}$ in an interval $\hat{I}$ containing all interpolation points. For each x in $\hat{I}$ and for each positive integer $k \leq n$ there exist $n + 1 - k$ distinct numbers $\xi_0, \xi_1, \ldots, \xi_{n-k}$ such that

7.33
$$x_j < \xi_j < x_{j+k}, \qquad j = 0, 1, \ldots, n - k$$

and a number η in $\hat{I}$ such that

7.34
$$f^{(k)}(x) - F^{(k)}(x) = \prod_{j=0}^{n-k} (x - \xi_j) \frac{f^{(n+1)}(\eta)}{(n+1-k)!}.$$

Error Formulas Based on Integrating the Remainder in Lagrangian Interpolation

It is easy to show that a numerical quadrature formula for the integration of a given function $f(x)$ between two consecutive interpolation points, say x_s and x_{s+1}, is definite if the method is based on Lagrangian interpolation in the distinct points $x_0, x_1, \ldots, x_n$. For, by (7.12) we have

7.35
$$\int_{x_s}^{x_{s+1}} f(x)dx - \int_{x_s}^{x_{s+1}} F(x)dx = \int_{x_s}^{x_{s+1}} Q(x)P(x)dx.$$

---

* It follows from (7.24) that the order of the method is exactly n.

But by (7.13) the function $Q(x)$ is one-signed in the interval $x_s \leqq x \leqq x_{s+1}$. Consequently, by the mean-value theorem for integrals we have

7.36
$$\int_{x_s}^{x_{s+1}} Q(x)P(x)dx = P(\xi_1) \int_{x_s}^{x_{s+1}} Q(x)dx$$

for some ξ_1 in $I = (x_s, x_{s+1})$. Thus by (7.14) we have

7.37
$$\int_{x_s}^{x_{s+1}} f(x)dx - \int_{x_s}^{x_{s+1}} F(x)dx = \frac{f^{(n+1)}(\xi)}{(n+1)!} \int_{x_s}^{x_{s+1}} Q(x)dx$$

for some ξ in I. This shows that the order of a one-step method for numerical quadrature is exactly n. Applying (7.37) for the case $f(x) = x^{n+1}$ we have

7.38
$$R[x^{n+1}] = \int_{x_s}^{x_{s+1}} Q(x)dx$$

so that (2.28) holds. Therefore the method is definite.

Let us now consider the error for Simpson's rule. Let the interpolation points be x_0, x_1, and x_2 and let $\hat{F}(x)$ be the unique polynomial of degree three or less such that*

7.39
$$\begin{cases} \hat{F}(x_0) = f(x_0) \\ \hat{F}(x_1) = f(x_1), \ \hat{F}'(x_1) = f'(x_1) \\ F(x_2) = f(x_2). \end{cases}$$

Since Simpson's rule is exact for any polynomial of degree three or less, we have

7.40
$$\int_{x_0}^{x_2} \hat{F}(x)dx = \frac{h}{3}[f(x_0) + 4f(x_1) + f(x_2)].$$

Moreover, as shown in Section 6.7 we have

7.41
$$f(x) - F(x) = (x - x_0)(x - x_1)^2(x - x_2)P(x)$$

for some continuous function $P(x)$ where for each x in $I : x_0 \leqq x \leqq x_2$, we have

7.42
$$P(x) = \frac{f^{(4)}(\xi)}{4!}$$

---

* See Section 6.7 concerning the existence and uniqueness of $\hat{F}(x)$.

for some ξ in (x_0, x_2). Therefore, since $(x - x_0)(x - x_1)^2(x - x_2)$ is one-signed in I we have by the mean-value theorem for integrals

7.43
$$\int_{x_0}^{x_2} f(x)dx - \frac{h}{3}[f(x_0) + 4f(x_1) + f(x_2)]$$

$$= P(\xi_1) \int_{x_0}^{x_2} (x - x_0)(x - x_1)^2(x - x_2)dx$$

$$= \frac{f^{(4)}(\xi)}{24} \int_{x_0}^{x_2} (x - x_0)(x - x_1)^2(x - x_2)dx$$

for some ξ_1 and ξ in (x_0, x_2). Applying (7.43) to the case $f(x) = x^4$ we have

7.44
$$R[x^4] = \int_{x_0}^{x_2} (x - x_0)(x - x_1)^2(x - x_2)dx$$

so that (2.28) holds.

A proof that the Newton-Cotes (closed) formulas are definite can be proved by integrating the remainder in Lagrangian interpolation and using integration by parts (see, for instance, Isaacson and Keller [1966] and Hayes and Rubin [1970]). Following Hayes and Rubin we let

7.45
$$\begin{cases} A(x) = \prod_{j=0}^{n-1} (x - x_j), & \text{if } n \text{ is odd} \\[2em] A(x) = \prod_{j=0}^{n} (x - x_j), & \text{if } n \text{ is even.} \end{cases}$$

We define

7.46
$$g(x) = \frac{f(x) - F(x)}{A(x)}.$$

As in the case of $P(x)$ we can show that $g(x)$ is continuous for all x in $I = [x_0, x_n]$, and is also continuously differentiable. In fact, it can be shown that if $f(x) \in C^{(n+1)}$ if n is odd and $C^{(n+2)}$ if n is even then

7.47
$$g'(x) = \begin{cases} \dfrac{f^{(n+1)}(\xi)}{(n+1)!}, & \text{if } n \text{ is odd} \\[1.5em] \dfrac{f^{(n+2)}(\xi)}{(n+2)!}, & \text{if } n \text{ is even} \end{cases}$$

for some ξ in (x_0, x_n). Since

7.48
$$f(x) - F(x) = g(x)A(x)$$

we have, upon integrating by parts,

7.49
$$\int_{x_0}^{x_n} [f(x) - F(x)]dx = \int_{x_0}^{x_n} g(x)A(x)dx$$

$$= [g(x)A^*(x)]_{x_0}^{x_n} - \int_{x_0}^{x_n} g'(x)A^*(x)dx$$

where

7.50
$$A^*(x) = \int_{x_0}^{x} A(x)dx.$$

It can also be shown that $A^*(x) \geq 0$ for $x_0 \leq x \leq x_n$ and that $A^*(x_n) = 0$ if n is even. Since $g(x_n) = 0$ if n is odd it follows that in all cases

7.51
$$\int_{x_0}^{x_n} [f(x) - F(x)]dx = -\int_{x_0}^{x_n} g'(x)A^*(x)dx$$

$$= -g'(\xi_1)\int_{x_0}^{x_n} A^*(x)dx$$

for some ξ_1 in (x_0, x_n), by the mean-value theorem for integrals. From this, from (7.47), and from the fact that the order is n for n even and $n + 1$ for n odd (see Exercise 16, Section 7.4) it follows at once that the method is definite.

Error Formulas Based on Peano's Theorem

We now describe a method based on the use of Peano's Theorem* for determining whether or not a given method is definite. Suppose we wish to represent the operator $L[f](x)$ (or $L[f]$), by $\tilde{L}[f](x)$ (or $\tilde{L}[f]$), based on the interpolation points $x_0, x_1, \ldots, x_n$ where $x_0 < x_1 < \cdots < x_n$. We let $\hat{I} = [\bar{a}, \bar{b}]$ be an interval containing the interpolation points and all values of x involved in $L[f](x)$. Thus, for example, for numerical differentiation we have $L[f](x) = f'(x)$ and $\hat{I}$ would contain x as well as the interpolation points. For numerical quadrature we have

$$L[f] = \int_{a}^{b} f(x)dx$$

and $\hat{I}$ would include the interval $[a, b]$ as well as the interpolation points.

We assume that the order of the method is p and that $f(x) \in C^{(p+1)}$ in $\hat{I}$. By

---

* See Davis [1963, pp. 69–75].

Taylor's Theorem with the integral form of the remainder we have (see Theorem A-3.66)

7.52 $$f(x) = f(\bar{a}) + (x - \bar{a})f'(\bar{a}) + \cdots + \frac{(x - \bar{a})^p}{p!} f^{(p)}(\bar{a}) + \int_{\bar{a}}^x (x - t)^p \frac{f^{(p+1)}(t)}{p!} dt.$$

Therefore, since the method is of order p we have

7.53 $$R[(x - \bar{a})^k](x) = L[(x - \bar{a})^k](x) - \tilde{L}[(x - \bar{a})^k](x) = 0, \qquad k = 0, 1, \ldots, p$$

and

7.54 $$R[f](x) = L[f](x) - \tilde{L}[f](x) = R\left[\int_{\bar{a}}^x (x - t)^p \frac{f^{(p+1)}(t)}{p!} dt\right](x)$$

$$= R\left[\int_{\bar{a}}^{\bar{b}} \phi_p(x - t) \frac{f^{(p+1)}(t)}{p!} dt\right](x)$$

$$= \int_{\bar{a}}^{\bar{b}} \{R[\phi_p(x - t)](x)\} \frac{f^{(p+1)}(t)}{p!} dt.$$

Here we define the functions

7.55 $$\phi_k(x) = \begin{cases} 0 & x \leqq 0 \\ x^k & x > 0 \end{cases}$$

for $k = 0, 1, \ldots$. Evidently $\phi_k(x)$ is continuous for all x if $k \geqq 1$; the function $\phi_0(x)$ is continuous except for a jump at $x = 0$. Similarly $\phi_k(x)$ is continuously differentiable for all x if $k \geqq 2$ and

7.56 $$\phi'_k(x) = k\phi_{k-1}(x).$$

The functions $\phi_1(x)$ and $\phi_0(x)$ are continuously differentiable for $x > 0$ and for $x < 0$. Hence $\phi'_0(x)$ and $\phi'_1(x)$ are piecewise continuous. Therefore, $R[\phi_p(x - t)](x)$ is a piecewise continuous function of t, for fixed x.

We now show that the method is definite if and only if $R[\phi_p(x - t)](x)$ is one-signed for all t in $\hat{I}$. If we apply (7.54) with $f(x) = x^{p+1}$ we have

7.57 $$R[x^{p+1}](x) = (p + 1) \int_{\bar{a}}^{\bar{b}} \{R[\phi_p(x - t)](x)\} dt.$$

Thus (2.28) holds if and only if

7.58 $$\int_{\bar{a}}^{\bar{b}} \{R[\phi_p(x - t)](x)\} f^{(p+1)}(t) dt = f^{(p+1)}(\xi) \int_{\bar{a}}^{\bar{b}} \{R[\phi_p(x - t)](x)\} dt$$

for some ξ in $(\bar{a}, \bar{b})$. But by the generalized mean-value theorem for integrals (Theorem A-3.40), this relation holds for every continuous function $f^{(p+1)}(x)$ if and only if the piecewise continuous function $R[\phi_p(x - t)](x)$ is one-signed for all t in $\hat{I}$.

We have thus shown that a method $\tilde{L}[f](x)$ of order p for representing $L[f](x)$ is definite if and only if

7.59 $$G(t; x) = R[\phi_p(x - t)](x)$$

is a one-signed function of t in $\hat{I}$. We now give some examples. First let us consider the case of linear interpolation in the points x_0 and $x_1 = x_0 + h$, where

7.60 $$L[f](x) = f(x), \quad \tilde{L}[f](x) = f(x_0) + \frac{x - x_0}{x_1 - x_0}[f(x_1) - f(x_0)].$$

If $x < x_0$ then we let $\bar{a} = x, \bar{b} = x_1$. Since $p = 1$, we consider

7.61 $R[\phi_1(x-t)](x) = \begin{cases} x - t - \left[(x_0 - t) + \dfrac{x - x_0}{x_1 - x_0}[x_1 - x_0] \right] = 0, & t \leq x \\[2mm] -\left[x_0 - t + \dfrac{x - x_0}{x_1 - x_0}[x_1 - x_0] \right] = t - x \geq 0, & x \leq t \leq x_0 \\[2mm] -\dfrac{x - x_0}{x_1 - x_0}(x_1 - t) \geq 0, & x_0 \leq t \leq x_1 \\[2mm] 0, & t \geq x_1. \end{cases}$

Thus $R[\phi_1(x - t)](x)$ is one-signed. If $x \in [x_0, x_1]$, then

7.62 $R[\phi_1(x-t)](x) = \begin{cases} x - t - \left[(x_0 - t) + \dfrac{x - x_0}{x_1 - x_0}(x_1 - x_0) \right] = 0, & t \leq x_0 \\[2mm] x - t - \left[\dfrac{x - x_0}{x_1 - x_0}(x_1 - t) \right] = \dfrac{(x_1 - x)(x_0 - t)}{x_1 - x_0} \leq 0, & x_0 \leq t \leq x \\[2mm] -\dfrac{(x - x_0)(x_1 - t)}{x_1 - x_0} \leq 0, & x \leq t \leq x_1 \\[2mm] 0, & t \geq x_1 \end{cases}$

and again $R[\phi_1(x - t)](x)$ is one-signed. Finally, if $x > x_1$ then

7.63 $R[\phi_1(x - t)](x) = \begin{cases} 0, & t \leq x_0 \\[2mm] \dfrac{(x_1 - x)(x_0 - t)}{(x_1 - x_0)} \geq 0, & x_0 \leq t \leq x_1 \\[2mm] x - t \geq 0, & x_1 \leq t \leq x \\[2mm] 0, & t \geq x. \end{cases}$

Thus in every case $R[\phi_1(x - t)](x)$ is one-signed, and we have verified that linear interpolation is definite.

As a second example, let us consider Simpson's rule

7.64
$$\int_{x_0}^{x_2} f(x)dx \sim \frac{h}{3}[f(x_0) + 4f(x_1) + f(x_2)].$$

We seek to show that $R[\phi_3(x - t)]$ is one-signed. If $t \leqq x_0$ we have

7.65
$$R[\phi_3(x - t)] = R[(x - t)^3] = 0$$

since Simpson's rule is exact for any polynomial of degree three or less. Also, if $t \geqq x_2$ we have $R[\phi_3(x - t)] = 0$. If $x_0 \leqq t \leqq x_2$, then we have

7.66
$$\int_{x_0}^{x_2} \phi_3(x - t)dx = \int_t^{x_2} (x - t)^3\,dx = \frac{(x_2 - t)^4}{4}.$$

Therefore, if $x_1 \leqq t \leqq x_2$ we have

7.67 $R[\phi_3(x - t)] = \dfrac{(x_2 - t)^4}{4} - \dfrac{h}{3}(x_2 - t)^3 = \dfrac{(x_2 - t)^3}{12}\{3(x_2 - t) - 4h\} \leqq 0.$

If $x_0 \leqq t \leqq x_1$, then*

7.68
$$R[\phi_3(x - t)] = \frac{(x_2 - t)^4}{4} - \frac{h}{3}[4(x_1 - t)^3 + (x_2 - t)^3]$$

$$= \frac{(x_0 - t)^3}{12}[h + 3(x_1 - t)] \leqq 0.$$

Hence $R[\phi_3(x - t)]$ is one-signed and we have verified that Simpson's rule is definite.

Let us now consider the numerical differentiation formula

7.69
$$f'(x) \sim \frac{f(x_1) - f(x_0)}{x_1 - x_0}.$$

Since

7.70
$$R[x^2](x) = L[x^2](x) - \tilde{L}[x^2](x) = 2x - (x_1 + x_0)$$

---

* This can be shown by noting that since $R[(x - t)^3] = 0$ we have

$$\frac{(x_2 - t)^4}{4} - \frac{(x_0 - t)^4}{4} = \frac{h}{3}[(x_0 - t)^3 + 4(x_1 - t)^3 + (x_2 - t)^3].$$

it follows that the order of the method is one unless $x = (x_0 + x_1)/2$. Let us first consider the case $x < x_0$. We have

7.71
$$R[\phi_1(x - t)](x) = \begin{cases} 0, & t < x \\ -1 < 0, & x < t < x_0 \\ -\dfrac{x_1 - t}{x_1 - x_0} \leq 0, & x_0 \leq t \leq x_1 \\ 0, & t > x_1. \end{cases}$$

Thus in this case $R[\phi_1(x - t)](x)$ is one-signed for all t and the method is definite.

Next we consider the case $x > x_1$. Here we have

7.72
$$R[\phi_1(x - t)](x) = \begin{cases} 0, & t < x_0 \\ \dfrac{t - x_0}{x_1 - x_0} \geq 0, & x_0 \leq t \leq x_1 \\ 1 > 0, & x_1 < t < x \\ 0, & t > x \end{cases}$$

and again the method is definite.

Next, we consider the case $x \in (x_0, x_1)$ where $x \neq (x_1 + x_0)/2$. We have

7.73
$$R[\phi_1(x - t)](x) = \begin{cases} 0, & t \leq x_0 \\ \dfrac{t - x_0}{x_1 - x_0} > 0, & x_0 < t < x \\ \dfrac{t - x_1}{x_1 - x_0} < 0, & x < t < x_1 \\ 0, & t \geq x_1. \end{cases}$$

Thus the function $R[\phi_1(x - t)](x)$ is not one-signed and hence the method is not definite. On the other hand, the reader should show that if $x = (x_0 + x_1)/2$ then $R[\phi_2(x - t)](x)$ is one-signed and the method is definite.

Let us now apply the numerical differentiation formula (7.69) to the case

7.74
$$f(x) = \begin{cases} 2x^3 - x^4, & 0 \leq x \leq 1 \\ 2x - 1, & 1 \leq x \leq 4 \end{cases}$$

with $x_0 = 0$, $x_1 = 4$. Evidently

7.75
$$f'(x) = \begin{cases} 6x^2 - 4x^3, & 0 \leq x \leq 1 \\ 2, & 1 \leq x \leq 4 \end{cases}$$

and

7.76
$$f''(x) = \begin{cases} 12x(1 - x), & 0 \leqq x \leqq 1 \\ 0, & 1 \leqq x \leqq 4 \end{cases}$$

so that $f(x) \in C^{(2)}[0, 4]$. Since $f(0) = 0$, $f(4) = 7$, and $f'(1) = 2$, we have

7.77
$$R[f](1) = 2 - \frac{7 - 0}{4} = \frac{1}{4}.$$

On the other hand,

7.78
$$R[x^2](1) = -2$$

and

7.79
$$R[x^2](1) \frac{f''(\xi)}{2} = -f''(\xi) \leqq 0$$

if $0 < \xi < 4$. Consequently, no matter what value of ξ we choose, we cannot have

7.80
$$R[f](1) = R[x^2](1) \frac{f''(\xi)}{2}.$$

Hence (2.28) cannot hold and the method is not definite.

EXERCISES 7.7

1. By using Taylor's theorem at the point a with the Lagrange form of the remainder for
$$I(x) = \int_a^x f(x)dx$$
   show that if $f(x) \in D^{(1)}(a, b)$, then
$$\int_a^b f(x)dx = (b - a)f(a) + \frac{(b - a)^2}{2} f'(\xi)$$
   for some ξ in the interval (a, b).

2. By using Taylor's theorem with the Lagrange form of the remainder for
$$I(x) = \int_a^x f(x)dx$$
   at the point $\alpha = (a + b)/2$, show that if $f(x) \in D^{(2)}[a, b]$, then
$$\int_a^b f(x)dx = (b - a)f\left(\frac{a + b}{2}\right) + \frac{f''(\tilde{\xi})}{24}(b - a)^3$$
   for some ξ in the interval (a, b).

3. Using Taylor's theorem with the Lagrange form of the remainder, show that

$$f''(a) = \frac{f(a + h) + f(a - h) - 2f(a)}{h^2} - \frac{h^2}{12} f^{(4)}(\xi)$$

for some ξ in $(a - h, a + h)$ if $f(x) \in D^{(4)}[a - h, a + h]$.

4. Apply (7.24) to the case of three evenly spaced points $x_0, x_1 = x_0 + h$, and $x_2 = x_0 + 2h$.

5. Carry out the details of the proof of (7.24) based on the use of (7.25).

6. Let $F(x)$ be the unique polynomial of degree n or less which agrees with $f(x)$ at the $n + 1$ distinct points $x_0, x_1, \ldots, x_n$. Let $f(x) \in C^{(n+2)}$ in the interval I spanned by the interpolation points and let $P(x)$ be continuous in I and satisfy (7.12), where

$$Q(x) = \prod_{j=0}^{n} (x - x_j).$$

Show that

$$P'(x) = \frac{f^{(n+2)}(\xi)}{(n + 2)!}$$

for some ξ in I. (Hint: Consider the function $\phi(z; x) = f(z) - F(z) - Q(z)[P(x) + (z - x)P'(x)]$.)

7. Let $f(x) \in C^{(n+3)}[a, b]$ and let $F(x)$ be the polynomial of degree n or less which agrees with $f(x)$ at the interpolation points $x_0, x_1, \ldots, x_n$ where $a \leqq x_0 < x_1 < \cdots < x_n \leqq b$. Let $P(x)$ be defined as in Exercise 6. Prove that $P(x) \in C^{(2)}[a, b]$. Also show that

$$P''(x) = \frac{f^{(n+3)}(\xi)}{(n + 3)!}$$

where $x_0 < \xi < x_n$.

8. Apply the method based on (7.25) to show that the second formula of (3.63) holds if $f(x) \in D^{(4)}[x_0, x_2]$.

9. Prove (7.47).

10. Show that the midpoint rule

$$\int_{x_0}^{x_2} f(x)dx \sim (x_2 - x_0)f((x_0 + x_2)/2)$$

is definite using the technique employed for Simpson's rule for the integration of the remainder in Lagrange interpolation.

11. Prove that $A^*(x) \geqq 0$ for the cases $n = 3, 4, 5$, where $A^*(x)$ is given by (7.50). Also show that in the case $n = 4$ we have $A^*(x_4) = 0$.

12. Derive the numerical quadrature formula Q.2.3.(-1) and determine a formula for its error by integrating the remainder of the appropriate Lagrange interpolation formula.

13. For each of the following methods determine whether the method is definite and, if so, find the error. (Assume in each case that $f(x)$ is well-behaved in the interval being considered.)

a) Interpolation in the three evenly spaced points x_0, $x_0 + h$, $x_0 + 2h$.

b) $f'(a) \sim \dfrac{f(a + h) - f(a - h)}{2h}$.

c) $f'\left(a + \dfrac{h}{2}\right) \sim \dfrac{1}{2h}[f(a + 2h) - f(a)]$

d) $f''(a) \sim \dfrac{f(a + h) + f(a - h) - 2f(a)}{h^2}$.

e) Simpson's second rule.

f) Q.2.3.0.

g) Q.2.3.1.

h) Q.2.3.(-1).

To test whether or not a method is definite, determine whether or not $R[\phi_p(x - t)]$ is one-signed in the relevant interval, where p is the order of the method.

14. Show that the formula

$$f'\left(a + \frac{h}{2}\right) \sim \frac{3}{2h}[f(a + h) - f(a)] - \tfrac{1}{4}[f'(a) + f'(a + h)]$$

is definite.

15. Show that the numerical quadrature formula Q.4.3.1 is definite.

16. Show that the numerical quadrature formula

$$\int_{x_0}^{x_1} f(x)dx \sim \frac{h}{2}[f(x_0) + f(x_1)] - \sum_{k=1}^{m} \frac{B_{2k}}{(2k)!} h^{2k}[f^{(2k-1)}(x_1) - f^{(2k-1)}(x_0)]$$

is definite.

17. Show that if $f(x) \in C^6[x - h, x + h]$ then

$$\frac{f(x + h) + f(x - h) - 2f(x)}{h^2} = f''(x) + \frac{h^2}{12}f^{(4)}(x) + \frac{h^4}{360}f^{(6)}(\xi)$$

for some ξ such that $x - h < \xi < x + h$.

18. Show that if $f(x) \in C^{(p+1)}$ and if $\tilde{L}[f](x)$ is a representation of $L[f](x)$ of order p, then

$$|R[f](x)| \leq GM_{p+1}$$

for some constant G independent of $f(x)$. Here $R[f](x) = L[f](x) - \tilde{L}[f](x)$ and M_{p+1} is a bound for $|f^{(p+1)}(x)|$ in an interval containing x and all interpolation points.

19. Find an expression for the error of the trapezoidal rule for a function $f(x)$ which belongs to $C^{(1)}$ in the interval of integration but not necessarily to $C^{(2)}$.

20. Find an expression for the error of Simpson's rule for a function $f(x)$ which belongs to $C^{(4)}$ in the interval of integration. Do the same for the cases $f(x) \in C^{(3)}$, $f(x) \in C^{(2)}$, and $f(x) \in C^{(1)}$.

21. Derive formulas for $f'(a)$ and $f''(a)$ in terms of $f(a)$, $f(a \pm h)$, $f(a \pm 2h)$ and show that they are definite. Find the error formulas. Assume that $f(x)$ is well-behaved in the interval $[a - 2h, a + 2h]$.

22. Give a bound in the error of (7.69) under the assumption that $f(x) \in C^{(2)}[x_0, x_1]$. Apply the formula to the case of (7.74) with $x_0 = 0$, $x_1 = 4$, and $x = 1$.

23. Show that the representation

$$\tilde{L}[f](a) = \frac{f(a + h) + f(a - h) - 2f(a)}{h^2} + \frac{f(a + h) - f(a - h)}{2h}$$

for the operator $L[f](a) = f''(a) + f'(a)$ is not definite. What is the order of the representation? Give a bound on the error in terms of $\max_{a-h \le x \le a+h} |f'''(x)|$.

7.8 NUMERICAL QUADRATURE—UNEQUAL INTERVALS

With the numerical quadrature formulas of Section 7.4, one is, in effect, determining the integral of an interpolating polynomial of degree n, if the number of interpolation points used is $n + 1$. The order of the method is normally n or in some cases $n + 1$. In this section we show that by using the same number of interpolation points but allowing them to be unevenly spaced we can either obtain a higher-order formula, as in the case of Gaussian quadrature formulas, or else we can obtain the same order but with equal weights, as in the case of Chebyshev quadrature formulas.

Gaussian Quadrature

We seek to represent the integral

8.1
$$\int_a^b f(x)dx$$

by the sum

8.2
$$\alpha_0 f(x_0) + \alpha_1 f(x_1) + \cdots + \alpha_n f(x_n)$$

where $x_0, x_1, \ldots, x_n$ are in the interval $[a, b]$. It is convenient to introduce the new independent variable u defined by the equivalent relations

8.3
$$u = \frac{2x - (b + a)}{b - a}, \qquad x = \frac{(b - a)u + (b + a)}{2}$$

so that the interval $[a, b]$ corresponds to the interval $[-1, 1]$. We seek to approximate the integral

8.4
$$\int_{-1}^1 \phi(u)du = \frac{2}{b - a}\int_a^b f(x)dx$$

by

8.5 $$\alpha_0\phi(u_0) + \alpha_1\phi(u_1) + \cdots + \alpha_n\phi(u_n)$$

where $u_0, u_1, \ldots, u_n$ lie in the interval $[-1, 1]$. Here

8.6 $$\phi(u) = f\left(\frac{(b-a)u + (b+a)}{2}\right).$$

Our procedure for determining the α_i and u_i is as follows. We determine the u_i as the roots of the polynomial equation

8.7 $$P_{n+1}(u) = 0$$

where for any nonnegative integer n we have

8.8 $$P_n(u) = \frac{d^n}{du^n}(u^2 - 1)^n.$$

Here

8.9 $$Q_n(u) = \frac{1}{2^n n!} P_n(u)$$

is the *Legendre polynomial* of degree n. As we shall later show, (8.7) has precisely $n + 1$ simple roots in $(-1, 1)$.

Having found the u_i we determine the α_i as follows. From the Lagrangian interpolation formula 6-(4.4) the unique polynomial $\Phi(u)$ of degree n or less which agrees with $\phi(u)$ at the points $u_0, u_1, \ldots, u_n$ is given by

8.10 $$\Phi(u) = \sum_{k=0}^{n} w_k(u)\phi(u_k)$$

where

8.11 $$w_k(u) = \prod_{\substack{j=0 \\ j \neq k}}^{n} \left(\frac{u - u_j}{u_k - u_j}\right).$$

Therefore we have

8.12 $$\int_{-1}^{1} \Phi(u)du = \sum_{k=0}^{n} \alpha_k\phi(u_k)$$

where

8.13 $$\alpha_k = \int_{-1}^{1} w_k(u)du.$$

Before seeking to justify the above procedure, we shall illustrate for the cases $n = 0, 1, 2$. If $n = 0$, we have

8.14
$$P_1(u) = \frac{d}{du}(u^2 - 1) = 2u;$$

hence $u_0 = 0$. Moreover, $w_0(x) = 1$ and $\alpha_0 = 2$. Thus we have

8.15
$$\int_{-1}^{1} \phi(u)du \sim 2\phi(0).$$

Evidently the method is of order *one*. Going back to the original integral we have

8.16
$$\int_{a}^{b} f(x)dx \sim (b - a)f\left(\frac{a + b}{2}\right)$$

which is the midpoint method.
  If $n = 1$, then

8.17
$$P_2(u) = \frac{d^2}{du^2}(u^2 - 1)^2 = 12u^2 - 4 = 0$$

or $u_0 = -\sqrt{\frac{1}{3}}$, $u_1 = \sqrt{\frac{1}{3}}$. Moreover,

8.18
$$\alpha_0 = \int_{-1}^{1}\left(\frac{u - u_1}{u_0 - u_1}\right)du = \frac{1}{u_0 - u_1}\left[\frac{u^2}{2} - u_1 u\right]_{-1}^{1} = 1.$$

Similarly, $\alpha_1 = 1$ and we have

8.19
$$\int_{-1}^{1} \phi(u)du \sim \phi(-\sqrt{\tfrac{1}{3}}) + \phi(\sqrt{\tfrac{1}{3}}).$$

  If $n = 2$, we have

8.20
$$P_3(u) = \frac{d^3}{du^3}(u^2 - 1)^3 = 120u^3 - 72u = 0$$

or $u_0 = -\sqrt{\frac{3}{5}}, u_1 = 0, u_2 = \sqrt{\frac{3}{5}}$. Moreover,

8.21
$$\begin{cases} \alpha_1 = \int_{-1}^{1} \dfrac{(u + \sqrt{\frac{3}{5}})(u - \sqrt{\frac{3}{5}})}{(-\sqrt{\frac{3}{5}})(\sqrt{\frac{3}{5}})} du = \frac{8}{9} \\[3mm] \alpha_0 = \int_{-1}^{1} \dfrac{(u - \sqrt{\frac{3}{5}})u}{(2\sqrt{\frac{3}{5}})(\sqrt{\frac{3}{5}})} du = \frac{5}{9} = \alpha_2. \end{cases}$$

Thus we have

8.22 $$\int_{-1}^{1} \phi(u)du \sim \tfrac{5}{9}\phi(-\sqrt{\tfrac{3}{5}}) + \tfrac{8}{9}\phi(0) + \tfrac{5}{9}\phi(\sqrt{\tfrac{3}{5}}).$$

Let us now determine the order of each of the above formulas. For the case $n = 0$, we have

8.23 $$\begin{cases} R[1] = \int_{-1}^{1} du - 2(1) = 2 - 2 = 0 \\[2mm] R[u] = \int_{-1}^{1} u\,du - 2(0) = 0 \\[2mm] R[u^2] = \int_{-1}^{1} u^2\,du - 2(0) = \tfrac{2}{3}. \end{cases}$$

If $n = 1$, we have

8.24 $$\begin{cases} R[1] = R[u] = R[u^2] = R[u^3] = 0, \quad \text{and} \\[2mm] R[u^4] = \int_{-1}^{1} u^4\,du - \tfrac{2}{9} = \tfrac{2}{5} - \tfrac{2}{9} = \tfrac{8}{45}. \end{cases}$$

Hence the order of approximation is *three*.
   If $n = 2$, then

8.25 $$\begin{cases} R[1] = R[u] = R[u^2] = R[u^3] = R[u^4] = R[u^5] = 0 \\[2mm] R[u^6] = \int_{-1}^{1} u^6\,du - \tfrac{10}{9}(\tfrac{27}{125}) = \tfrac{8}{175}. \end{cases}$$

Hence the method is of order *five*. In general we shall show that with $n + 1$ points we obtain an order of approximation of $2n + 1$ using Gaussian quadrature.
   We now seek to show that the order of accuracy is indeed $2n + 1$. To do this we show that the Gaussian quadrature formula with $n + 1$ points is exact for any polynomial $\Phi(u)$ of degree $2n + 1$ or less. Let $P_{n+1}(u)$ be the polynomial of degree $n + 1$ given by (8.8). Then there exist* unique polynomials $S(u)$ and $T(u)$ of degree n or less such that

8.26 $$\Phi(u) = P_{n+1}(u)S(u) + T(u).$$

---

* This can be shown by repeated subtractions of powers of x times $P_{n+1}(u)$ from $\Phi(u)$, starting with x^n. (See Exercise 8, Section 5.2.)

Because of the choice of the $\alpha_0, \alpha_1, \ldots, \alpha_n$ we have

8.27
$$\sum_{i=0}^{n} \alpha_i T(u_i) = \int_{-1}^{1} T(u)du$$

since the Gaussian quadrature formula is certainly exact for any polynomial of degree n or less. Therefore,

8.28
$$\int_{-1}^{1} \Phi(u)du = \int_{-1}^{1} P_{n+1}(u)S(u)du + \int_{-1}^{1} T(u)du$$

and

8.29
$$\sum_{i=0}^{n} \alpha_i \Phi(u_i) = \sum_{i=0}^{n} \alpha_i P_{n+1}(u_i)S(u_i) + \sum_{i=0}^{n} \alpha_i T(u_i)$$
$$= \sum_{i=0}^{n} \alpha_i T(u_i)$$

since $P_{n+1}(u_i) = 0$, $i = 0, 1, \ldots, n$. Evidently

8.30
$$\int_{-1}^{1} \Phi(u)du = \sum_{i=0}^{n} \alpha_i \Phi(u_i)$$

provided

8.31
$$\int_{-1}^{1} P_{n+1}(u)S(u)du = 0.$$

But since $S(u)$ is a polynomial of degree n or less we can find coefficients $\beta_0, \beta_1, \ldots, \beta_n$ so that

8.32
$$S(u) = \beta_0 P_0(u) + \beta_1 P_1(u) + \cdots + \beta_n P_n(u).$$

To find the β_i, simply subtract a multiple, say β_n, of $P_n(u)$ from $S(u)$ to eliminate the u^n term, then subtract a multiple of $P_{n-1}(u)$ from $S(u) - \beta_n P_n(u)$ to eliminate u^{n-1}, etc. We can show (8.31) provided we can show that

8.33
$$\int_{-1}^{1} P_m(u)P_n(u)du = 0, \qquad m \neq n.$$

To prove (8.33) we first show that for each $n \geq 0$, $y = P_n(u)$ satisfies the Legendre differential equation

8.34
$$(1 - u^2)y'' - 2uy' + n(n + 1)y = 0.$$

This can perhaps be most easily shown as follows. Assume

8.35
$$y = \sum_{k=0}^{\infty} c_k u^k.$$

Then

8.36
$$\begin{cases} 2uy' = 2 \sum_{k=0}^{\infty} kc_k u^k, \\[2mm] y'' = \sum_{k=2}^{\infty} k(k-1)c_k u^{k-2} = \sum_{k=0}^{\infty} (k+1)(k+2)c_{k+2} u^k \\[2mm] u^2 y'' = \sum_{k=0}^{\infty} k(k-1)c_k u^k. \end{cases}$$

Therefore we get

8.37 $\displaystyle\sum_{k=0}^{\infty} u^k \{(k+1)(k+2)c_{k+2} - [k(k-1) + 2k - n(n+1)]c_k\} = 0.$

Equating coefficients of u^k to zero we have

8.38
$$c_{k+2} = -\frac{(n-k)(n+k+1)}{(k+1)(k+2)}c_k, \qquad k = 0, 1, 2, \ldots .$$

Evidently $c_{n+2} = c_{n+4} = \cdots = 0$. Thus if n is even, a polynomial solution of the differential equation can be found by letting $c_0 = 1$, $c_1 = 0$, while if n is odd, we can let $c_0 = 0$, $c_1 = 1$. Thus, for example, we have

8.39
$$\begin{cases} n = 0, \quad c_0 = 1, \quad c_1 = 0, \quad P_0(u) = 1 \\[1mm] n = 1, \quad c_0 = 0, \quad c_1 = 1, \quad P_1(u) = u \\[1mm] n = 2, \quad c_0 = 1, \quad c_1 = 0, \quad c_2 = -3, \quad P_2(u) = 1 - 3u^2 \\[1mm] n = 3, \quad c_0 = 0, \quad c_1 = 1, \quad c_2 = 0, \quad c_3 = -\tfrac{5}{3}, \quad P_3(u) = u - \tfrac{5}{3}u^3, \end{cases}$$

etc. Evidently if n is even, the corresponding solution is an even polynomial, while if n is odd, the solution is an odd polynomial.

Let us now verify that $P_n(u)$ is indeed a solution of (8.34). By the binomial theorem we have

8.40 $(u^2 - 1)^n = u^{2n} - \dbinom{n}{1}u^{2n-2} + \cdots + (-1)^s \dbinom{n}{s}u^{2n-2s} + \cdots + (-1)^n$

and hence

8.41
$$\frac{1}{n!}\frac{d^n}{du^n}(u^2 - 1)^n = \binom{2n}{n}u^n - \binom{n}{1}\binom{2n-2}{n}u^{n-2} + \cdots$$
$$+ (-1)^s\binom{n}{s}\binom{2n-2s}{n}u^{n-2s} + \cdots.$$

Now let $k = n - 2s$. Evidently either k and n are both even or both odd. We have

8.42
$$\frac{1}{n!}\frac{d^n}{du^n}(u^2 - 1)^n = \begin{cases} \displaystyle\sum_{\substack{k=0 \\ k \text{ even}}}^{n} d_k u^k, & \text{if } n \text{ is even} \\[2em] \displaystyle\sum_{\substack{k=1 \\ k \text{ odd}}}^{n} d_k u^k, & \text{if } n \text{ is odd} \end{cases}$$

where

8.43
$$d_k = (-1)^{(n-k)/2}\frac{1}{\left(\frac{n-k}{2}\right)!\left(\frac{n+k}{2}\right)!}\frac{(n+k)!}{k!}.$$

Therefore

8.44
$$d_{k+2} = -\frac{(n-k)(n+k+1)}{(k+1)(k+2)}d_k$$

which is the same recurrence relation as (8.38).

If n is even and if we let $c_0 = d_0$ and $c_1 = 0$, then the polynomial

8.45
$$\frac{1}{n!}\frac{d^n}{du^n}(u^2 - 1)^n$$

satisfies the differential equation (8.34), by (8.38) and (8.44). Similarly, if n is odd and if we let $c_0 = 0$ and $c_1 = d_1$, then a similar result holds.

From (8.34) we have for any m and n

8.46
$$\frac{d}{du}[(1 - u^2)P_n'(u)] + n(n + 1)P_n(u) = 0$$

and

8.47
$$\frac{d}{du}[(1 - u^2)P_m'(u)] + m(m + 1)P_m(u) = 0.$$

Therefore,

8.48 $\int_{-1}^{1} P_m(u) \left[\dfrac{d}{du}[(1 - u^2)P_n'(u)] \right] du + \int_{-1}^{1} n(n + 1)P_m(u)P_n(u)du = 0$

and, integrating by parts, we get

8.49 $-\int_{-1}^{1} (1 - u^2)P_n'(u)P_m'(u)du + \int_{-1}^{1} n(n + 1)P_m(u)P_n(u)du = 0.$

Similarly,

8.50 $-\int_{-1}^{1} (1 - u^2)P_n'(u)P_m'(u)du + \int_{-1}^{1} m(m + 1)P_m(u)P_n(u)du = 0;$

and (8.33) follows. Hence (8.31) holds by (8.32).

We now show that all roots of $P_n(u) = 0$ lie in the interval $(-1, 1)$ and are simple. If any root, say α, were repeated, then we would have $P_n(\alpha) = P_n'(\alpha) = 0$. If $-1 < \alpha < 1$, then by (8.34), since $\alpha \neq \pm 1$, we have $P_n''(\alpha) = 0$. By repeated differentiation of (8.34) we can show that all derivatives of $P_n(u)$ vanish for $u = \alpha$; hence the polynomial $P_n(u)$ vanishes identically which contradicts (8.8), since by (8.41) the coefficient of u^n does not vanish.

We now show there is at least one root of $P_n(u) = 0$ in the interval $(-1, 1)$ for $n \geq 1$. Otherwise, $P_n(u)$ would be one-signed and

8.51 $\int_{-1}^{1} P_n(u)P_0(u)du = \int_{-1}^{1} P_n(u)du \neq 0,$

a contradiction. Suppose now there are $m < n$ roots $u_1, u_2, \ldots, u_m$ of $P_n(u) = 0$ in the interval $(-1, 1)$. Then $P_n(u)$ and $W_m(u) = (u - u_1)(u - u_2) \cdots (u - u_m)$ have the same sign everywhere in the interval $[-1, 1]$ or else have opposite signs everywhere. But this implies that

8.52 $\int_{-1}^{1} P_n(u)W_m(u)du \neq 0.$

On the other hand, since $W_m(u)$ is a polynomial of degree less than n it follows from (8.31) that

8.53 $\int_{-1}^{1} P_n(u)W_m(u)du = 0$

and we have a contradiction. Thus there are precisely n simple roots of $P_n(u) = 0$ in the interval $(-1, 1)$.

An Error Formula for Gaussian Quadrature

We have shown that the Gaussian quadrature formula with $n + 1$ points is of order $2n + 1$. We now show that the Gaussian quadrature formula corresponding to each n is *definite* and hence we have by (2.28)

8.54
$$R[\phi] = R[u^{2n+2}] \frac{\phi^{(2n+2)}(\xi)}{(2n+2)!}$$

for some ξ in the interval $(-1, 1)$. To do this let us consider the polynomial $\hat{\Phi}(u)$ such that for $i = 0, 1, \ldots, n$ we have

8.55
$$\begin{cases} \hat{\Phi}(u_i) = \phi(u_i) \\ \hat{\Phi}'(u_i) = \phi'(u_i). \end{cases}$$

Evidently, as in Section 6.7, we have

8.56
$$\phi(u) - \hat{\Phi}(u) = \prod_{i=0}^{n} (u - u_i)^2 P(u)$$

for some continuous function $P(u)$ such that

8.57
$$P(u) = \frac{\phi^{(2n+2)}(\xi)}{(2n+2)!}$$

for some $\xi \in (-1, 1)$. Since the Gaussian quadrature formula is exact for any polynomial of degree $2n + 1$ or less, we have

8.58
$$\tilde{L}[\phi] = \tilde{L}[\hat{\Phi}] = L[\hat{\Phi}]$$

and

8.59
$$R[\phi] = L[\phi] - \tilde{L}[\phi] = L[\phi - \hat{\Phi}] = \int_{-1}^{1} [\phi(u) - \hat{\Phi}(u)]du$$

$$= \int_{-1}^{1} \left(\prod_{i=0}^{n} (u - u_i)^2 \right) P(u)du$$

$$= P(\xi_1) \int_{-1}^{1} \prod_{i=0}^{n} (u - u_i)^2 \, du = \frac{\phi^{(2n+2)}(\xi)}{(2n+2)!} \int_{-1}^{1} \prod_{i=0}^{n} (u - u_i)^2 \, du$$

by the mean-value theorem for integrals (Theorem A-3.36), where ξ_1 and ξ lie in $(-1, 1)$. Letting $\phi(u) = u^{2n+2}$ we get

8.60
$$R[u^{2n+2}] = \int_{-1}^{1} \prod_{i=0}^{n} (u - u_i)^2 \, du$$

and hence (2.28) holds and the method is definite.

In the special cases $n = 0, 1, 2$ we have (see (8.23), (8.24), (8.25), etc.)

8.61
$$\begin{cases} \int_{-1}^{1} \phi(u)du = 2\phi(0) + \frac{1}{3}\phi''(\xi) \\ \int_{-1}^{1} \phi(u)du = \phi\left(-\sqrt{\frac{1}{3}}\right) + \phi\left(\sqrt{\frac{1}{3}}\right) + \frac{1}{135}\phi^{(4)}(\xi) \\ \int_{-1}^{1} \phi(u)du = \frac{5}{9}\phi\left(-\sqrt{\frac{3}{5}}\right) + \frac{8}{9}\phi(0) + \frac{5}{9}\phi\left(\sqrt{\frac{3}{5}}\right) + \frac{1}{15,750}\phi^{(6)}(\xi). \end{cases}$$

In general it can be shown that the error with Gaussian quadrature based on $n + 1$ points for (8.4) is

8.62
$$E_n = \frac{2^{2n+3}((n+1)!)^4}{(2n+3)((2n+2)!)^3}\phi^{(2n+2)}(\xi).$$

For the integral (8.1) the error is

8.63
$$E_n = \frac{(b-a)^{2n+3}((n+1)!)^4}{(2n+3)((2n+2)!)^3}f^{(2n+2)}(\xi)$$

(see Davis and Rabinowitz [1967]).

Numerical Example

Let us compute the integral

8.64
$$\int_{1}^{2} e^x \, dx \doteq 4.67078$$

using Gaussian quadrature with one, two, and three points. Evidently, if we let

8.65
$$u = 2x - 3, \qquad x = \frac{u+3}{2}$$

we have

8.66
$$\int_{1}^{2} e^x \, dx = \frac{1}{2}\int_{-1}^{1} e^{(u+3)/2} \, du.$$

Hence we seek to evaluate

8.67
$$\int_{-1}^{1} \phi(u)du$$

where

8.68
$$\phi(u) = \tfrac{1}{2}e^{(u+3)/2}.$$

First, let $n = 0$. We have, by (8.15)

8.69
$$\int_1^2 e^x\,dx \sim e^{1.5} \doteq 4.48169.$$

Thus the error is 0.18909. The bounds for the error given by (8.63) are

8.70
$$\tfrac{1}{24}e^{1.0} \qquad \text{and} \qquad \tfrac{1}{24}e^{2.0}$$

or

8.71
$$0.11326 \qquad \text{and} \qquad 0.30788.$$

Next, for the case $n = 1$ we have by (8.19)

8.72
$$\int_1^2 e^x\,dx \sim \tfrac{1}{2}[e^{x_0} + e^{x_1}]$$

where

8.73
$$x_0 = \frac{3 - \sqrt{\tfrac{1}{3}}}{2} \doteq 1.21132, \quad x_1 \doteq \frac{3 + \sqrt{\tfrac{1}{3}}}{2} = 1.78868.$$

Therefore,

8.74
$$\tfrac{1}{2}[e^{x_0} + e^{x_1}] \doteq \tfrac{1}{2}[3.35791 + 5.98155] = 4.66973.$$

The error is 0.00105. The bounds given for the error by (8.63) are

8.75
$$\tfrac{1}{4320}e^{1.0} \qquad \text{and} \qquad \tfrac{1}{4320}e^{2.0}$$

or

8.76
$$0.00063 \qquad \text{and} \qquad 0.00171.$$

For the case $n = 2$ we have by (8.22)

8.77
$$\int_1^2 e^x\,dx \sim \tfrac{5}{18}e^{x_0} + \tfrac{8}{18}e^{x_1} + \tfrac{5}{18}e^{x_2}$$

where

8.78 $x_0 = \dfrac{3 - \sqrt{\tfrac{3}{5}}}{2} = 1.11270, \quad x_2 = 1.50000, \quad x_3 = \dfrac{3 + \sqrt{\tfrac{3}{5}}}{2} = 1.88730.$

Therefore,

8.79 $\displaystyle\int_1^2 e^x\,dx \sim \tfrac{5}{18}(3.04256) + \tfrac{8}{18}(4.48169) + \tfrac{5}{18}(6.60152) = 4.67077$

which agrees with (8.64) to within 0.00001. The bounds for the error given by
(8.63) are

8.80 $\displaystyle\frac{e^{1.0}}{2,016,000}$ and $\displaystyle\frac{e^{2.0}}{2,016,000}$

or

8.81   0.00000135   and   0.00000367.

Chebyshev Quadrature

We now seek a numerical quadrature formula with $n + 1$ points $x_0, x_1, \ldots, x_n$
which has an order of accuracy of at least $n + 1$ and which has equal weights. As
in the case of Gaussian quadrature, if we wish to evaluate the integral (8.1) we
introduce the change of variable (8.3) and obtain the integral (8.4). We seek to
determine $u_0, u_1, \ldots, u_n$ and w so that the quadrature formula

8.82 $\displaystyle\int_{-1}^1 \phi(u)\,du \sim w[\phi(u_0) + \phi(u_1) + \cdots + \phi(u_n)]$

is exact for any polynomial of degree p or less where p is as large as possible.
   In any case, if we require that (8.82) be exact for $\phi(u) = 1, u, u^2, \ldots, u^{n+1}$ we
have

8.83 $\displaystyle w = \frac{2}{n + 1}$

and

8.84 $\begin{cases} 0 = \displaystyle\sum_{i=0}^n u_i^k, & k = 1, 2, \ldots, n + 1, k \text{ odd} \\[2ex] \dfrac{n + 1}{k + 1} = \displaystyle\sum_{i=0}^n u_i^k, & k = 1, 2, \ldots, n + 1, k \text{ even.} \end{cases}$

   Suppose now that $u_0, u_1, \ldots, u_n$ are the roots of the polynomial equation

8.85 $G_n(u) = a_0 u^{n+1} + a_1 u^n + \cdots + a_{n+1} = 0.$

Let us define the numbers

8.86
$$s_k = \sum_{i=0}^{n} u_i^k, \qquad k = 0, 1, 2, \ldots .$$

By 5-(7.26), the numbers $s_0, s_1, \ldots, s_{n+1}$ satisfy the conditions

8.87
$$\begin{cases} n + 1 = s_0 \\ 0 = a_0 s_1 + a_1 \\ 0 = a_0 s_2 + a_1 s_1 + 2a_2 \\ 0 = a_0 s_3 + a_1 s_2 + a_2 s_1 + 3a_3 \\ \vdots \\ 0 = a_0 s_{n+1} + a_1 s_n + \cdots + (n+1)a_{n+1} = 0. \end{cases}$$

Our procedure then is to determine $s_0, s_1, \ldots, s_{n+1}$ by

8.88
$$\begin{cases} s_k = \dfrac{n+1}{k+1}, & \text{if } k \text{ is even,} \\ s_k = 0, & \text{if } k \text{ is odd,} \end{cases}$$

and then find the a_k from (8.87) letting $a_0 = 1$. We then find the roots $u_0, u_1, \ldots, u_n$ of (8.85).

For the case $n = 0$ we have $s_0 = 1$. Since $s_1 = 0$, we have, by (8.87) with $a_0 = 1$

8.89
$$a_1 = 0.$$

Hence (8.85) becomes

8.90
$$u = 0$$

which has the single root $u_0 = 0$. Our quadrature formula becomes

8.91
$$\int_{-1}^{1} \phi(u)\,du \sim 2\phi(0)$$

which is the same as the Gaussian quadrature formula corresponding to $n = 0$. Thus the order of accuracy is one in this case.

For the case $n = 1$ we have $s_0 = 2$, $s_1 = 0$, $s_2 = \frac{2}{3}$. Hence by (8.87) we have, with $a_0 = 1$,

8.92
$$a_1 = 0, \quad a_2 = -\tfrac{1}{3}.$$

We then solve

8.93
$$u^2 - \tfrac{1}{3} = 0$$

obtaining

8.94
$$u_0 = -\sqrt{\tfrac{1}{3}}, \quad u_1 = \sqrt{\tfrac{1}{3}}$$

and we have

8.95
$$\int_{-1}^{1} \phi(u)du \sim \phi(-\sqrt{\tfrac{1}{3}}) + \phi(\sqrt{\tfrac{1}{3}})$$

which is again the same as the Gaussian quadrature formula corresponding to $n = 1$. Thus the order of accuracy is 3 in this case.

The reader should show that for $n = 2$ the method is

8.96
$$\int_{-1}^{1} \phi(u)du \sim \tfrac{2}{3}[\phi(-\sqrt{\tfrac{1}{2}}) + \phi(0) + \phi(\sqrt{\tfrac{1}{2}})]$$

and for $n = 3$ it is

8.97
$$\int_{-1}^{1} \phi(u)du \sim \tfrac{1}{2}[\phi(u_0) + \phi(u_1) + \phi(u_2) + \phi(u_3)]$$

where u_0, u_1, u_2, and u_3 satisfy the equation

8.98
$$u^4 - \tfrac{2}{3}u^2 + \tfrac{1}{45} = 0.$$

Thus

8.99 $u_0 = -0.79465, \quad u_1 = -0.18759, \quad u_2 = 0.18759, \quad u_3 = 0.79465.$

It can be shown (see Bernstein [1937]) that the abscissas for Chebyshev quadrature are real for the cases

8.100
$$n = 0, 1, 2, 3, 4, 5, 6, 8;$$

otherwise, at least some are complex. Consequently, Chebyshev quadrature is only useful for the eight indicated values of n.

We now show that if n is one of the values given by (8.100) then the order p of the Chebyshev method is given by

8.101
$$p = \begin{cases} n + 1, & \text{if } n \text{ is even} \\ n + 2, & \text{if } n \text{ is odd.} \end{cases}$$

Thus for n even the order of the Chebyshev method is the same as that of the closed

Newton-Cotes method with $n + 1$ points, namely, Q.n.n.0.* On the other hand, if n is odd, then the order of the Chebyshev method is two higher than the order of the Newton-Cotes method Q.n.n.0.

By the construction of the method we already know that

8.102 $R[u^k] = 0, \qquad k = 0, 1, \ldots, n + 1.$

If n is even, then

8.103 $R[u^{n+2}] = \dfrac{2}{n + 3} - \dfrac{2}{n + 1} s_{n+2}.$

But by 5-(7.27) we have

8.104 $s_{n+2} = -[a_2 s_n + a_4 s_{n-2} + \cdots + a_n s_2]$

and, by (8.88),

8.105 $R[u^{n+2}] = \dfrac{2}{n + 3} + 2\left[\dfrac{a_2}{n + 1} + \dfrac{a_4}{n - 1} + \cdots + \dfrac{a_n}{3}\right].$

Thus, for the cases $n = 0, 2, 4$, we have

8.106 $R[u^{n+2}] = \begin{cases} \frac{2}{3}, & n = 0 \\ \frac{1}{15}, & n = 2 \\ \frac{13}{756}, & n = 4. \end{cases}$

Thus the order is precisely $n + 1$ for $n = 0, 2, 4$. The reader should show that this is also the case for $n = 6, 8$.

If n is odd, then $R[u^{n+2}] = 0$ and hence the order is at least $n + 2$. Moreover,

8.107 $R[u^{n+3}] = \dfrac{2}{n + 4} - \dfrac{2}{n + 1} s_{n+3}.$

But by 5-(7.27) we have

8.108 $s_{n+3} = -[a_2 s_{n+1} + a_4 s_{n-1} + \cdots + a_{n+1} s_2],$

and, by (8.88)

8.109 $R[u^{n+3}] = \dfrac{2}{n + 4} + 2\left[\dfrac{a_2}{n + 2} + \dfrac{a_4}{n} + \cdots + \dfrac{a_{n+1}}{3}\right].$

---

* Except for the case $n = 0$ where Q.0.0.0 does not exist.

Thus for the cases $n = 1$, and 3 we have

8.110
$$R[u^{n+3}] = \begin{cases} \frac{8}{45}, & n = 1 \\ \frac{32}{945}, & n = 3. \end{cases}$$

Thus the order is exactly $n + 2$ for $n = 1, 3$. The reader should show that this is also the case for $n = 5$.

It can also be shown that each Chebyshev quadrature formula is definite. (See, for instance, Hildebrand [1956].) We give here a sketch of the argument. First one defines the continuous function $\pi(u)$ in the interval $[-1, 1]$ by

8.111
$$\pi(u) = \frac{\phi(u) - \Phi(u)}{Q(u)}$$

where $\Phi(u)$ is the unique polynomial of degree n or less which agrees with $\phi(u)$ at the interpolation points and where

8.112
$$Q(u) = \prod_{j=0}^{n} (u - u_j).$$

As we have seen in Section 6.5,

8.113
$$\pi(u) = \frac{\phi^{(n+1)}(\xi)}{(n+1)!}$$

for some ξ in $(-1, 1)$. It can also be shown* that $\pi(u) \in C^{(2)}(-1, 1)$ and moreover,

8.114
$$\begin{cases} \pi'(u) = \dfrac{\phi^{(n+2)}(\xi_1)}{(n+2)!} \\[2mm] \pi''(u) = \dfrac{2\phi^{(n+3)}(\xi_2)}{(n+3)!} \end{cases}$$

for some ξ_1 and ξ_2 in $(-1, 1)$. Here we assume that $\phi(u) \in C^{(n+3)}(-1, 1)$. For a proof of (8.114) see Ralston [1963]. See also Hayes and Rubin [1970] for a proof for the case of $\pi'(u)$.

By (8.111) we have

8.115
$$\int_{-1}^{1} \phi(u)du - \int_{-1}^{1} \Phi(u)du = \int_{-1}^{1} Q(u)\pi(u)du = \int_{-1}^{1} G_n(u)\pi(u)du$$

---

* In Section 7.7 we proved the existence and continuity of $\pi'(u)$.

where $G_n(u)$ is given by (8.85) with $a_0 = 1$. Integrating by parts we have

8.116 $$\int_{-1}^{1} G_n(u)\pi(u)du = [H_n(u)\pi(u)]_{-1}^{1} - \int_{-1}^{1} H_n(u)\pi'(u)du$$

where

8.117 $$H_n(u) = \int_{-1}^{u} G_n(u)du.$$

We now show that

8.118 $$H_n(1) = H_n(-1) = 0.$$

But from (8.87) and (8.88) it follows that

8.119 $$a_k = 0, \qquad k = 1, 2, \ldots, n + 1, k \text{ odd}.$$

Hence (8.118) certainly holds if n is even. If n is odd, then by (8.85) and (8.119) we have

8.120 $$\frac{1}{2}\int_{-1}^{1} G_n(u)du = \frac{a_0}{n + 2} + \frac{a_2}{n} + \cdots + a_{n+1}$$

$$= \frac{1}{n + 1}[a_0 s_{n+1} + a_2 s_{n-1} + \cdots + a_{n+1} s_0] = 0$$

by the last equation of 5-(7.26).
From (8.116) we have

8.121 $$\int_{-1}^{1} G_n(u)\pi(u)du = -\int_{-1}^{1} H_n(u)\pi'(u)du.$$

Since

8.122 $$H_n(u) = \begin{cases} \dfrac{u^2 - 1}{2}, & n = 0 \\[3mm] \dfrac{u^2(u^2 - 1)}{4} & n = 2 \\[3mm] \dfrac{(u^2 - 1)(24u^4 - 6u^2 + 1)}{144}, & n = 4 \end{cases}$$

it follows that if $n = 0, 2, 4$, $H_n(u)$ is one-signed in I and hence by the mean-value theorem for integrals

8.123
$$\int_{-1}^{1} G_n(u)\pi(u)du = - \pi'(\xi) \int_{-1}^{1} H_n(u)du$$

for some ξ in $(-1, 1)$. It is then easy to show that the method is definite. The reader should show that this is also the case for $n = 6$ and 8.

If n is odd, we integrate by parts once again obtaining

8.124
$$\int_{-1}^{1} G_n(u)\pi(u)du = [-K_n(u)\pi'(u)]_{-1}^{1} + \int_{-1}^{1} K_n(u)\pi''(u)du$$

where

8.125
$$K_n(u) = \int_{-1}^{u} H_n(u)du.$$

We verify that

8.126 $H_1(u) = \dfrac{u^2 - u}{3}$, $K_1(u) = \frac{1}{12}(u^2 - 1)^2$, $K_1(-1) = K_1(1) = 0$

8.127 $H_3(u) = \dfrac{u^5}{5} - \dfrac{2}{9}u^3 + \dfrac{u}{45}$, $K_3(u) = \frac{1}{90}(u^2 - 1)^2(3u^2 + 1)$,

$$K_3(-1) = K_3(1) = 0.$$

Since $K_1(u)$ and $K_3(u)$ are one-signed, we can also show that the method is definite for $n = 1, 3$. This can also be done for the remaining case, namely $n = 5$.

In Table 8.128 we give the abscissas and error formulas for the Chebyshev formulas corresponding to $n = 0, 1, 2, 3, 4$, i.e., for 1, 2, 3, 4, and 5 points. For a more complete table of the abscissas see Hildebrand [1956].

8.128 Table Abscissas and Errors for Chebyshev Quadrature

n	Weight	Abscissas	$G_n(u)$	Error
0	2	$u_0 = \quad 0$	u	$\dfrac{1}{3}\phi''(\xi)$
1	1	$u_0 = -0.57735$ $u_1 = \quad 0.57735$	$u^2 - \dfrac{1}{3}$	$\dfrac{1}{135}\phi^{(4)}(\xi)$
2	$\dfrac{2}{3}$	$u_0 = -0.70711$	$u^3 - \dfrac{1}{2}u$	$\dfrac{1}{360}\phi^{(4)}(\xi)$

Continued

n	Weight	Abscissas	$G_n(u)$	Error
		$u_1 = \quad 0$		
		$u_2 = \quad 0.70711$		
3	$\dfrac{1}{2}$	$u_0 = -0.79465$	$u^4 - \dfrac{2}{3}u^2 + \dfrac{1}{45}$	$\dfrac{2}{42{,}525}\phi^{(6)}(\xi)$
		$u_1 = -0.18759$		
		$u_2 = \quad 0.18759$		
		$u_3 = \quad 0.79465$		
4	$\dfrac{2}{5}$	$u_0 = -0.83250$	$u^5 - \dfrac{5}{6}u^3 + \dfrac{7}{72}u$	$\dfrac{13}{544{,}320}\phi^{(6)}(\xi)$
		$u_1 = -0.37454$		
		$u_2 = \quad 0$		
		$u_3 = \quad 0.37454$		
		$u_4 = \quad 0.83250$		

EXERCISES 7.8

1. Find the Gaussian quadrature formulas for 4 and 5 points. Verify in each case that the order of the method is $2n + 1$ where the number of points is $n + 1$.

2. Verify that the Legendre polynomials of degree n satisfy the Legendre differential equation (8.34) for $n = 0(1)5$.

3. Find the polynomials $S(u)$ and $T(u)$ of degree two or less such that $u^5 = P_3(u)S(u) + T(u)$ where $P_3(u)$ is given by (8.8). Also find β_0, β_1, and β_2 such that $S(u) = \beta_0 P_0(u) + \beta_1 P_1(u) + \beta_2 P_2(u)$. Verify directly that

$$\int_{-1}^{1} P_3(u)S(u)\,du = 0.$$

4. Verify that

$$\int_{-1}^{1} P_n(u)P_m(u) = 0$$

for the following cases: $n = 2, m = 3$; $n = 2, m = 4$; $n = 3, m = 4$.

5. Derive error formulas for Gaussian quadrature for 4 and 5 points.

6. Compute

$$\int_{1}^{2} x^{-1}\,dx$$

by Gaussian quadrature with $n + 1$ interpolation points for $n = 0(1)3$. Compare the results with the exact value of the integral (namely, $\log 2 = 0.693147$). Also compute the errors and compare with theoretical upper and lower bounds for the error. Do the same for

$$\int_{-1}^{1} (1 + x^2)^{-1} \, dx.$$

7. For $n = 0, 1, 2, \ldots$ let a_n be the coefficient of u^n for $Q_n(u)$, the Legendre polynomial of degree n. Show that $a_0 = a_1 = 1$ and $a_{n+1} = (2n + 1)a_n/(n + 1), n = 1, 2, \ldots$.

8. For $n = 0, 1, 2, \ldots$ let $Q_n(u)$ be the Legendre polynomial of degree n. Show that $Q_0(u) = 1$, $Q_1(u) = u$, and $(n + 1)Q_{n+1}(u) - (2n + 1)uQ_n(u) + nQ_{n-1}(u) = 0$.

9. Using (8.9) and (8.8) and repeated integration by parts show that

$$\int_{-1}^{1} Q_n(u)^2 \, du = \frac{2}{2n + 1}.$$

(See Ford [1963], p. 194.)

10. Verify the values given in Table 8.128 for the case $n = 4$.

11. Apply the Chebyshev quadrature formula with three points for the integral (8.64). Compare the actual error with theoretical upper and lower bounds.

12. Compute

$$\int_{-1}^{1} \frac{dx}{1 + x^2}$$

using Chebyshev quadrature with 4 and 5 points. Compare the actual error with the theoretical bounds.

13. Verify (8.96) and (8.97).

14. Find the Chebyshev quadrature formulas corresponding to $n = 5, 6,$ and 8 and determine the order in each case.

15. Compute $R[x^{n+2}]$ for the Chebyshev quadrature formulas corresponding to $n = 6$ and 8. Compute $R[x^{n+3}]$ for the formula corresponding to $n = 5$.

16. Show that the Chebyshev quadrature formulas corresponding to $n = 5, n = 6,$ and $n = 8$ are definite.

17. Show that some of the abscissas for Chebyshev integration based on 8 points are complex.

SUPPLEMENTARY DISCUSSION
Section 7.1

For a more extensive treatment of methods for numerical quadrature, including the treatment of improper integrals and integrals with singularities, see Davis and Rabinowitz [1967].

Section 7.3

For a more extensive table of coefficients for numerical differentiation for $f'(x)$ and for higher derivatives, see Kopal [1955, Appendix III] or Bickley [1941].

Section 7.7

Various other devices based on the integration by parts of the remainder in Lagrange interpolation can be used—see, for instance, Hildebrand [1956]. See also Section 7.8.

A proof that the Newton-Cotes formulas are definite is given by Steffensen [1950].

The representation (7.54) is a consequence of Peano's theorem (see Davis and Rabinowitz [1967]). The function $(p!)^{-1}\{R[\phi_p(x-t)](x)\}$ is the *Peano kernel*.

Sard [1963] proposed criteria for optimum numerical quadrature formulas based on the use of Peano's theorem. Connections between these formulas and numerical quadrature formulas based on splines are shown by Schoenberg [1964].

Section 7.8

An extensive table of the coefficients for Gaussian quadrature formulas is given by Stroud and Secrest [1966].

Other quadrature formulas can be obtained for the integral

$$\int_a^b w(x)f(x)dx$$

where $w(x) > 0$ for $x \in [a, b]$. The function $w(x)$ is known as the *weight function*. If $w(x) \equiv 1$ we obtain the ordinary Gaussian quadrature formulas sometimes referred to as Legendre-Gauss quadrature (Ralston [1965]). Other numerical quadrature formulas of Gauss type correspond to different weight functions. Thus Jacobi-Gauss quadrature corresponds to the weight function $w(x) = (1 - x)^\alpha(1 + x)^\beta$ where $\alpha > -1$, $\beta > -1$. Chebyshev-Gauss quadrature corresponds to the weight function $w(x) = (1 - x^2)^{-1/2}$.

Gaussian quadrature formulas can be derived from numerical quadrature formulas based on integration formulas corresponding to the Hermite interpolation formula (see Hildebrand [1956, p. 317]).

CHAPTER 8

ORDINARY DIFFERENTIAL EQUATIONS

8.1 INTRODUCTION

In this chapter and the two subsequent chapters we shall study numerical methods for solving problems involving ordinary differential equations. We shall be primarily concerned with the single first-order equation

1.1
$$y' = \frac{dy}{dx} = f(x, y).$$

We shall, however, also consider the case of a single equation of order r of the form

1.2
$$y^{(r)} = \frac{d^r y}{dx^r} = f(x, y, y', \ldots, y^{(r-1)})$$

and the case of a system of n simultaneous first-order equations of the form

1.3
$$y_i' = f_i(x, y_1, y_2, \ldots, y_n), \qquad i = 1, 2, \ldots, n.$$

Many of the traditional undergraduate courses on ordinary differential equations are concerned almost entirely with "closed-form" solutions of a single equation of first order or higher order. In such courses, there is a danger that the student may not understand that the number of cases where such closed-form solutions is possible is very small. He may also not understand that in most cases a solution exists, even when a closed-form solution does not exist, and, moreover, the solution can usually be found to a high degree of accuracy by numerical methods such as those described below. In fact, even when a closed-form solution does exist, it may be less convenient to use than a numerical method.

Consider, for instance, the differential equation

1.4
$$y' = 2xy + \frac{1}{1 + x^2}.$$

The general solution is

1.5
$$y = e^{x^2} \int \frac{e^{-t^2}}{1 + t^2} \, dt + c e^{x^2}$$

where c is an arbitrary constant. The solution through the point (x_0, y_0) is given by

1.6
$$y = e^{x^2} \int_{x_0}^{x} \frac{e^{-t^2}}{1 + t^2} \, dt + y_0 e^{x^2 - x_0^2}.$$

As we shall see, the numerical evaluation of the integral appearing in (1.6) by numerical quadrature is less convenient than the numerical solution of (1.4) by some of the schemes which we shall describe later in this chapter.

For a given differential equation or system of equations there will, in general, be many solutions. For example, for the equation

1.7
$$y' = Ay,$$

where A is a constant, the function

1.8
$$y = ce^{Ax}$$

is a solution for any value of the constant c. Thus we have a "one-parameter family" of solutions. For the second-order equation

1.9
$$y'' = A^2 y$$

the function

1.10
$$y = c_1 e^{Ax} + c_2 e^{-Ax}$$

is a solution for any values of the constants c_1 and c_2. In this case we have a "two-parameter family" of solutions.

Most problems arising in practice, which involve ordinary differential equations, have been formulated with the desire to predict some quantity. For example, one might be interested in knowing how far a ball will go if thrown at a given angle of elevation and initial velocity. To do this one may construct a mathematical model and predict the solution of the physical problem by solving the mathematical model involving a differential equation. Clearly, a mathematical model will be most useful if it has a unique solution. In some cases, if the problem has a small number of different solutions, a knowledge of these solutions might provide some helpful information. However, mathematical models for which there is either no solution or an infinite number of solutions are seldom very useful. Consequently, in formulating and solving problems involving ordinary differential equations one should be familiar with the relevant existence and uniqueness theorems.

Existence and uniqueness theorems are considered in Section 8.2. Under quite general conditions the *initial-value problem*

1.11
$$\begin{cases} y' = \dfrac{dy}{dx} = f(x, y) \\ y(x_0) = y_0 \end{cases}$$

where x_0, y_0 and the function $f(x, y)$ are given, has a unique solution in the interval $[x_0, \bar{x}]$ for some $\bar{x} > x_0$. Moreover, if $f(x, y)$ is analytic at (x_0, y_0), then for some $\delta > 0$ there exists a function $y(x)$ which is analytic on $[x_0 - \delta, x_0 + \delta]$ and which satisfies (1.11).

As an example, let us consider the problem defined by

1.12
$$\begin{cases} y' = Ay \\ y(x_0) = y_0 \end{cases}$$

where A is a constant. The general solution of the differential equation is given by (1.8). Letting $x = x_0$ and $y = y_0$ we get

1.13
$$c = y_0 e^{-Ax_0}$$

which upon substitution in (1.8) gives us

1.14
$$y = y_0 e^{A(x - x_0)}.$$

This function is, of course, analytic at x_0.

For an rth-order equation of the form (1.2) one can, in general, specify r conditions, each involving y and some of its derivatives at $s \leqq r$ values of the abscissa, say $x_0, x_1, \ldots, x_{s-1}$. However, in this chapter and in Chapter 9 we shall assume that $y, y', \ldots, y^{(r-1)}$ are prescribed at a single value of x, say x_0. Similarly, for the system (1.3) we shall assume that $y_1, y_2, \ldots, y_n$ are specified at the point x_0. In each case a unique solution can be shown to exist in the interval $[x_0, \bar{x}]$ for some $\bar{x} > x_0$ provided that certain conditions hold which are analogous to those assumed for the case of a single first-order equation.

In Section 8.3 we consider two methods based on the use of series which can be used to solve (1.11) if $f(x, y)$ is analytic. In Section 8.4 we develop an integral equation formulation of (1.11) and consider the Picard method of successive approximation. This method is primarily of theoretical interest rather than of practical use.

In Sections 8.5–8.10 we consider three classes of methods, namely: methods based on numerical quadrature (Sections 8.6–8.8); methods based on the use of Taylor's series (Section 8.9); and methods based on the use of numerical differentiation (Section 8.10). The simple Euler method considered in Section 8.5 belongs to all three classes.

The solution of higher-order equations and systems of first-order equations is considered in Section 8.11. A typical computer program for solving a system of first-order equations is described in Section 8.12.

In Chapter 9 we consider more advanced topics related to ordinary differential equations including stability, convergence, and accuracy. In Chapter 10 we discuss more complicated problems involving ordinary differential equations including: two-point boundary value problems where the equation is of second or higher order and conditions are prescribed for the solution at two points; and eigenvalue problems involving a parameter (eigenvalue) which is chosen so that an appropriate two-point boundary value problem can be solved.

EXERCISES 8.1

1. Verify that (1.6) is the solution of (1.4) satisfying the condition $y(x_0) = y_0$.

2. Show that if $y(x)$ is a solution of $y' + P(x)y = Q(x)$ then for some constant of c we have

$$y(x) = e^{-\int P(x)dx} \int Q e^{\int P(x)dx}\, dx + c e^{-\int P(x)dx}.$$

Use this formula to derive (1.5).

3. Verify that (1.10) is a solution of (1.9). Show that if $A \neq 0$ then (1.10) is the most general solution of (1.9). (Hint: let $v = y' - Ay$.) What about the case $A = 0$?

4. Verify that if m_1, m_2, and m_3 are roots of

$$am^3 + bm^2 + cm + d = 0$$

where a, b, c, and d are constants, then

$$y = c_1 e^{m_1 x} + c_2 e^{m_2 x} + c_3 e^{m_3 x}$$

is a solution of

$$ay''' + by'' + cy' + dy = 0$$

for any constants c_1, c_2, and c_3.

8.2 EXISTENCE AND UNIQUENESS

In this section we discuss existence and uniqueness for the initial-value problem (1.11). We shall assume that $f(x, y)$ is continuous in a region $R: x_0 \leq x \leq \hat{x}$, $|y - y_0| \leq \beta$ for some $\hat{x} > x_0$ and for some $\beta > 0$. We shall normally also assume that $f(x, y)$ satisfies a Lipschitz condition in R. This condition is defined as follows.

2.1 Definition. The function $f(x, y)$ satisfies a *Lipschitz condition in y* in a region R if there exists a constant L such that

2.2 $$|f(x, y) - f(x, z)| \leq L|y - z|$$

for any two points (x, y) and (x, z) in R with the same abscissa such that the segment joining (x, y) and (x, z) lies in R.

For example, the function $f(x, y) = x^2 + y^2$ satisfies a Lipschitz condition in the region $0 \leq x \leq 1, 0 \leq y \leq 2$ with $L = 4$ since

2.3 $$|f(x, y) - f(x, z)| = |y + z||y - z| \leq 4|y - z|.$$

In general, if $f_y(x, y)$ exists and is bounded in R then $f(x, y)$ satisfies a Lipschitz condition in R, for by the mean-value theorem we have

2.4
$$f(x, y) - f(x, z) = (y - z)f_y(x, c)$$

where c lies between y and z. Thus if

2.5
$$|f_y(x, y)| \leqq L$$

in R, then (2.2) holds.

In the example just considered we have

2.6
$$f_y(x, y) = 2y$$

and

2.7
$$|f_y(x, y)| \leqq 4$$

in the region $0 \leqq x \leqq 1, 0 \leqq y \leqq 2$.

We now prove

2.8 Lemma. If $f(x, y)$ is continuous for $x_0 \leqq x \leqq \hat{x}$ and for $|y - y_0| \leqq \beta$ for some $\hat{x} > x_0$ and for some $\beta > 0$, then there exists $\bar{x}$ and M such that

2.9
$$x_0 < \bar{x} \leqq \hat{x}, \qquad M \geqq |f(x_0, y_0)|$$

and such that

2.10
$$|f(x, y)| \leqq M$$

in R:

2.11
$$\begin{cases} x_0 \leqq x \leqq \bar{x} \\ |y - y_0| \leqq M(\bar{x} - x_0). \end{cases}$$

Proof. The determination of $\bar{x}$ is illustrated in Fig. 2.13. It is clearly sufficient to let M be a bound on $|f(x, y)|$ for $x_0 \leqq x \leqq \hat{x}$, $|y - y_0| \leqq \beta$ and to let

2.12
$$\bar{x} = \min\left(\hat{x}, x_0 + \frac{\beta}{M}\right).$$

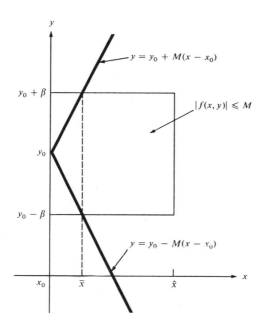

2.13 Fig. Determination of $\bar{x}$.

As an example, let us consider the problem defined by

2.14 $$y' = f(x, y) = y^2, \qquad y(0) = 1.$$

We remark that the exact solution of (2.14) is

2.15 $$y = \frac{1}{1 - x}.$$

For any $\hat{x} > 0$ and $\beta > 0$ we have

2.16 $$M = \max_{\substack{0 \le x \le \hat{x} \\ |y-1| \le \beta}} |f(x, y)| = (1 + \beta)^2.$$

Thus, since $\hat{x}$ can be chosen arbitrarily large, we have, by (2.12),

2.17 $$\bar{x} = \frac{\beta}{M} = \frac{\sqrt{M} - 1}{M}.$$

The largest value of $\bar{x}$ for which this is possible for some $M > 1$ corresponds to $M = 4$ and is given by

2.18 $$(\bar{x})_{\text{max}} = \tfrac{1}{4}.$$

We verify that for $0 \leqq x \leqq \frac{1}{4}$, $|y - 1| \leqq 1$ we have

2.19 $$|f(x, y)| \leqq 4.$$

For any $\bar{x} < \frac{1}{4}$ we can find the smallest value of M such that (2.10) holds in $R: 0 \leqq x \leqq \bar{x}$, $|y - 1| \leqq M\bar{x}$, as follows. We seek the smallest value of M such that

2.20 $$(1 + \bar{x}M)^2 \leqq M.$$

In the case $\bar{x} = 0.1$ we get

2.21 $$M = 40 - \sqrt{1500} \doteq 1.270.$$

Suppose that a solution $y(x)$ of (1.11) exists for $x \in [x_0, \bar{x}]$ for some $\bar{x} > x_0$ and that, for some M, (2.10) holds for R given by (2.11). We show that if

2.22 $$|y(x) - y_0| \leqq M(\bar{x} - x_0)$$

for all $x \in [x_0, \bar{x}]$, then for $x \in [x_0, \bar{x}]$ we have

2.23 $$|y(x) - y_0| \leqq M(x - x_0).$$

For if we consider the function

2.24 $$w(x) = y(x) - [y_0 + M(x - x_0)],$$

we have

2.25 $$w(x_0) = 0$$

and

2.26 $$w'(x) = y'(x) - M$$
$$= f(x, y(x)) - M$$
$$\leqq 0.$$

This follows since, for $x \in [x_0, \bar{x}]$, (2.22), and hence (2.10), holds and we have

2.27 $$|f(x, y(x))| \leqq M.$$

For $x \in [x_0, \bar{x}]$ we have, by the mean-value theorem,

2.28 $$w(x) \leqq 0$$

and so

2.29 $y(x) \leqq y_0 + M(x - x_0).$

Similarly, we can show that

2.30 $y(x) \geqq y_0 - M(x - x_0)$

and (2.23) follows.

We now state without proof the following result.

2.31 Theorem. For some $\hat{x} > x_0$ and for some $\beta > 0$ let $f(x, y)$ be continuous in R:

2.32 $x_0 \leqq x \leqq \hat{x}, \qquad |y - y_0| \leqq \beta$

and let $M > 0$ such that

2.33 $|f(x, y)| \leqq M$

in R. Let

2.34 $\bar{x} = \min\left(\hat{x}, x_0 + \dfrac{\beta}{M}\right).$

There exists a function $y(x)$, continuously differentiable in $x_0 \leqq x \leqq \bar{x}$ which satisfies (1.11). If $f(x, y)$ satisfies a Lipschitz condition in y in R then the solution is unique.

For a proof of Theorem 2.31 see Birkhoff and Rota [1962]. See also Ford [1933], and Henrici [1962].

It can be shown that any solution $y(x)$ satisfies the condition (2.22) for $x \in [x_0, \bar{x}]$. (See Exercise 10.) Hence, by the previous argument, (2.23) holds for $x \in [x_0, \bar{x}]$.

In the example (2.14), we have $|f(x, y)| \leqq 4$ for $0 \leqq x \leqq \frac{1}{4}$ and for $|y - 1| \leqq 1$. Moreover, $f(x, y)$ satisfies a Lipschitz condition in y in that region with Lipschitz constant $L = 4$. Hence the existence of a unique solution is guaranteed by Theorem 2.31.

On the other hand let us consider the example

2.35 $y' = y^{1/3}, \qquad y(0) = 0.$

It is easy to verify that

2.36 $|f(x, y)| \leqq 1$

for $0 \leqq x \leqq 1$ and for

2.37 $|y| \leqq 1.$

Consequently a solution exists by Theorem 2.31 for $0 \leqq x \leqq 1$. However, $f(x, y)$ does not satisfy a Lipschitz condition in y since

2.38 $$\frac{|f(x, y) - f(x, z)|}{|y - z|} = \frac{|y^{1/3} - z^{1/3}|}{|y - z|} = \frac{1}{|y^{2/3} + y^{1/3}z^{1/3} + z^{2/3}|}$$

can become arbitrarily large for y and z sufficiently small.

It is easy to show that for any a such that $0 \leqq a \leqq 1$ the function

2.39 $$y_a(x) = \begin{cases} 0, & 0 \leqq x \leqq a \\ (\tfrac{2}{3}(x - a))^{3/2}, & a \leqq x \leqq 1 \end{cases}$$

is a solution. Hence there are many solutions of (2.35).

We also state without proof the following result (see, for instance, Birkhoff and Rota [1962]).

2.40 Theorem. If $f(x, y)$ is analytic at (x_0, y_0) then for δ sufficiently small there exists a unique function $y(x)$ which is analytic for $x_0 - \delta \leqq x \leqq x_0 + \delta$ and which satisfies (1.11). If $f(x, y)$ is analytic for all x and y, then any solution of (1.11) is analytic.

Since the function $f(x, y) = y^2$ is analytic for all x and y it follows that the unique solution of (2.14) which we know exists in $I : 0 \leqq x \leqq \tfrac{1}{4}$ is analytic in I. As a matter of fact, by (2.15) the solution is given by

2.41 $y(x) = 1 + x + x^2 + \cdots .$

One can verify that this solution is actually valid for $0 \leqq x < 1$.

EXERCISES 8.2

1. Show that the function $f(x, y) = xy + y^3$ satisfies a Lipschitz condition in y in the region $x^2 + y^2 \leqq 1$, and determine a bound for the Lipschitz constant.

2. Show that the initial-value problem $y' = x + y^2, y(0) = 1$ has a unique solution in the interval $0 \leqq x \leqq \alpha$, where α is the positive real root of $4\alpha^3 + 4\alpha - 1 = 0$, $(\alpha \doteq 0.237)$.

3. For the initial-value problem $y' = f(x, y) = x + y^2, y(0) = 1$, find the smallest value of M such that $|f(x, y)| \leqq M$ in the region $0 \leqq x \leqq 0.1, 1 - 0.1M \leqq y \leqq 1 + 0.1M$.

4. Consider the initial-value problem $y' = x + y^3, y(0) = 2$. Find $\bar{x}$ such that $\bar{x} > 0$ and such that a unique solution exists in the interval $[0, \bar{x}]$. (You do *not* need to find the largest possible $\bar{x}$.)

5. Prove Theorem 2.31 under the assumption that $f(x, y)$ satisfies a Lipschitz condition in y in R. Assume that a continuously differentiable function $y(x)$ is a solution of (1.11) if and

only if $y(x)$ satisfies the integral equation

$$y(x) = y_0 + \int_{x_0}^{x} f(x, y(x))dx.$$

(See Section 8.4.) Let $y_0(x) \equiv y_0$ and let

$$y_{n+1}(x) = y_0 + \int_{x_0}^{x} f(x, y_n(x))dx, \qquad n = 0, 1, 2, \ldots.$$

Show that the sequence $y_0(x), y_1(x), \ldots$ converges uniformly in $[x_0, \bar{x}]$ to a continuously differentiable function $y(x)$ which satisfies (1.11). Show that if $z(x)$ is any solution of (1.11) then $z(x) \equiv y(x)$ for $x \in [x_0, \bar{x}]$. (See Ford [1933], pp. 86–89.)

6. By repeated application of Theorem 2.31 show that there exists a unique solution of the initial-value problem $y' = y^2, y(0) = 1$ in the interval $[0, \frac{1}{2}]$. How would you apply Theorem 2.31 to show the existence of a unique solution on $[0, 1)$?

7. Using Theorem 2.31 prove the following theorem: if, for some $\hat{x} > x_0$, $f(x, y)$ is defined and continuous on $[x_0, \hat{x}]$ and for all y and if for some constant L we have

$$|f(x, y) - f(x, y^*)| \leq L|y - y^*|$$

for any x in $[x_0, \hat{x}]$ and for any y and y^* then there exists a unique solution of (1.11) on $[x_0, \hat{x}]$.

8. Use the preceding result to prove the existence of a unique solution on $[x_0, \hat{x}]$ of the initial-value problem

$$\begin{cases} y' = P(x)y + Q(x) \\ y(x_0) = y_0 \end{cases}$$

where $P(x)$ and $Q(x)$ are continuous on $[x_0, \hat{x}]$.

9. Show that the initial-value problem $y' = \sqrt{y^2 - 1}$, $y(0) = 1$ has more than one solution in the interval $[0, 1]$.

10. Under the hypotheses of Theorem 2.31, show that any solution $y(x)$ satisfies

$$|y(x) - y_0| \leq M(\bar{x} - x_0) \text{ for } x \in [x_0, \bar{x}].$$

8.3 ANALYTIC METHODS

As indicated above, in most cases a closed-form solution of the differential equation (1.1) does not exist. Also, as indicated by the example (1.4), even when a closed-form solution does exist, its use may be less efficient than the use of numerical methods. In spite of this, every reasonable effort should be made to find a closed-form solution, if it exists, since useful information about the solution can often be obtained even if the actual evaluation of the solution is obtained by other methods. Methods for finding closed-form solutions are given in most texts on differential equations. A systematic approach to the question of whether or not a given differential equation can be solved in closed form is given by Cohen [1931]. An extensive list of solvable differential equations is given by Kamke [1948].

Infinite Series Solution: Direct Determination of Taylor's Series

As we have seen (Theorem 2.40), if $f(x, y)$ is analytic at (x_0, y_0), there exists a unique solution $y(x)$ of (1.11) which is analytic at x_0. Therefore, $y(x)$ has derivatives of all orders and

3.1 $$y(x) = y(x_0) + (x - x_0)y'(x_0) + \frac{(x - x_0)^2}{2} y''(x_0) + \cdots$$

in a neighborhood of x_0. Using (1.11), we can evaluate as many derivatives of $y(x)$ as we please. Indeed, since $y(x_0) = y_0$ we have

3.2 $$y'(x_0) = f(x_0, y_0)$$

3.3 $$y''(x_0) = [f_x(x, y) + y'f_y(x, y)]_{\substack{x = x_0 \\ y = y_0}}$$

$$= f_x(x_0, y_0) + f(x_0, y_0)f_y(x_0, y_0).$$

Higher derivatives can be evaluated in a similar way. In actual practice, however, one usually computes the derivatives recursively rather than computing the various partial derivatives. This is illustrated in the following example:

3.4 $$y' = x^2 + y^2, \qquad y(0) = 1.$$

By successive differentiation we have

3.5
$$y' = x^2 + y^2, \qquad\qquad y_0' = x_0^2 + y_0^2 = 1$$
$$y'' = 2x + 2yy', \qquad\qquad y_0'' = 2x_0 + 2y_0y_0' = 2$$
$$y''' = 2 + 2yy'' + 2y'^2, \qquad y_0''' = 2 + 2y_0y_0'' + 2y_0'^2 = 8$$
$$y^{(4)} = 2yy''' + 6y'y'', \qquad y_0^{(4)} = 2y_0y_0''' + 6y_0'y_0'' = 28.$$

Here $y_0 = y(x_0)$, $y_0' = y'(x_0)$, etc. Substituting in (3.1), we obtain

3.6 $$y(x) = 1 + x + 2\frac{x^2}{2!} + 8\frac{x^3}{3!} + 28\frac{x^4}{4!} + \cdots$$

$$= 1 + x + x^2 + \tfrac{4}{3}x^3 + \tfrac{7}{6}x^4 + \cdots.$$

For any suitable value of x we can evaluate y directly. For example, if $x = 0.1$, we have

3.7 $$y(0.1) = 1 + 0.1 + 0.01 + \tfrac{4}{3}(0.001) + \tfrac{7}{6}(0.0001) + \cdots \doteq 1.111.$$

On the other hand, if we let $x = 0.5$, then many more terms will be required before we can get three decimal places of accuracy. If we let $x = 1$, we find that the series does not converge.

We now show that the series converges in the interval $0 \leqq x \leqq \sqrt[3]{6}/2 \doteq 0.909$. To do this, consider the initial-value problem

3.8
$$\begin{cases} v' = \alpha v^2, & \alpha > 0 \\ v(0) = 1 \end{cases}$$

whose solution is

3.9
$$v = \frac{1}{1 - \alpha x} = 1 + \alpha x + \alpha^2 x^2 + \cdots,$$

the above series converging for $0 \leqq x < \alpha^{-1}$. Evidently, we have

3.10
$$\begin{aligned} v' &= \alpha v^2, & v'_0 &= \alpha \\ v'' &= \alpha(2vv'), & v''_0 &= 2\alpha^2 \\ v''' &= \alpha(2(v')^2 + 2vv''), & v'''_0 &= 6\alpha^3 \\ v^{(4)} &= \alpha(2vv''' + 6v'v''), & v_0^{(4)} &= 24\alpha^4 \end{aligned}$$

etc. Evidently each derivative of v will exceed the corresponding derivative of y, at $x = 0$, provided $6\alpha^3 \geq 8$, i.e., if $\alpha \geq 2/\sqrt[3]{6} \doteq 1.101$. Thus, since the series (3.9) for v converges for $0 \leqq x \leqq \sqrt[3]{6}/2$, it follows from Theorem A-2.15 that the series (3.6) also converges for the same range.

Actually, the series for $y(x)$ converges for a larger range in x. However, we can easily show that it does not converge if $x \geqq 1$. Indeed, consider the initial-value problem

3.11
$$\begin{cases} w' = w^2 \\ w(0) = 1. \end{cases}$$

One can easily verify that every derivative of $w(x)$ at $x = 0$ does not exceed the corresponding derivative of y. Hence, since

3.12
$$w = \frac{1}{1 - x} = 1 + x + x^2 + \cdots,$$

and since the series for $w(x)$ does not converge for $x \geq 1$, it follows from Theorem A-2.15 that the series (3.6) does not converge for $x \geqq 1$.

As another example, let us consider the initial-value problem

3.13
$$\begin{cases} y' = -2xy^2 \\ y(0) = 1. \end{cases}$$

Here the solution is

3.14
$$y = \frac{1}{1 + x^2} = 1 - x^2 + x^4 - x^6 + \cdots.$$

This series would be obtained if one used the Taylor series method described above. The radius of convergence of the series is 1. For any x' such that $0 < x' < 1$ we can use the series to get $y(x')$. Then we can use the Taylor series method with $x_0 = x'$ and get a new series which we could evaluate for some $x'' > x'$. Continuing in this way, we could obtain y for any desired value of x. This process is an example of "analytic continuation."

Infinite Series Solution: The Method of Undetermined Coefficients

The method of undetermined coefficients is sometimes easier to use than the Taylor series method. We write (3.1) in the form

3.15
$$y(x) = c_0 + c_1(x - x_0) + c_2(x - x_0)^2 + \cdots.$$

We seek to evaluate the c_k, not by computing $y^{(k)}(x_0)$, but rather by substituting in the differential equation and equating like powers of $(x - x_0)$. For example, let us again consider the example (3.4). For $x_0 = 0$ we have

3.16
$$y = c_0 + c_1 x + c_2 x^2 + c_3 x^3 + \cdots$$
$$y' = c_1 + 2c_2 x + 3c_3 x^2 + \cdots.$$

Substituting in (3.4) we get

3.17
$$c_1 + 2c_2 x + 3c_3 x^2 + 4c_4 x^3 + 5c_5 x^4 + \cdots = x^2 + (c_0 + c_1 x + c_2 x^2 + \cdots)^2$$
$$= c_0^2 + 2c_0 c_1 x + (1 + 2c_0 c_2 + c_1^2)x^2$$
$$+ (2c_0 c_3 + 2c_1 c_2)x^3 + (2c_0 c_4 + 2c_1 c_3 + c_2^2)x^4 + \cdots.$$

Equating coefficients of $1, x, x^2, x^3, \ldots$ we obtain

3.18 $\quad c_1 = c_0^2, \quad 2c_2 = 2c_0 c_1, \quad 3c_3 = 1 + 2c_0 c_2 + c_1^2, \quad 4c_4 = 2c_0 c_3 + 2c_1 c_2$

etc. Since $y(0) = 1$, then $c_0 = 1$ and we have

3.19
$$c_1 = 1, \qquad c_2 = 1, \qquad c_3 = \tfrac{4}{3}, \qquad c_4 = \tfrac{7}{6},$$

etc. Substituting in (3.15) we have

3.20
$$y = 1 + x + x^2 + \tfrac{4}{3}x^3 + \tfrac{7}{6}x^4 + \cdots$$

which agrees with the result obtained using the method of Taylor's series.

It is easy to see that both the Taylor's series method and the method of undetermined coefficients can be carried out for initial-value problems involving higher-order equations or systems of first-order equations. For a single higher-order equation of the form (1.2) one would be given $y_0, y_0', \ldots, y_0^{(r-1)}$. Hence $y_0^{(r)}$ could be evaluated. By differentiation of (1.2) one could get higher derivatives. The method of undetermined coefficients could also be carried out if $f(x, y, y', \ldots, y^{(r-1)})$ were analytic. For a system of first-order equations the extension of both procedures is obvious.

There are many disadvantages for methods based on infinite series for solving ordinary differential equations. In the first place, the further one wishes to proceed from the starting value x_0 the more terms, in general, will be required. In certain cases, one will have to proceed in several steps as shown in the example (3.13). If more than a few terms are required, the labor of repeated differentiation may be very great in the Taylor's series method, and the manipulations with power series may be very cumbersome in the method of undetermined coefficients. However, computer programs have been developed for the automatic manipulation of mathematical symbols (including symbolic differentiation and operations on power series) and such programs may eventually make the infinite series method feasible for general use.

EXERCISES 8.3

1. Find the first five terms of the power series expansion about $x = 0$ for the solution of $y' = -y$, $y(0) = 1$ using the method of Taylor's series. Obtain values of $y(0.1)$ and $y(0.2)$ correct to four decimal places.

2. Consider the initial-value problem $y' = x + y^2$, $y(0) = 1$. Find the first five terms of the power series expansion about $x = 0$ for the solution using the method of Taylor's series. Use the series to find $y(0.1)$, $y(0.2)$, $y(0.3)$, and $y(0.4)$. Also, with the value obtained at $x = 0.1$, use the Taylor's series expansion about $x = 0.1$ to obtain $y(0.2)$.

3. Use the method of Taylor's series to find the solution of $y' = x + y(1 + x)^{-1}$, $y(0) = 1$, correct to four decimal places for $x = 0(0.1)0.5$.

4. Find upper and lower bounds for the radius of convergence of the power series expansion of Exercise 2.

5. Use the method of undetermined coefficients to find the first five terms of the power series expansion about $x = 0$ of the solution of the initial-value problem of Exercise 2.

6. Solve the initial-value problem $y' = y^2$, $y(0) = 1$ by the method of undetermined coefficients and verify that one obtains the power series for the solution $y = (1 - x)^{-1}$.

7. By the method of Taylor's series and also by the method of undetermined coefficients obtain a power series solution for the initial-value problem $y'' + y = 0$, $y(0) = 1$, $y'(0) = 0$.

8. Consider the initial-value problem $y' = xy + y^2$, $y(0) = 1$. Find the first five terms of the Taylor's series expansion for $y(x)$ about $x = 0$ and evaluate $y(0.1)$ and $y(0.2)$.

9. Apply the method of Taylor's series to the initial-value problem $y'' = y' + xy^2$, $y(0) = 1$, $y'(0) = 2$, and evaluate the solution at $x = 0.2$.

10. By the method of Taylor's series and also by the method of undetermined coefficients obtain a power series solution to the initial-value problem $y' = z$, $z' = -y$, $y(0) = 1$, $z(0) = 0$.

8.4 INTEGRAL EQUATION FORMULATION—
THE PICARD METHOD OF SUCCESSIVE APPROXIMATIONS

The initial-value problem (1.11) can be formulated in terms of the integral equation

4.1
$$y(x) = y_0 + \int_{x_0}^{x} f(t, y(t))dt,$$

for, by the fundamental theorem of integral calculus, if $y(x)$ satisfies (1.11) then

4.2
$$y(x) - y(x_0) = \int_{x_0}^{x} y'(t)dt = \int_{x_0}^{x} f(t, y(t))dt$$

and (4.1) holds since $y(x_0) = y_0$. Conversely, if $y(x)$ satisfies (4.1) then (1.11) holds since $y(x_0) = y_0$ and since the derivative of the right member of (4.2) is $f(x, y(x))$.

We now consider the following special case of the Picard method of successive approximation defined by*

4.3
$$\begin{cases} y^{[0]}(x) = y_0 \\ y^{[k+1]}(x) = y_0 + \int_{x_0}^{x} f(t, y^{[k]}(t))dt, \qquad k = 0, 1, \ldots . \end{cases}$$

Suppose now that for some $\bar{x} > x_0$ and for some $M \geq |f(x_0, y_0)|$ we have

4.4 $|f(x, y)| \leq M$,   for $x_0 \leq x \leq \bar{x}$, $|y - y_0| \leq M(\bar{x} - x_0)$.

By Theorem 2.31 problem (1.11) has a solution defined for $x_0 \leq x \leq \bar{x}$. One can easily show that each $y^{[k]}(x)$ lies in the triangle bounded by the lines

4.5 $x = \bar{x}$, $y = y_0 \pm M(x - x_0)$

and hence certainly lies in the region $x_0 \leq x \leq \bar{x}$, $|y - y_0| \leq M(\bar{x} - x_0)$. Moreover, if $f(x, y)$ satisfies a Lipschitz condition for $x_0 \leq x \leq \bar{x}$, $|y - y_0| \leq M(\bar{x} - x_0)$, then the sequence $y^{[0]}(x)$, $y^{[1]}(x), \ldots$ converges to the unique solution. (See, for instance, Ford [1933].)

---

* Here $y^{[k]}(x)$ denotes the $(k + 1)$st function in the process (4.3) and should be distinguished from $y^{(k)}(x)$, the kth derivative of $y(x)$. In the general case of the Picard method $y^{[0]}(x)$ can be more general than the constant function. (See Ford [1933].)

Let us apply the Picard method to obtain an approximate solution of the initial-value problem (3.4). We let $y^{[0]}(x) = 1$ and we obtain

4.6
$$y^{[1]}(x) = 1 + \int_0^x (t^2 + 1)dt = 1 + x + \frac{x^3}{3}$$

and

4.7
$$y^{[2]}(x) = 1 + \int_0^x \left(t^2 + \left(1 + t + \frac{t^3}{3} \right)^2 \right) dt$$

$$= 1 + x + x^2 + \tfrac{2}{3}x^3 + \tfrac{1}{6}x^4 + \tfrac{2}{15}x^5 + \tfrac{1}{63}x^7.$$

This agrees with the result obtained using the method of Taylor's series up to the first three terms. Subsequent iterations would produce agreement to more terms.

Let us now consider a discrete analog of (4.1) obtained by evaluating the integral using the trapezoidal rule. We choose a positive integer N and we let

4.8
$$x_n = x_0 + nh, \qquad n = 0, 1, \ldots, N$$

where

4.9
$$h = \frac{\bar{x} - x_0}{N} = \frac{x_N - x_0}{N}.$$

We also let y_n represent an approximate value of the solution at $x = x_n$. Applying the trapezoidal rule to (4.1) we have

4.10
$$y_n = y_0 + h \left[\tfrac{1}{2}f(x_0, y_0) + \sum_{s=1}^{n-1} f(x_s, y_s) + \tfrac{1}{2}f(x_n, y_n) \right], \qquad n = 1, 2, \ldots, N.$$

Analogous to (4.3) we have

4.11
$$\begin{cases} y_n^{[0]} = y_0 \\ y_n^{[k+1]} = y_0 + h \left[\tfrac{1}{2}f(x_0, y_0) + \sum_{s=1}^{n-1} f(x_s, y_s^{[k]}) + \tfrac{1}{2}f(x_n, y_n^{[k]}) \right], \qquad \begin{matrix} n = 1, 2, \ldots, N, \\ k = 0, 1, 2, \ldots. \end{matrix} \end{cases}$$

We now show that (4.10) has a unique solution provided that $f(x, y)$ satisfies a Lipschitz condition in y for $x_0 \leq x \leq \bar{x}$, $\quad |y - y_0| \leq M(\bar{x} - x_0)$ and if

4.12
$$\frac{hL}{2} < 1.$$

Here L is any positive number such that

4.13 $|f(x, y) - f(x, z)| \leqq L|y - z|$

for any x, y, and z such that

4.14 $x_0 \leqq x \leqq \bar{x}$, $|y - y_0| \leqq M(\bar{x} - x_0)$, $|z - y_0| \leqq M(\bar{x} - x_0)$.

First, let us consider the equation of (4.10) corresponding to $n = 1$, namely,

4.15 $y_1 = y_0 + h[\tfrac{1}{2}f(x_0, y_0) + \tfrac{1}{2}f(x_1, y_1)] = \phi(y_1)$.

Let I be the interval

4.16 $y_0 - M(\bar{x} - x_0) \leqq y \leqq y_0 + M(\bar{x} - x_0)$.

Because of (4.12) it follows that if y and z belong to I, then

4.17 $|\phi(y) - \phi(z)| \leqq \mu|y - z|$

for some $\mu < 1$. Moreover, if $y \in I$ then $\phi(y) \in I$ since, by (4.15) and (4.4), we have

4.18 $|\phi(y) - y_0| \leqq Mh$.

Consequently, by Corollary 4-7.31, there exists a unique solution of (4.15). Moreover, the sequence defined by

4.19 $\begin{cases} y_1^{[0]} = y_0 \\ y_1^{[k+1]} = y_0 + \dfrac{h}{2}[f(x_0, y_0) + f(x_1, y_1^{[k]})], \qquad k = 0, 1, 2, \ldots \end{cases}$

converges to this solution. It is evident, then, that for the iterative method (4.11) the sequence $y_1^{[0]}, y_1^{[1]}, \ldots$ is the same as that defined by (4.19) and hence converges to y_1.

Having established the existence of y_1 we can write the equation of (4.10) corresponding to $n = 2$ in the form

4.20 $y_2 = y_0 + h[\tfrac{1}{2}f(x_0, y_0) + f(x_1, y_1) + \tfrac{1}{2}f(x_2, y_2)]$
              $= y_1 + h[\tfrac{1}{2}f(x_1, y_1) + \tfrac{1}{2}f(x_2, y_2)]$.

The existence of a unique solution y_2 of (4.20) can be shown in the same way as we showed that (4.15) has a unique solution. Continuing this process we can show that (4.10) has a unique solution.

Now let us show that the sequence $y_2^{[0]}, y_2^{[1]}, \ldots$ defined by the iterative process (4.11) converges to y_2. Evidently we have for $k = 0, 1, \ldots,$

4.21
$$\begin{cases} y_1^{[k+1]} = y_0 + h[\tfrac{1}{2}f(x_0, y_0) + \tfrac{1}{2}f(x_1, y_1^{[k]})] \\ y_2^{[k+1]} = y_0 + h[\tfrac{1}{2}f(x_0, y_0) + f(x_1, y_1^{[k]}) + \tfrac{1}{2}f(x_2, y_2^{[k]})] \end{cases}$$

and hence

4.22
$$\begin{cases} |y_1^{[k+1]} - y_1| \leqq \dfrac{hL}{2}|y_1^{[k]} - y_1| \\[2mm] |y_2^{[k+1]} - y_2| \leqq hL|y_1^{[k]} - y_1| + \dfrac{hL}{2}|y_2^{[k]} - y_2|. \end{cases}$$

It is easy to show by induction that

4.23
$$\begin{cases} |y_1^{[k]} - y_1| \leqq \left(\dfrac{hL}{2}\right)^k |y_1^{[0]} - y_1| \\[2mm] |y_2^{[k]} - y_2| \leqq \left(\dfrac{hL}{2}\right)^k |y_2^{[0]} - y_2| + \left(\dfrac{hL}{2}\right)^k 2k|y_1^{[0]} - y_1|. \end{cases}$$

Therefore, by (4.12), the convergence follows.

By similar methods we can show successively that for $n = 3, 4, \ldots, N$ we have

4.24
$$\lim_{k \to \infty} y_n^{[k]} = y_n.$$

As an example, let us consider the initial-value problem (3.4) with $h = 0.1$. We shall consider the range $0 \leqq x \leqq 0.5$. The following results were obtained using (4.11).

$k\backslash x$	0	0.1	0.2	0.3	0.4	0.5
0	1.000000	1.000000	1.000000	1.000000	1.000000	1.000000
1	1.000000	1.100500	1.203000	1.309500	1.422000	1.542500
2	1.000000	1.111055	1.246470	1.411070	1.610414	1.850984
3	1.000000	1.112222	1.254129	1.437869	1.679597	2.001076
4	1.000000	1.112352	1.255346	1.443861	1.700787	2.062554
5	1.000000	1.112366	1.255527	1.445059	1.706429	2.084270
6	1.000000	1.112368	1.255553	1.445281	1.707785	2.091089
7	1.000000	1.112368	1.255557	1.445319	1.708088	2.093047
8	1.000000	1.112368	1.255557	1.445326	1.708152	2.093572
9	1.000000	1.112368	1.255558	1.445327	1.708165	2.093706
10	1.000000	1.112368	1.255558	1.445327	1.708167	2.093739
11	1.000000	1.112368	1.255558	1.445327	1.708168	2.093746
12	1.000000	1.112368	1.255558	1.445327	1.708168	2.093748
13	1.000000	1.112368	1.255558	1.445327	1.708168	2.093749
14	1.000000	1.112368	1.255558	1.445327	1.708168	2.093749

We remark that the solution thus obtained is the same, to within rounding errors, as one would obtain by iterating over one step to get y_1, then iterating over two steps, which actually amounts to iterating over the second step, to get y_2, etc. This method is equivalent to the modified Euler method which is discussed in Section 8.6. However, the modified Euler method requires many fewer arithmetic operations.

If instead of the trapezoidal rule of integration we had used the rectangle rule, then, instead of (4.10), we would have the system

4.25 $$y_n = y_0 + h \sum_{s=0}^{n-1} f(x_s, y_s), \qquad n = 1, 2, \ldots, N.$$

If we used the iterative method

4.26
$$\begin{cases} y_n^{[0]} = y_0 \\ y_n^{[k+1]} = y_0 + h \sum_{s=0}^{n-1} f(x_s, y_s^{[k]}), \qquad k = 0, 1, 2, \ldots, \\ \qquad\qquad\qquad\qquad\qquad\qquad\qquad n = 1, 2, \ldots, N, \end{cases}$$

then the method would converge in precisely N steps. If we used the slightly different method

4.27
$$\begin{cases} y_n^{[0]} = y_0 \\ y_n^{[k+1]} = y_0 + h \sum_{s=0}^{n-1} f(x_s, y_s^{[k+1]}) \end{cases}$$

then the method would converge in one step. Actually, of course, it is more efficient to write (4.25) in the equivalent form

4.28 $$y_n = y_{n-1} + hf(x_{n-1}, y_{n-1}), \qquad n = 1, 2, \ldots, N.$$

The method then becomes equivalent to the Euler method which we consider in the next section.

EXERCISES 8.4

1. Verify that the function $y(x) = e^{-x}$ satisfies the integral equation (4.1) if $f(x, y) = -y$, $x_0 = 0$, and $y_0 = 1$.

2. Carry out the Picard method of successive approximation for the problem $y' = -y$, $y(0) = 1$, and verify that one obtains the solution $y(x) = e^{-x}$. Let $y^{[0]}(x) = 1$.

3. For the initial-value problem (3.4) find $y^{[3]}(x)$ using the Picard method where $y^{[0]}(x) \equiv 1$, and where $y^{[1]}(x)$ and $y^{[2]}(x)$ are given by (4.6) and (4.7), respectively.

4. Carry out two iterations of the Picard method for the initial-value problem $y' = x + y^2$, $y(0) = 1$.

5. Show that the sequence $y_3^{[0]}, y_3^{(1)}, \ldots$ defined by the iterative process (4.11) converges to y_3 provided (4.12) holds.

6. Prove (4.23).

7. For $k = 1, 2, 3$, verify the calculations given in the text for the problem (3.4) with $h = 0.1$ and for the range $0 \leq x \leq 0.5$. Also, iterate over one step to get $y(0.1)$, then over the second step to get $y(0.2)$, etc. Compare the work involved with the two procedures.

8.5 THE EULER METHOD

The Euler method is probably the simplest of all numerical methods for solving ordinary differential equations. Even though the method is seldom used in practice because of its poor accuracy, nevertheless, a study of its properties is instructive and serves as a basis for the analysis of more accurate methods.

Derivation of the Euler Method

One procedure for deriving the basic formula for the Euler method is to use the Taylor's series method of the previous section, with only the first two terms considered. Thus we have (read Section 9.2 of Appendix A)

5.1 $$y(x) \sim y(x_0) + (x - x_0)y'(x_0).$$

From (1.11) we obtain

5.2 $$y(x) \sim y_0 + (x - x_0)f(x_0, y_0).$$

Applying (5.2) for a value x_1 reasonably close to x_0 and letting $y_1 \doteq y(x_1)$ we obtain

5.3 $$y_1 = y_0 + (x_1 - x_0)f(x_0, y_0).$$

Geometrically, the point (x_1, y_1) lies on the tangent to the solution curve which passes through the point (x_0, y_0) (see Fig. 5.4). In general, the larger $x_1 - x_0$ the larger the deviation between the solution curve and the tangent line.

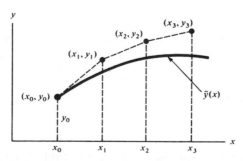

5.4 Fig. The Euler Method.

The geometric interpretation of the Euler method given above suggests an alternate derivation of (5.3) based on the direct use of the differential equation (5.1). We replace the derivative dy/dx by the forward difference quotient $(y_1 - y_0)/(x_1 - x_0)$ and equate to $f(x_0, y_0)$, obtaining

5.5
$$\frac{y_1 - y_0}{x_1 - x_0} = f(x_0, y_0)$$

which is equivalent to (5.3).

Having found y_1 by (5.3) or (5.5), we next determine y_2, an approximate value for $y(x_2)$, using (5.3), with x_0 replaced by x_1 and y_0 replaced by y_1. We have

5.6
$$y_2 = y_1 + (x_2 - x_1)f(x_1, y_1).$$

In general, for any set $x_0 < x_1 < x_2 < \cdots$ we can determine $y_1 \doteq y(x_1)$, $y_2 \doteq y(x_2), \ldots$ using

5.7
$$y_{n+1} = y_n + (x_{n+1} - x_n)f(x_n, y_n).$$

If, as is frequently the case, the abscissas x_i are chosen with equal spacing, we have $x_1 - x_0 = x_2 - x_1 = \cdots = h$, for some step size $h > 0$ and

5.8
$$y_{n+1} = y_n + hf(x_n, y_n), \qquad n = 0, 1, 2, \ldots.$$

If one connects consecutive points of the set $(x_0, y_0), (x_1, y_1), (x_2, y_2), \ldots$ by straight lines, one obtains a polygonal line (see Fig. 5.4). This polygonal line is frequently referred to as an "Euler polygon" or as an "Euler-Cauchy Polygon." The Euler polygons can be used to prove existence of a solution of (1.11) if $f(x, y)$ is continuous in the domain of interest. (See, for instance, Bieberbach [1930].) They can also be used to prove existence and uniqueness if $f(x, y)$ is continuous and satisfies a Lipschitz condition in y. (See, for instance, Henrici [1962].)

Numerical Example

Let us now apply the Euler method to the initial-value problem

5.9
$$y' = -y, \qquad y(0) = 1,$$

whose solution is

5.10
$$\bar{y}(x) = e^{-x}.$$

(From this point on we shall use $\bar{y}(x)$ for the *exact* solution in order to distinguish it from the *approximate* solutions under consideration.)

Euler's method (5.8) becomes

5.11 $y_{n+1} = y_n + h(-y_n) = (1 - h)y_n.$

The calculations up to $x = 0.4$, with $h = 0.1$, are shown below.

n	x_n	y_n	$-hy_n$	e^{-x_n}	$e^{-x_n} - y_n$
0	0	1.0000	−0.1000	1.0000	0
1	0.1	0.9000	−0.0900	0.9048	0.0048
2	0.2	0.8100	−0.0810	0.8187	0.0087
3	0.3	0.7290	−0.0729	0.7408	0.0118
4	0.4	0.6561	—	0.6703	0.0142

With $h = 0.2$, we have

n	x_n	y_n	$-hy_n$	e^{-x_n}	$e^{-x_n} - y_n$
0	0	1.0000	−0.2000	1.0000	0
1	0.2	0.8000	−0.1600	0.8187	0.0187
2	0.4	0.6400	—	0.6703	0.0303

Evidently the error increases with x. Moreover, the error is approximately twice as great with $h = 0.2$ as with $h = 0.1$. This suggests that the error may be approximately proportional to h. Later in this section we shall show that this is indeed the case.

Convergence and Accuracy of the Euler Method in a Special Case

We now consider the application of the Euler method to the problem

5.12
$$\begin{cases} y' = Ay \\ y(0) = y_0 \end{cases}$$

where A is a constant. Evidently the exact solution of (5.12) is given by

5.13 $\bar{y}(x) = y_0 e^{Ax}.$

We seek to show that the solution obtained using the Euler method converges to the exact solution (5.13) as the step size h tends to zero. We also seek a bound on the error involved in using the Euler method for a given h.

For given h we let $y_h(x)$ denote the solution obtained using the Euler method with step size h. We assume that for values of x such that x/h is not an integer we use linear interpolation. Thus, by (5.8), $y_h(x)$ is defined as follows

5.14
$$\begin{cases} y_h(0) = y_0 \\ y_h(x + h) = y_h(x) + hAy_h(x), \qquad \text{if } x/h \text{ is an integer,} \end{cases}$$

5.15 $y_h(x) = \dfrac{x_{n+1} - x}{h} y_h(x_n) + \dfrac{x - x_n}{h} y_h(x_{n+1})$,   if x/h is not an integer.

Here

5.16 $x_k = kh$, $k = 0, 1, 2, \ldots$

and n is the largest integer such that

5.17 $x_n \leqq x$.

By (5.14) we have $y_h(x_1) = (1 + Ah)y_0$, $y_h(x_2) = (1 + Ah)y_h(x_1) = (1 + Ah)^2 y_0$, and in general

5.18 $y_h(x_n) = (1 + Ah)^n y_0$, $n = 0, 1, 2, \ldots$.

We remark that if one evaluates (5.18) for any given n one obtains the same result, except for rounding errors, as one obtains by carrying out the Euler method for n steps. Of course, the analytic solution (5.18) of (5.14), though useful for theoretical purposes, is not of practical use for numerical calculation since the exact solution of the differential equation problem (5.12) is itself known explicitly.

The second equation of (5.14) can be written in the form

5.19 $\Delta y_h(x) = hAy_h(x)$

where $\Delta y_h(x) = y_h(x + h) - y_h(x)$ is the first (forward) difference of $y_h(x)$. Equations such as (5.19) which involve differences are often referred to as "difference equations." In fact, (5.19) is a linear difference equation with constant coefficients, and, like a linear differential equation with constant coefficients, it can be solved explicitly.

Before continuing our study of the convergence of the Euler method, let us examine the behavior of the solution (5.18) of the difference equation (5.14) for values of x tending to infinity such that x/h is an integer. In the case $A < 0$ if $|Ah| < 1$, then both the solution (5.13) of the differential equation and the solution (5.18) of the difference equation approach zero as $x \to \infty$. Similarly, if $A > 0$, the solution of the differential equation and the solution of the difference equation both approach infinity as $x \to \infty$ provided $y_0 > 0$. This similarity between the behavior of the solution of the differential equation and that of the difference equation is related to the *stability* of the Euler method and will be discussed in detail in Chapter 9.

We now prove

5.20 Lemma. If $|hA| < 1$, then for $n = 0, 1, 2, \ldots$ we have

5.21
$$|\bar{y}(x_n) - y_h(x_n)| \leq \begin{cases} \dfrac{h}{2} A^2 |y_0| x_n e^{Ax_n}, & \text{if } A \geq 0 \\[3mm] \dfrac{h}{2} A^2 |y_0| x_n e^{Ax_{n-1}}, & \text{if } A < 0. \end{cases}$$

Here $\bar{y}(x_n)$ and $y_h(x_n)$ are given by (5.13) and (5.18), respectively.

Proof. For convenience we let

5.22
$$\alpha = Ah.$$

Evidently we have

5.23
$$\bar{y}(x_n) - y_h(x_n) = [e^{Ax_n} - (1 + hA)^n]y_0$$
$$= [e^{\alpha n} - (1 + \alpha)^n]y_0.$$

By Taylor's theorem we have

5.24
$$e^{\alpha} = 1 + \alpha + \frac{\alpha^2}{2} e^{\xi}$$

where ξ lies between 0 and α. Thus we have

5.25
$$\begin{cases} 0 < e^{\alpha} - (1 + \alpha) < \dfrac{\alpha^2}{2} e^{\alpha}, & \text{if } \alpha \geq 0 \\[3mm] 0 < e^{\alpha} - (1 + \alpha) < \dfrac{\alpha^2}{2}, & \text{if } \alpha < 0. \end{cases}$$

Therefore since

5.26 $e^{\alpha n} - (1 + \alpha)^n = (e^{\alpha} - (1 + \alpha))(e^{(n-1)\alpha} + e^{(n-2)\alpha}(1 + \alpha) + \cdots + (1 + \alpha)^{n-1})$

it follows that

5.27
$$\begin{cases} 0 < e^{\alpha n} - (1 + \alpha)^n < n\dfrac{\alpha^2}{2} e^{n\alpha}, & \text{if } \alpha \geq 0 \\[3mm] 0 < e^{\alpha n} - (1 + \alpha)^n < n\dfrac{\alpha^2}{2} e^{(n-1)\alpha}, & \text{if } \alpha < 0. \end{cases}$$

The result (5.21) follows from (5.23) and (5.27).

We are now in a position to prove

5.28 Theorem. If $|hA| < 1$ and if

5.29 $x \in [x_n, x_{n+1}]$

then

5.30 $|\bar{y}(x) \div y_h(x)| \leq \begin{cases} \dfrac{h}{2} A^2 |y_0| e^{A x_{n+1}} \left\{ x_{n+1} + \dfrac{h}{4} \right\}, & \text{if } A \geqq 0 \\[3mm] \dfrac{h}{2} A^2 |y_0| e^{A x_{n-1}} \cdot \left\{ x_{n+1} + \dfrac{h}{4} e^{Ah} \right\}, & \text{if } A < 0. \end{cases}$

Proof. By (5.15) it follows that

5.31 $|\bar{y}(x) - y_h(x)| \leqq \max \Big(|\bar{y}(x_n) - y_h(x_n)|, \ |\bar{y}(x_{n+1}) - y_h(x_{n+1})| \Big) + E$

where E is the error in linear interpolation for $\bar{y}(x)$ in the points x_n and x_{n+1}, i.e., E is given by

5.32 $E = \bar{y}(x) - \left[\dfrac{x_{n+1} - x}{h} \bar{y}(x_n) + \dfrac{x - x_n}{h} \bar{y}(x_{n+1}) \right].$

By Theorem 6-3.9, we have, since $\bar{y}''(x) = A^2 y_0 e^{Ax}$,

5.33 $|E| \leqq \begin{cases} \dfrac{h^2}{8} A^2 |y_0| e^{A x_{n+1}}, & \text{if } A \geqq 0 \\[3mm] \dfrac{h^2}{8} A^2 |y_0| e^{A x_n}, & \text{if } A < 0. \end{cases}$

Evidently, by (5.21),

5.34 $\max \Big(|\bar{y}(x_n) - y_h(x_n)|, \ |\bar{y}(x_{n+1}) - y_h(x_{n+1})| \Big)$

$\leqq \begin{cases} \dfrac{h}{2} A^2 |y_0| x_{n+1} e^{A x_{n+1}}, & \text{if } A \geqq 0 \\[3mm] \dfrac{h}{2} A^2 |y_0| x_{n+1} e^{A x_{n-1}}. & \text{if } A < 0. \end{cases}$

The result (5.30) now follows from (5.31), (5.34), and (5.33).

Let us now apply the error bound (5.21) to the example given earlier in this section. With $h = 0.1$, $A = -1$, $y_0 = 1$, and $x = 0.4$, we have

5.35 $|\bar{y}(0.4) - y_h(0.4)| \leqq \dfrac{h}{2}(0.4)e^{-(0.3)} = (0.02)e^{-0.3} \doteq 0.014816$

which agrees closely with the actual error of 0.0142.

Suppose now that $x = 0.35$. Then

5.36 $$y_h(0.35) = \tfrac{1}{2}(y_h(0.3) + y_h(0.4)) \doteq 0.6926,$$

and $\bar{y}(0.35) \doteq 0.7047$. Thus the actual error is

5.37 $$\bar{y}(0.35) - y_h(0.35) \doteq 0.0121.$$

From (5.30) we have, since $x_n = 0.3$,

5.38 $$|\bar{y}(0.35) - y_h(0.35)| \leq (0.05)e^{-0.2}(0.4 + (0.025)(e^{-0.1}))$$
$$\doteq 0.0173.$$

We now show that a result analogous to Theorem 5.28 can be proved for a much more general class of problems.

5.39 Theorem. Let $f(x, y)$ satisfy all of the hypotheses of Theorem 2.31 including the Lipschitz condition and let $\bar{y}''$ be continuous for $x_0 \leq x \leq \bar{x}$ where $\bar{y}(x)$ is the unique solution of (1.11). Let $y_h(x)$ be the solution obtained by the Euler method as defined by (5.14) and (5.15) with $hAy_h(x)$ replaced by $hf(x, y_h(x))$. If $x_{n+1} \in [x_0, \bar{x}]$ and if $x \in [x_n, x_{n+1}]$ where

5.40 $$x_n = x_0 + nh, \qquad n = 0, 1, 2, \ldots$$

then we have

5.41 $$|\bar{y}(x) - y_h(x)| \leq \frac{h}{2} M_2 \left\{ \frac{e^{L(x_{n+1} - x_0)} - 1}{L} + \frac{h}{4} \right\}.$$

Here

5.42 $$M_2 = \max_{x_0 \leq x \leq \bar{x}} |\bar{y}''(x)|$$

and L is any positive* constant such that

5.43 $$|f(x, y) - f(x, z)| \leq L|y - z|$$

for any two points (x, y) and (x, z) with the same abscissa in the region $x_0 \leq x \leq \bar{x}$, $|y - y_0| \leq M(\bar{x} - x_0)$.

Proof. Let

5.44 $$y_n = y_h(x_n), \qquad \bar{y}_n = \bar{y}(x_n), \qquad n = 0, 1, 2, \ldots .$$

---

* If (5.43) holds with $L = 0$ then the right member of (5.41) can be replaced by
$$(h/2)M_2((x_{n+1} - x_0) + h/4).$$

Evidently we have for $n = 0, 1, 2, \ldots$

5.45
$$y_{n+1} = y_n + hf(x_n, y_n)$$

and

5.46
$$\bar{y}_{n+1} = \bar{y}_n + hf(x_n, \bar{y}_n) + E$$

where

5.47
$$|E| \leqq \frac{h^2}{2} M_2.$$

This follows since, by Taylor's theorem, we have

5.48
$$\bar{y}_{n+1} - \bar{y}_n - hf(x_n, \bar{y}_n) = \bar{y}(x_n + h) - \bar{y}(x_n) - h\bar{y}'(x_n)$$
$$= \frac{h^2}{2} \bar{y}''(c)$$

for some c with $x_n < c < x_{n+1}$. By (5.45), (5.46), and (5.43) we have

5.49
$$|e_{n+1}| \leqq (1 + hL)|e_n| + |E|$$

where we let

5.50
$$e_n = \bar{y}_n - y_n, \qquad n = 0, 1, 2, \ldots .$$

It is easy to verify that

5.51
$$|e_n| \leqq \frac{(1 + hL)^n - 1}{hL} |E|.$$

Since $1 + hL \leqq e^{hL}$ it follows that

5.52
$$|e_n| \leqq \frac{e^{L(x_n - x_0)} - 1}{hL} |E| \leqq \frac{(e^{L(x_n - x_0)} - 1)}{L} \frac{h}{2} M_2.$$

The rest of the proof is analogous to the proof of Theorem 5.28 where we note that the error in linear interpolation for $\bar{y}(x)$ is bounded by $h^2 M_2/8$.

We remark that in order that $\bar{y}''(x)$ be continuous in $x_0 \leqq x \leqq \bar{x}$ it is clearly sufficient for $f_x(x, y)$ and $f_y(x, y)$ to exist and be continuous for

5.53
$$x_0 \leqq x \leqq \bar{x}, \qquad |y - y_0| \leqq M(\bar{x} - x_0).$$

This follows since

5.54
$$\bar{y}''(x) = f_x(x, \bar{y}(x)) + \bar{y}'(x) f_y(x, \bar{y}(x)).$$

From Theorem 5.39 it follows that for all x we have

5.55
$$\lim_{h \to 0} y_h(x) = \bar{y}(x).$$

Moreover, for fixed x, the difference

5.56
$$e_h(x) = \bar{y}(x) - y_h(x)$$

tends to zero like the first power of the step size h. Unfortunately, this means that in order to reduce the error involved in using a step size h by a factor of m, one would have to reduce h by a factor of m. Because of this, the Euler method is very seldom used in practice. In later sections we shall consider methods where the error is proportional to h^p for some integer $p > 1$. With such methods much greater accuracy can usually be achieved for a given step size than with the Euler method.

EXERCISES 8.5

1. Find $y(0.1)$ and $y(0.2)$ for the problem (3.4) using the Euler method with $h = 0.1$. Find the error for the two values by comparing these values with the results obtained by the methods of Section 8.3.

2. In the preceding problem obtain $y(0.05)$, $y(0.1)$, $y(0.15)$, and $y(0.2)$ by the Euler method with $h = 0.05$. Compare the errors at $x = 0.1$ and $x = 0.2$ with those corresponding to the case $h = 0.1$.

3. Carry out the calculations of Exercises 1 and 2 for the problem $y' = x + y^2$, $y(0) = 1$.

4. Verify (5.27) for the cases $n = 5$, $\alpha = \frac{1}{2}$, and $n = 5$, $\alpha = -\frac{1}{2}$.

5. Verify (5.51).

6. Apply Theorem 5.28 to the case $y' = -y$, $y(0) = 1$, $h = 0.1$, $x = 0.4$.

7. Apply Theorem 5.39 to the example $y' = -y$, $y(0) = 1$, $h = 0.1$, $x = 0.4$. Also consider the case $x = 0.35$.

8. Apply Theorem 5.39 to the problem (3.4) with $h = 0.1$ and $x = 0.2$. Compare your result with the exact value of $|\bar{y}(0.2) - y_h(0.2)|$.

9. Apply Theorem 5.39 to the problem $y' = x + y^2$, $y(0) = 1$, with $h = 0.1$ and $x = 0.2$. Compare with the exact value of $|\bar{y}(0.2) - y_h(0.2)|$.

10. Write out the details of the proof of Theorem 5.39.

8.6 METHODS BASED ON NUMERICAL QUADRATURE

In the previous section we have seen that the Euler method, though extremely simple, is too inaccurate to be of practical use. In this section we show that greater accuracy can be obtained using methods based on numerical quadrature formulas.

We assume throughout that we are seeking approximate values $y_1, y_2, \ldots$ for $\bar{y}(x_1), \bar{y}(x_2), \ldots$, respectively, where $\bar{y}(x)$ is the exact solution of (1.11) and where

6.1
$$x_n = x_0 + nh, \qquad n = 0, 1, 2, \ldots \; .$$

A method based on numerical quadrature can be described by the procedure for finding y_{n+1}, given $y_n, y_{n-1}, \ldots$. (Normally, the procedure used to find the first few values $y_1, y_2, \ldots, y_s$ for some s is different from the general procedure.)

The Euler method can be derived by using the rectangle rule to evaluate the integral

6.2
$$\bar{y}(x_{n+1}) - \bar{y}(x_n) = \int_{x_n}^{x_{n+1}} f(t, \bar{y}(t))dt = \int_{x_n}^{x_{n+1}} g(t)dt$$

where, as before, we let

6.3
$$g(t) = f(t, \bar{y}(t)).$$

From the rectangle rule we have

6.4
$$\int_{x_n}^{x_{n+1}} g(t)dt \sim hg(x_n) = hf(x_n, \bar{y}(x_n)).$$

The formula

6.5
$$y_{n+1} = y_n + hf(x_n, y_n)$$

for the Euler method is obtained from (6.2) by replacing $\bar{y}(x_{n+1})$ by y_{n+1}, $\bar{y}(x_n)$ by y_n and by representing the integral by $hf(x_n, y_n)$.

The Modified Euler Method

The modified Euler method is based on the trapezoidal rule where we represent the integral in (6.2) by

6.6
$$\frac{h}{2}[g(x_n) + g(x_{n+1})] = \frac{h}{2}[f(x_n, \bar{y}(x_n)) + f(x_{n+1}, \bar{y}(x_{n+1}))].$$

if we replace $\bar{y}(x_{n+1})$ and $\bar{y}(x_n)$ in (6.2) by y_{n+1} and y_n, respectively, and if we represent the integral by

6.7
$$\frac{h}{2}[f(x_n, y_n) + f(x_{n+1}, y_{n+1})]$$

we get

6.8
$$y_{n+1} = y_n + \frac{h}{2}[f(x_n, y_n) + f(x_{n+1}, y_{n+1})]$$

which is the formula for the modified Euler method.

In general, (6.8) defines a nonlinear equation in which y_{n+1} appears implicitly. In the special case where $f(x, y)$ is a linear function of y of the form

6.9
$$f(x, y) = A(x)y + B(x)$$

we can solve (6.8) explicitly for y_{n+1} obtaining

6.10
$$y_{n+1} = \frac{y_n + \frac{h}{2}[A(x_n)y_n + B(x_n) + B(x_{n+1})]}{1 - \frac{h}{2}A(x_{n+1})}.$$

As in Section 8.4, we can show that if $f(x, y)$ satisfies a Lipschitz condition in y in the region of interest and if h is sufficiently small, then (6.8) has a unique solution. Moreover, this solution can be found by the iterative process.

6.11
$$\begin{cases} y_{n+1}^{[0]} = y_n \\ y_{n+1}^{[k+1]} = y_n + \frac{h}{2}[f(x_n, y_n) + f(x_{n+1}, y_{n+1}^{[k]})], \quad k = 0, 1, \dots, \\ \qquad\qquad n = 0, 1, \dots. \end{cases}$$

The reader should show that convergence also holds if we let $y_{n+1}^{[0]}$ be determined by the Euler method, i.e., if we let

6.12
$$y_{n+1}^{[0]} = y_n + hf(x_n, y_n).$$

Let us illustrate this by the example

6.13
$$\begin{cases} y' = -y + x + \frac{1}{2} \\ y(0) = 1. \end{cases}$$

We seek to determine $y_1 = y(0.1)$ by the modified Euler method with step size $h = 0.1$.

We first note that since $f(x, y)$ is a linear function of y we could solve (6.8) explicitly for y_1 obtaining, by (6.10),

6.14 $$y_1 = \frac{1 + (0.05)((-1)(1) + 0.5 + 0.6)}{1 - (0.05)(-1)} = \frac{1 + 0.005}{1 + 0.05} = \frac{1.005}{1.05} \doteq 0.95714.$$

On the other hand, with the iterative procedure (6.11) we have $y_1^{[0]} = 1.0$. Subsequent calculations are given by the following table:

n	$y_1^{[k]}$	$f(0, y_0)$	$f(0.1, y_1^{[k]})$
0	1.0000	−0.500	−0.400
1	0.955	−0.500	−0.355
2	0.95725	−0.500	−0.35725
3	0.95714	−0.500	−0.35714
4	0.95714	—	—

Thus after 4 iterations the value of $y_1^{[k]}$ agrees with the exact value to five decimal places.

Using methods described in the next section, one can show that there is no appreciable loss of accuracy if, instead of iterating (6.11) until convergence is reached, we simply iterate three times. If we use the starting value $y_{n+1}^{[0]}$ given by (6.12) we need iterate only twice. With a still better estimate of $y_{n+1}^{[0]}$, described later in this section, we need iterate only once.

We now define the *Heun method* and the *corrected Heun method* as follows. The Heun method is defined by

6.15
$$\begin{cases} y_{n+1}^{(P)} = y_n + hf(x_n, y_n) \\ y_{n+1} = y_n + \dfrac{h}{2}[f(x_n, y_n) + f(x_{n+1}, y_{n+1}^{(P)})]. \end{cases}$$

The corrected Heun method is defined by

6.16
$$\begin{cases} y_{n+1}^{(P)} = y_n + hf(x_n, y_n) \\ y_{n+1}^{(C)} = y_n + \dfrac{h}{2}[f(x_n, y_n) + f(x_{n+1}, y_{n+1}^{(P)})] \\ y_{n+1} = y_n + \dfrac{h}{2}[f(x_n, y_n) + f(x_{n+1}, y_{n+1}^{(C)})]. \end{cases}$$

In example (6.13) we have

6.17
$$\begin{cases} y_1^{(P)} = y_0 + h[-y_0 + x_0 + \tfrac{1}{2}] = 0.950 \\ y_1 = y_0 + \dfrac{h}{2}[(-y_0 + x_0 + \tfrac{1}{2}) + (-y_1^{(P)} + x_1 + \tfrac{1}{2})] \\ = 1 - 0.0425 = 0.9575 \end{cases}$$

for the Heun method. For the corrected Heun method we have $y_1^{(C)} = 0.9575$ and

6.18 $y_1 = y_0 + \dfrac{h}{2}[(-y_0 + x_0 + \frac{1}{2}) + (-y_1^{(C)} + x_1 + \frac{1}{2})] \doteq 0.95713.$

Subsequent iterations would yield

6.19 $0.95715, \qquad 0.95714, \qquad 0.95714, \ldots .$

We note that $y_1 = 0.95714$ is the solution correct to 5 decimal places of equation (6.8), with $n = 0$. Evidently the value given by the corrected Heun method is extremely close to this value.

The Midpoint Method

The modified Euler method is said to be a *closed method* since the approximate expression for the integral in (6.2) involves $g(x_{n+1})$. The associated quadrature formula is closed in this case. If such is not the case, then we have an *open method*. Thus the Euler method which is based on the rectangle rule, an open quadrature formula, is an open method. Another open method is based on the midpoint rule of integration. If we write

6.20 $\bar{y}(x_{n+1}) - \bar{y}(x_{n-1}) = \displaystyle\int_{x_{n-1}}^{x_{n+1}} f(t, y(t))dt = \int_{x_{n-1}}^{x_{n+1}} g(t)dt,$

then we represent the integral by

6.21 $\displaystyle\int_{x_{n-1}}^{x_{n+1}} g(t)dt \sim 2hg(x_n) = 2hf(x_n, \bar{y}(x_n)).$

If we replace $\bar{y}(x_{n+1})$ and $\bar{y}(x_{n-1})$ by y_{n+1} and y_{n-1}, respectively, and if we replace the integral by $2hf(x_n, y_n)$ we obtain the formula for the midpoint method

6.22 $y_{n+1} = y_{n-1} + 2hf(x_n, y_n), \qquad n = 1, 2, \ldots .$

Evidently (6.22) cannot be used to find y_1. One could, however, determine y_1 by a different procedure such as the Euler method or the modified Euler method. Thus, for example, suppose we wish to solve (6.13) by the midpoint method with the step size $h = 0.1$. We use the modified Euler method and compute, as above,

6.23 $y_1 = y(0.1) \doteq 0.95714.$

Using (6.22) we get

6.24 $y_2 = y_0 + 2hf(0.1, y_1) \doteq 1 + 2(0.1)(-0.35714)$

$= 0.928572.$

The midpoint method is easier to apply than the modified Euler method but has the following disadvantages: first, the quadrature error, while of the same order in h, is four times as great; second, a special starting procedure is required; third, the method often suffers from numerical instability, as will be shown in Chapter 9.

Other Methods Based on Numerical Quadrature

We now consider methods based on the use of the quadrature formulas considered in Section 7.4. Given the integers q, M, and m with $q > 0$, $M \geqq m \geqq 0$ we define the method $Q(m).q.M$ by

6.25
$$y_{n+1} - y_{n+1-q} = h \sum_{i=m}^{M} \beta_i f(x_{n+1-i}, y_{n+1-i}).$$

In this chapter we shall assume that the β_i are the same as those in the numerical quadrature formula $Q.q.M.m$ given by 7-(4.49). However, the "associated quadrature formula," referred to below, for method $Q(m).q.M.$ is not $Q.q.M.m$.

For a *closed* method based on numerical quadrature we have $m = 0$, and we designate the method by $QC.q.M$. Such methods are defined by

6.26 $$y_{n+1} - y_{n+1-q} = h\beta_0 f(x_{n+1}, y_{n+1}) + h \sum_{i=1}^{M} \beta_i f(x_{n+1-i}, y_{n+1-i}).$$

The most important class of *open* methods corresponds to the case $m = 1$. Such methods are defined by (6.26) with $\beta_0 = 0$. We designate the method $Q(1).q.M$ by $QO.q.M$.

For the methods considered so far, we have the following designations:

the Euler method   QO.1.1

the modified Euler method   QC.1.1

the midpoint method   QO.2.1

(The midpoint method is also QO.2.2 and QC.2.1.)

We define the *order* of a method based on numerical quadrature as the largest integer p such that

6.27 $$(R[y])_n = y_{n+1} - y_{n+1-q} - h \sum_{i=0}^{M} \beta_i y'_{n+1-i} = 0$$

for any polynomial $y(x)$ of degree p or less. The *quadrature error* $(QE[y])_n$ corresponding to a given function $y(x)$ is defined by

6.28 $$(QE[y])_n = (R[(x - \alpha)^{p+1}])_n \frac{y^{(p+1)}(x_n)}{(p+1)!}.$$

Here α can be chosen for convenience, the quadrature error being independent of α.

For the Euler method, we note that (6.27) holds for $y = 1$ and for $y = x - x_n$ but not for $y = (x - x_n)^2$. Since

6.29 $$(R[(x - x_n)^2])_n = h^2$$

we have

6.30 $$(QE[y])_n = \frac{h^2}{2} y''(x_n).$$

It is easy to show that the order of a method based on numerical quadrature is one higher than the order of the associated quadrature formula. We assume that the latter is definite (see Section 7.7) so that if its order is s, then its error is

6.31 $$Kh^{s+2}g^{(s+1)}(\xi)$$

for some constant K and for some ξ in the interval $x_{n+1-N} < x < x_{n+1}$, where $N = \max(M, q)$ for the method QC.q.M or QO.q.M. Here $g(x)$ is the function being integrated. It can easily be shown that

6.32 $$(QE[y])_n = Kh^{p+1}y^{(p+1)}(x_n)$$

where $p = s + 1$ is the order of the method (6.25).

From the results of Chapter 7 on numerical quadrature we have

6.33 Theorem. The order of method QO.q.M is at least M. The order of method QC.q.M is at least $M + 1$.

For example, let us consider the Euler method. The order of accuracy of the rectangle rule

6.34 $$\int_{x_n}^{x_{n+1}} g(x)dx \sim hg(x_n)$$

is $s = 0$ and by Table 7-4.51*

6.35 $$\int_{x_n}^{x_{n+1}} g(x)dx - hg(x_n) = \tfrac{1}{2}h^2g'(\xi)$$

for some $\xi \in (x_n, x_{n+1})$. Therefore the order of the Euler method is $p = 1$ and the quadrature error is given by (6.30).

---

* Note that even though the coefficients β_i for the Euler method are the same as those for the backward rectangle rule Q.1.1.1, the error (6.31) corresponds to that of the (forward) rectangle rule Q.1.0.0. In fact, Q.1.0.0 is the "associated quadrature formula" for method Q.0.1.1.

Tables 6.36 and 6.37 give the coefficients for several open and closed methods, respectively, based on numerical quadrature. In each case the quadrature error is given. The argument for the derivative of $y(x)$ is omitted. The formulas are the same as those of Table 7-4.51 for the same q, M, and m except that for the quadrature error $f^{(p)}$ is replaced by $(-1)^p y^{(p+1)}$ in each case.

6.36 Table   Table of Coefficients for Open Methods Based on Numerical Quadrature.

For the q-step open method QO.q.M involving the M values $f(x_i, y_i)$, $i = n, n-1, \ldots,$ $n - (M - 1)$ we have

$$y_{n+1} - y_{n+1-q} = h \sum_{i=1}^{M} \beta_i f(x_{n+1-i}, y_{n+1-i}).$$

For the case $M = 4$ the coefficients β_i refer to the points x_{n-i+1} as indicated.

$$
\begin{array}{ccccc}
\beta_4 & \beta_3 & \beta_2 & \beta_1 & \\
\hline
x_{n-3} & x_{n-2} & x_{n-1} & x_n & x_{n+1}
\end{array}
$$

M	β_1	β_2	β_3	β_4	Quadrature Error	Name	Symbol
Single-step $(q = 1)$	1	1	—	—	—	$\frac{1}{2}h^2 y^{(2)}$	Euler	QO.1.1
$y_{n+1} - y_n$	2	$\frac{3}{2}$	$-\frac{1}{2}$	—	—	$\frac{5}{12}h^3 y^{(3)}$	Predictor for improved Heun	QO.1.2
	3	$\frac{23}{12}$	$-\frac{16}{12}$	$\frac{5}{12}$	—	$\frac{3}{8}h^4 y^{(4)}$	—	QO.1.3
	4	$\frac{55}{24}$	$-\frac{59}{24}$	$\frac{37}{24}$	$-\frac{9}{24}$	$\frac{251}{720}h^5 y^{(5)}$	Adams-Moulton Predictor	QO.1.4
Double-step $(q = 2)$	1	2	—	—	—	$\frac{1}{3}h^3 y^{(3)}$	Midpoint	QO.2.1
$y_{n+1} - y_{n-1}$	2	2	0	—	—	$\frac{1}{3}h^3 y^{(3)}$	Midpoint	QO.2.2
	3	$\frac{7}{3}$	$-\frac{2}{3}$	$\frac{1}{3}$	—	$\frac{1}{3}h^4 y^{(4)}$	—	QO.2.3
	4	$\frac{8}{3}$	$-\frac{5}{3}$	$\frac{4}{3}$	$-\frac{1}{3}$	$\frac{29}{90}h^5 y^{(5)}$	—	QO.2.4
Three-step $(q = 3)$	2	$\frac{3}{2}$	$\frac{3}{2}$	—	—	$\frac{3}{4}h^3 y^{(3)}$	—	QO.3.2
$y_{n+1} - y_{n-2}$	3	$\frac{9}{4}$	0	$\frac{3}{4}$	—	$\frac{3}{8}h^4 y^{(4)}$	—	QO.3.3
	4	$\frac{21}{8}$	$-\frac{9}{8}$	$\frac{15}{8}$	$-\frac{3}{8}$	$\frac{27}{80}h^5 y^{(5)}$	—	QO.3.4
Four-step $(q = 4)$	3	$\frac{8}{3}$	$-\frac{4}{3}$	$\frac{8}{3}$	—	$\frac{14}{45}h^5 y^{(5)}$	Milne-Simpson Predictor	QO.4.3
$y_{n+1} - y_{n-3}$	4	$\frac{8}{3}$	$-\frac{4}{3}$	$\frac{8}{3}$	0	$\frac{14}{45}h^5 y^{(5)}$	Milne-Simpson Predictor	QO.4.4

As we shall see in Section 8.7, one can (without any serious loss of accuracy) obtain y_{n+1} for a closed method by first determining $y_{n+1}^{[0]}$ by an open method which has the same order as the closed method and then applying the closed formula once. The value $y_{n+1}^{[0]}$ is called the "predicted value" and is denoted by $y_{n+1}^{(P)}$. The value obtained using the closed method is called the "corrected value" and is sometimes denoted by $y_{n+1}^{(C)}$.

6.37 Table Table of Coefficients for Closed Methods Based on Numerical Quadrature.
For the q-step closed methods QC.q.M involving the $M + 1$ values $f(x_i, y_i), i = n + 1, n, \ldots,$
$n - (M - 1)$ we have

$$y_{n+1} - y_{n+1-q} = h \sum_{i=0}^{M} \beta_i f(x_{n+1-i}, y_{n+1-i}).$$

For the case $M = 4$ the coefficients β_i refer to the points x_{n-i+1} as indicated.

$$
\begin{array}{ccccc}
\beta_4 & \beta_3 & \beta_2 & \beta_1 & \beta_0 \\
x_{n-3} & x_{n-2} & x_{n-1} & x_n & x_{n+1}
\end{array}
$$

	M	β_0	β_1	β_2	β_3	β_4	Quadrature Error	Name	Symbol
Single-step	0	1	—	—	—	—	$-\frac{1}{2}h^2 y^{(2)}$	Backward Euler	QC.1.0
($q = 1$)	1	$\frac{1}{2}$	$\frac{1}{2}$	—	—	—	$-\frac{1}{12}h^3 y^{(3)}$	Modified Euler	QC.1.1
$y_{n+1} - y_n$	2	$\frac{5}{12}$	$\frac{8}{12}$	$-\frac{1}{12}$	—	—	$-\frac{1}{24}h^4 y^{(4)}$	Method Based on Five-eight Rule	QC.1.2
	3	$\frac{9}{24}$	$\frac{19}{24}$	$-\frac{5}{24}$	$\frac{1}{24}$	—	$-\frac{19}{720}h^5 y^{(5)}$	Adams-Moulton Corrector	QC.1.3
Double-step	0	2	—	—	—	—	$-2h^2 y^{(2)}$	—	QC.2.0
($q = 2$)	1	0	2	—	—	—	$\frac{1}{3}h^3 y^{(3)}$	Midpoint Method	QC.2.1
$y_{n+1} - y_{n-1}$	2	$\frac{1}{3}$	$\frac{4}{3}$	$\frac{1}{3}$	—	—	$-\frac{1}{90}h^5 y^{(5)}$	Milne-Simpson Corrector	QC.2.2
	3	$\frac{1}{3}$	$\frac{4}{3}$	$\frac{1}{3}$	0	—	$-\frac{1}{90}h^5 y^{(5)}$	Milne-Simpson Corrector	QC.2.3
Three-step	1	$-\frac{3}{2}$	$\frac{9}{2}$	—	—	—	$\frac{9}{4}h^3 y^{(3)}$	—	QC.3.1
($q = 3$)	2	$\frac{3}{4}$	0	$\frac{9}{4}$	—	—	$-\frac{3}{8}h^4 y^{(4)}$	—	QC.3.2
$y_{n+1} - y_{n-2}$	3	$\frac{3}{8}$	$\frac{9}{8}$	$\frac{9}{8}$	$\frac{3}{8}$	—	$-\frac{3}{80}h^5 y^{(5)}$	Method Based on Simpson's Second Rule	QC.3.3
Four-step	2	$\frac{8}{3}$	$-\frac{16}{3}$	$\frac{20}{3}$	—	—	$-\frac{8}{3}h^4 y^{(4)}$	—	QC.4.2
($q = 4$)	3	0	$\frac{8}{3}$	$-\frac{4}{3}$	$\frac{8}{3}$	—	$\frac{14}{45}h^5 y^{(5)}$	Milne-Simpson Predictor	QC.4.3
$y_{n+1} - y_{n-3}$	4	$\frac{28}{90}$	$\frac{128}{90}$	$\frac{48}{90}$	$\frac{128}{90}$	$\frac{28}{90}$	$-\frac{8}{945}h^7 y^{(7)}$	—	QC.4.4

The Heun method given by (6.15) uses QO.1.1 as a predictor and QC.1.1 as a corrector. (The corrected Heun method (6.16) is the same as the Heun method except that there is one additional application of the corrector.) We now consider two additional variants of the Heun method, namely the *improved Heun method* and the *Heun-midpoint* method. The improved Heun method is based on QO.1.2 as the predictor and the modified Euler method QC.1.1 as the corrector. The formulas are

6.38
$$
\begin{cases}
y_{n+1}^{(P)} = y_n + h[\frac{3}{2}f(x_n, y_n) - \frac{1}{2}f(x_{n-1}, y_{n-1})] \\
y_{n+1} = y_n + h[\frac{1}{2}f(x_{n+1}, y_{n+1}^{(P)}) + \frac{1}{2}f(x_n, y_n)]
\end{cases}
$$

The Heun-midpoint method uses the midpoint method QO.2.1 as the predictor and the modified Euler method QC.1.1 as the corrector. The formulas are

6.39
$$\begin{cases} y_{n+1}^{(P)} = y_{n-1} + 2hf(x_n, y_n) \\ y_{n+1} = y_n + h[\tfrac{1}{2}f(x_{n+1}, y_{n+1}^{(P)}) + \tfrac{1}{2}f(x_n, y_n)]. \end{cases}$$

For both of these methods a special starting procedure must be used to get y_1.

Suppose, in example (6.13), that the value $y_1 \doteq 0.95713$ has been obtained using the corrected Heun method. The value of y_2 is, by (6.10), given by

6.40
$$y_2 = \frac{y_1 + (0.05)[-y_1 + x_1 + x_2 + 1]}{1.05} \doteq 0.92788.$$

For the improved Heun method we have

6.41
$$\begin{cases} y_2^{(P)} = 0.95713 + (0.1)[\tfrac{3}{2}(-0.35713) - \tfrac{1}{2}(-0.50000)] \doteq 0.92856 \\ y_2 = 0.95713 + (0.1)[\tfrac{1}{2}(-0.35713) + \tfrac{1}{2}(-0.22856)] \doteq 0.92785. \end{cases}$$

Subsequent iterations would yield $0.92788, 0.92788, \ldots$. For the Heun-midpoint method we have

6.42
$$\begin{cases} y_2^{(P)} = 1.00000 + (0.2)(-0.35713) \doteq 0.92857 \\ y_2 = 0.95713 + (0.1)[\tfrac{1}{2}(-0.35713) + \tfrac{1}{2}(-0.22857)] \doteq 0.92785. \end{cases}$$

Subsequent iterations would yield $0.92788, 0.92788, \ldots$.

The improved Heun method is one of a family of predictor-corrector methods. More accurate methods include QO.1.3/QC.1.2 and the Adams-Moulton method QO.1.4/QC.1.3 given by

6.43
$$\begin{cases} y_{n+1}^{(P)} = y_n + \dfrac{h}{24}[55f(x_n, y_n) - 59f(x_{n-1}, y_{n-1}) + 37f(x_{n-2}, y_{n-2}) \\ \qquad\qquad\qquad - 9f(x_{n-3}, y_{n-3})] \\ y_{n+1} = y_n + \dfrac{h}{24}[9f(x_{n+1}, y_{n+1}^{(P)}) + 19f(x_n, y_n) \\ \qquad\qquad\qquad - 5f(x_{n-1}, y_{n-1}) + f(x_{n-2}, y_{n-2})]. \end{cases}$$

Evidently to use the Adams-Moulton method one must determine y_1, y_2, and y_3 by a different procedure.

One can also consider a family of methods based on two-step quadrature formulas. It turns out that QC.2.1 is the midpoint method and hence is actually open as well as closed. The next method in the family is QC.2.2 which corresponds to Simpson's rule. The corresponding open method, namely QO.2.3, has a quadrature error of $O(h^4)$ which is less than that of QC.2.2. Thus, if we wish to avoid more than one application of the corrector we would have to use as a predictor

QO.1.4, QO.2.4, QO.3.4, or QO.4.3 (which is the same as QO.4.4). The latter method is attractive in that it corresponds to the smallest value of M. The method QO.4.3/QC.2.2 is known as the "Milne-Simpson method" and is defined by

6.44
$$\begin{cases} y_{n+1}^{(P)} = y_{n-3} + \frac{4h}{3}[2f(x_n, y_n) - f(x_{n-1}, y_{n-1}) + 2f(x_{n-2}, y_{n-2})] \\ y_{n+1} = y_{n-1} + \frac{h}{3}[f(x_{n+1}, y_{n+1}^{(P)}) + 4f(x_n, y_n) + f(x_{n-1}, y_{n-1})]. \end{cases}$$

Evidently to use the Milne-Simpson method one must determine y_1, y_2, and y_3 by a different procedure as in the case of the Adams-Moulton method.

An inspection of the quadrature errors for QC.2.2 and QC.1.3 reveals that the error of the Milne-Simpson method is less than one-half that of the Adams-Moulton method. Moreover, the method involves fewer calculations. However, as we shall see in Chapter 9, the Milne-Simpson method is subject to a mild form of instability which sometimes is manifested by the occurrence of wild oscillations in the numerical solution. This difficulty, which does not exist with the Adams-Moulton method, can usually be controlled, but nevertheless it does make the Milne-Simpson method less attractive than it might otherwise be.

EXERCISES 8.6

1. Consider the modified Euler method as applied to the initial-value problem $y' = x + e^x y$, $y(0) = 1$, with $h = 0.1$. Find $y(0.1)$ exactly.

2. For the initial-value problem $y' = xy + y^2$, $y(0) = 1$, determine $y(0.2)$ using two steps of the Heun method with $h = 0.1$ and also using one step of the Heun method with $h = 0.2$. Compute the error at $x = 0.2$ in each case. (See Exercise 8, Section 8.3.) Comment on the behavior of the error as $h \to 0$.

3. Consider the initial-value problem $y' = x + y^2$, $y(0) = 1$.

   a) Find an exact value of $y(0.1)$ as determined by the modified Euler method with $h = 0.1$.
   b) Find $y(0.1)$ using the iterative process (6.11) with $y_1^{[0]}$ determined by the Euler method. Identify the values produced by the Heun method and by the corrected Heun method.
   c) Determine $y(0.2)$ using $h = 0.1$ and also using $h = 0.05$. Use the Heun-midpoint method for all steps except the first where the corrected Heun method should be used. Determine the error in $y(0.2)$ in each case (see Exercise 2, Section 8.3 for the exact solution) and verify that it is approximately proportioned to h^2.
   d) With $y(0.1)$ determined by the corrected Heun method with $h = 0.1$, find $y(0.2)$ and $y(0.3)$ by the improved Heun method with $h = 0.1$.
   e) Using exact values for $y(0.1)$, $y(0.2)$, and $y(0.3)$, find $y(0.4)$ by the Milne-Simpson method. Also find $y(0.4)$ by the Adams-Moulton method and compare the errors. (See Exercise 2, Section 8.3, for the exact values.)

4. Consider the method (6.25) which is based on the numerical quadrature formula
$$\int_{x_{n+1-q}}^{x_{n+1}} g(x)dx \sim h \sum_{i=0}^{M} \beta_i g(x_{n+1-i}).$$

Let s be the order of the quadrature formula and let K be the constant such that for any sufficiently differentiable function $g(x)$ we have

$$\int_{x_{n+1-q}}^{x_{n+1}} g(x)dx - h \sum_{i=0}^{M} \beta_i g(x_{n+1-i}) = Kh^{s+2}g^{(s+1)}(\xi)$$

for some ξ in the interval $x_{n+1-q} \leqq x \leqq x_{n+1}$. Show that the quadrature error for (6.25) is of order $p = s + 1$ and has quadrature error

$$(QE[y])_n = Kh^{p+1}y^{(p+1)}(x_n).$$

5. Verify the formulas for the quadrature errors given in Tables 6.36 and 6.37 for the following methods: QO.1.2, QO.1.4, QO.2.3, QO.3.3, QO.4.3, QC.1.2, QC.1.3, QC.3.3, and QC.4.2.
6. Relate the quadrature errors given in Tables 6.36 and 6.37 to the errors given in Table 7-4.51.

8.7 ERROR ESTIMATION FOR PREDICTOR-CORRECTOR METHODS

Throughout this discussion we assume that we are interested in finding an approximate solution of (1.11) in an interval $[x_0, \bar{x}]$ for some $\bar{x} > x_0$ where for some $M \geqq |f(x_0, y_0)|$ we have

7.1 $|f(x, y)| \leqq M$

in $R: x_0 \leqq x \leqq \bar{x}, |y - y_0| \leqq M(\bar{x} - x_0)$. We also assume that sufficiently many partial derivatives of $f(x, y)$ exist and are continuous in R so that all appropriate derivatives of $\bar{y}(x)$ are continuous on $[x_0, \bar{x}]$. As a matter of fact, for any solution $y(x)$ of the differential equation, the corresponding derivatives are bounded for values of x for which $(x, y(x))$ is in R.

We shall let

7.2 $F_{i,j} = \max \left| \dfrac{\partial^{i+j} f(x, y)}{\partial x^i \partial y^j} \right|, \qquad i, j = 0, 1, \ldots$

for $(x, y) \in R$ and

7.3 $M_k = \max_{x_0 \leqq x \leqq \bar{x}} |\bar{y}^{(k)}(x)|, \qquad k = 0, 1, \ldots,$

where $\bar{y}(x)$ is the exact solution of (1.11). Evidently $f(x, y)$ satisfies a Lipschitz condition in y with

7.4 $L = F_{0,1}.$

Moreover, we have

7.5 $\begin{cases} M_1 = F_{0,0} \\ M_2 \leqq F_{1,0} + F_{0,0}F_{0,1}. \end{cases}$

Similar bounds can be given for M_3, M_4, etc. These bounds are also valid for any solution $y(x)$ of the differential equation if we let

7.6
$$M_k[y] = \max_{x \in S} |y^{(k)}(x)|$$

where S is the set of all values of x such that $(x, y(x)) \in R$.

We now discuss the accuracy of various predictor-corrector methods based on numerical quadrature. We shall show that if the predictor is sufficiently good, the result obtained using the predictor and one application of the corrector is essentially as accurate as that obtained by solving the corrector equation exactly for y_{n+1}. Thus, the accuracy of such a predictor-corrector method depends on the accuracy of the corrector alone.

We have already, in Section 8.6, defined the *order* of a method based on a numerical quadrature formula and the *quadrature error*. Normally, we shall be interested in the quadrature error corresponding to the exact solution, $\bar{y}(x)$, of (1.11), which is given by

7.7
$$(QE[\bar{y}])_n.$$

Usually be shall designate $(QE[\bar{y}])_n$ by $(QE)_n$ or QE when no confusion will arise.

Single-step Error

The *single-step error* can be defined for a somewhat wider class of methods than those based on numerical quadrature. We define the single-step error in a method as

7.8
$$(SSE[\bar{y}])_n = \bar{y}_{n+1} - \tilde{y}_{n+1}$$

where $\bar{y}_{n+1}$ refers to the exact solution of (1.11) and $\tilde{y}_{n+1}$ is defined to be the value of y_{n+1} produced by the method if the values $\bar{y}_n, \bar{y}_{n-1}, \dots$ were used instead of $y_n, y_{n-1}, \dots$. Thus, for instance, with the improved Heun method (6.38) we have

7.9
$$\begin{cases} \tilde{y}_{n+1}^{(P)} = \bar{y}_n + h[\tfrac{3}{2}f(x_n, \bar{y}_n) - \tfrac{1}{2}f(x_{n-1}, \bar{y}_{n-1})] \\ \tilde{y}_{n+1} = \bar{y}_n + h[\tfrac{1}{2}f(x_{n+1}, \tilde{y}_{n+1}^{(P)}) + \tfrac{1}{2}f(x_n, \bar{y}_n)]. \end{cases}$$

Normally we shall designate $(SSE[\bar{y}])_n$ by SSE_n or SSE when no confusion will arise.

Let us now determine the single-step error for various variants of the Heun method. For the Heun method itself we have

7.10
$$\begin{cases} \tilde{y}_{n+1}^{(P)} = \bar{y}_n + hf(x_n, \bar{y}_n) \\ \tilde{y}_{n+1} = \bar{y}_n + \dfrac{h}{2}[f(x_n, \bar{y}_n) + f(x_{n+1}, \tilde{y}_{n+1}^{(P)})]. \end{cases}$$

Moreover,

7.11
$$\bar{y}_{n+1} = \bar{y}_n + hf(x_n, \bar{y}_n) + QE^* + O(h^3)$$

and

7.12
$$\bar{y}_{n+1} = \bar{y}_n + \frac{h}{2}[f(x_n, \bar{y}_n) + f(x_{n+1}, \bar{y}_{n+1})] + QE + O(h^4).$$

Here QE^* and QE refer to the quadrature errors for the predictor (the Euler method) and the corrector (the modified Euler method), respectively, i.e.,

7.13
$$QE^* = \frac{h^2}{2}\bar{y}''(x_n), \quad QE = -\frac{h^3}{12}\bar{y}^{(3)}(x_n).$$

We have assumed that $\bar{y}^{(4)}(x)$ exists and is continuous and hence bounded in the interval of interest. From this it follows that, if $\xi \in (x_n, x_{n+1})$, then

7.14
$$\begin{cases} \dfrac{h^2}{2}\bar{y}''(\xi) = \dfrac{h^2}{2}\bar{y}''(x_n) + O(h^3) \\[2mm] -\dfrac{h^3}{12}\bar{y}^{(3)}(\xi) = -\dfrac{h^3}{12}y'''(x_n) + O(h^4). \end{cases}$$

Evidently, we have

7.15
$$|\bar{y}_{n+1} - \tilde{y}_{n+1}^{(P)} - QE^*| = O(h^3)$$

and

7.16
$$|\bar{y}_{n+1} - \tilde{y}_{n+1} - QE| \leqq \frac{hL}{2}|\bar{y}_{n+1} - \tilde{y}_{n+1}^{(P)}| + O(h^4)$$

$$\leqq \frac{hL}{2}|QE^*| + O(h^4).$$

Thus

7.17
$$|\bar{y}_{n+1} - \tilde{y}_{n+1}| \leqq |QE| + \frac{hL}{2}|QE^*| + O(h^4)$$

$$\leqq \frac{h^3}{12}M_3 + \frac{h^3 L}{4}M_2 + O(h^4).$$

If we were to solve the corrector equation exactly so that

7.18
$$\tilde{y}_{n+1} = \bar{y}_n + \frac{h}{2}[f(x_n, \bar{y}_n) + f(x_{n+1}, \tilde{y}_{n+1})]$$

then we would have

7.19
$$|\bar{y}_{n+1} - \tilde{y}_{n+1} - QE| \leq \frac{hL}{2}|\bar{y}_{n+1} - \tilde{y}_{n+1}| + O(h^4)$$

and

7.20
$$\bar{y}_{n+1} - \tilde{y}_{n+1} = QE + O(h^4) = -\frac{h^3}{12}\bar{y}'''(x_n) + O(h^4).$$

Thus, the single-step error for the Heun method is of the same order as that of the modified Euler method.

The reader should show that with the corrected Heun method we have

7.21
$$\bar{y}_{n+1} - \tilde{y}_{n+1} = QE + O(h^4).$$

He should also show that this result holds for the improved Heun and for the Heun-midpoint method. More generally, he should show that if the predictor and the corrector are of the same order, say p, then the single-step error of the corresponding predictor-corrector method is equal to the quadrature error of the corrector to within $O(h^{p+2})$.

For the improved Heun method, it can be shown that

7.22
$$\bar{y}_{n+1} - \tilde{y}_{n+1} = QE + O(h^4)$$
$$= -\tfrac{1}{12}h^3\bar{y}^{(3)}(x_n) + O(h^4)$$

and

7.23
$$\bar{y}_{n+1} - \tilde{y}^{(P)}_{n+1} = QE^* + O(h^4)$$
$$= \tfrac{5}{12}h^3\bar{y}^{(3)}(x_n) + O(h^4).$$

Therefore, we have, upon elimination of $\bar{y}^{(3)}(x_n)$

7.24
$$\bar{y}_{n+1} = \tfrac{5}{6}\tilde{y}_{n+1} + \tfrac{1}{6}\tilde{y}^{(P)}_{n+1} + O(h^4)$$

and

7.25
$$\bar{y}_{n+1} - \tilde{y}_{n+1} = -\tfrac{1}{6}(\tilde{y}_{n+1} - \tilde{y}^{(P)}_{n+1}) + O(h^4).$$

Thus we can estimate the single-step error in terms of the difference between the predicted and the corrected values $\tilde{y}^{(P)}_{n+1}$ and $\tilde{y}_{n+1}$. However, in practice, one seldom has $\tilde{y}_{n+1}$ and $\tilde{y}_{n+1}^{(P)}$ available, but, rather, one has y_{n+1} and $y_{n+1}^{(P)}$. It seems reasonable to use the approximate formula

7.26 $$\bar{y}_{n+1} - \tilde{y}_{n+1} \sim -\tfrac{1}{6}(y_{n+1} - y_{n+1}^{(P)}).$$

A similar analysis for the Heun-midpoint method gives

7.27 $$\bar{y}_{n+1} - \tilde{y}_{n+1} = -\tfrac{1}{5}(\tilde{y}_{n+1} - \tilde{y}_{n+1}^{(P)}) + O(h^4)$$

and we use the approximate formula

7.28 $$\bar{y}_{n+1} - \tilde{y}_{n+1} \sim -\tfrac{1}{5}(y_{n+1} - y_{n+1}^{(P)}).$$

Accumulated Error

We now consider the *accumulated error* which we define by

7.29 $$e_n = \bar{y}_n - y_n, \qquad n = 0, 1, 2, \ldots.$$

We prove

7.30 Theorem. Let $y_1, y_2, \ldots, y_{N^*-1}$ be chosen so that

7.31 $$|\bar{y}_n - y_n| \leqq \delta, \qquad n = 1, 2, \ldots, N^* - 1$$

and for $n = N^* - 1, N^*, N^* + 1, \ldots$ let y_{n+1} be determined by the predictor-corrector method

7.32
$$
\begin{cases}
y_{n+1}^{(P)} = y_{n+1-q^*} + h \sum_{i=1}^{M^*} \beta_i^* f(x_{n+1-i}, y_{n+1-i}) \\[2mm]
y_{n+1} = y_{n+1-q} + h \sum_{i=1}^{M} \beta_i f(x_{n+1-i}, y_{n+1-i}) + h\beta_0 f(x_{n+1}, y_{n+1}^{(P)}).
\end{cases}
$$

Here we let

7.33 $$N = \max(q, M), \qquad N^* = \max(q^*, M^*).$$

We assume $N^* \geqq N$. Let the order of the predictor and of the corrector be p and let the quadrature errors for the predictor and the corrector be

7.34 $$(QE^*)_n = K^* h^{p+1} \bar{y}^{(p+1)}(x_n)$$

and

7.35 $$(QE)_n = K h^{p+1} \bar{y}^{(p+1)}(x_n),$$

respectively. Then for $n = 0, 1, 2, \ldots, (\bar{x} - x_0)/h$, we have

7.36 $|e_n| \leq \left\{ \delta + \dfrac{(x_n - x_0)}{q} h^p M_{p+1}[|K| + Lh|\beta_0|K^*|] \right\} e^{L(x_n - x_0)(B + hL|\beta_0|B^*)}$

where L is the Lipschitz constant for $f(x, y)$ and

7.37 $B = \displaystyle\sum_{i=0}^{M} |\beta_i|, \qquad B^* = \sum_{i=1}^{M^*} |\beta_i^*|.$

Proof. We give a proof for the Milne-Simpson method where we have

7.38 $\begin{cases} q = 2, \ M = 2, \ N = 2, \ \beta_0 = \frac{1}{3}, \ \beta_1 = \frac{4}{3}, \ \beta_2 = \frac{1}{3}, \ B = 2 \\ p = 4, \ K = -\frac{1}{90}, \ QE = -\frac{1}{90}h^5 y^{(5)} \end{cases}$

7.39 $\begin{cases} q^* = 4, \ M^* = 3, \ N^* = 4, \ \beta_1^* = \frac{8}{3}, \ \beta_2^* = -\frac{4}{3}, \ \beta_3^* = \frac{8}{3}, \ B^* = \frac{20}{3} \\ p^* = 4, \ K^* = \frac{14}{45}, \ QE^* = \frac{14}{45}h^5 y^{(5)}. \end{cases}$

We seek to show that

7.40 $|e_n| \leq \left\{ \delta + \dfrac{x_n - x_0}{2} h^4 M_5 [\frac{1}{90} + \frac{14}{135}hL] \right\} e^{L(x_n - x_0)(2 + (20/9)hL)}.$

For $n = 3, 4, \ldots$ we have

7.41 $\begin{cases} y_{n+1}^{(P)} = y_{n-3} + h\{\frac{8}{3}f(x_n, y_n) - \frac{4}{3}f(x_{n-1}, y_{n-1}) + \frac{8}{3}f(x_{n-2}, y_{n-2})\} \\ y_{n+1} = y_{n-1} + h\{\frac{4}{3}f(x_n, y_n) + \frac{1}{3}f(x_{n-1}, y_{n-1})\} + \dfrac{h}{3}f(x_{n+1}, y_{n+1}^{(P)}) \end{cases}$

and

7.42 $\begin{cases} \bar{y}_{n+1} = \bar{y}_{n-3} + h\{\frac{8}{3}f(x_n, \bar{y}_n) - \frac{4}{3}f(x_{n-1}, \bar{y}_{n-1}) + \frac{8}{3}f(x_{n-2}, \bar{y}_{n-2})\} \\ \qquad\qquad + \frac{14}{45}h^5\bar{y}^{(5)}(\xi^*) \\ \bar{y}_{n+1} = \bar{y}_{n-1} + h\{\frac{4}{3}f(x_n, \bar{y}_n) + \frac{1}{3}f(x_{n-1}, \bar{y}_{n-1})\} + \dfrac{h}{3}f(x_{n+1}, \bar{y}_{n+1}) \\ \qquad\qquad - \frac{1}{90}h^5\bar{y}^{(5)}(\xi) \end{cases}$

for some ξ^* and ξ in the interval (x_{n-3}, x_{n+1}). Therefore,

7.43 $|e_{n+1}| \leq |e_{n-1}| + hL\{\frac{4}{3}|e_n| + \frac{1}{3}|e_{n-1}|\} + \dfrac{hL}{3}|\bar{y}_{n+1} - y_{n+1}^{(P)}| + \overline{QE}$

and

7.44 $|\bar{y}_{n+1} - y_{n+1}^{(P)}| \leq |e_{n-3}| + hL\{\frac{8}{3}|e_n| + \frac{4}{3}|e_{n-1}| + \frac{8}{3}|e_{n-2}|\} + \overline{QE^*}$

where

7.45 $\overline{QE^*} = (\frac{14}{45})h^5 M_5$ and $\overline{QE} = (\frac{1}{90})h^5 M_5$.

Thus we have

7.46 $|e_{n+1}| - |e_{n-1}| \leq |e_n|\{hL[\frac{4}{3} + \frac{8}{9}hL]\} + |e_{n-1}|\{hL[\frac{1}{3} + \frac{4}{9}hL]\}$

$$+ |e_{n-2}|\{hL(\frac{8}{9}hL)\} + |e_{n-3}|\left\{\frac{hL}{3}\right\} + \left\{\overline{QE} + \frac{hL}{3}\overline{QE^*}\right\}.$$

Similar inequalities can be derived for

7.47 $|e_{n-1}| - |e_{n-3}|, \qquad |e_{n-3}| - |e_{n-5}|, \qquad \cdots, \qquad |e_{n+1-2s}| - |e_{n+1-2(s+1)}|$

where $n - 2s \geq 3, n - 2s < 5$. Summing these inequalities and replacing $n + 1$ by n we have

7.48 $$|e_n| \leq Q \sum_{k=0}^{n-1} |e_k| + R$$

for $n = 4, 5, \ldots$, where

7.49 $$Q = hL(B + \frac{1}{3}B^*hL) = hL(2 + \frac{20}{9}hL)$$

and where

7.50 $$R = \delta + \left[\overline{QE} + \frac{hL}{3}\overline{QE^*}\right]\left(\frac{n}{2}\right)$$

$$= \delta + \frac{(x_n - x_0)}{2}h^4 M_5[\frac{1}{90} + \frac{14}{135}hL].$$

We now prove

7.51 Lemma. If Q and R are nonnegative numbers, if $r \geq 1$, and if

7.52 $$|e_n| \leq Q \sum_{k=0}^{n-1} |e_k| + R$$

for $n = r, r + 1, \ldots$, then for $n = r, r + 1, \ldots$, we have

7.53
$$|e_n| \leq \left(R + Q \sum_{k=0}^{r-1} |e_k|\right)(1 + Q)^{n-r}$$

$$\leq \left(R + Q \sum_{k=0}^{r-1} |e_k|\right) e^{Q(n-r)}.$$

Proof. We use mathematical induction. If $n = r$, then by (7.52) we have

7.54
$$|e_r| \leq Q \sum_{k=0}^{r-1} |e_k| + R$$

so that (7.53) holds. Suppose (7.53) is true, with n replaced by j, for $j = r, r + 1, \ldots, n$. Then by (7.52) we have

7.55
$$|e_{n+1}| \leq Q \sum_{k=0}^{n} |e_k| + R = \left(Q \sum_{k=0}^{r-1} |e_k| + R\right) + Q \sum_{k=r}^{n} |e_k|$$

$$\leq \left(Q \sum_{k=0}^{r-1} |e_k| + R\right)\left\{1 + Q \sum_{k=r}^{n} (1 + Q)^{k-r}\right\}.$$

But if $Q > 0$, then

7.56 $1 + Q \sum_{k=r}^{n} (1 + Q)^{n-r} = 1 + Q \dfrac{1 - (1 + Q)^{n+1-r}}{(-Q)} = (1 + Q)^{n+1-r}$

and hence (7.53) holds for $n + 1$ for the case $Q > 0$. (If $Q = 0$ the result (7.53) is obviously true.) Since $1 + x \leq e^x$ for any $x \geq 0$, the lemma follows.

Since (7.48) holds for $n = 0, 1, \ldots$, and since $e_0 = 0$ we can apply Lemma 7.51 with $r = 1$ and obtain

7.57
$$|e_n| \leq Re^{Q(n-1)} \leq Re^{Qn}.$$

The result (7.40) follows from (7.49) and (7.50). This completes the proof of Theorem 7.1 for the case of the Milne-Simpson method. The reader should work out the details of the proof for the general case.

To illustrate the application of Theorem 7.30, let us consider the example

7.58
$$\begin{cases} y' = -y \\ y(0) = 1. \end{cases}$$

Suppose that the Milne-Simpson method is used with $h = 0.1$ and that we desire a bound on $e_{10} = \bar{y}_{10} - y_{10}$. We assume that exact values of y_1, y_2, y_3 have been obtained. Evidently since $\bar{y}(x) = e^{-x}$ we have

7.59
$$M_5 = \max_{0 \leq x \leq 1} |e^{-x}| = 1.$$

Moreover, $L = 1$ and $\delta = 0$. Therefore,

7.60
$$|e_{10}| \leq \tfrac{1}{2}h^4\left\{\frac{1}{90} + \frac{h}{3}\left(\frac{14}{45}\right)\right\}e^{(2 + (h/3)(20/3))}$$

$$= \frac{(0.1)^4}{2}\left\{\frac{1}{90} + \frac{1.4}{135}\right\}e^{(20/9)} \doteq 0.0000099.$$

EXERCISES 8.7

1. Under the assumptions stated at the beginning of Section 8.7, find a bound for M_3 in terms of the $F_{i,j}$. For the problem

$$y' = x + y^2$$

$$y(0) = 1$$

and for the region $R : 0 \leq x \leq 0.2, |y - 1| \leq 0.5$ find $F_{0,0}, F_{1,0}, F_{0,1}, F_{2,0}, F_{1,1}, F_{0,2}$, and also M_1, M_2, M_3.

2. Consider the predictor-corrector method QO.1.2/QC.2.2. How many applications of the corrector are necessary so that the order of the single-step error is of the same order as though the corrector equation were solved exactly? How many applications are necessary so that the single-step error is the same, to within $0(h^6)$, as that of the corrector?

3. Prove that (7.21) holds for the corrected Heun, improved Heun, and Heun-midpoint methods.

4. Show that if the predictor and the corrector are of the same order, say p, then the single-step error of the corresponding predictor-corrector method is equal to the quadrature error of the corrector to within $0(h^{p+2})$.

5. Find the single-step error for the Adams-Moulton method and for the Milne-Simpson method.

6. Find the single-step error for the method involving the Euler method as a predictor followed by one correction with method QC.1.1 followed by one correction with the Adams-Moulton corrector.

7. Apply Theorem 7.30 to the following methods:

Predictor	Corrector
QO.1.2	QC.1.1
QO.2.1	QC.1.1
QO.1.4	QC.1.3
QO.1.3	QC.1.2
QO.3.4	QC.3.3

Give numerical values for $\bar{y}_{10} - y_{10}$ for the case $h = 0.1$ and for the example (7.58). Assume that the starting values are exact.

8. Work out the analog of Theorem 7.30 when the corrector equation is solved exactly.

9. Prove Theorem 7.30 in the general case.

8.8 A NUMERICAL EXAMPLE

Let us again consider the example

8.1
$$\begin{cases} y' = -y \\ y(0) = 1 \end{cases}$$

whose solution is

8.2
$$\bar{y}(x) = e^{-x}.$$

We shall apply several of the methods described above for two steps with $h = 0.1$. We have already given a theoretical analysis for some of the methods for this problem in Section 8.6.

From the tables we have

8.3
$$\begin{cases} \bar{y}_1 = \bar{y}(x_1) \doteq 0.904837 \\ \bar{y}_2 = \bar{y}(x_2) \doteq 0.818731. \end{cases}$$

For the Euler method we have

n	x_n	$\bar{y}_n$	y_n	$e_n = \bar{y}_n - y_n$	$\tilde{y}_n$	$(SSE)_n =$ $\bar{y}_{n+1} - \tilde{y}_{n+1}$
0	0	1.0	1.0	0	1.0	0.004837
1	0.1	0.904837	0.9	0.004837	0.9	0.004378
2	0.2	0.818731	0.81	0.008731	0.814353	—

The sum of the single-step errors is

8.4
$$(SSE)_0 + (SSE)_1 \doteq 0.009215$$

which agrees closely with

8.5
$$e_2 = \bar{y}_2 - y_2 \doteq 0.008731.$$

We note that the estimate

8.6
$$(SSE)_n \sim \frac{h^2}{2} y''(x_n) = \frac{h^2}{2} e^{-x_n}$$

is reasonably close since

8.7
$$\begin{cases} (SSE)_0 \sim \dfrac{h^2}{2} e^0 = 0.005 \\[2mm] (SSE)_1 \sim \dfrac{h^2}{2} e^{-0.1} \doteq (0.005)(0.904837) \doteq 0.00452. \end{cases}$$

For the Heun method we have

n	x_n	$\bar{y}_n$	$y_n^{(P)}$	y_n	$\bar{y}_n - y_n$	$\tilde{y}_n^{(P)}$	$\tilde{y}_n$	$(SSE)_n =$ $\bar{y}_{n+1} - \tilde{y}_{n+1}$
0	0	1.0	—	1.0	0	—	1.0	−0.000163
1	0.1	0.904837	0.9	0.905	−0.000163	0.9	0.905	−0.000146
2	0.2	0.818731	0.8145	0.819025	−0.000294	0.814353	0.818877	—

The sum of the single-step errors is

8.8 $$(SSE)_0 + (SSE)_1 \doteq -0.000309$$

which agrees closely with

8.9 $$\bar{y}_2 - y_2 \doteq -0.000294$$

We note that the estimate

8.10 $$(SSE)_n \sim -\frac{h^3}{12} y'''(x_n) + \frac{h}{2} f_y(x_n, y_n)(SSE^*)_n$$

$$\sim -\frac{h^3}{12} y'''(x_n) - \frac{h^3}{4} y''(x_n)$$

$$= -\frac{h^3}{6} e^{-x_n}$$

for the single-step error is reasonably good since

8.11
$$\begin{cases} (SSE)_0 \sim -\dfrac{h^3}{6} e^0 \doteq -0.000167 \\[3mm] (SSE)_1 \sim -\dfrac{h^3}{6} e^{-0.1} \doteq 0.000151. \end{cases}$$

Let us now consider the procedure based on the Heun method for the first step followed by the midpoint method for the second step. We have

n	x_n	$\bar{y}_n$	y_n	$\bar{y}_n - y_n$	$\tilde{y}_n$	$(SSE)_n =$ $\bar{y}_{n+1} - \tilde{y}_{n+1}$
0	0	1.0	1.0	0	1.0	−0.000163
1	0.1	0.904837	0.905	−0.000163	0.905	−0.000302
2	0.2	0.818731	0.819	−0.000269	0.819033	—

Since the midpoint method is a two-step method we compare $(SSE)_1 \doteq -0.000302$ with $\bar{y}_2 - y_2 \doteq -0.000269$ and find reasonably close agreement.

To estimate $(SSE)_1$ we use

8.12
$$(SSE)_1 \sim \frac{h^3}{3} y'''(x_1) = -\frac{h^3}{3} e^{-0.1}$$

$$\doteq -0.000302$$

which agrees with the exact value to three significant figures.
For the corrected Heun method we have

n	x_n	$\bar{y}_n$	$y_n^{(P)}$	$y_n^{(C)}$	y_n	$\bar{y}_n - y_n$	$\tilde{y}_n^{(P)}$	$\tilde{y}_n^{(C)}$	$\tilde{y}_{n+1}$	$(SSE)_n =$ $\bar{y}_{n+1} - \tilde{y}_{n+1}$
0	0	1.0	—	—	1.0	0	—	—	—	0.000087
1	0.1	0.904837	0.9	0.905	0.90475	0.000087	0.9	0.905	0.90475	0.000080
2	0.2	0.818731	0.814275	0.818799	0.818573	0.000158	0.814353	0.818877	0.818651	—

The sum of the single-step errors

8.13
$$(SSE)_0 + (SSE)_1 \doteq 0.000167$$

agrees reasonably closely with

8.14
$$\bar{y}_2 - y_2 \doteq 0.000158$$

To estimate the single-step errors we use the estimate

8.15
$$(SSE)_n \sim -\frac{h^3}{12} y'''(x_n) = \frac{h^3}{12} e^{-x_n}$$

and we have

8.16
$$\begin{cases} (SSE)_0 \sim \dfrac{h^3}{12} e^0 \doteq 0.000083 \\[2mm] (SSE)_1 \sim \dfrac{h^3}{12} e^{-0.1} \doteq 0.000075 \end{cases}$$

which agree closely with the exact values.
If we had used the Heun-midpoint method in the second step we would have obtained

8.17
$$y_2^{(P)} \doteq 0.819050, \qquad y_2 \doteq 0.818560$$

and

8.18
$$e_2 = \bar{y}_2 - y_2 \doteq 0.000171.$$

Moreover

8.19 $\bar{y}_2^{(P)} \doteq 0.819032, \qquad \tilde{y}_2 \doteq 0.818643$

and

8.20 $(SSE)_1 = \bar{y}_2 - \tilde{y}_2 \doteq 0.000088.$

The sum of the single-step errors

8.21 $(SSE)_0 + (SSE)_1 \doteq 0.000175$

is close to e_2. Our estimate for $(SSE)_1$ would be the same as (8.16) and is close to $(SSE)_1$. We also note that

8.22 $-\tfrac{1}{5}(y_2 - y_2^{(P)}) \doteq 0.000098$

which agrees closely with $(SSE)_1$.

If we had used the improved Heun method in the second step we would have obtained

8.23 $y_2^{(P)} \doteq 0.819038, \qquad y_2 \doteq 0.818561$

and

8.24 $e_2 = \bar{y}_2 - y_2 \doteq 0.000170.$

Moreover

8.25 $\bar{y}_2^{(P)} \doteq 0.819111, \qquad \tilde{y}_2 \doteq 0.818640$

and

8.26 $(SSE)_1 = \bar{y}_2 - \tilde{y}_2 \doteq 0.000091.$

The sum of the single-step errors

8.27 $(SSE)_0 + (SSE)_1 \doteq 0.000178$

is again close to e_2. Our estimate of $(SSE)_1$ would be the same as (8.16) and is close. We also note that

8.28 $-\tfrac{1}{6}(y_2 - y_2^{(P)}) \doteq 0.000079$

which agrees reasonably closely with $(SSE)_1$.

For the modified Euler method we can compute y_1 and y_2 accurately using (6.10) obtaining

8.29
$$\begin{cases} y_1 = \left(\dfrac{1 - \dfrac{h}{2}}{1 + \dfrac{h}{2}}\right) y_0 \doteq 0.904762 \\[2em] y_2 = \left(\dfrac{1 - \dfrac{h}{2}}{1 + \dfrac{h}{2}}\right) y_1 \doteq 0.818594. \end{cases}$$

The reader should verify that these results can be obtained to the accuracy indicated using the Euler method as a predictor followed by four iterations.

We obtain the following results

n	x_n	$\bar{y}_n$	y_n	$\bar{y}_n - y_n$	$\tilde{y}_n$	$(SSE)_n = \bar{y}_{n+1} - \tilde{y}_{n+1}$
0	0	1.0	1.0	0	1.0	0.000075
1	0.1	0.904837	0.904762	0.000075	0.904762	0.000069
2	0.2	0.818731	0.818594	0.000137	0.818662	—

The sum of the single-step errors

8.30 $(SSE)_0 + (SSE)_1 \doteq 0.000144$

agrees closely with $\bar{y}_2 - y_2 \doteq 0.000137$. Moreover, the single-step errors agree closely with the estimated values which are given by (8.16).

EXERCISES 8.8

1. Extend the computations given in the text to $x = 0.3$.
2. Verify the statement following (8.29).
3. Carry out an analysis similar to that given in the text for the Adams-Moulton method with $h = 0.1$, and continue until $x = 0.5$. Use the Runge-Kutta method for the starting values. Do the same for the Milne-Simpson method.

8.9 RUNGE-KUTTA METHODS

Nearly all of the methods of the previous section required a knowledge of y at several starting values $x_0 + h, x_0 + 2h, \ldots, x_0 + rh$ in addition to the given value of $y(x_0)$. For example, for the Adams-Moulton method y is required at three additional points. The only methods which we have so far considered which do not require additional starting values are the Euler method, the modified Euler method, the Heun method, and the corrected Heun method, all of which are relatively inaccurate. The object of this section is to develop methods of relatively high accuracy, with single-step errors of the order of h^5, which are "self-starting," i.e., require only a knowledge of y at one point.

Let us suppose we wish to solve the differential equation $y' = f(x, y)$ and we are given $y(a)$. In order to find $y(a + h)$ we could use the Taylor series method

9.1 $$y(a + h) = y(a) + hy'(a) + \frac{h^2}{2} y''(a) + \frac{h^3}{3!} y'''(a) + \cdots$$

$$= y(a) + hf + \frac{h^2}{2}[f_x + ff_y] + \frac{h^3}{6}[f_{xx} + 2ff_{xy} + f^2f_{yy} + ff_y^2 + f_xf_y] + \cdots$$

where f and its derivatives are evaluated at $(a, y(a))$. As we have already noted, Euler's method is obtained by considering two terms of the Taylor series. We now show that the Heun method yields a value of $y(a + h)$ which agrees with the first three terms of the Taylor series to within terms of order h^3.

Indeed, by (7.10) the value of $y(a + h)$ given by the Heun method is

9.2 $$y^{(H)}(a + h) = y(a) + \frac{h}{2}[f(a, y(a)) + f(a + h, y(a) + hf(a, y(a)))].$$

Expanding the right side of (9.2) in a Taylor series in two variables we have

9.3 $$y^{(H)}(a + h) = y(a) + \frac{h}{2}[f + f + hf_x + hff_y + 0(h^2)]$$

$$= y(a) + hf + \frac{h^2}{2}[f_x + ff_y] + 0(h^3),$$

which agrees with the right member of (9.1) to within terms of order h^3.

The formulas for the Heun method can be written in the following form

9.4 $$y^{(H)}(a + h) = y(a) + \alpha_1\Delta_1 + \alpha_2\Delta_2$$

where

9.5 $$\begin{cases} \Delta_1 = hf(a, y(a)) \\ \Delta_2 = hf(a + \rho h, y(a) + \rho\Delta_1) \end{cases}$$

and where $\rho = 1, \alpha_1 = \alpha_2 = \frac{1}{2}$. However, other values of α_1, α_2, and ρ are possible. Thus we have

9.6 $$\alpha_1\Delta_1 + \alpha_2\Delta_2 = h(\alpha_1 + \alpha_2)f + \alpha_2[\rho h^2f_x + \rho h^2ff_y] + 0(h^3).$$

Equating this, to within terms of order h^3, to the right side of (9.2) and equating coefficients of h and of h^2 we get

9.7
$$\begin{cases} \alpha_1 + \alpha_2 = 1 \\ \alpha_2 \rho = \frac{1}{2}. \end{cases}$$

One solution of the above equations is $\alpha_1 = \alpha_2 = \frac{1}{2}, \rho = 1$. However, we could let $\alpha_2 = 1, \alpha_1 = 0$, and $\rho = \frac{1}{2}$. This would give the formula

9.8
$$y(a + h) = y(a) + hf(a + \tfrac{1}{2}h, y(a) + \tfrac{1}{2}hf)$$

which is similar to the midpoint method with step-size $h/2$. There are infinitely many possibilities; for each $\rho \neq 0$ one could compute the corresponding α_1 and α_2.

The above idea can be generalized to obtain more accurate formulas. Let us consider the following method

9.9
$$y(a + h) = y(a) + \alpha_1 \Delta_1 + \alpha_2 \Delta_2 + \alpha_3 \Delta_3$$

where

9.10
$$\begin{cases} \Delta_1 = hf(a, y(a)) \\ \Delta_2 = hf(a + \rho h, y(a) + \rho \Delta_1) \\ \Delta_3 = hf(a + (\sigma + \tau)h, y(a) + \sigma \Delta_1 + \tau \Delta_2). \end{cases}$$

Expanding $\Delta_1, \Delta_2, \Delta_3$ in Taylor's series we get

9.11
$$\begin{cases} \Delta_1 = hf \\ \Delta_2 = hf + \rho h^2 [f_x + ff_y] + \frac{\rho^2 h^3}{2} [f_{xx} + 2ff_{xy} + f^2 f_{yy}] + 0(h^4) \\ \Delta_3 = hf + (\sigma + \tau)h^2 [f_x + ff_y] + \frac{(\sigma + \tau)^2 h^3}{2} [f_{xx} + 2ff_{xy} + f^2 f_{yy}) \\ \qquad + h^3 \rho \tau [f_x f_y + ff_y^2] + 0(h^4). \end{cases}$$

Substituting in (9.9) and equating coefficients of h, h^2, and h^3 of the resulting expression with those of (9.1), we obtain

9.12
$$\begin{cases} \alpha_1 + \alpha_2 + \alpha_3 = 1 \\ \alpha_2 \rho + \alpha_3 (\sigma + \tau) = \frac{1}{2} \\ \left(\alpha_2 \frac{\rho^2}{2} + \alpha_3 \frac{(\sigma + \tau)^2}{2} \right) [f_{xx} + 2ff_{xy} + f^2 f_{yy}] + \alpha_3 \rho \tau [f_x f_y + ff_y^2] \\ \qquad = \frac{1}{6} [f_{xx} + 2ff_{xy} + f^2 f_{yy} + ff_y^2 + f_x f_y]. \end{cases}$$

For the above conditions to hold for all functions $f(x, y)$ we must have

9.13
$$\begin{cases} \alpha_1 + \alpha_2 + \alpha_3 = 1 \\ \alpha_2\rho + \alpha_3(\sigma + \tau) = \frac{1}{2} \\ \alpha_2\frac{\rho^2}{2} + \alpha_3\frac{(\sigma + \tau)^2}{2} = \frac{1}{6} \\ \alpha_3\rho\tau = \frac{1}{6}. \end{cases}$$

Eliminating α_2 and α_3 from the second, third, and fourth equations, we get

9.14
$$\tau = \frac{\lambda(\rho - \lambda)}{(3\rho - 2)\rho}$$

where $\lambda = \sigma + \tau$. For any value of ρ such that $\rho \neq 0$ and $\rho \neq \frac{2}{3}$ one can assign any value of λ different from zero and ρ. One can then compute τ and $\alpha_3 = \frac{1}{6}(\rho\tau)^{-1}$, $\alpha_2 = \rho^{-1}[\frac{1}{2} - \alpha_3\lambda]$, $\alpha_1 = 1 - \alpha_2 - \alpha_3$. If $\rho = \frac{2}{3}$, then we can find a solution provided $\lambda = 0$ or $\frac{2}{3}$. If $\rho = \frac{2}{3}$ and $\lambda = 0$, then τ can have any nonzero value and $\alpha_3 = \frac{1}{4}\tau^{-1}$, $\alpha_2 = \frac{3}{4}$, $\alpha_1 = \frac{1}{4}(1 - \tau^{-1})$. If $\rho = \frac{2}{3}$ and $\lambda = \frac{2}{3}$, then τ may have any nonzero value and $\alpha_3 = \frac{1}{4}\tau^{-1}$, $\alpha_2 = \frac{3}{4} - \alpha_3$, $\alpha_1 = \frac{1}{4}$.

An interesting special case is $\rho = \frac{1}{2}$, $\lambda = 1$. We get $\tau = 2$, $\sigma = -1$, $\alpha_3 = \frac{1}{6}$, $\alpha_2 = \frac{2}{3}$, $\alpha_1 = \frac{1}{6}$. This leads to the following formula which is sometimes referred to as *Kutta's third-order rule*

9.15
$$\begin{cases} \Delta_1 = hf(a, y(a)) \\ \Delta_2 = hf\left(a + \frac{h}{2}, y(a) + \frac{1}{2}\Delta_1\right) \\ \Delta_3 = hf(a + h, y(a) + 2\Delta_2 - \Delta_1) \end{cases}$$

9.16
$$y(a + h) = y(a) + \frac{1}{6}[\Delta_1 + 4\Delta_2 + \Delta_3].$$

Evidently, as noted by Kunz [1957], Kutta's third-order rule is analogous to Simpson's rule since when applied to the differential equation $y' = f(x)$ it leads to Simpson's rule with step size $h/2$ for integrating

$$\int_a^{a+h} f(t)dt.$$

Runge-Kutta formulas of higher order can be derived by the above methods. However, the labor involved and the complexity increases very rapidly. It is almost prohibitively difficult to get a formula with an error of higher order than

$O(h^5)$.   Methods with error of order $O(h^5)$, which are often called *fourth-order Runge-Kutta methods*, appear to be by far the most popular.   One of the most frequently used fourth-order method is given by

9.17 $$y(a + h) = y(a) + \tfrac{1}{6}(\Delta_1 + 2\Delta_2 + 2\Delta_3 + \Delta_4)$$

where

9.18
$$\begin{cases} \Delta_1 = hf(a, y(a)) \\ \Delta_2 = hf(a + \tfrac{1}{2}h, y(a) + \tfrac{1}{2}\Delta_1) \\ \Delta_3 = hf(a + \tfrac{1}{2}h, y(a) + \tfrac{1}{2}\Delta_2) \\ \Delta_4 = hf(a + h, y(a) + \Delta_3). \end{cases}$$

The single-step error for this method is $O(h^5)$, but the complete expression for the error is extremely complicated (see, for instance, Ralston [1965]).

As an example, let us consider the problem of solving $y' = -y$, $y(0) = 1$. Letting $h = 0.1$ we have

9.19
$$\begin{cases} \Delta_1 = (0.1)(-1) = -0.1 \\ \Lambda_2 = (0.1)(-[1 + \tfrac{1}{2}(-0.1)]) = 0.1(-0.95) = -0.095 \\ \Delta_3 = (0.1)(-[1 + \tfrac{1}{2}(-0.095)]) = 0.1(-0.9525) = -0.09525 \\ \Delta_4 = (0.1)(-[1 + (-0.09525)]) = 0.1(-0.90475) = -0.090475. \end{cases}$$

9.20 $y(0.1) = 1 + \tfrac{1}{6}[(-0.1) + 2(-0.095) + 2(-0.09525) + (-0.09475)]$

$= 1 + \tfrac{1}{6}(-0.570975) = 1 - 0.0951625 = 0.9048375.$

This agrees closely with the value of the exact solution at $x = 0.1$, which, from tables, is 0.904837 to six decimals.   Using the first five terms of the Taylor's series solution we also obtain 0.904837.

In future discussion, by the term *Runge-Kutta method* we shall mean the method defined by (9.17) and (9.18).   This method has the same order of accuracy as the Milne-Simpson method and the Adams-Moulton method.   It has the advantage of being self-starting.   It has the following disadvantages: it requires four evaluations of $f(x, y)$ per step rather than two as in the predictor-corrector methods; the formula for the error is very complicated; and it is difficult to estimate the single-step error.   An ideal arrangement seems to be to use the Runge-Kutta method to compute starting values and then to use a predictor-corrector method. This procedure permits one to change step size conveniently when appropriate. This is discussed in more detail in Section 8.12.

EXERCISES 8.9

1. Find $y(0.1)$ by the Runge-Kutta third-order method (9.15)–(9.16) for the problem $y' = -y$, $y(0) = 1$, with $h = 0.1$.

2. Find the Runge-Kutta third-order method (9.9)–(9.10) such that $\rho = 1$, $\lambda = 2$. Apply the method to find $y(0.1)$ for $y' = -y$, $y(0) = 1$, with $h = 0.1$.

3. Carry out the derivation of the Runge-Kutta fourth-order method (9.17)–(9.18).

4. For the problem $y' = -y$, $y(0) = 1$, find an analytic formula for $y(h)$ based on the Runge-Kutta method (9.17)–(9.18) and compare with e^{-h}.

5. In the initial-value problem $y' = -y$, $y(0) = 1$, find $y(0.1)$ and $y(0.2)$ using the Runge-Kutta method with $h = 0.1$. Compare the results with those obtained in Exercise 1, Section 8.3. Using the exact value of $y(0.2)$, compute $y(0.3)$ by the Runge-Kutta method. Then determine $y(0.4)$ by the Adams-Moulton method.

6. In the preceding exercise determine $y(0.4)$ by the Adams-Moulton method based on the exact values of $y(0.1)$ and $y(0.2)$ and on the value of $y(0.3)$ obtained by the Runge-Kutta method with $h = 0.1$.

7. In the initial-value problem $y' = x + y^2$, $y(0) = 1$ determine $y(0.3)$ by the Runge-Kutta method with $h = 0.1$ and also with $h = 0.05$, using the value of $y(0.2)$ obtained from Exercise 2, Section 8.3. Compute the error for $y(0.3)$ in each case and verify that it is approximately proportional to h^4.

8. In the problem of the preceding exercise, with the values of $y(0.1)$ and $y(0.2)$ obtained from Exercise 2, Section 8.3, and with $y(0.3)$ determined by the Runge-Kutta method with $h = 0.1$, find $y(0.6)$ using the Adams-Moulton method with $h = 0.1$.

8.10 METHODS BASED ON NUMERICAL DIFFERENTIATION

In this section we consider a class of methods based on the numerical differentiation formulas of Section 7.3. Given the integers q and N such that

10.1 $$N \geq 1, \qquad 0 \leq q \leq N$$

we can represent $y'(x_{n+1-q})$ as a linear combination of $y_{n+1}, y_n, \ldots, y_{n+1-N}$ using an appropriate numerical differentiation formula. Thus we have

10.2 $$y'_{n+1-q} \sim \frac{1}{h} \sum_{i=0}^{N} \alpha_i y_{n+1-i}$$

where the α_i are independent of h and n. Moreover, as shown in Section 7.3 the coefficient α_0 does not vanish. Therefore, the method defined by

10.3 $$f(x_{n+1-q}, y_{n+1-q}) = \frac{1}{h} \sum_{i=0}^{N} \alpha_i y_{n+1-i}$$

affords a procedure for finding y_{n+1} provided the values $y_n, y_{n-1}, \ldots, y_{n+1-N}$ are available. Such a method is a *method based on numerical differentiation*.

If $q \neq 0$, the method is said to be *open* and we can solve (10.3) explicitly for y_{n+1} obtaining

10.4
$$y_{n+1} = \frac{h}{\alpha_0} f(x_{n+1-q}, y_{n+1-q}) - \sum_{i=1}^{N} \frac{\alpha_i}{\alpha_0} y_{n+1-i}.$$

Otherwise, if $q = 0$, the method is said to be *closed* and we have the implicit equation

10.5
$$y_{n+1} = \frac{h}{\alpha_0} f(x_{n+1}, y_{n+1}) - \sum_{i=1}^{N} \frac{\alpha_i}{\alpha_0} y_{n+1-i}.$$

However, if $f(x, y)$ satisfies a Lipschitz condition in y, then (10.5) has a unique solution for

10.6
$$\frac{hL}{\alpha_0} < 1$$

where L is the Lipschitz constant. Moreover, as in the case of methods based on numerical quadrature, the solution can be found by an iterative process where very few iterations are required.

As in the case of methods based on numerical quadrature, the first few values of y_n, namely $y_1, y_2, \ldots, y_{N-1}$, must be determined by a special routine. If one uses a predictor-corrector method based on (10.3) with a predictor which involves N^* previous values of y_n where $N^* \geq N$, then $y_1, y_2, \ldots, y_{N^*-1}$ would have to be found by a special procedure.

Given q and N we designate the method based on the numerical differentiation formula for y'_{n+1-q} in terms of $y_{n+1}, y_n, \ldots, y_{n+1-N}$ as *method D.q.N.* Thus, for example, D.1.1 is the Euler method, and D.1.2 is the midpoint method. The backward Euler method is D.0.1.

The *order* of a method based on numerical differentiation is the largest integer p such that

10.7
$$(R[y])_n = y_{n+1} + \sum_{i=1}^{N} \frac{\alpha_i}{\alpha_0} y_{n+1-i} - \frac{h}{\alpha_0} y'_{n+1-i} = 0$$

for any polynomial of degree p or less. From the results of Section 7.3 it follows that the order of method D.q.N is N. We define the *differentiation error* $(DE[y])_n$ corresponding to a given function $y(x)$ by

10.8
$$(DE[y])_n = (R[(x - \alpha)^{p+1}])_n \frac{y^{(p+1)}(x_n)}{(p+1)!}.$$

The constant α can be chosen for convenience, the differentiation error being independent of α.

Since each numerical differentiation formula is definite it follows that the error term for the numerical differentiation formula corresponding to method D.q.N is

10.9 $Kh^p y^{(p+1)}(\xi)$

for some ξ in (x_{n+1-N}, x_{n+1}). Consequently by 7-(2.28) we have

10.10 $(DE[y])_n = -\dfrac{K}{\alpha_0} h^{p+1} y^{(p+1)}(x_n).$

For example, with the Euler method, which is method D.1.1, the corresponding numerical differentiation formula is

10.11 $y'_n \sim \dfrac{y_{n+1} - y_n}{h}$

and the error term is

10.12 $-\tfrac{1}{2} h y''(\xi)$

for some ξ in the interval (x_n, x_{n+1}). Thus by (10.10) we have, since $\alpha_0 = 1$, $\alpha_1 = -1$,

10.13 $(DE[y])_n = \tfrac{1}{2} h^2 y''(x_n)$ (Euler method).

For the backward Euler method we have

10.14 $(DE[y])_n = -\tfrac{1}{2} h^2 y''(x_n)$ (backward Euler method).

The midpoint method, which is method D.1.2, corresponds to the numerical differentiation formula

10.15 $y'_n \sim \dfrac{y_{n+1} - y_{n-1}}{2h}$

and the error term is

10.16 $-\tfrac{1}{6} h^2 y^{(3)}(\xi)$

for some ξ in the interval (x_{n-1}, x_{n+1}). Consequently, since $\alpha_0 = \tfrac{1}{2}$, we have, by (10.10),

10.17 $(DE[y])_n = \tfrac{1}{3} h^3 y^{(3)}(x_n)$ (midpoint method).

For an open method based on numerical differentiation the single-step error is simply equal to the differentiation error to within $O(h^{p+2})$, where p is the order of the method, provided $y \in C^{(p+2)}$ in the interval of interest. Thus we have

10.18 $$(SSE[y])_n = (DE[y])_n + O(h^{p+2}).$$

(We recall that $(DE[y])_n = O(h^{p+1})$ and that $p = N$.) This result is also true for a closed method since $f(x, y)$ satisfies a Lipschitz condition in y.

Evidently for the Euler method we have

10.19 $$(SSE[y])_n = \tfrac{1}{2}h^2 y^{(2)}(x_n) + O(h^3) \qquad \text{(Euler method)}$$

and for the backward Euler method

10.20 $\quad (SSE[y])_n = -\tfrac{1}{2}h^2 y^{(2)}(x_n) + O(h^3) \qquad$ (backward Euler method).

Table 10.23 gives the coefficients for several methods based on numerical differentiation. Included are closed methods with $q = 0$ and open methods where $q = 1$ and $q = 2$ (see (10.4) and (10.5)). The differentiation error $(DE[y])_n$ is given in each case.

For example, for method D.0.3 we have, by (10.5),

10.21 $$y_{n+1} = \tfrac{6h}{11} f(x_{n+1}, y_{n+1}) + \tfrac{18}{11}y_n - \tfrac{9}{11}y_{n-1} + \tfrac{2}{11}y_{n-2}$$

and the differentiation error is

10.22 $$(DE[y])_n = -\tfrac{3}{22}h^4 y^{(4)}.$$

10.23 Table   Methods Based on Numerical Differentiation.

	N	α_0	α_1	α_2	α_3	α_4	$(DE[y])_n$	Name	Symbol
$q = 0$ (closed)	1	1	-1	—	—	—	$-\tfrac{1}{2}h^2 y^{(2)}$	Backward Euler	D.0.1
	2	$\tfrac{3}{2}$	-2	$\tfrac{1}{2}$	—	—	$-\tfrac{2}{3}h^3 y^{(3)}$		D.0.2
	3	$\tfrac{11}{6}$	-3	$\tfrac{3}{2}$	$-\tfrac{1}{3}$	—	$-\tfrac{3}{22}h^4 y^{(4)}$		D.0.3
	4	$\tfrac{25}{12}$	-4	3	$-\tfrac{4}{3}$	$\tfrac{1}{4}$	$-\tfrac{12}{125}h^5 y^{(5)}$		D.0.4
$q = 1$	1	1	-1	—	—	—	$\tfrac{1}{2}h^2 y^{(2)}$	Euler	D.1.1
	2	$\tfrac{1}{2}$	0	$-\tfrac{1}{2}$	—	—	$\tfrac{1}{3}h^3 y^{(3)}$	Midpoint	D.1.2
	3	$\tfrac{1}{3}$	$\tfrac{1}{2}$	-1	$\tfrac{1}{6}$	—	$\tfrac{1}{4}h^4 y^{(4)}$		D.1.3*
	4	$\tfrac{1}{4}$	$\tfrac{5}{6}$	$-\tfrac{3}{2}$	$\tfrac{1}{2}$	$-\tfrac{1}{12}$	$\tfrac{1}{5}h^5 y^{(5)}$		D.1.4*
$q = 2$	1	1	-1	—	—	—	$\tfrac{3}{2}h^2 y^{(2)}$		D.2.1
	2	$-\tfrac{1}{2}$	2	$-\tfrac{3}{2}$	—	—	$\tfrac{2}{3}h^3 y^{(3)}$		D.2.2*
	3	$-\tfrac{1}{6}$	1	$-\tfrac{1}{2}$	$-\tfrac{1}{3}$	—	$\tfrac{1}{4}h^4 y^{(4)}$		D.2.3*
	4	$-\tfrac{1}{12}$	$\tfrac{2}{3}$	0	$-\tfrac{2}{3}$	$\tfrac{1}{12}$	$\tfrac{2}{5}h^5 y^{(5)}$		D.2.4*

* These methods are unstable and should not be used for numerical computation.

It can be seen from Tables 10.23, 6.36, and 6.37 that one can get single-step errors using methods based on numerical differentiation which are comparable with those obtained by methods based on numerical quadrature and involving the same number of points. However, as we shall see in Chapter 9, methods based on numerical differentiation are unstable for large values of N. They are generally less satisfactory than methods based on numerical quadrature.

We remark that methods based on numerical differentiation and methods based on numerical quadrature are special cases of linear multistep methods which are considered in Chapter 9. Such methods are defined by

10.24
$$\sum_{i=0}^{N} \alpha_i y_{n+1-i} = h \sum_{i=0}^{N} \beta_i f(x_{n+1-i}, y_{n+1-i})$$

where the α_i and β_i are independent of h and where $\alpha_0 \neq 0$ and $|\alpha_N| + |\beta_N| \neq 0$. If all α_i vanish except α_0 and one other α_i, say α_k, then we have a method based on numerical quadrature. (It can be shown that α_k must equal $-\alpha_0$.) If all β_i vanish except one, then we have a method based on numerical differentiation.

EXERCISES 8.10

1. Show that the order of method D.q.N is N.

2. Verify the formulas given in Table 10.23 for the differentiation error for methods D.1.1, D.2.1, D.0.2, D.1.2, D.2.2.

3. Apply method D.2.2 to solve $y' = -y$, $y(0) = 1$, with $h = 0.1$. Use the Euler method to determine the starting value $y(0.1)$. Integrate as far as $x = 0.6$. Comment on the behavior of the numerical solution. Do the same if the exact value is used for $y(0.1)$.

4. Apply the predictor-corrector method D.1.2/D.0.2 to solve $y' = -y$, $y(0) = 1$ with $h = 0.1$. Integrate to $x = 0.5$. Use the corrected Heun method for the starting value. Do the same for method D.1.3.

5. Derive formulas for methods D.0.5, D.1.5, and D.2.5. Also find the differentiation error in each case.

6. Consider the initial-value problem $y' = -y$, $y(0) = 1$. Obtain $y(0.1)$, $y(0.2)$, $y(0.3)$ by the Runge-Kutta method. Then find $y(1)$ using method D.1.3 with $h = 0.1$. Do the same for method D.0.4. With method D.0.4, solve (10.5) exactly for y_{n+1} at each step.

8.11 HIGHER-ORDER EQUATIONS
AND SYSTEMS OF FIRST-ORDER EQUATIONS

Most of the methods which we have been considering so far can be used for initial-value problems involving systems of first-order equations. Before showing this, however, we first show that an initial-value problem involving a single equation of higher order can normally be reduced to an initial-value problem involving a system of first-order equations. Thus, if we are given the initial-value problem

11.1
$$y^{(n)} = f(x, y, y', \ldots, y^{(n-1)})$$

where

11.2 $\qquad y(x_0) = y_0, \qquad y'(x_0) = y_0', \ldots, \qquad y^{(n-1)}(x_0) = y_0^{(n-1)}$

and where $y_0, y_0', \ldots, y_0^{(n-1)}$ are given, we introduce the new dependent variables $y_1(x), y_2(x), \ldots, y_n(x)$, where

11.3 $\qquad\qquad y_1 = y, \qquad y_2 = y', \ldots, \qquad y_n = y^{(n-1)}.$

Evidently the functions $y_1(x), y_2(x), \ldots, y_n(x)$ satisfy the system

11.4
$$\begin{cases} y_1' = y_2 \\ y_2' = y_3 \\ \quad\vdots \\ y_{n-1}' = y_n \\ y_n' = f(x, y_1, y_2, \ldots, y_n). \end{cases}$$

Moreover, the functions also satisfy the initial conditions

11.5 $\qquad y_1(x_0) = y_0, \qquad y_2(x_0) = y_0', \ldots, \qquad y_n(x_0) = y_0^{(n-1)}.$

As an example, consider the initial-value problem

11.6 $\qquad\qquad\qquad\qquad y'' = -y$

where

11.7 $\qquad\qquad\qquad y(0) = 1, \qquad y'(0) = 0.$

Letting $y_1(x) = y, \; y_2(x) = y'$ we get the new initial-value problem

11.8 $\qquad\qquad\qquad\quad \begin{cases} y_1' = y_2 \\ y_2' = -y_1 \end{cases}$

with the initial conditions

11.9 $\qquad\qquad\qquad y_1(0) = 1, \qquad y_2(0) = 0.$

We remark that the general solution of (11.6) is

11.10 $\qquad\qquad\qquad y = A \cos x + B \sin x.$

Hence the initial conditions imply that $A = 1$, $B = 0$, so that

11.11 $y(x) = \cos x.$

We can easily verify that if $y_1(x) = y(x) = \cos x$, $y_2(x) = y'(x) = -\sin x$, then (11.8) and (11.9) are satisfied.

In a similar manner, one can handle a system of higher-order equations. For example, the system

11.12 $\begin{cases} y'' + yz = 0 \\ z' + 2yz = 4 \end{cases}$

with initial conditions

11.13 $y(0) = 1, \qquad y'(0) = 0, \qquad z(0) = 3$

can be reduced to the system

11.14 $\begin{cases} y_1' = 4 - 2y_1y_2 \\ y_2' = y_3 \\ y_3' = -y_1y_2 \end{cases}$

where

11.15 $y_1(0) = 3, \qquad y_2(0) = 1, \qquad y_3(0) = 0$

by letting

11.16 $y_1 = z, \qquad y_2 = y, \qquad y_3 = y'.$

Let us now consider the general first-order system

11.17 $\begin{cases} y_1'(x) = f_1(x, y_1, y_2, \ldots, y_n) \\ y_2'(x) = f_2(x, y_1, y_2, \ldots, y_n) \\ \vdots \\ y_n'(x) = f_n(x, y_1, y_2, \ldots, y_n) \end{cases}$

with initial conditions

11.18 $y_1(x_0) = (y_1)_0, \qquad y_2(x_0) = (y_2)_0, \ldots, \qquad y_n(x_0) = (y_n)_0$

where $(y_1)_0, (y_2)_0, \ldots, (y_n)_0$ are given.

It is convenient to introduce the vector notation

11.19
$$\vec{y}(x) = \begin{pmatrix} y_1(x) \\ y_2(x) \\ \vdots \\ y_n(x) \end{pmatrix}, \quad \vec{y}'(x) = \begin{pmatrix} y_1'(x) \\ y_2'(x) \\ \vdots \\ y_n'(x) \end{pmatrix},$$

etc. Evidently the system (11.17) can be written in the form

11.20
$$\vec{y}' = \vec{f}(x, \vec{y})$$

where

11.21
$$\vec{f}(x, \vec{y}) = \begin{pmatrix} f_1(x, y_1, y_2, \ldots, y_n) \\ f_2(x, y_1, y_2, \ldots, y_n) \\ \vdots \\ f_n(x, y_1, y_2, \ldots, y_n) \end{pmatrix} = \begin{pmatrix} f_1(x, \vec{y}) \\ f_2(x, \vec{y}) \\ \vdots \\ f_n(x, \vec{y}) \end{pmatrix}.$$

The initial conditions (11.18) can be written in the form

11.22
$$\vec{y}(x_0) = \vec{y}_0$$

where

11.23
$$\vec{y}_0 = \begin{pmatrix} (y_1)_0 \\ (y_2)_0 \\ \vdots \\ (y_n)_0 \end{pmatrix}.$$

In order to discuss existence and uniqueness it is convenient to introduce the (Euclidean) norm, $\|\vec{v}\|$, of a vector $\vec{v}$ given by

11.24
$$\vec{v} = \begin{pmatrix} v_1 \\ v_2 \\ \vdots \\ v_n \end{pmatrix}.$$

as

11.25
$$\|\vec{v}\| = (|v_1|^2 + |v_2|^2 + \cdots + |v_n|^2)^{1/2}.$$

The (vector) function $\vec{f}(x, \vec{y})$ satisfies a *Lipschitz condition* in $\vec{y}$ in a region R of the variables $x, y_1, y_2, \ldots, y_n$ if for some constant L we have

11.26 $\|\vec{f}(x, \vec{y}) - \vec{f}(x, \vec{z})\| \leq L\|\vec{y} - \vec{z}\|$

for any x, $\vec{y}$, and $\vec{z}$ such that $(x, y_1, y_2, \ldots, y_n)$ and $(x, z_1, z_2, \ldots, z_n)$ are in R. We can define limits, continuity, and derivatives for functions defined in R in an obvious way.

We now state without proof the following theorems (see Birkhoff and Rota [1962] and Henrici [1963]).

11.27 Theorem. (Existence and Uniqueness). For some $\hat{x} > x_0$ and for some $\delta > 0$ let $\vec{f}(x, \vec{y})$ be continuous in R:

11.28 $x_0 \leq x \leq \hat{x}, \qquad \|\vec{y} - \vec{y}_0\| \leq \delta$

and let

11.29 $\|\vec{f}(x, \vec{y})\| \leq M$

in R. Let

11.30 $\bar{x} = \min\left(\hat{x}, x_0 + \dfrac{\delta}{M}\right).$

There exists a function $\vec{y}(x)$, continuously differentiable in $[x_0, \bar{x}]$ which satisfies (11.20) and (11.22). If, moreover, $\vec{f}(x, \vec{y})$ satisfies a Lipschitz condition in $\vec{y}$ in R, then the solution is unique.

11.31 Theorem. If, for some $\hat{x} > x_0$, $\vec{f}(x, \vec{y})$ is defined and continuous for $x_0 \leq x \leq \hat{x}$ and for all $\vec{y}$, and if for some constant L we have

11.32 $\|\vec{f}(x, y) - \vec{f}(x, \vec{y}^*)\| \leq L\|\vec{y} - \vec{y}^*\|$

for $x \in [x_0, \hat{x}]$ and for all $\vec{y}$ and $\vec{y}^*$, then there exists a unique solution of (11.20) and (11.22) for the interval $[x_0, \hat{x}]$.

11.33 Theorem. If $\vec{f}(x, \vec{y})$ is an analytic function of $x, y_1, y_2, \ldots, y_n$ at $x_0, (y_1)_0$, $(y_2)_0, \ldots, (y_n)_0$ then for δ sufficiently small there exists a unique function $\vec{y}(x)$ which is analytic for $x_0 - \delta \leq x \leq x_0 + \delta$ and which satisfies (11.20) and (11.22). If $\vec{f}(x, \vec{y})$ is analytic for all x and $\vec{y}$ then any solution of (11.20) and (11.22) is analytic.

The extension of methods which we have been considering can be carried out directly using the vector notation. For example, for the Euler method we have

11.34 $\vec{y}_{k+1} = \vec{y}_k + h\vec{f}(x_k, \vec{y}_k).$

In expanded form (for the case $n = 2$) this is equivalent to

11.35
$$\begin{cases} (y_1)_{k+1} = (y_1)_k + hf_1(x_k, (y_1)_k, (y_2)_k) \\ (y_2)_{k+1} = (y_2)_k + hf_2(x_k, (y_1)_k, (y_2)_k). \end{cases}$$

In a similar way the Heun method can be written in the form

11.36
$$\begin{cases} \vec{y}_{k+1}^{(P)} = \vec{y}_k + h\vec{f}(x_k, \vec{y}_k) \\ \vec{y}_{k+1} = \vec{y}_k + \dfrac{h}{2}[\vec{f}(x_k, \vec{y}_k) + \vec{f}(x_{k+1}, \vec{y}_{k+1}^{(P)})]. \end{cases}$$

The Runge-Kutta method becomes

11.37
$$\vec{y}_{k+1} = \vec{y}_k + \vec{\Delta}_k$$

where

11.38
$$\vec{\Delta}_k = \tfrac{1}{6}(\vec{\Delta}_k^{(1)} + 2\vec{\Delta}_k^{(2)} + 2\vec{\Delta}_k^{(3)} + \vec{\Delta}_k^{(4)})$$

and

11.39
$$\begin{cases} \vec{\Delta}_k^{(1)} = h\vec{f}(x_k, \vec{y}_k) \\ \vec{\Delta}_k^{(2)} = h\vec{f}\left(x_k + \dfrac{h}{2}, \vec{y}_k + \tfrac{1}{2}\vec{\Delta}_k^{(1)}\right) \\ \vec{\Delta}_k^{(3)} = h\vec{f}\left(x_k + \dfrac{h}{2}, \vec{y}_k + \tfrac{1}{2}\vec{\Delta}_k^{(2)}\right) \\ \vec{\Delta}_k^{(4)} = h\vec{f}(x_k + h, \vec{y}_k + \vec{\Delta}_k^{(3)}). \end{cases}$$

EXERCISES 8.11

1. Consider the initial-value problem $y'' = y' + xy^2$, $y(0) = 1$, $y'(0) = 2$.
   a) Determine an equivalent first-order system.
   b) Find both dependent variables in the first-order system at $x = 0.2$ by the Taylor's series method. Compare with the result obtained in Exercise 9, Section 8.3.
   c) With $h = 0.1$, find both of the dependent variables at $x = 0.2$ using the corrected Heun method for the first step and the Heun-midpoint method for the second step.

2. In the problem of Exercise 1 find $\bar{x}$ such that a unique solution exists for $0 \leq x \leq \bar{x}$.

3. Using Theorem 11.27, prove Theorem 11.31.

4. Use Theorem 11.31 to prove the existence of a unique solution over the interval $[x_0, \hat{x}]$ of the initial-value problem
$$y'' = P(x)y' + Q(x)y + R(x)$$
$$y(x_0) = y_0$$
$$y'(x_0) = y_0'$$
where $P(x)$, $Q(x)$, and $R(x)$ are continuous on $[x_0, \hat{x}]$.

5. Consider the initial-value problem $y_1' = x^2 + y_1^2 + y_2^2$, $y_2' = x^2 y_1^2 y_2^2$ with $y_1(0) = 1$, $y_2(0) = 2$. Find $\bar{x}$ such that $\bar{x} > x_0$ and such that a unique solution exists for $[0, \bar{x}]$.

8.12 THE USE OF HIGH-SPEED COMPUTERS

In this section we describe a procedure for solving a system of first-order ordinary differential equations using a high-speed digital computer. Several programs have been written on the basis of this procedure (see, for instance, Raney [1961] and Kincaid [1969]). A similar procedure is described by Conte [1962].

Suppose we have $\bar{y}(x)$ for $x = a$ but that we know $\bar{y}(x)$ for fewer than 3 of the abscissas $a - h, a - 2h, a - 3h$. We then determine $\bar{y}(a + h)$ by the Runge-Kutta method by the formulas (see (11.37) and (11.38))

12.1 $\bar{y}(a + h) = \bar{y}(a) + \frac{1}{6}(\vec{\Delta}^{(1)}(a) + 2\vec{\Delta}^{(2)}(a) + 2\vec{\Delta}^{(3)}(a) + \vec{\Delta}^{(4)}(a))$

where

12.2
$$\begin{cases} \vec{\Delta}^{(1)}(a) = h\vec{f}(a, \bar{y}(a)) \\[2mm] \vec{\Delta}^{(2)}(a) = h\vec{f}\left(a + \frac{h}{2}, \bar{y}(a) + \frac{1}{2}\vec{\Delta}^{(1)}(a)\right) \\[2mm] \vec{\Delta}^{(3)}(a) = h\vec{f}\left(a + \frac{h}{2}, \bar{y}(a) + \frac{1}{2}\vec{\Delta}^{(2)}(a)\right) \\[2mm] \vec{\Delta}^{(4)}(a) = h\vec{f}(a + h, \bar{y}(a) + \vec{\Delta}^{(3)}(a)). \end{cases}$$

One could, of course, use the Runge-Kutta method throughout the computation, but this would involve computing $\vec{f}(x, y)$ four times per step. We can reduce this to two times per step without loss of accuracy using the Adams-Moulton method, provided we have $\bar{y}(x)$ available for $x = a - h$, $a - 2h$, and $a - 3h$ as well as $x = a$. We use the formulas

12.3
$$\begin{cases} \bar{y}^{(P)}(a + h) = \bar{y}(a) + \frac{h}{24}[55\vec{f}(a, \bar{y}(a)) - 59\vec{f}(a - h, \bar{y}(a - h)) \\[2mm] \qquad + 37\vec{f}(a - 2h, \bar{y}(a - 2h)) - 9\vec{f}(a - 3h, \bar{y}(a - 3h))] \\[2mm] \bar{y}(a + h) = \bar{y}(a) + \frac{h}{24}[9\vec{f}(a + h, \bar{y}(a + h)) + 19\vec{f}(a, \bar{y}(a)) \\[2mm] \qquad - 5\vec{f}(a - h, \bar{y}(a - h)) + \vec{f}(a - 2h, \bar{y}(a - 2h))]. \end{cases}$$

As the calculation proceeds, we monitor the difference between the predicted values of $y(a + h)$ and the corrected values. Actually, if there are n equations, we use the "maximum norm"

12.4 $\Delta = \|\bar{y}(a + h) - \bar{y}^{(P)}(a + h)\|_\infty = \max\limits_{i=1,2,\ldots,n} |y_i(a + h) - y_i^{(P)}(a + h)|.$

12.5 Fig. The Solution of a System of Ordinary Differential Equations.

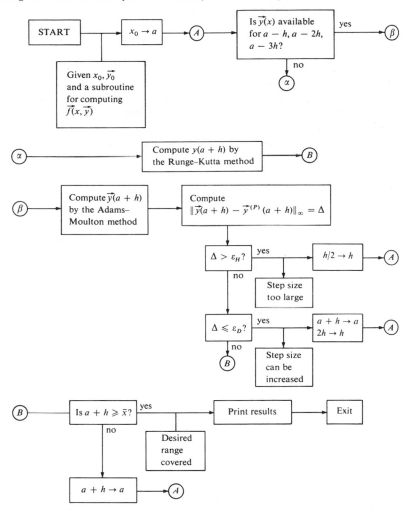

If Δ exceeds some prescribed quantity, say ε_H, then we halve the mesh size, re-compute starting values by the Runge-Kutta method, with the reduced mesh size, and then proceed with the Adams-Moulton method. If Δ is less than some prescribed quantity, say ε_D, then we double the mesh size, obtain new starting values by the Runge-Kutta method, if necessary, and proceed.

A flow diagram for the procedure described above is given in Fig. 12.5. One is given x_0, the initial conditions $\vec{y}_0$, a subroutine for computing $\vec{f}(x, \vec{y})$ for any x

and $\bar{y}$, and $\bar{x}$. Here $\bar{x}$ is the upper end of the interval for which the solution is desired. One is also given the quantities ε_H and ε_D.

To begin with we let $a = x_0$. Since $\bar{y}(x)$ is not available for $a - h$, we compute $\bar{y}(a + h)$ by the Runge-Kutta method. We also compute $\bar{y}(a + 2h)$, $\bar{y}(a + 3h)$. At this point we have sufficiently many values of $\bar{y}(x)$ to use the Adams-Moulton method. After each use of the method we compute Δ by (12.4) and decide whether to halve the interval size (if $\Delta > \varepsilon_H$) or to double it (if $\Delta < \varepsilon_D$), or to leave it alone. If the interval size is halved, it is necessary to use the Runge-Kutta method for three steps with the smaller interval size before returning to the Adams-Moulton method. If the interval size is doubled, then there may be sufficiently many values $a - 2h$, $a - 4h$, and $a - 6h$ for which $\bar{y}(x)$ is available so that the use of the Runge-Kutta method is not necessary.

Partial Double-precision

An important feature of the program is the use of partial double-precision arithmetic to control the accumulation of rounding errors. Basically, the procedure is as follows. Each component of $\bar{y}$ is stored as a double-precision floating-point number. However, in computing $\bar{f}(x, \bar{y})$ one uses only the most significant half of each component of $\bar{y}$. Only when $\bar{y}$ itself is to be modified is the addition done in double-precision. Thus, the procedure requires only a relatively small amount of extra machine time. However, the reduction in the accumulation of rounding errors is almost as great as though one used complete double-precision throughout.

We illustrate the procedure in Table 12.9 for the Euler method for the initial-value problem $y' = y$, $y(0) = 1$, whose solution is $\bar{y}(x) = e^x$. The formula for the Euler method is

12.6 $y(x + h) = y(x) + hf(x, y) = y(x) + hy(x).$

Let us assume we are using a computer with a basic word length of three decimal digits for the mantissa of a floating-point number. In Procedure I we use single-precision throughout. For example, to go from $x = 0.06$ to $x = 0.07$ we have

12.7 $y(0.06) = 1.06 + 0.0106 = 1.06 \times 10^0 + 1.06 \times 10^{-2}.$

But in floating-point addition, the machine would replace 1.06×10^{-2} by 0.010×10^0, shifting right and dropping bits in order to obtain the same exponents. The result of the addition would be

$$1.07 \times 10^0.$$

On the other hand, with our partial double-precision scheme, Procedure II, we use only the most significant half of y to compute $f(x, y)$, but when we add $hf(x, y)$ to y we use double-precision. Thus, in order to go from $x = 0.06$ to $x = 0.07$ we first compute $f(x, y) = y = 1.06$. We then multiply by $h = 0.01$ in floating-point obtaining 0.0106. We extend this number to double-precision, i.e., to 0.0106000 and add in floating-point double-precision to 1.06150 obtaining 1.07210.

It can be seen that the result obtained by Procedure II for $x = 0.2$ is about 2 percent greater than that obtained by Procedure I. Moreover, if complete precision had been used, we would have obtained

12.8 $$y(0.20) = (1.01)^{20} \doteq 1.22019.$$

Thus, Procedure II has an error of 0.00069 as compared with an error of 0.02019 for Procedure I. Consequently there is a considerable improvement in accuracy as a result of using Procedure II.

12.9 Table Numerical Solution of $y' = y$, $y(0) = 1$ by the Euler Method with $h = 0.01$.

	Procedure I		Procedure II (Partial Double-precision)		
x	y	hy	y	$\{y\}$	$h\{y\}$
0	1.00	0.0100	1.00000	1.00	0.0100
0.01	1.01	0.0101	1.01000	1.01	0.0101
0.02	1.02	0.0102	1.02010	1.02	0.0102
0.03	1.03	0.0103	1.03030	1.03	0.0103
0.04	1.04	0.0104	1.04060	1.04	0.0104
0.05	1.05	0.0105	1.05100	1.05	0.0105
0.06	1.06	0.0106	1.06150	1.06	0.0106
0.07	1.07	0.0107	1.07210	1.07	0.0107
0.08	1.08	0.0108	1.08280	1.08	0.0108
0.09	1.09	0.0109	1.09360	1.09	0.0109
0.10	1.10	0.0110	1.10450	1.10	0.0110
0.11	1.11	0.0111	1.11550	1.11	0.0111
0.12	1.12	0.0112	1.12660	1.12	0.0112
0.13	1.13	0.0113	1.13780	1.13	0.0113
0.14	1.14	0.0114	1.14910	1.14	0.0114
0.15	1.15	0.0115	1.16050	1.16	0.0116
0.16	1.16	0.0116	1.17210	1.17	0.0117
0.17	1.17	0.0117	1.18380	1.18	0.0118
0.18	1.18	0.0118	1.19560	1.19	0.0119
0.19	1.19	0.0119	1.20750	1.20	0.0120
0.20	1.20		1.21950		

$\{y\}$ is the single-precision representation of y.

EXERCISES 8.12

1. Carry out a partial double-precision calculation as in the text for the initial-value problem

$$y' = x + y^2$$

with $y(0) = 2$, $h = 0.02$, from $x = 0$ to 0.2. Do the same using Procedure I.

2. Write a program based on the algorithm given in the text and apply it to the system $y_1' = -y_2$, $y_2' = y_1$ with $y_1(0) = 1$, $y_2(0) = 0$. Let the initial step size be $h = \pi/20$, and let $\varepsilon_H = 10^{-6}$, $\varepsilon_D = 10^{-12}$.

SUPPLEMENTARY DISCUSSION

Section 8.3

Milne [1953] gave a formula for the size of the interval of convergence of the Taylor series solution in terms of certain characteristic parameters for $f(x, y)$.

Section 8.4

The symbolic manipulations needed to carry out the method of undetermined coefficients can, for certain functions $f(x, y)$, be carried out on the computer using the program SYMBAL described by Engeli [1969].

Section 8.6

Methods based on single-step closed quadrature formulas are known as *Adams-Moulton methods*. Methods based on single-step open quadrature formulas are known as *Adams-Bashforth methods*. Methods based on double-step open methods are known as *Nyström methods* (Henrici [1962]).

Section 8.7

The single-step error is equivalent to what Dahlquist [1956] refers to as the "local truncation error."

A bound for the accumulated error for a more general class of methods is given by Henrici [1962, Chapter 5]. A statement of the result is given in Section 9.6.

Our discussion of the accumulated error is based on the assumption that the quadrature formulas for the predictor and the corrector are *definite* (see Section 7.7).

Milne [1953] estimated $y^{(p+1)}(x_n)$ in terms of the difference between the predicted and the corrected values of y_{n+1} for certain predictor-corrector methods.

Section 8.11

In the case of second-order differential equations of the form $y'' = f(x, y)$ special methods can be used which do not involve deriving an equivalent first-order system. See, for instance, Henrici [1962, Chapter 6].

For a thorough discussion of numerical methods for solving systems of ordinary differential equations see Henrici [1963].

Section 8.12

The method of partial double precision was proposed by Young [1955c]. Other methods to control the growth of round-off error are described by Gill [1951] and by Blum [1957]. An analysis of the use of partial double precision based on the use of statistics is given by Henrici [1962].

APPENDIX A

MATHEMATICAL PRELIMINARIES

A.1 INTRODUCTION

In this appendix we give many of the mathematical facts which are used in the text. Most of the material can be found in standard texts. However, the exact statement of a given theorem frequently varies from one book to another; hence it may be useful to give the exact form which we use. In some cases, proofs are given in the text.

A.2 SETS, SEQUENCES, AND SERIES

Sets of Real Numbers

Let $\mathbf{R}$ denote the set of real numbers, $\mathbf{R}^+$ the set of positive real numbers and $\mathbf{R}^-$ the set of negative real numbers.

2.1 Theorem. Given a set $S \subseteq \mathbf{R}$ such that for some α each element $x \in S$ does not exceed α, then there exists $\beta \in \mathbf{R}$ such that $\beta \geq x$ for all $x \in S$ and such that if $\beta' \geq x$ for all $x \in S$ then $\beta' \geq \beta$. The number β is called the *least upper bound*, or *supremum*, of S, and we write

2.2 $$\beta = \sup_{x \in S} x.$$

Similarly, if for some γ we have $\gamma \leq x$ for all $x \in S$, then there exists a *greatest lower bound* or *infimum*

2.3 $$\delta = \inf_{x \in S} x$$

of S which is the number δ such that $\delta \leq x$ for all $x \in S$ and if $\delta' \leq x$ for all $x \in S$ then $\delta' \leq \delta$.

2.4 Definition. Given a set $S \subseteq \mathbf{R}$, if $\sup_{x \in S} x$ exists and belongs to S, then we say that $\max_{x \in S} x$ exists and

2.5 $$\max_{x \in S} x = \sup_{x \in S} x.$$

Similarly, if $\inf_{x \in S} x$ exists and belongs to S, then we say that $\min_{x \in S} x$ exists and

2.6 $$\min_{x \in S} x = \inf_{x \in S} x.$$

2.7 Definition. Given $x_0 \in \mathbf{R}$ and $\delta > 0$ the set of all x such that

$$|x - x_0| < \delta$$

is a *neighborhood* of x_0.

We use the following notation to describe various kinds of intervals in $\mathbf{R}$.

2.8
$$\begin{cases}
[a, b] = \{x : a \leqq x \leqq b \text{ or } b \leqq x \leqq a\} \\
[a, b) = \{x : a \leqq x < b \text{ or } b < x \leqq a\} \\
(a, b] = \{x : a < x \leqq b \text{ or } b \leqq x < a\} \\
(a, b) = \{x : a < x < b \text{ or } b < x < a\}.
\end{cases}$$

(Here the notation $\{x : a \leqq x \leqq b\}$ means the set of all x such that $a \leqq x \leqq b$.)

Sequences

2.9 Definition. The sequence $\{x_i\} = x_1, x_2, \ldots$ of real numbers *converges* (to a limit x) if there exists a number $x \in \mathbf{R}$ such that given any $\varepsilon > 0$ there exists N such that for all $n > N$

$$|x - x_n| < \varepsilon.$$

2.10 Theorem. If the sequence $\{x_i\}$ is a non-decreasing (non-increasing) sequence which has a finite upper (lower) bound, then the sequence converges.

2.11 Definition. A sequence $\{x_i\}$ is a *Cauchy sequence* if, given any $\varepsilon > 0$, there exists N such that for all m and n such that $m > N$ and $n > N$ we have

$$|x_n - x_m| < \varepsilon.$$

2.12 Theorem. If the sequence $\{x_i\}$ is a Cauchy sequence, then it converges.

2.13 Theorem. If for some a and b each element of the sequence $\{x_i\}$ satisfies the condition $x_i \in [a, b]$, then there exists a convergent subsequence. That is, there exist positive integers $n_1, n_2, \ldots$ such that

$$1 \leqq n_1 < n_2 < \cdots$$

and such that the sequence $\{y_i\}$ converges, where $y_1 = x_{n_1}, y_2 = x_{n_2}, \ldots$.

Series of Real Numbers

2.14 Definition. The series

$$\sum_{i=1}^{\infty} u_i$$

converges to a limit s if the sequence of partial sums

$$s_1 = u_1$$
$$s_2 = u_1 + u_2$$
$$s_3 = u_1 + u_2 + u_3$$
$$\vdots$$

converges to s.

2.15 Theorem. If the series $\sum_{i=1}^{\infty} u_i$ converges and if

$$0 \leqq v_i \leqq u_i, \qquad i = 1, 2, \ldots$$

then the series $\sum_{i=1}^{\infty} v_i$ converges.

2.16 Theorem. Let $u_1, u_2, \ldots$ be nonnegative numbers such that for some N

2.17 $$u_{n+1} \leqq u_n, \qquad n = N, N+1, \ldots$$

and such that

$$\lim_{n \to \infty} u_n = 0.$$

Then the series

2.18 $$\sum_{i=1}^{\infty} (-1)^{n+1} u_i = u_1 - u_2 + u_3 - u_4 + \cdots$$

converges to a limit s. Moreover,

2.19 $$\left| s - \sum_{i=1}^{n} u_i \right| \leqq u_{n+1}$$

for $n \geqq N$.

A.3 FUNCTIONS OF A REAL VARIABLE

Limits

3.1 Definition. Given a function $f(x)$ defined in a neighborhood of x_0 (except possibly at x_0) we say that $\lim_{x \to x_0} f(x)$ exists and

3.2 $$\lim_{x \to x_0} f(x) = A$$

if given any $\varepsilon > 0$, there exists $\delta > 0$ such that for any x satisfying

3.3 $$0 < |x - x_0| < \delta$$

we have

3.4
$$|f(x) - A| < \varepsilon.$$

In the next definition $x \to x_0^+$ means "x approaches x_0 from the right" and $x \to x_0^-$ means "x approaches x_0 from the left." This terminology is not to be confused with $x \in \mathbf{R}^+$ (or $x \in \mathbf{R}^-$) which means x is a positive real number (or negative real number).

3.5 Definition. If $f(x)$ is defined for $I : (x_0, x_0 + \delta]$ for some $\delta > 0$, then $\lim_{x \to x_0+} f(x)$ exists and

3.6
$$\lim_{x \to x_0 +} f(x) = A$$

if, given any $\varepsilon > 0$, there exists $\delta > 0$ such that for all $x \in I$ we have

3.7
$$|f(x) - A| < \varepsilon.$$

A similar definition can be given for

3.8
$$\lim_{x \to x_0 -} f(x).$$

3.9 Definition. If $f(x)$ is defined in a neighborhood of x_0 (except possibly at x_0), then

3.10
$$\overline{\lim_{x \to x_0}} f(x) = \lim_{\delta \to 0} \left\{ \sup_{0 < |x - x_0| < \delta} f(x) \right\}$$

if the limit exists. The limit exists if, for some $\delta > 0$, the set of all $f(x)$ such that $0 < |x - x_0| < \delta$ has an upper bound. Similarly, we define

3.11
$$\underline{\lim_{x \to x_0}} f(x) = \lim_{\delta \to 0} \left\{ \inf_{0 < |x - x_0| < \delta} f(x) \right\}$$

if the limit exists.

3.12 Theorem. If

3.13
$$\overline{\lim_{x \to x_0}} f(x) = \underline{\lim_{x \to x_0}} f(x),$$

then $\lim_{x \to x_0} f(x)$ exists and

3.14
$$\lim_{x \to x_0} f(x) = \overline{\lim_{x \to x_0}} f(x) = \underline{\lim_{x \to x_0}} f(x).$$

Continuity

3.15 Definition. Given a function $f(x)$ defined in a neighborhood of x_0 we say that $f(x)$ is *continuous* at x_0 if $\lim_{x \to x_0} f(x)$ exists and

3.16
$$\lim_{x \to x_0} f(x) = f(x_0).$$

If $f(x)$ is continuous at each point of the interval $[a, b]$, we say that

3.17
$$f(x) \in C[a, b].$$

Similarly, $f(x) \in C[a, b)$, $C(a, b]$, or $C(a, b)$ if $f(x)$ is continuous in $[a, b)$, $(a, b]$, or (a, b), respectively.

3.18 Definition. The function $f(x)$, defined on the interval $I = [a, b]$, is *uniformly continuous* on I if, for every $\varepsilon > 0$, there exists a $\delta > 0$ such that for any two points x and x' in I satisfying

3.19
$$|x - x'| < \delta$$

we have

3.20
$$|f(x) - f(x')| < \varepsilon.$$

(It should be noted that δ is independent of x and x'.)

3.21 Theorem. If $f(x)$ is a continuous function defined on the interval $I = [a, b]$ then $f(x)$ is uniformly continuous on I.

3.22 Theorem. If $f(x)$ is a continuous function defined on the interval $I = [a, b]$, then for some ξ and η in I we have

3.23
$$f(\xi) = \max_{a \leq x \leq b} f(x), \qquad f(\eta) = \min_{a \leq x \leq b} f(x).$$

3.24 Theorem. If $f(x) \in C[a, b]$ and if $f(a)f(b) < 0$, then for some $\xi \in (a, b)$ we have

3.25
$$f(\xi) = 0.$$

3.26 Definition. A function $f(x)$ is *piecewise continuous* on the interval $I = [a, b]$ if $f(x)$ is defined and continuous for all but a finite number of points $x_1, x_2, \ldots, x_n$ in I. At each such point the limit

3.27
$$\lim_{x \to x_i+} f(x)$$

exists unless $x_i = b$ and the limit

3.28
$$\lim_{x \to x_i -} f(x)$$

exists unless $x_i = a$.

Integration

3.29 Definition. Given a function $f(x)$ defined on the interval $I = [a, b]$ we say that $f(x)$ is *integrable* over I, that $\int_a^b f(x)dx$ exists, and

3.30
$$\int_a^b f(x)dx = A$$

if, given any $\varepsilon > 0$, there exists $\delta > 0$ such that for any $x_1, x_2, \ldots, x_n$ satisfying

3.31
$$a = x_1 < x_2 < x_3 < \cdots < x_n = b$$

with

3.32
$$|x_{i+1} - x_i| < \delta, \qquad i = 1, 2, \ldots, n - 1$$

and for any $\xi_1, \xi_2, \ldots, \xi_{n-1}$ satisfying

3.33
$$\xi_i \in [x_i, x_{i+1}], \qquad i = 1, 2, \ldots, n - 1$$

we have

3.34
$$\left| A - \sum_{i=1}^{n-1} (x_{i+1} - x_i)f(\xi_i) \right| < \varepsilon.$$

3.35 Definition. If $f(x)$ is piecewise continuous on $[a, b]$, then $f(x)$ is integrable.

3.36 Theorem (Mean-value theorem for integrals). If $f(x)$ is continuous and $g(x)$ is piecewise continuous on $[a, b]$ and if $g(x) \geqq 0$ (or $g(x) \leqq 0$) on $[a, b]$, then

3.37
$$\int_a^b f(x)g(x)dx = f(\xi) \int_a^b g(x)dx$$

for some $\xi \in [a, b]$.

3.38 Corollary. If $f(x)$ is continuous on $[a, b]$, then

3.39
$$\int_a^b f(x)dx = f(\xi)(b - a)$$

for some $\xi \in [a, b]$.

3.40 Theorem. Let $g(x)$ be piecewise continuous function on the interval $I = [a, b]$. Then the relation

3.41
$$\int_a^b f(x)g(x)dx = f(\xi) \int_a^b g(x)dx$$

holds for all $f(x) \in C[a, b]$ and for some $\xi \in I$ if and only if $g(x)$ is one-signed in I (i.e., if $g(x) \geq 0$ for all $x \in I$ or $g(x) \leq 0$ for all $x \in I$).

Differentiation

3.42 Definition. Given a function $f(x)$ defined in a neighborhood of x_0, we say that $f(x)$ is *differentiable* at x_0 if

3.43
$$\lim_{x \to x_0} \frac{f(x) - f(x_0)}{x - x_0}$$

exists. If the limit exists, we let

3.44
$$f'(x_0) = \lim_{x \to x_0} \frac{f(x) - f(x_0)}{x - x_0}.$$

If $f(x)$ is differentiable (and hence continuous) at all points of the interval $[a, b]$, we say that

3.45
$$f(x) \in D^{(1)}[a, b].$$

If $f'(x)$ is continuous on the interval, then we say that

3.46
$$f(x) \in C^{(1)}[a, b].$$

Similarly, we say that $f(x) \in D^{(1)}[a, b)$ if $f'(x)$ exists on $[a, b)$, that $f(x) \in D^{(1)}(a, b]$ if $f'(x)$ exists on $(a, b]$, etc.

3.47 Theorem (Rolle's theorem). If $f(x) \in C[a, b]$ and $f(x) \in D^{(1)}(a, b)$ and if

3.48
$$f(a) = f(b) = 0,$$

then for some $\xi \in (a, b)$ we have

3.49
$$f'(\xi) = 0.$$

3.50 Theorem (Mean-value theorem). If $f(x) \in C[a, b]$ and $f(x) \in D^{(1)}(a, b)$, then for some $\xi \in (a, b)$ we have

3.51
$$f(b) - f(a) = (b - a)f'(\xi).$$

3.52 Theorem (L'Hospital's rule). Let $f(x) \in C^{(1)}[a, b]$ and $g(x) \in C^{(1)}[a, b]$ and let $c \in (a, b)$. If

3.53
$$g'(x) \neq 0, \qquad a \leqq x \leqq b, \qquad x \neq c$$

and if

3.54
$$f(c) = g(c) = 0,$$

then

3.55
$$\lim_{x \to c} \frac{f(x)}{g(x)} = \lim_{x \to c} \frac{f'(c)}{g'(c)},$$

provided the latter limit exists.

3.56 Theorem (Fundamental theorem of integral calculus). Let $f(x) \in C[a, b]$ and let $\phi(x)$ be defined for $x \in [a, b]$ by

3.57
$$\phi(x) = \int_a^x f(x)dx.$$

Then $\phi(x) \in C^{(1)}[a, b]$ and

3.58
$$\phi'(x) = f(x).$$

3.59 Theorem. If $f(x) \in C^{(1)}[a, b]$, then

3.60
$$\int_a^b f'(x)dx = f(b) - f(a).$$

3.61 Definition. If $f(x)$ is continuous and has continuous derivatives of all orders up to and including the n-th on the interval $I = [a, b]$, then we say that

3.62
$$f(x) \in C^{(n)}[a, b].$$

If $f(x) \in C^{(n-1)}[a, b]$ and $f^{(n)}(x)$ exists on I, then we say that

3.63
$$f(x) \in D^{(n)}[a, b].$$

3.64 Theorem (Taylor's theorem with the Lagrange form of the remainder). If $f(x) \in C^{(n)}[a, b]$ and if $f(x) \in D^{(n+1)}(a, b)$, then

3.65 $f(b) = f(a) + (b-a)f'(a) + \cdots + \dfrac{(b-a)^n}{n!}f^{(n)}(a) + \dfrac{(b-a)^{n+1}}{(n+1)!}f^{(n+1)}(\xi)$

for some $\xi \in (a, b)$.

3.66 Theorem (Taylor's theorem (with the integral form of remainder)).   If $f(x) \in C^{(n+1)}(a, b)$, then

3.67 $f(b) = f(a) + (b-a)f'(a) + \cdots + \dfrac{(b-a)^n}{n!}f^{(n)}(a) + \displaystyle\int_a^b \dfrac{(b-t)^n}{n!}f^{(n+1)}(t)dt.$

3.68 Definition.   The function $f(x)$ is *analytic* at x_0 if it is defined in a neighborhood N of x_0 and has continuous derivatives of all orders in N, and if for all $x \in N$ we have

3.69 $f(x) = f(x_0) + (x-x_0)f'(x_0) + \dfrac{(x-x_0)^2}{2!}f''(x_0) + \cdots.$

The function is analytic on an interval I if it is analytic at each point of I. The series (3.69) is the *Taylor series for $f(x)$*.

O and o Notation

3.70 Definition.   If for some constant K we have

3.71 $|f(x)| \leq K|g(x)|$

for $|x - a|$ sufficiently small, then we say that

3.72 $f(x) = O(g(x))$

as $x \to a$.   On the other hand, if

3.73 $\overline{\lim\limits_{x \to a}}\left|\dfrac{f(x)}{g(x)}\right| = 0,$

then we say that

3.74 $f(x) = o(g(x))$

as $x \to a$.

Sequences and Series of Functions

3.75 Definition.   The sequence of functions $s_1(x), s_2(x), \ldots$ converges to $s(x)$ if

3.76 $\lim\limits_{n \to \infty} s_n(x) = s(x).$

The sequence *converges uniformly* on the interval $I = [a, b]$ if, given any $\varepsilon > 0$, there exists N such that for all $n > N$ we have

3.77 $$|s_n(x) - s(x)| < \varepsilon$$

for all $x \in I$. (It should be noted that N is independent of x.)

3.78 Definition. The series

3.79 $$\sum_{i=1}^{\infty} u_i(x)$$

converges to $s(x)$ if the sequence of partial sums $\{s_n(x)\}$, where

3.80 $$s_n(x) = \sum_{i=1}^{n} u_i(x),$$

converges to $s(x)$. The convergence of the series is uniform on the interval $I = [a, b]$ if the convergence of the sequence is uniform.

3.81 Theorem. If the series $\sum_{i=1}^{\infty} u_i(x)$ converges uniformly to $s(x)$ on the interval $I = [a, b]$, and if each $u_i(x)$ is continuous on I then $s(x)$ is continuous on I.

3.82 Definition. A *power series* is a series of the form

3.83 $$\sum_{k=0}^{\infty} a_k x^k.$$

3.84 Theorem. If a power series converges for $\hat{x} \neq 0$, then it converges for all x such that

$$|x| < |\hat{x}|.$$

Moreover, for any positive δ such that $\delta < |\hat{x}|$, the convergence is uniform in the interval

$$|x| \leq |\hat{x}| - \delta.$$

3.85 Definition. The *radius of convergence* r of the power series (3.83) is defined by

3.86 $$r = \frac{1}{\varlimsup_{k \to \infty} \sqrt[k]{|a_k|}}.$$

3.87 Theorem. If r is the radius of convergence of the power series (3.83), then (3.83) converges for $|x| < r$ and diverges for $|x| > r$.

3.88 Theorem. If r is the radius of convergence of the power series (3.83), then

3.89
$$f(x) = \sum_{k=0}^{\infty} a_k x^k$$

defines a function which is analytic for $|x| < r$. Moreover, if $|x| < r$, then

3.90
$$\begin{cases} f'(x) = \sum_{k=1}^{\infty} k a_k x^{k-1} \\ \\ f''(x) = \sum_{k=2}^{\infty} k(k-1) a_k x^{k-2}, \end{cases}$$

etc., and if $|a| < r, |b| < r$, then

3.91
$$\int_a^b f(x)dx = \sum_{k=0}^{\infty} a_k \left(\frac{b^{k+1} - a^{k+1}}{k+1} \right).$$

A.4 FUNCTIONS OF TWO REAL VARIABLES

4.1 Definition. Given a point (x_0, y_0), for each $r > 0$ the circle

$$(x - x_0)^2 + (y - y_0)^2 < r^2$$

is a *neighborhood* of (x_0, y_0).

4.2 Definition. A *domain D* is a connected set of points such that for each point (x_0, y_0) of D there is a neighborhood of (x_0, y_0) contained in D.

4.3 Definition. A set S of points is *open* if for any point (x_0, y_0) of S there is a neighborhood of (x_0, y_0) each of whose points belongs to S.

4.4 Definition. A point (x_0, y_0) is a *limit point* of a set S if, given any $\varepsilon > 0$, there exists a point (x, y) (which may be (x_0, y_0) if $(x_0, y_0) \in S$) such that

$$(x - x_0)^2 + (y - y_0)^2 < \varepsilon^2.$$

4.5 Definition. A set S of points is *closed* if every limit point of S belongs to S.

4.6 Theorem (Taylor's theorem in two variables). If the function $f(x, y)$ is defined in a neighborhood N of (x_0, y_0) and has continuous partial derivatives of all orders up to and including the n-th in N, and if the partial derivatives of order $n + 1$ exist in N, then for any (x, y) in N we have

4.7 $\quad f(x, y) = \sum\limits_{i=0}^{n} \sum\limits_{j=0}^{i} \dfrac{(x - x_0)^j (y - y_0)^{i-j}}{j!(i-j)!} \dfrac{\partial^i f(x_0, y_0)}{\partial x^j \partial y^{i-j}}$

$$+ \sum\limits_{j=0}^{n+1} \dfrac{(x - x_0)^j (y - y_0)^{n+1-j}}{j!(n+1-j)!} \dfrac{\partial^{n+1} f(x_0 + \theta(x - x_0), y_0 + \theta(y - y_0))}{\partial x^j \partial y^{n+1-j}}$$

for some θ such that $0 < \theta < 1$.

4.8 Definition. The function $f(x, y)$ is *analytic* at a point (x_0, y_0) if it is defined in a neighborhood N of (x_0, y_0) and has continuous partial derivatives of all orders in N, and if for all $(x, y) \in N$ we have

4.9 $\qquad\qquad f(x, y) = \sum\limits_{i=0}^{\infty} \sum\limits_{j=0}^{i} \dfrac{(x - x_0)^i (y - y_0)^{i-j}}{j!(i-j)!} \dfrac{\partial^i f(x_0, y_0)}{\partial x^i \partial y^{i-j}}.$

The function is analytic in a domain D if it is analytic at each point of D. The series (4.9) in the *Taylor series for* $f(x, y)$.

A.5 FUNCTIONS OF A COMPLEX VARIABLE

Let $\mathbf{C}$ denote the set of complex numbers z of the form

5.1 $\qquad\qquad\qquad\qquad z = x + iy$

where $x, y \in \mathbf{R}$ and where

5.2 $\qquad\qquad\qquad\qquad i^2 = -1.$

If (5.1) holds then we let

5.3 $\qquad\qquad\qquad \operatorname{Re} z = x, \quad \operatorname{Im} z = y.$

The *modulus*, $|z|$, of $z \in \mathbf{C}$ is given by

5.4 $\qquad\qquad\qquad |z| = \sqrt{(\operatorname{Re} z)^2 + (\operatorname{Im} z)^2}.$

If $z_1, z_2 \in \mathbf{C}$ then $z_1 = z_2$ if and only if

5.5 $\qquad\qquad\qquad \begin{cases} \operatorname{Re} z_1 = \operatorname{Re} z_2 \\ \operatorname{Im} z_1 = \operatorname{Im} z_2. \end{cases}$

Elementary Arithmetic Operations with Complex Numbers

If $z_1 = x_1 + iy_1$ and $z_2 = x_2 + iy_2$ where $x_1, y_1, x_2, y_2 \in \mathbf{R}$, then

5.6 $\qquad\qquad\qquad z_1 \pm z_2 = (x_1 \pm x_2) + i(y_1 \pm y_2).$

5.7 $$z_1 z_2 = (x_1 x_2 - y_1 y_2) + i(x_1 y_2 + x_2 y_1)$$

and

5.8 $$\frac{z_1}{z_2} = \frac{(x_1 x_2 + y_1 y_2) + i(-x_1 y_2 + x_2 y_1)}{x_2^2 + y_2^2}, \qquad \text{if } z_2 \neq 0.$$

Sets, Sequences, and Series

5.9 Definition. Given $z_0 \in \mathbf{C}$ and $\delta > 0$ the set of all z such that

$$|z - z_0| < \delta$$

is a *neighborhood* of z_0.

Definitions 2.7 and 2.9 and Theorems 2.12 and 2.13 concerning sequences of real numbers and Definition 2.14 concerning series of real numbers apply also to sequences and series of complex numbers. In the hypothesis of Theorem 2.13 the condition that each element of the sequence lie in $[a, b]$ should be replaced by the condition that each element lie in the circle

$$|z - z_0| < r$$

for some $z_0 \in \mathbf{C}$ and for some $r \in \mathbf{R}^+$.

5.10 Definition. Given a function $f(z)$ of a complex variable defined in a neighborhood of z_0 (except possibly at z_0) we say that $\lim_{z \to z_0} f(z)$ exists and

5.11 $$\lim_{z \to z_0} f(z) = A$$

if, given any $\varepsilon > 0$, there exists $\delta > 0$ such that for any z satisfying

$$0 < |z - z_0| < \delta$$

we have

$$|f(z) - A| < \varepsilon.$$

5.12 Definition. Given a function $f(z)$ of a complex variable defined in a neighborhood of z_0 we say that $f(z)$ is *continuous* at z_0 if $\lim_{z \to z_0} f(z)$ exists and

5.13 $$\lim_{z \to z_0} f(z) = f(z_0).$$

If $f(z)$ is continuous at each point of a domain D, we say that $f(z)$ is continuous in D.

5.14 Definition. Given a function $f(z)$ of a complex variable defined in a neighborhood of z_0, we say that $f(z)$ is *differentiable* at z_0 if

5.15
$$\lim_{z \to z_0} \frac{f(z) - f(z_0)}{z - z_0}.$$

If $f(z)$ is differentiable at each point of a domain D we say that $f(z)$ is differentiable in D.

5.17 Definition. The function $f(z)$ of a complex variable is *analytic* at z_0 if $f(z)$ is defined and differentiable at each point in a neighborhood of z_0. The function $f(z)$ is analytic in a domain D if it is analytic at each point of D.

5.18 Theorem. If $f(z)$ is analytic at z_0 then, for some neighborhood N of z_0, $f(z)$ has continuous derivatives of all orders in N and

5.19
$$f(z) = f(z_0) + (z - z_0)f'(z_0) + \frac{(z - z_0)^2}{2!}f''(z_0) + \cdots$$

for all z such that $|z - z_0| < \delta$, for some $\delta > 0$.

5.20 Theorem Cauchy-Riemann equations). Let $f(z) = u(x, y) + iv(x, y)$, where $u(x, y)$ and $v(x, y)$ are real-valued functions of the real variables x and y, and where $z = x + iy$. If $f(z)$ is analytic at $z_0 = x_0 + iy_0$, where $x_0, y_0 \in \mathbf{R}$, then for each point (x, y) in some neighborhood of (x_0, y_0) we have

5.21
$$\begin{cases} \dfrac{\partial u(x, y)}{\partial x} = \dfrac{\partial v(x, y)}{\partial y} \\[2mm] \dfrac{\partial v(x, y)}{\partial x} = -\dfrac{\partial u(x, y)}{\partial y}. \end{cases}$$

5.22 Theorem. Let $f(z)$ be analytic in a domain R and let S be the boundary of R. Then

5.23
$$\max_{z \in R + S} |f(z)| \leq \max_{z \in S} |f(z)|.$$

Moreover, if $f(z)$ does not vanish in R we have

5.24
$$\min_{z \in R + S} |f(z)| \geq \min_{z \in S} |f(z)|.$$

5.25 Definition. A *power series* is a series of the form

5.26
$$\sum_{k=0}^{\infty} a_k z^k.$$

5.27 Theorem. If a power series converges for $\hat{z} \neq 0$ then it converges for all z such that

$$|z| < |\hat{z}|.$$

Moreover, for any positive δ such that $\delta < |\hat{z}|$ the convergence is uniform for

$$|z| \leqq |\hat{z}| - \delta.$$

5.28 Definition. The *radius of convergence r* of the power series (5.26) is defined by

5.29
$$r = \frac{1}{\lim\limits_{k \to \infty} \sqrt[k]{|a_k|}}.$$

5.30 Theorem. If r is the radius of convergence of the power series (5.26) then (5.26) converges for $|z| < r$ and diverges for $|z| > r$.

5.31 Theorem. If r is the radius of convergence of the power series (5.26) then

5.32
$$f(z) = \sum_{k=0}^{\infty} a_k z^k$$

defines a function which is analytic for $|z| < r$. Moreover, if $|z| < r$ then

5.33
$$\begin{cases} f'(z) = \sum_{k=1}^{\infty} k a_k z^{k-1} \\ f''(z) = \sum_{k=2}^{\infty} k(k-1) a_k z^{k-2}, \end{cases}$$

etc.

A.6 EXTREMA OF A FUNCTION OF SEVERAL VARIABLES

Let $f(x_1, x_2, \ldots, x_n)$ be continuous and have continuous partial derivatives of orders one and two in a neighborhood of $(x_1^{(0)}, x_2^{(0)}, \ldots, x_n^{(0)})$. Then a *necessary* condition that $f(x_1, x_2, \ldots, x_n)$ have a relative extremum (a relative minimum or a relative maximum) at $(x_1^{(0)}, x_2^{(0)}, \ldots, x_n^{(0)})$ is that

6.1
$$\frac{\partial f(x_1^{(0)}, x_2^{(0)}, \ldots, x_n^{(0)})}{\partial x_i} = 0, \qquad i = 1, 2, \ldots, n.$$

A *sufficient* condition for a relative minimum (maximum) is that the matrix

6.2
$$A = \begin{bmatrix} \alpha_{11} & \alpha_{12} & \cdots & \alpha_{1n} \\ \alpha_{21} & \alpha_{22} & \cdots & \alpha_{2n} \\ \vdots & & & \\ \alpha_{n1} & \alpha_{n2} & \cdots & \alpha_{nn} \end{bmatrix}$$

where

6.3 $$\alpha_{ij} = \frac{\partial^2 f(x_1^{(0)}, x_2^{(0)}, \ldots, x_n^{(0)})}{\partial x_i \, \partial x_j}, \qquad i,j = 1, 2, \ldots, n$$

is positive definite (negative definite). See Chapter 11 for the definition of a positive definite matrix.

A.7 LINEAR ALGEBRA

Consider the system of n linear algebraic equations

7.1
$$\begin{cases} a_{11}u_1 + a_{12}u_2 + \cdots + a_{1n}u_n = b_1 \\ a_{21}u_1 + a_{22}u_2 + \cdots + a_{2n}u_n = b_2 \\ \vdots \\ a_{n1}u_1 + a_{n2}u_2 + \cdots + a_{nn}u_n = b_n \end{cases}$$

with n unknowns $u_1, u_2, \ldots, u_n$.

7.2 Theorem (Cramer's rule). If the determinant‡

7.3 $$\Delta = \det A$$

$$= \det \begin{bmatrix} a_{11} & a_{12} & \cdots & a_{1n} \\ a_{21} & a_{22} & \cdots & a_{2n} \\ \vdots & & & \\ a_{n1} & a_{n2} & \cdots & a_{nn} \end{bmatrix}$$

does not vanish then the system (7.1) has a unique solution which is given by

7.4 $$u_i = \frac{\Delta_i}{\Delta}, \qquad i = 1, 2, \ldots, n.$$

Here, for each i, Δ_i is the determinant of the matrix formed from A by replacing a_{1i} by b_1, a_{2i} by $b_2, \ldots, a_{ni}$ by b_n.

The *homogeneous system* corresponding to (7.1) is obtained from (7.1) by letting

7.5 $$b_1 = b_2 = \cdots = b_n = 0.$$

The homogeneous system always has the *trivial solution*.

---

‡ We assume that the reader is familiar with the definition of a determinant.

7.6 $$u_1 = u_2 = \cdots = u_n = 0.$$

Any other solution is said to be *nontrivial*.

7.7 Theorem. The homogeneous system, corresponding to (7.1) has a nontrivial solution if and only if

7.8 $$\Delta = 0.$$

7.9 Theorem. The system (7.1) has a unique solution if and only if the corresponding homogeneous system has no solution except the trivial solution.

A.8 POLYNOMIALS

8.1 Definition. Given the real or complex numbers $a_0, a_1, \ldots, a_n$, where $a_0 \neq 0$, the function

8.2 $$P(x) = a_0 x^n + a_1 x^{n-1} + \cdots + a_n$$

is a *polynomial of degree n*. (If $n = 0$, the function

8.3 $$P(x) = a_0$$

is a polynomial of degree zero whether or not $a_0 = 0$.)

8.4 Definition. The number α is a *zero* of *multiplicity m* of the polynomial $P(x)$ of degree n if

8.5 $$\begin{cases} P(\alpha) = P'(\alpha) = \cdots = P^{(m-1)}(\alpha) = 0 \\ P^{(m)}(\alpha) \neq 0. \end{cases}$$

8.6 Theorem (The fundamental theorem of algebra). Any polynomial of degree one or more has at least one zero.

8.7 Theorem. A polynomial of degree $n \geq 0$ has exactly n zeros, provided that a zero of multiplicity m is counted as m zeros.

A.9 MISCELLANEOUS

9.1 (*Principle of Mathematical Induction*). If a certain proposition involving the integer variable n is true for $n = 1$ and can be shown to be true for $n + 1$ provided it is true for all integers less than $n + 1$, then it is true for all n.

As an example, consider the proposition that

$$S_n = 1 + 2 + \cdots n = \frac{n(n+1)}{2}$$

for $n = 1, 2, \ldots$. By direct verification the proposition is true for $n = 1$ since $S_1 = 1$. If the proposition is true for n then

$$S_{n+1} = S_n + n + 1$$

$$= \frac{n(n + 1)}{2} + n + 1$$

$$= \frac{(n + 1)(n + 2)}{2}$$

$$= \frac{(n + 1)((n + 1) + 1)}{2}$$

so that the proposition is true for $n + 1$. Hence the proposition is true in general.

9.2 (The symbol "$\sim$"). The symbol $\sim$ is used rather loosely throughout this book in the following three senses:

a) Asymptotic equality. For example, we write

$$f(x) \sim g(x) \qquad \text{as } x \to a,$$

if and only if

$$\lim_{x \to a} \frac{f(x)}{g(x)} = 1.$$

b) Approximately equal to (in some sense). The symbol "$\doteq$" is used to mean "is approximately equal to," also. For example, we might write

$$e \doteq 2.71828$$

(a numerical approximation) and

$$e^x \sim 1 + x + \frac{x^2}{2!} + \frac{x^3}{3!}$$

(an algebraic approximation).

c) Is represented by. For example, in 2-(4.4) we write

$$a \sim a_{59}a_{58}a_{57} \cdots a_2 a_1 a_0.$$

APPENDIX B

REVISION OF THEOREM 7.36

Replace the material between the asterisks on pages 125 and 126 by the following:

7.36. Theorem. Let $f(x)$ be a continuous function in $I = [\bar{x} - \delta, \bar{x} + \delta]$ where $\delta > 0$, and let $\bar{x}$ be a root of (1.1) in I. If the iterative method (7.4) is consistent with (1.1), if $\phi(x) \in C^{(1)}$ in I, and if $|\phi'(x)| < 1$ in I, except possibly at $\bar{x}$, then the sequence $x_0, x_1, x_2, \ldots$ defined by (7.4) converges to $\bar{x}$ provided $x_0 \in I$.

Proof. Since the iterative method is consistent, we have

7.37
$$\bar{x} = \phi(\bar{x}).$$

By the mean-value theorem

7.38
$$x_{n+1} - \bar{x} = \phi(x_n) - \phi(\bar{x}) = (x_n - \bar{x})\phi'(\xi)$$

where ξ lies between x_n and $\bar{x}$. Hence we have

7.39
$$|x_{n+1} - \bar{x}| \le |x_n - \bar{x}| \le \ldots \le |x_0 - \bar{x}|.$$

Thus all of the x_n lie in the interval I. The sequence of numbers $|x_0 - \bar{x}|, |x_1 - \bar{x}|, \ldots$ is a monotone nonincreasing sequence, bounded below by zero, and therefore converges. If it converges to zero, then $x_n \to \bar{x}$. Otherwise, $|x_n - \bar{x}| \to \eta > 0$, and $|x_n - \bar{x}| \ge \eta$ for all n.

Let M be the maximum of $|\phi'(x)|$ for I': $\eta/2 \le |x - \bar{x}| \le \delta$. By the continuity of $\phi'(x)$, and the fact that $|\phi'(x)| < 1$ for all $x \in I'$, we have $M < 1$. Moreover, since

7.40
$$\phi(x_n) - \phi(\bar{x}) = \int_{\bar{x}}^{x_n} \phi'(x)dx$$

it follows that

$$|x_{n+1} - \bar{x}| = |\phi(x_n) - \phi(\bar{x})| \le M\left(|x_n - \bar{x}| - \frac{\eta}{2}\right) + \frac{\eta}{2}$$

$$= M|x_n - \bar{x}| + \frac{\eta}{2}(1 - M) \le \frac{1 + M}{2}|x_n - \bar{x}|$$

since $\eta \le |x_n - \bar{x}|$. Therefore,

$$|x_n - \bar{x}| \le \left(\frac{1 + M}{2}\right)^n |x_0 - \bar{x}|.$$

For n large enough we have $|x_n - \bar{x}| < \eta$, which contradicts the assumption that $|x_n - \bar{x}|$ $\geq \eta$ for all n. Therefore, $x_n - \bar{x} \to 0$ as $n \to \infty$ and the method converges.

We remark that there can be only one root of (1.1) in I. If $\bar{x}'$ were another root, then by consistency $\bar{x}' = \phi\,(\bar{x}')$. If $x_0 = \bar{x}'$ then $x_1 = x_2 = \ldots = \bar{x}'$. Since the sequence converges to $\bar{x}'$ and must also converge to $\bar{x}$, it follows that $\bar{x}' = \bar{x}$. This completes the proof.

APPENDIX C

REVISION OF THEOREM 8.12

Replace the material between the asterisks on page 129 by the following:

8.12 Theorem. Given $\bar{x}$ and $r > 0$ let $g(x)$ and $h(x)$ be continuously differentiable in $I = [\bar{x} - r, \bar{x} + r]$. Let $\bar{x}$ be a root of $g(x) - h(x) = 0$, and let $|g'(x)| \geq \alpha$ and $|h'(x)| \leq \beta$ in I, where $\alpha > 0$, $\beta \geq 0$, and $\beta/\alpha < 1$. For any $a \in I$, there exists $b \in I$ such that $g(b) = h(a)$. Moreover, the following method converges to $\bar{x}$ for any $x_0 \in I$: given x_n we let x_{n+1} be any number in I such that $g(x_{n+1}) = h(x_n)$.

Proof. By the continuity of $g'(x)$ in I we have either $g'(x) \geq \alpha$ for all x in I or else $g'(x) \leq -\alpha$ for all x in I. For any $a \in I$ with $a \neq \bar{x}$, we have, by the mean-value theorem, and from the fact that $g(\bar{x}) = h(\bar{x})$.

$$g(a) - h(a) = (a - \bar{x})[g'(\xi) - h'(\xi)]$$

and

$$g(\bar{x} + (\bar{x} - a)) - h(a) = g(\bar{x} + (\bar{x} - a)) - g(\bar{x}) + h(\bar{x}) - h(a)$$
$$= (\bar{x} - a)[g'(\xi_1) + h'(\xi_2)]$$

where ξ, ξ_1, and ξ_2 lie in I. Evidently

$$\Delta = [g(a) - h(a)][g(\bar{x} + (\bar{x} - a)) - h(a)] = -(\bar{x} - a)^2[g'(\xi) - h'(\xi)][g'(\xi_1) + h'(\xi_2)].$$

Since $|h'(x)| \leq \beta$ for all $x \in I$ and since either $g'(x) \geq \alpha$ for all $x \in I$ or else $g'(x) \leq -\alpha$ for all $x \in I$, where $\beta < \alpha$, it follows that

$$\Delta < 0.$$

Hence the continuous function $g(x) - h(a)$ changes sign in I and therefore has a zero in I. Thus, for any $a \in I$ there exists a root of $g(x) - h(a) = 0$ in I.

It now follows that the method defined above can be carried out. By the mean-value theorem we have

BIBLIOGRAPHY

Pages on which a reference is cited are listed in italic numerals at the end of each reference.

Achieser, N. I. [1956], *Theory of Approximation*, Frederick Ungar Pub. Co., New York; *328*.

Adams, Duane, A. [1967], "A stopping criterion for polynomial root finding," *Comm. Assoc. Comput. Mach.* 10, 655–658; *212*.

Ahlberg, J. H., E. N. Nilson, and J. L. Walsh [1967], *The Theory of Splines and Their Applications*, Academic Press, Inc., New York; *296*.

Ahlfors, L. V. [1966], *Complex Analysis* (second edition), McGraw-Hill Book Co., New York; *169*.

Aitken, A. C. [1932], "An interpolation by iteration of proportional parts, without the use of differences," *Proc. Edinburgh Math. Soc.* 3, Series 2, 56–76; *264*.

Allen, D. N. de G. [1954], *Relaxation Methods*, McGraw-Hill Book Co., New York; *997, 998*.

Antosiewicz, Henry A., and Walter Gautschi [1962], "Numerical methods in ordinary differential equations," Chapter 9 in *A Survey of Numerical Analysis*, edited by John Todd, McGraw-Hill Book Co., New York.

Arms, R. J., L. D. Gates, and B. Zondek [1956], "A method of block iteration," *J. Soc. Indust. Appl. Math.* 4, 220–229; *1074*.

Bareiss, E. H. [1960], "Resultant procedure and the mechanization of the Graeffe process," *J. Assoc. Comput. Mach.* 7, 346–386; *242*.

Bareiss, E. H. [1967], "The numerical solution of polynomial equations and the resultant procedure," Chapter 10 of Ralston and Wilf [1967]; *242*.

Batschelet, von Eduard [1952], "Über die numerische Auflösung von Randwertproblemen bei elliptischen partiellen Differentialgleichungen," *ZAMP* III, 165–193; *998*.

Bauer, F. L. [1963], "Optimally scaled matrices," *Numer. Math.* 5, 73–87; *815*.

Bauer, F. L., and C. T. Fike [1960], "Norms and exclusion theorems," *Numer. Math.* 2, 137–141; *947*.

Bauer, F. L., H. Rutishauser, and E. Stiefel [1963], "New aspects of numerical quadrature," *Proc. of Symposia in Applied Math.*, Vol. XV, Amer. Math. Soc., Providence, Rhode Island; *383, 384*.

Beckman, F. S. [1960], "The solution of linear equations by the conjugate gradient method," Chapter 4 of Ralston and Wilf [1960]; *1074*.

Bellman, R. [1960], *Introduction to Matrix Analysis*, McGraw-Hill Book Co., New York; *754*.

Bernstein, S. [1937], "Sur les formules de quadrature de Cotes et de Tchebycheff," *C. R. de l'Academie des Sciences de l'URSS* 14, 323–326; *414*.

Bickley, W. G. [1941], "Formulae for numerical differentiation," *Math. Gazette* 25, 19–27; *420*.

Bieberbach, L. [1930], *Theorie der Differentialgleichungen*, Springer, Berlin; *442*.

Birkhoff, G., C. de Boor, B. Swartz, and B. Wendroff [1966], "Rayleigh-Ritz approximation by piecewise cubic polynomials," *SIAM J. Numer. Anal.* 3, 188–203; *666*.

Birkhoff, Garrett, and Gian-Carlo Rota [1962], *Ordinary Differential Equations*, Ginn and Co., Boston; *429, 430, 486*.

Birkhoff, Garrett, and Saunders MacLane [1953], *A Survey of Modern Algebra*, The MacMillan Co., New York.

Birkhoff, Garrett, Richard S. Varga, and David Young [1962], "Alternating direction implicit methods," in *Advances in Computers*, Vol. 3, edited by F. Alt and M. Rubinoff, Academic Press, New York, 189–273; *1047, 1055, 1060, 1061, 1074*.

Birkhoff, G., David M. Young, and E. H. Zarantenello [1951], "Effective conformal transformation of smooth, simply connected domains," *Proc. Nat. Acad. Sci.* 37, 411–414.

Blair, A., N. Metropolis, J. von Neumann, A. H. Taub, and M. Tsingori [1959], "A study of a numerical solution of a two-dimensional hydrodynamical problem," *Math. Tables Aids Comput.* 13, 145–184; *1074*.

Blanch, G. [1964], "Numerical evaluation of continued fractions," *SIAM Review* 6, 383–421; *329*.

Blum, E. K. [1957], "A modification of the Runge-Kutta fourth-order method," Numerical Note NN–80, Ramo-Wooldridge Corp., Los Angeles; *492*.

Borosh, I., and A. S. Fraenkel [1966], "Exact solutions of linear equations with rational coefficients by congruence techniques," *Math. of Comp.* 20, 107–112; *68, 885, 888*.

Bramble, J. H., and B. E. Hubbard [1963], "A theorem on error estimation for finite difference analogues of the Dirichlet problem for elliptic equations," in *Contributions to Differential Equations*, Vol. II, John Wiley and Sons, New York; *992*.

Bramble, J. H., and B. E. Hubbard [1964], "On a finite difference analogue of an elliptic boundary problem which is neither diagonally dominant nor of nonnegative type," *Jour. of Math. and Phys.* 43, 117–132; *614*.

Bronson, R. [1969], *Matrix Methods, an Introduction*, Academic Press, New York; *717, 731, 753, 755*.

Businger, P. A. [1968], "Matrices which can be optimally scaled," *Numer. Math.* 12, 346–348; *817*.

Businger, P. A. [1969], "Reducing a matrix to Hessenberg form," *Math. of Comp.* 23, 819–822; *924*.

Businger, P. A. [1971], "Monitoring the numerical stability of Gaussian elimination," *Numer. Math.* 16, 360–361; *806*.

Butcher, J. C. [1965], "A modified multistep method for the numerical integration of ordinary differential equations," *J. Assoc. Comput. Mach.* 12, 124–135.

Champagne, W. P. [1964], "On finding roots of polynomials by hook or by crook," Master's thesis, The University of Texas at Austin, August 1964; also TNN–37, Computation Center, The University of Texas at Austin, 1964; *177, 216, 217, 235, 237, 238*.

Chartres, B. A. [1966], "Automatic controlled precision calculations," *J. Assoc. Comput. Mach.* 13, 386–403; *68*.

Chase, P. E. [1962], "Stability properties of predictor-corrector methods for ordinary differential equations," *J. Assoc. Comput. Mach.* 9, 457–468.

Cheney, E. W. [1966], *Introduction to Approximation Theory*, McGraw-Hill Book Co., New York.

Ciarlet, P., M. Schultz, and R. Varga [1967], "Numerical methods of high-order accuracy for nonlinear boundary value problems," *Numer. Math.* 9, 394–430; *665, 666*.

Clenshaw, C. W. [1962], "Chebyshev series for mathematical functions," *Mathematical Tables* 5, Nat. Physical Lab., London; *343*.

Cody, W. J. [1967], "The influence of machine design on numerical algorithms," *Proceedings Spring Joint Computer Conference 1967*, 305–309; *69*.

Cohen, Abraham [1931], *An Introduction to the Lie Theory of One-Parameter Groups*, Stechert and Co., New York; *431*.

Collatz, L. [1933], "Bemerkungen zur Fehlerabschätzung für das Differenzenverfahren bei partiellen Differentialgleichungen," *Z. Angew. Math. Mech.* 13, 56–57; *576, 612, 960*.

Collatz, L. [1960], *The Numerical Treatment of Differential Equations*, 3rd ed., Springer-Verlag, Berlin; *665, 666*.

Collatz, L. [1966], *Functional Analysis and Numerical Mathematics*, Academic Press, New York; *761*.

Conte, S. D. [1962], "The computation of satellite orbit trajectories," in *Advances in Computers*, Vol. 3, edited by Franz L. Alt and Morris Rubinoff, Academic Press: New York, 1–76; *488*.

Conte, S. D., and David M. Young [1957], "Eigenvalues in modern industry III: Problems involving differential operators," *Proceedings of the Joint N.Y.U.-I.B.M. Symposium on Digital Computing in the Aircraft Industry, New York, Jan. 31, 1957–Feb. 28, 1957*; *659, 661*.

Control Data Corp. [1969], "*Control Data 6400/6500/6600 Computer Systems Reference Manual*, Publication 60100000, Control Data Corp., Minneapolis, Minn.; *45, 52*.

Courant, R., and D. Hilbert [1962], *Methods of Mathematical Physics*, Vol. II, Interscience Publishers, New York; *952*.

Courant, R., and F. John [1965], *Introduction to Calculus and Analysis*, Interscience Publishers, New York.

Crank, J., and P. Nicolson [1947], "A practical method for numerical evaluation of solutions of partial differential equations of the heat-conduction type," *Proc. Cambridge Philos. Soc.* 43, 50–67; *1078*.

Cullen, C. G. [1966], *Matrices and Linear Transformations*, Addison-Wesley, Reading, Mass.; *681*.

Cuthill, E. H., and R. S. Varga [1959], "A method of normalized block iteration," *J. Assoc. Comput. Mach.* 6, 236–244; *1064*.

Dahlquist, G. [1956], "Convergence and stability in the numerical integration of ordinary differential equations," *Math. Scand.* 4, 33–53; *492, 502, 513, 516, 522, 523, 574*.

Davis, Phillip J. [1963], *Interpolation and Approximation*, Blaisdell Publishing Company, New York; *288*.

Davis, Phillip J., and Phillip Rabinowitz [1967], *Numerical Integration*, Blaisdell Publishing Company, Waltham, Mass.; *420, 421*.

Dejon, Bruno, and Peter Henrici [1969], *Constructive Aspects of the Fundamental Theorem of Algebra*. (Proceedings of a Symposium conducted by the IBM Research Laboratory, Zurich-Ruschlikon, Switzerland, June 5–7, 1967), Wiley-Interscience, New York; *241*.

Delves, L. M., and J. N. Lyness [1967], "A numerical method for locating the zeros of an analytic function," *Math. Comp.* 21, 543–560; *245*.

DeVogelaere, R. [1958], "Over-relaxations," Abstract No. 539–53, *Amer. Math. Soc. Notices* 5, 147; *1068*.

Douglas, Jim, Jr. [1955], "On the numerical integration of $\partial^2 u/\partial x^2 + \partial^2 u/\partial y^2 = \partial u/\partial t$ by implicit methods," *J. Soc. Indust. Appl. Math.* 3, 42–65; *1097*.

Douglas, Jim, Jr. [1961], "A survey of numerical methods for parabolic differential equations," in *Advances in Computers*, 2, edited by F. L. Alt, Academic Press, New York, 1–54; *1098*.

Douglas, J., Jr., and J. E. Gunn [1962], "Alternating direction methods for parabolic systems in *m* space variables," *J. Assoc. Comput. Mach.* 9, 450–456; *1098*.

Douglas, J., Jr., and J. E. Gunn [1964], "A general formulation of alternating direction methods," *Numer. Math.* 6, 428–453; *1098*.

Douglas, J., Jr., and H. Rachford [1956], "On the numerical solution of heat conduction problems in two and three space variables," *Trans. Amer. Math. Soc.* 82, 421–439; *1098*.

Downing, J. A. [1966], "The automatic construction of contour plots with applications to numerical analysis," Master's thesis, The University of Texas at Austin; also TNN–58, Computation Center, The University of Texas at Austin; *162, 174*.

DuFort, E. C., and S. P. Frankel [1953], "Stability conditions in the numerical treatment of parabolic differential equations," *Math. of Comp.* (formerly *Math. Tables Aids Comput.*) 7, 135–152; *1085*.

Ehrlich, Louis W. [1964], "The block symmetric successive overrelaxation method," *SIAM J.* 12, 807–826; *1068*.

Eidson, Harold D., Jr. [1969], "The convergence of Richardson's finite-difference analogue for the heat equation," Master's thesis, The University of Texas at Austin; also TNN–90, Computation Center, The University of Texas at Austin; *1085*.

Engeli, Max E. [1969], "User's manual for the formula manipulation language SYMBAL," Report TRM-8.01, Computation Center, The University of Texas at Austin; *492*.

Faddeev, D. K., and V. N. Faddeeva [1963], *Computational Methods of Linear Algebra*, W. H. Freeman and Co., San Francisco; *721, 725, 747, 767, 823*.

Fanett, Mary [1963], "Application of the Remes algorithm to a problem in rational approximation," Master's thesis, The University of Texas at Austin; also TNN–23, Computation Center, The University of Texas at Austin; *317*.

Fejér, L. [1907], "Untersuchungen über Fouriersche Reihen," *Math. Annalen* 58, 51–69.

Fike, C. T. [1968], *Computer Evaluation of Mathematical Functions*, Prentice-Hall, Englewood Cliffs, N.J.; *68, 343*.

Flanders, D., and G. Shortley [1950], "Numerical determination of fundamental modes," *J. Appl. Physics* 21, 1326–1332; *1074*.

Ford, Lester R. [1933], *Differential Equations*, McGraw-Hill Book Co., New York; *389, 429, 431, 436, 643*.

Forsythe, G. E. [1957], "Generation and use of orthogonal polynomials for data-fitting with a digital computer," *J. Soc. Indust. Appl. Math.* 5, 74–88; *323, 324*.

Forsythe, G. E. [1966], "How do you solve a quadratic equation?" Tech. Report CS40, Stanford University; *92*.

Forsythe, G. E. [1967], "What is a satisfactory quadratic equation solver?" Tech. Report CS74, Stamford University; *92*.

Forsythe, G. E. [1969], "Solving a quadratic equation on a computer," *The Mathematical Sciences—A Collection of Essays*, edited by the National Research Council's Committee on Support of Research in the Mathematical Sciences, published for the National Academy of Sciences—National Research Council by the M.I.T. Press, Cambridge, Mass.; *92*.

Forsythe, G. E. [1970], "Pitfalls in computation," *Amer. Math. Monthly* 77, 931–956.

Forsythe, G. E., and C. Moler [1967], *Computer Solution of Linear Algebraic Equations*, Prentice-Hall, Englewood Cliffs, N.J.; *810, 812, 813, 814, 815, 816, 827, 833*.

Forsythe, G. E., and W. R. Wasow [1960], *Finite Difference Methods for Partial Differential Equations*, John Wiley and Sons, Inc., New York; *998, 1073, 1098*.

Fox, L. [1944], "Solution by relaxation methods of plane potential problems with mixed boundary conditions," *Quart. Appl. Math.* 2, 251–257; *993*.

Fox, L. [1947], "Some improvements in the use of relaxation methods for the solution of ordinary and partial differential equations," *Proc. Roy. Soc.*, London, Ser. A, Vol. 190, 31–59; *577, 666*.

Fox, L. [1957], *Numerical Solution of Two-point Boundary Value Problems in Ordinary Differential Equations*, Clarendon Press, Oxford; *665*.

Fox, L. [1962], *Numerical Solution of Ordinary and Partial Differential Equations*, Addison-Wesley Pub. Co., Reading, Mass.; *993, 998*.

Fox, L. [1965], *An Introduction to Numerical Linear Algebra*, Oxford Univ. Press, New York.

Fox, L., and D. F. Mayers [1968], *Computing Methods for Scientists and Engineers*, Clarendon Press, Oxford; *826, 834*.

Francis, J. G. F. [1961-1962], "The QR transformation, Parts I and II," *Computer Jour.* 4, 265-271 and 332-345; *923, 932, 934*.

Frank, W. L. [1958], "Computing eigenvalues of complex matrices by determinant evaluation and by methods of Danilewski and Wielandt," *J. Soc. Indust. Appl. Math.* 6, 378-392; *946*.

Frank, Werner [1960], "Solution of linear systems by Richardson's method," *J. Assoc. Comput. Mach.* 7, 274-286; *1068*.

Franklin, J. N. [1968], *Matrix Theory*, Prentice-Hall, Englewood Cliffs, N.J.; *773, 786*.

Fraser, M., and N. Metropolis [1968], "Algorithms in unnormalized arithmetic III. Matrix inversion," *Numer. Math.* 12, 416-428; *68*.

Friedman, B. [1956], *Principles and Techniques of Applied Mathematics*, John Yiley and Sons, Inc., New York; *753*.

Friedman, B. [1957], "The iterative solution of elliptic difference equations," A.E.C. Research and Development Report NYO-7698, Institute of Mathematical Sciences, New York University; *1074*.

Gantmacher, F. R. [1960], *The Theory of Matrices*, Vols. I and II (translated by K. A. Hirsch), Chelsea, New York; *739, 744, 747, 911*.

Garabedian, P. R. [1956], "Estimation of the relaxation factor for small mesh size," *Math. of Comp.* (formerly *Math. Tables Aids Comput.*); 10, 183-185; *1098*.

Geiringer, H. [1949], "On the solution of systems of linear equations by certain iterative methods," *Reissner Anniversary Volume*, University of Michigan Press, Ann Arbor, Mich., 365-393; *1073*.

Gerschgorin, S. [1930], "Fehlerabschätzung für das Differenzenverfahren zur Lösung partieller Differentialgleichungen," *Z. Agnew Math. Mech.* 10, 373-383; *576, 666, 970, 998*.

Gerschgorin, S. [1931], "Über die Abgrenzung der Eigenwerte einer Matrix," *Izv. Akad. Nauk SSSR*, Ser. fiz.-mat. 6, 749-754; *891*.

Gilbert, J. D. [1970], *Elements of Linear Algebra*, International Textbook Co., Scranton, Pa.

Gill, S. [1951], "A process for the step-by-step integration of differential equations in an automatic digital computing machine," *Proc. Cambridge Phil. Soc.* 47, 96-108; *492*.

Givens, Wallace [1953], "A method of computing eigenvalues and eigenvectors suggested by classical results on symmetric matrices," Chapter 17 of *Simultaneous Linear Equations and the Determination of Eigenvalues*, edited by L. J. Paige and Olga Taussky, National Bureau of Standards, Applied Mathematics Series No. 29, Washington, D.C.; *245*.

Givens, J. W. [1954], "Numerical computation of the characteristic values of a real symmetric matrix," Oak Ridge National Laboratory Report ORNL-1574; *10, 900, 909*.

Goldstine, H. H., F. J. Murray, and J. von Neumann [1959], "The Jacobi method for real symmetric matrices," *J. Assoc. Comput. Mach.* 6, 59-96; *896, 899*.

Golub, Gene H. [1959], "The use of Chebyshev matrix polynomials in the iterative solution of linear systems compared with the method of successive overrelaxation," doctoral thesis, University of Illinois.

Golub, G. H., and R. S. Varga [1961], "Chebyshev semi-iterative methods, successive over-relaxation iterative methods, and second order Richardson iterative methods," Parts I and II, *Numer. Math.* 3, 147–168; *1066, 1069, 1074.*

Gragg, William B., and Hans J. Stetter [1964], "Generalized multistep predictor-corrector methods," *J. Assoc. Comput. Mach.* 11, 188–209; *574.*

Greenstadt, J. [1960], "The determination of the characteristic roots of a matrix by the Jacobi method," Chapter 7 of Ralston and Wilf [1960]; *893, 896.*

Gregory, R. T. [1953], "Computing eigenvalues and eigenvectors of a symmetric matrix on the ILLIAC," *Math. of Comp.* 7, 215–220; *896.*

Gregory, R. T. [1957], "A method of deriving numerical differentiation formulas," *Amer. Math. Monthly* 64, 79–82.

Gregory, R. T. [1960], "Defective and derogatory matrices," *SIAM Review* 2, 134–139; *748.*

Gregory, R. T. [1963], *Numeral Systems*, Wm. C. Brown Book Co., Dubuque, Iowa (now Kendall/Hunt Publishing Company).

Gregory, R. T. [1966], "On the design of the arithmetic unit of a fixed-word-length computer from the standpoint of computational accuracy," *I.E.E.E. Trans. on Electronic Computers*, EC–15, 255–257; *69.*

Gregory, R. T., and D. L. Karney [1969], *A Collection of Matrices for Testing Computational Algorithms*, John Wiley and Sons, Inc., New York.

Griffith, H. W. [1971], "Preliminary investigations using interval arithmetic in the numerical evaluation of polynomials," doctoral dissertation, The University of Texas at Austin; also CNA–8, Center for Numerical Analysis, The University of Texas at Austin; *68.*

Guilinger, Willis H., Jr. [1965], "The Peaceman-Rachford method for small mesh increments," *J. of Math. Anal. and Appl.* 11, 261–277.

Habetler, G. J., and E. L. Wachspress [1961], "Symmetric successive overrelaxation in solving difference equations," *Math. of Comp.* 15, 356–362; *1068.*

Hageman, L. A., and R. B. Kellogg [1968], "Estimating optimum overrelaxation parameters," *Math. of Comp.* 22, 60–68; *1037.*

Hansen, Eldon [1969], "Cyclic composite multistep predictor-corrector methods," *Proc. of 24th National ACM Conference, New York*, 135–139; *574.*

Hansen, E. R. (editor) [1969], *Topics in Interval Analysis*, Clarendon Press, Oxford; *68, 574.*

Hart, J. F., E. W. Cheney, C. L. Lawson, H. J. Maehly, C. K. Mesztenyi, J. R. Rice, H. C. Thatcher, Jr., and C. Witzgall [1968], *Computer Approximations*, John Wiley and Sons, Inc., New York; *317, 326, 328, 329, 343.*

Hartree, D. R. [1952]. *Numerical Analysis* (second edition 1958), Clarendon Press, Oxford; *11, 174, 343.*

Hayes, D. R., and L. Rubin [1970], "A proof of the Newton-Cotes quadrature formulas with error term," *Amer. Math. Monthly* 77, 1065–1072; *389, 392, 416.*

Henrici, Peter [1957]. "Theoretical and experimental studies on the accumulation of error in the numerical solution of initial-value problems for systems of ordinary differential equations," *Proc. of the International Conference on Information Processing*, UNESCO, Paris, 15–20 June 1959 (published in 1960), 36–44.

Henrici, P. [1957a], unpublished lecture notes. Department of Mathematics, University of California, Los Angeles; *564, 569, 573.*

Henrici, Peter [1958], "On the speed of convergence of cyclic and quasicyclic Jacobi methods for computing eigenvalues of Hermitian matrices," *J. SIAM* 6, 144–162; *896.*

Henrici, P. [1960], "Estimating the best over-relaxation factor," Report NN-144, Ramo-Wooldridge Technical Memo, Los Angeles, Calif.; *1037*.

Henrici, Peter [1962], *Discrete Variable Methods in Ordinary Differential Equations*, John Wiley and Sons, Inc., New York; *389, 429, 442, 492, 498, 511, 513, 516, 520, 542, 543, 546, 563, 574*.

Henrici, P. [1963], *Error Propagation for Difference Methods*, John Wiley, New York; *486, 492*.

Henrici, Peter [1967], "Quotient-difference algorithms," Chapter 2 of Ralston and Wilf [1967]; *242*.

Hestenes, M. R., and E. Stiefel [1952], "Method of conjugate gradients for solving linear systems," *J. Res. Nat. Bur. Standards* 49, 409–436; *1071*.

Hildebrand, F. B. [1956], *Introduction to Numerical Analysis*, McGraw-Hill Book Co., Inc., New York; *416, 418, 420, 421*.

Hodge, W. V. D., and D. Pedoe [1947], *Methods of Algebraic Geometry*, Vol. I, Cambridge, at the University Press; *748*.

Hohn, F. E. [1958], *Elementary Matrix Algebra*, Macmillan, New York; *711, 734*.

Householder, Alston S. [1953], *Principles of Numerical Analysis*, McGraw-Hill Book Co., Inc., New York; *245*.

Householder, A. S. [1958], "The approximate solution of matrix problems," *J. Assoc. Comput. Mach.* 5, 205–243.

Householder, A. S. [1958a], "Unitary triangularization of a nonsymmetric matrix," *J. Assoc. Comput. Mach.* 5, 339–342; *901*.

Householder, A. S. [1964], *The Theory of Matrices in Numerical Analysis*, Blaisdell Publishing Co., New York; *202, 816, 901*.

Householder, A. S. [1970], *The Numerical Treatment of a Single Nonlinear Equation*, McGraw-Hill Book Co., New York; *173, 241, 245*.

Howell, J. A. [1971], "Algorithm 406. Exact solution of linear equations using residue arithmetic," *Comm. Assoc. Comput. Mach.* 14, 180–184; *888*.

Howell, J. A., and R. T. Gregory [1969a], "Solving systems of linear algebraic equations using residue arithmetic," Report TNN-82 (revised), Computation Center, The University Of Texas at Austin.

Howell, J. A., and R. T. Gregory [1969b], "An algorithm for solving linear algebraic equations using residue arithmetic, Parts I and II," *BIT* 9, 200–224 and 324–337; *888*.

Howell, J. A., and R. T. Gregory [1970], "Solving linear equations using residue arithmetic—Algorithm II," *BIT* 10, 23–37; *68, 874, 876, 888*.

Isaacson, Eugene, and Herbert B. Keller [1966], *Analysis of Numerical Methods*, John Wiley and Sons, New York; *390, 392, 1082, 1088*.

Jackson, Dunham [1930], *The Theory of Approximation*, American Math. Soc., Providence, R.I.

Jackson, Dunham [1941], *Fourier Series and Orthogonal Polynomials*, The Carus Mathematical Monographs, No. 6, Math. Assoc. of America; *330*.

Jenkins, M. A. [1969], "Three-stage variable-shift iterations for the solution of polynomial equations with *a posteriori* error bounds for the zeros," doctoral dissertation, Stanford University; also Tech. Report CS 138, Computer Science Department, Stanford University; *243*.

Jenkins, M. A., and J. F. Traub [1967], "An algorithm for an automatic general polynomial solver," Tech. Report No. CS 71, Computer Science Department, Stanford University; *243, 245*.

Jenkins, M. A., and J. F. Traub [1968, 1970], "A three-stage variable-shift iteration for polynomial zeros and its relation to generalized Rayleigh iteration," CS 107, Stanford Univ. (1968), *Numer. Math.* 14 (1970), 252–263; *243*.

John, F. [1952], "On integration of parabolic equations by difference methods," *Comm. Pure Appl. Math.* 5, 155–211.

John, F. [1967], *Lectures on Advanced Numerical Analysis*, Gordon and Breach, New York; *761, 769, 770*.

Juncosa, M. L., and David M. Young [1953], "On the order of convergence of solutions of a difference equation to a solution of the diffusion equation," *J. Soc. Indust. Appl. Math.* 1, 111–135; *1098*.

Juncosa, M. L., and David M. Young [1954], "On the convergence of a solution of difference equation to a solution of the equation of diffusion," *Proc. Amer. Math. Soc.* 5, 168–174; *1080, 1098*.

Juncosa, M. L., and David M. Young [1957], "On the Crank-Nicolson procedure for solving parabolic partial differential equations," *Proc. Cambridge Philos. Soc.* 53, part 2, 448–461; *1088*.

Kahan, W. [1958], "Gauss-Seidel methods of solving large systems of linear equations," doctoral thesis, University of Toronto; *1029, 1094*.

Kamke, E. [1948], *Differentialgleichungen Lösungsmethoden und Lösungen*, 1, Chelsea Pub. Co., New York; *431*.

Keller, Herbert B. [1968], *Numerical Methods for Two-point Boundary-value Problems*, Blaisdell, Waltham, Mass.; *665, 666*.

Kincaid, David [1969], "Solution of N simultaneous first-order differential equations," Program Writeup UTD2–01–CC07, The University of Texas Computation Center; *488*.

Kolman, B. [1970], *Elementary Linear Algebra*, Macmillan, New York.

Kopal, Zdeněk [1955], *Numerical Analysis*, John Wiley and Sons, Inc., New York; *420*.

Krause, E. F. [1970], *Introduction to Linear Algebra*, Holt, Rinehart and Winston, New York; *693, 694, 695, 699, 713, 715, 785*.

Kublanovskaya, V. N. [1961], "On some algorithms for the solution of the complete eigenvalue problem," *Zh. vych. mat.* 1, 555–570; *921*.

Kunz, K. S. [1957], *Numerical Analysis*, McGraw-Hill Book Co., New York; *276, 476*.

Kusmin, R. O. [1931], "Zur Theorie der mechanischen Quadraturen," *Nachr. Poly. Inst. Leningrad* 33, 5–14; *371*.

Lanczos, Cornelius [1956], *Applied Analysis*, Prentice-Hall, Englewood Cliffs, N.J.; *10*.

Lax, P. D., and R. D. Richtmyer [1956], "Survey of the stability of linear finite difference equations," *Comm. Pure Appl. Math.* 9, 267–293.

Lees, Milton [1966], "Discrete methods for nonlinear two-point boundary value problems," in *Numerical Solution of Partial Differential Equations*, edited by James Bramble, Academic Press, New York, 59–72; *665*.

Lehmer, D. H. [1961], "A machine method for solving polynomial equations," *J. Assoc. Comput. Mach.* 8, 151–161; *202*.

Leutert, Werner [1952], "On the convergence of unstable approximate solutions of the heat equation to the exact solution," *J. Math. Physics* 30, 245–251; *1098*.

Lindamood, G. E. [1964], "Numerical analysis in residue number systems," Univ. of Maryland Computer Science Report TR–64–7, College Park, Maryland; *68, 888*.

MacLane, S., and G. Birkhoff [1967], *Algebra*, 3rd edition, Macmillan, New York.

Maehly, H. J. [1959], "Rational approximations for transcendental functions," *Proc. of the International Conference on Information Processing*, UNESCO, Paris, 15-20 June 1959 (published in 1960), 57-62; *329*.

Marcus, M., and H. Minc [1964], *A Survey of Matrix Theory and Matrix Inequalities*, Allyn and Bacon, Inc., Boston; *816*.

Marden, Morris [1966], *Geometry of Polynomials*, Amer. Math. Soc., Providence, Rhode Island; *245*.

Martin, R. S., C. Reinsch, and J. H. Wilkinson [1968], "Householder's tridiagonalization of a symmetric matrix," *Numer. Math.* 11, 181-195.

Matula, D. W. [1969], "Towards an abstract mathematical theory of floating-point arithmetic," *Proc. Spring Joint Computer Conference*, IFIPS 34, 765-772; *68*.

McClellan, M. T. [1971], "The exact solution of systems of linear equations with polynomial coefficients," *Proc. of the Second Symposium on Symbolic and Algebraic Manipulation*, Assoc. for Comput. Mach., March 23-25, 1971, Los Angeles, Calif.; *68, 888*.

McDonald, A. E. [1970], "A multiplicity-independent, global iteration for meromorphic functions," doctoral dissertation, The University of Texas at Austin; also TNN-98, Computation Center, The University of Texas at Austin; *206, 210, 244, 245*.

McDowell, Leland K. [1967], "Variable successive overrelaxation," Report No. 244, Department of Computer Sciences, University of Illinois; *1068*.

McKeeman, W. M. [1962], "Algorithm 135, 'Crout with equilibration and iteration'," *Comm. Assoc. Comput. Mach.* 5, 553-555; *819*.

Milne, W. E. [1949], *Numerical Calculus*, Princeton University Press, Princeton, New Jersey.

Milne, W. E. [1953], *Numerical Solution of Differential Equations*, John Wiley and Sons, Inc., New York; *492*.

Milne, W. E., and R. R. Reynolds [1959], "Stability of a numerical solution of differential equations," *J. Assoc. Comput. Mach.* 6, 196-203; *574*.

Mitchell, B. E. [1953], "Normal and diagonalizable matrices," *Amer. Math. Monthly* 60, 94-96; *742*.

Montel, P. [1910], "Leçons sur les séries de polynomes à une variable complexe," Borel Monograph, Paris; *296*,

Moore, R. [1966], *Interval Analysis*, Prentice-Hall, Englewood Cliffs, N.J.; *68*.

Mouradoglou, A. J. [1967], "Numerical studies on the convergence of the Peaceman-Rachford alternating direction implicit method," Master's thesis, The University of Texas at Austin; also TNN-67, Computation Center, The University of Texas at Austin; *1035, 1061*.

Moursund, D. G. [1967], "Optimal starting values for Newton-Raphson calculation of $\sqrt{x}$," *Comm. Assoc. Comput. Mach.* 10, 430-432; *68*.

Muller, D. E. [1956], "A method for solving algebraic equations using an automatic computer," *Math. of Comp.* (formerly *Math. Tables Aids Comput.*) 10, 208-215; *195*.

Naiser, Lou Ann [1967], "The QR algorithm applied to Hessenberg matrices," Master's thesis, The University of Texas at Austin; also TNN-66, Computation Center, The University of Texas at Austin; *932, 934*.

Newman, M. [1967], "Solving equations exactly," *J. Res. Nat. Bur. Stds.* 17B, 171-179; *68, 868, 869, 881, 886, 888*.

Noble, Ben [1969], *Applied Linear Algebra*, Prentice-Hall, Inc., Englewood Cliffs, N.J.; *809, 817*.

O'Brien, George G., Morton A. Hyman, and Sidney Kaplan [1951], "A study of the numerical solution of partial differential equations," *J. Math. Phys.* 29, 223-252; *1079, 1098*.

Ortega, J. [1967], "The Givens-Householder method for symmetric matrices," Chapter 4 of Ralston and Wilf [1967]; *901, 938*.

Ortega, James M., and Werner C. Rheinboldt [1970], *Iterative Solution of Nonlinear Equations in Several Variables*, Academic Press, New York; *173, 174*.

Osborne, E. E. [1960], "On pre-conditioning of matrices," *J. Assoc. Comput. Mach.* 7, 338–335; *950*.

Ostrowski, A. M. [1954], "On the linear iteration procedures for symmetric matrices," *Rend. Mat. e Appl.* 13, 140–163; *1029, 1068, 1073*.

Ostrowski, A. M. [1958–1959], "On the convergence of the Rayleigh quotient iteration for the computation of the characteristic roots and vectors," (in six parts), *Arch. Rat. Mech. Anal.*: I, vol. 1, pp. 233–241; II, vol. 2, pp.423–428; III, vol. 3, pp. 325–340; IV, vol. 3, pp. 341–347; V, vol. 3, pp. 472–481; VI, vol. 4, pp. 153–165; *245*.

Ostrowski, A. M. [1966], *Solution of Equations and Systems of Equations*, Academic Press, New York; *163, 173, 174, 549*.

Padé, H. [1892], "Sur la représentation approchée d'une fonction par des fractions rationnelles," thèse, Ann. de l'Ec. Nor. (3) 9; *326*.

Parlett, B. N. [1964], "Laguerre's method applied to the matrix eigenvalue problem," *Math. of Comp.* 18, 143–145; *242*.

Parlett, B. N. [1966], "Singular and invariant matrices under the QR transformation," *Math. of Comp.* 20, 611–615; *930*.

Parlett, B. N. [1967], "The LU and QR algorithms," Chapter 5 of Ralston and Wilf [1967]; *924, 930, 932, 936*.

Parlett, B. N. [1968], "Global convergence of the basic QR algorithm on Hessenberg matrices," *Math. of Comp.* 22, 803–817; *930*.

Parlett, B. N., and C. Reinsch [1969], "Balancing a matrix for calculation of eigenvalues and eigenvectors," *Numer. Math.* 13, 293–304; *950*.

Parter, Seymour V. [1959], "On 'two-line' iterative methods for the Laplace and biharmonic difference equations," *Numer. Math.* 1, 240–252; *1074*.

Parter, Seymour V. [1961], "'Multi-line' iterative methods for elliptic difference equations and fundamental frequencies," *Numer. Math.* 3, 305–319; *1074*.

Parter, Seymour V. [1965], "On estimating the 'rates of convergence' of iterative methods for elliptic difference equations," *Trans. Amer. Math. Soc.* 114, 320–354; *1074*.

Peaceman, D. W., and H. H. Rachford, Jr. [1955], "The numerical solution of parabolic and elliptic differential equations," *J. Soc. Indust. Appl. Math.* 3, 28–41; *1041, 1050, 1098*.

Peetre, J., and V. Thomee [1967], "On the rate of convergence for discrete initial-value problems," *Math. Scand.* 21, 159–176; *1098*.

Perlis, S. [1952], *Theory of Matrices*, Addison-Wesley, Reading, Mass.; *687, 713*.

Poole, William G. [1965], "Numerical experiments with several iterative methods for solving partial differential equations," Master's thesis, The University of Texas at Austin; also TNN–49, Computation Center, The University of Texas at Austin; *1071*.

Poole, W. G. [1970], "A geometric convergence theory for the QR, Rayleigh quotient, and power iterations," Computer Center Technical Report No. 41, Univ. of Calif., Berkeley; *941*.

Pope, D. A., and C. Tompkins [1957], "Maximizing functions of rotations," *J. Assoc. Comput. Mach.* 4, 459–466; *896*.

Price, H., and R. S. Varga [1962], "Recent numerical experiments comparing successive overrelaxation iterative methods with alternating direction implicit methods," Report No. 91, Gulf Research and Development Company, Pittsburgh, Pa.; *1061*.

Ralston, A. [1963], "On differentiating error terms," *Amer. Math. Monthly* 71, 187–189; *389*.

Ralston, Anthony [1965], *A First Course in Numerical Analysis*, McGraw-Hill Book Co., New York; *202, 242, 245, 327, 328, 329, 389, 421, 477*.

Ralston, Anthony, and Herbert S. Wilf [1960], *Mathematical Methods for Digital Computers*, Vol. I, John Wiley and Sons, Inc., New York.

Ralston, Anthony, and Herbert S. Wilf [1967], *Mathematical Methods for Digital Computers*, Vol. II, John Wiley and Sons, Inc., New York.

Raney, J. L. [1961], "Solution of N simultaneous differential equations by the Adams-Moulton method using a Runge-Kutta starter and partial double-precision arithmetic," Program Writeup UTD2–02–003, The University of Texas Computation Center; *488*.

Redish, K. A., and W. Ward [1971], "Environment enquires for numerical analysis," *SIGNUM Newsletter*, Assoc. for Comput. Mach. 6, 10–15; *92*.

Reich, E. [1949], "On the convergence of the classical iterative method of solving linear simultaneous equations," *Ann. Math. Statist.* 20, 448–451; *1073*.

Remes, E. [1934], "Sur un procédé convergent d'approximations successives pour déterminer les polynomes d'approximation," *Comptes Rendus*, 198, 2063–2065; *309*.

Remes, E. [1934a], "Sur le calcul effectif des meilleures approximations de Tchebichef," *Comptes Rendus* 199, 337–340; *309, 317*.

Rice, John R. [1964], *The Approximation of Functions I. Linear Theory*, Addison-Wesley, Boston; *317*.

Richardson, L. F. [1910], "The approximate arithmetical solution by finite differences of physical problems involving differential equations, with application to the stresses in a masonry dam," *Philos. Trans. Roy. Soc. London*, Ser. A, vol. 210, 307–357, and *Proc. Roy. Soc. London*, Ser. A, vol 83, 335–336; *577, 614, 1067, 1083*.

Richtmyer, Robert D., and K. W. Morton [1967], *Difference Methods for Initial-value Problems* (second edition), Interscience Publishers, New York, Second Edition; *1098*.

Romberg, W. [1955], "Vereinfachte numerische Integration," Det. Kong. Norske Videnskaber Selskab Forhandlinger, Band, 23, Nr. 7, Trondheim.

Ruhe, Axel [1970], "An algorithm for numerical determination of the structure of a general matrix," *BIT* 10, 196–216; *941*.

Runge, C. [1901], "Über empirische Funktionen und die Interpolation zwischen äquidistanten Ordinaten," *Zeit. für. Math. und Phys.*, XLVL, 229; *296*.

Rutishauser, H. [1957], *Der Quotienten-Differenzen-Algorithmus*, Birkhauser, Verlag, Basel; *242*.

Rutishauser, H. [1958], "Solution of eigenvalue problems with the LR transformation," in *Further Contributions to the Solution of Simultaneous Linear Equations and the Determination of Eigenvalues*, Nat. Bur. of Std. AMS 49; *921*.

Salzer, H. E. [1956], "Osculatory extrapolation and a new method for the numerical integration of differential equations," *J. Franklin Inst.* 262, 111–119; *574*.

Sard, A. [1963], "Linear approximation," *Mathematical Surveys, No. 9*, American Mathematical Society, Providence, R.I.; *421*.

Saul'yev, V. K. [1964], *Integration of Equations of Parabolic Type by the Method of Nets*, translated by G. J. Tee, Pergamon Press, New York; *1098*.

Schmidt, Jochen W., and Harmut Dressel [1967], "Fehlerabschätzungen bei Polynomgleichungen mit dem Fixpunktsatz von Brouwer," *Numer. Math.* 10, 42–50; *245*.

Schoenberg, I. J. [1964], "Spline interpolation and best quadrature formulas," *Bull. Amer. Math. Soc.* 70, 143–148; *421*.

Schröder, E. [1870], "Über unendlich viele Algorithmen zur Auflösung der Gleichungen," *Math. Ann.* 2, 317–365. English translation by G. W. Stewart, III, ORNL Translations No. 1851; *243, 245.*

Shanks, D. [1955], "Non-linear transformations of divergent and slowly convergent sequences," *J. Math. and Phys.* 34, 1–42; *174.*

Sheldon, J. [1955], "On the numerical solution of elliptic difference equations," *Math. of Comp.* (formerly *Math. Tables Aids Comput.*) 9, 101–112; *1068.*

Sheldon, J. W. [1959], "On the spectral norms of several iterative processes," *J. Assoc. for Comput. Mach.* 6, 494–505; *1069.*

Shortley. G. [1953], "Use of Tchebycheff polynomial operators in the numerical solution of boundary value problems," *J. Appl. Phys.* 32, 243–255; *1074.*

Simeunovic, D. M. [1967], "Les limites des modules des zéros des polynomes et des séries de Taylor," *Mat. Bech.* 4, 209–303; *245.*

Snyder, Martin A. [1966], *Chebyshev Methods in Numerical Approximation,* Prentice Hall, Inc., Englewood Cliffs, N.J.; *343.*

Southwell, R. V. [1946], *Relaxation Methods in Theoretical Physics,* Oxford University Press, New York; *1027.*

Steffensen, J. F. [1927], *Interpolation* (second edition 1950), Chelsea Publishing Co., New York; *296, 421.*

Stein, P., and R. Rosenberg [1948], "On the solution of linear simultaneous equations by iteration," *J. London Math. Soc.* 23, 111–118; *1021, 1073.*

Stewart, G. W. [1968], "Translation of the paper by E. Schröder [1870], 'On infinitely many algorithms for solving equations,'" Oak Ridge National Laboratory Report No. ORNL–tr–1851, Oak Ridge, Tennessee; *245.*

Stewart, G. W. [1969], "On some methods for solving equations related to Schröder's iterations," unpublished manuscript; *245.*

Stewart, G. W. [1970a], "Algorithm 384, Eigenvalues and eigenvectors of a real symmetric matrix," *Comm. Assoc. Comput. Mach.* 13, 369–371; *913.*

Stewart, G. W. [1970b], "Incorporating origin shifts into the QR algorithm for symmetric tridiagonal matrices," *Comm. Assoc. Comput. Mach.* 13, 365–367; *913.*

Stiefel, E. [1952], "Über einige Methoden der Relaxationrechnung," *Z. angew. Math. Phys.* 3, 1–33; *1071.*

Stiefel, E. L. [1959], "Numerical methods of Tchebycheff approximation," in *On Numerical Approximation,* edited by R. E. Langer, Univ. of Wisconsin Press, Madison, Wisconsin, 217–232; *309, 310, 316.*

Stiefel, E. L. [1959a], "Über diskrete und lineare Tschebyscheff-Approximationen," *Numer. Math.* 1, 1–28; *317.*

Strang. Gilbert [1970], "The finite element method and approximation theory," in *Numerical Solution of Partial Differential Equations,* II (SYNSPADE, 1970) edited by Bert Hubbard, Academic Press, New York; *998.*

Stroud, A. H., and Don Secrest [1966], *Gaussian Quadrature Formulas,* Prentice-Hall, Englewood Cliffs, N.J.; *421.*

Szabó, S., and R. Tanaka [1967]; *Residue Arithmetic and Its Applications to Computer Technology,* McGraw-Hill, New York; *68, 835, 837, 842.*

Takahasi, H., and Y. Ishibashi [1961], "A new method for 'exact calculation' by a digital computer," Information Processing in Japan 1, 28–42; *68, 888.*

Taussky, O. [1949], "A recurring theorem on determinants," *Amer. Math. Monthly* 56, 672–676; *1073*.

Thomas, L. H. [1949], "Elliptic problems in linear difference equations over a network," Watson Scientific Computing Laboratory, Columbia University, New York; *587*.

Thomason, John M. [1968], "Stabilizing averages for multistep methods of solving ordinary differential equations," Master's thesis, The University of Texas at Austin; also TNN–83, Computation Center, The University of Texas at Austin; *574*.

Thrall, R. M., and L. Tornheim [1957], *Vectors, Spaces and Matrices*, John Wiley & Sons, New York; *1047*.

Timlake, W. P. [1965], "On an algorithm of Milne and Reynolds," *BIT* 5, 276–281; *574*.

Todd, John [1963], *Introduction to the Constructive Theory of Functions*, Academic Press, Inc., New York; *309*.

Traub, J. F. [1964], *Iterative Methods for the Solution of Equations*, Prentice-Hall, Inc., Englewood Cliffs, N.J.; *120, 142, 146, 159, 173, 174*.

Traub, J. F. [1966a], "A class of globally convergent iteration functions for the solution of polynomial equations," *Math. Comp.* 20, 113–138; *243*.

Traub, J. F. [1966b], "Proof of global convergence of an iterative method for calculating complex zeros of a polynomial," *Notices Amer. Math. Soc.* 13, 117; *243*.

Traub, J. F. [1967], "The calculation of zeros of polynomials and analytic functions," Proceedings of Symposia in Applied Mathematics, Vol. 19, *Mathematical Aspects of Computer Science*, Amer. Math. Soc., Providence, Rhode Island, 138–152; *243*.

Tropper, A. M. [1969], *An Introduction to Linear Algebra*, American Elsevier, New York; *686, 704*.

Uspensky, J. V. [1948], *Theory of Equations*, McGraw-Hill, New York; *785, 787*.

Varah, James M. [1967], "The computation of bounds for the invariant subspaces of a general matrix operator," Tech. Report No. C.S. 66, Computer Science Dept., Stanford University; *941*.

Varga, Richard S. [1957], "A comparison of the successive overrelaxation method and semi-iterative methods using Chebyshev polynomials," *J. Soc. Indust. Appl. Math.* 5, 39–46; *1068, 1074*.

Varga, R. S. [1959], "Orderings of the successive overrelaxation scheme," *Pacific Jour. Math.* 9, 925–939; *1074*.

Varga, Richard S. [1960], "Factorization and normalized iterative methods," in *Boundary Problems in Differential Equations*, edited by R. E. Langer, University of Wisconsin Press, Madison, 121–142; *1064, 1074*.

Varga, Richard S. [1962], *Matrix Iterative Analysis*, Prentice-Hall, Inc., Englewood Cliffs, New Jersey; *816, 997, 998, 1013, 1014, 1021, 1029, 1037, 1073, 1074, 1098*.

Varga, R. S. [1970], *Functional Analysis and Approximation Theory in Numerical Analysis*, Proceedings of the Regional Conference Sponsored by the National Science Foundation at Boston University, Boston, Mass.; *666*.

Viswanathan, R. V. [1957], "Solution of Poisson's equation by relaxation method—normal gradient specified on curved boundaries," *Math. of Comp.* 11, 67–78; *993*.

Wachspress, E. L. [1957], "CURE: A generalized two-space-dimension multigroup coding for the IBM-704," Report KAPL-1724, Knolls Atomic Power Laboratory, Schenectady, New York; *1074*.

Wachspress, B. L. [1966], *Iterative Solution of Elliptic Systems and Applications to the Neutron Diffusion Equations of Reactor Physics*, Prentice-Hall, Inc., Englewood Cliffs, N.J.; *735, 1050, 1073, 1074.*

Walsh, J. L. [1931], "The existence of rational functions of best approximation," *Trans. Amer. Math. Soc.* 33, 668–689; *328.*

Warlick, Charles H. [1955], "Convergence rates of numerical methods for solving $\partial^2 u/\partial x^2 + (k/\rho)\partial u/\partial\rho + \partial^2 u/\partial\rho^2 = 0$," Master's thesis, The University of Maryland; *1037.*

Warlick, Charles H., and David M. Young [1970], "*A priori* methods for the determination of the optimum relaxation factor for the successive overrelaxation method," TNN–105, Computation Center, The University of Texas at Austin; *1037.*

Wedderburn, J. H. M. [1934], *Lectures on Matrices*, Amer. Math. Soc. Colloquium Publications, Vol. 27; *753.*

Wendroff, B. [1966], *Theoretical Numerical Analysis*, Academic Press, New York; *826, 828.*

Werner, H. [1912], "Die konstruktive Ermittlung der Tschebyscheff—Approximierenden im Bereich der rationalen Funktionen," *Arch. Rational Mech. Anal.* 11, 368–384; *328.*

Widder, David V. [1947], *Advanced Calculus*, Prentice-Hall, Englewood Cliffs, N.J.; *995.*

Widlund, O. B. [1966], "On the rate of convergence of an alternating direction implicit method in a noncommutative case," *Math. Comp.* 20, 500–515; *1074.*

Widlund, O. B. [1968], "On the rate of convergence for parabolic difference schemes, I," *Proc. Amer. Math. Soc. Symposium for Applied Mathematics 21*, Durham, North Carolina; *1098.*

Widlund, O. B. [1971], "On the rate of convergence for parabolic difference schemes, II," *Comm. Pure Appl. Math.* 23, 79–96; *1098.*

Wilf, Herbert S. [1960], "The numerical solution of polynomial equations," Chapter 21 of Ralston and Wilf [1960]; *210.*

Wilkinson, Belinda M. [1969], "A polyalgorithm for finding roots of polynomial equations," Master's thesis, The University of Texas at Austin; also TNN–93, Computation Center, The University of Texas at Austin; *177, 218, 229, 235.*

Wilkinson, J. H. [1954], "The calculation of the latent roots and vectors of matrices on the pilot model of the A.C.E.," *Proc. Camb. Phil. Soc.* 50, 536–566; *915.*

Wilkinson, J. H. [1960], "Householder's method for the solution of the algebraic eigenproblem," *Comp. Jour.* 3, 23–27; *901.*

Wilkinson, J. H. [1961], "Error analysis of direct methods of matrix inversion," *J. Assoc. Comput. Mach.* 8, 281–330; *813.*

Wilkinson, J. H. [1962], "Householder's method for symmetric matrices," *Numer. Math.* 4, 354–361; *901.*

Wilkinson, J. H. [1963], *Rounding Errors in Algebraic Processes*, Prentice-Hall, Inc., Englewood Cliffs, N.J.; *45, 46, 52, 53, 191, 213, 216, 230, 833, 889, 938, 939, 944.*

Wilkinson, J. H. [1965], *The Algebraic Eigenvalue Problem*, Clarendon Press, Oxford; *53, 745, 804, 808, 813, 826, 832, 891, 896, 915, 919, 923, 924, 932, 941, 946, 948, 950.*

Wilkinson, J. H. [1965a], "Convergence of the *LR, QR,* and related algorithms," *The Computer Jour.* 8, 77–84.

Wilkinson, J. H. [1967], "The solution of ill-conditioned linear equations," Chapter 3 of Ralston and Wilf [1967]; *806, 826, 828, 832, 834.*

Wilkinson, J. H. [1968], see Martin, Reinsch, and Wilkinson [1968]; *901.*

WPA [1944], "Tables of Lagrangian interpolation coefficients," *Mathematical Tables Project*, Works Progress Administration, Federal Works Agency; *271.*

Young, David M. [1950], "Iterative methods for solving partial difference equations of elliptic type," doctoral thesis, Harvard University; *1073, 1074.*

Young, David M. [1954], "Iterative methods for solving partial difference equations of elliptic type," *Trans. Amer. Math. Soc.* 76, 92–111; *1023, 1025, 1073.*

Young, David M. [1954a], "On Richardson's method for solving linear systems with positive definite matrices," *J. Math. Phys.* XXXII, 243–255; *1074.*

Young, David M. [1955], "Gill's method for solving ordinary differential equations," Numerical Note NN–4, Ramo-Wooldridge Corp., Los Angeles, Calif.

Young, David M. [1961], "The numerical solution of elliptic and parabolic partial differential equations," in *Numerical Analysis*, Vol. VI, edited by E. F. Beckenbach, McGraw-Hill Book Co., New York, 283–298.

Young, David M. [1962], "The numerical solution of elliptic and parabolic partial differential equations," in *Survey of Numerical Analysis*, edited by John Todd, McGraw-Hill Book Co., New York, 380–438; *1098, 1099.*

Young, David M. [1971], "On the consistency of linear stationary iterative methods," *SIAM Jour. of Numer. Analysis* 9, 89–96; *174.*

Young, David M. [1971a], *Iterative Solution of Large Linear Systems*, Academic Press, New York; *174, 755, 1013, 1014, 1025, 1029, 1037, 1073, 1098.*

Young, David M. [1971b], "Second-degree iterative methods for the solution of large linear systems," *Jour. of Approx. Theory* 5, 137–148; *1068.*

Young, David M. [1971c], "A bound for the optimum relaxation factor for the successive overrelaxation method," *Numer. Math.* 16, 408–413; *1037.*

Young, David M., and John H. Dauwalder [1965], "Discrete representations of partial differential operators," in *Error in Digital Computation*, Vol. 2, edited by L. B. Rall, John Wiley and Sons, Inc., New York, 181–217; *666, 989.*

Young, David M., and John H. Dauwalder [1971], "Discrete representations of partial differential operators—II," unpublished manuscript; *990.*

Young, David M., and Louis Ehrlich [1956], "On the numerical solution of linear and nonlinear parabolic equations on the Ordvac," Interim Tech. Report No. 18, Office of Ordnance Research Contract DA–36–034–ORD–1486, University of Maryland.

Young, David M., and Louis Ehrlich [1960], "Some numerical studies of iterative methods for solving elliptic difference equations," in *Boundary Problems in Differential Equations*, edited by R. E. Langer, The University of Wisconsin Press, Madison, 143–162; *1061.*

Young, David M., and Thurman Frank [1962], "A survey of computer methods for solving elliptic and parabolic partial differential equations," *Bulletin of the International Computation Center* 2, Rome, Italy, 3–61; also TNN–20, Computation Center, The University of Texas at Austin.

Young David M., L. D. Gates, Jr., R. J. Arms, and D. F. Eliezer [1955], "The computation of an axially symmetric free boundary problem on NORC," U.S. Naval Proving Ground Report No. 1413, Dahlgren, Va.; *997.*

Young, David M., and David R. Kincaid [1969], "Norms of the successive overrelaxation method and related methods," TNN–94, Computation Center, The University of Texas at Austin.

Young, David M., and Alvis E. McDonald [1969], "On the surveillance and control of number range and accuracy in numerical computation," in *Information Processing 68*, North-Holland Publishing Company, Amsterdam, 145–152; *87, 92.*

Young, D. M., Jr., A. E. McDonald, H. E. Eidson, and B. M. Wilkinson [1971], "Matrix eigenvalue methods for solving polynomial equations," in *Progress*; *177, 228, 235, 238, 240*.

Young, David M., and Harry Shaw [1955], "Ordvac solutions of $\partial^2 u/\partial x^2 + \partial^2 u/\partial y^2 + (k/y)\partial u/\partial y = 0$ for boundary value problems and problems of mixed type," Interim Tech. Report No. 14, Office of Ordnance Research Contract DA-36-034-ORD-1486, Univ. of Maryland; *1037*.

Young, David M., and Charles H. Warlick [1953], "On the use of Richardson's method for the numerical solution of Laplace's equation on the ORDVAC," Ballistic Research Labs. Memorandum Report No. 707, Aberdeen Proving Ground, Maryland; *1067*.

Young, David M., Mary F. Wheeler, and James A. Downing [1965], "On the use of the modified successive overrelaxation method with several relaxation factors," in *Proc. of IFIP Congress 65*, edited by W. A. Kalenich, Spartan Books, Inc., Washington, D.C., 177–182; *1068*.

Zelinsky, D. [1968], *A First Course in Linear Algebra*, Academic Press, New York; *730*.

INDEX

DATE DUE

This item is Due on
or before Date shown.